Plant Elicitor Peptides

Shachi Singh • Rajesh Mehrotra
Editors

Plant Elicitor Peptides

New Tool for Sustainable Agriculture

Springer

Editors
Shachi Singh
Department of Botany, MMV
Banaras Hindu University
Varanasi, Uttar Pradesh, India

Rajesh Mehrotra
Department of Biological Sciences
BITS Pilani, K. K. Birla Goa Campus
Zuarinagar, Goa, India

ISBN 978-981-97-6373-3 ISBN 978-981-97-6374-0 (eBook)
https://doi.org/10.1007/978-981-97-6374-0

This Springer imprint is published by the registered company Springer Nature Singapore Pte Ltd.
The registered company address is: 152 Beach Road, #21-01/04 Gateway East, Singapore 189721, Singapore

Preface

Plant innate immunity can be triggered by external stimuli, called elicitors. The elicitors bind to the receptor site on plasma membrane and trigger the signal cascades for the activation of metabolic pathways. Several groups of elicitor molecules have been identified to stimulate plant growth and development as well as provide protection against biotic and abiotic stress. Among various groups of elicitor signals, peptides have emerged as an important regulator of plant immunity. Generally, the peptide signals originate from external sources, such as pathogens and pests; however, recent findings have revealed that endogenous molecules originating from host plant can also trigger plant immunity. Identification of these endogenous peptide elicitors in various plant families has generated interest in exploring their potential toward sustainable agriculture.

This book provides a comprehensive overview on plant innate immunity triggered by elicitors. It explains how peptide signals can help in providing protection against environmental stress and can act as an eco-friendly crop protectant. The chapters deal with investigation of plant elicitor peptides, understanding the processes controlled by peptide signals, methods of isolation and their application in agriculture. Future strategies as well as challenges associated with the utilization of these immunity inducers are also discussed. The content of this book is not only limited to peptide elicitors but also covers other groups of compounds involved in plant protection.

Molecular and functional characterization of the elicitors will be useful in the development of elicitation-based technology, including large-scale production and commercialization. This book serves as an essential resource for researchers and industries working in the field of environmental stress management of crop plants. We are thankful to all our contributors for accepting our invitation and working hard to compose this book. We hope you will find this compilation worth reading.

Varanasi, India — Shachi Singh
Zuarinagar, Goa, India — Rajesh Mehrotra

Contents

1 **Plant Elicitor Peptides as Amplifiers of Immune Responses Against Biotic Stressors** . . . 1
Nalika P. Ranatnuge and W. P. Thisali Hasara

2 **Peptide Elicitors for Defense Against Abiotic Stress** . . . 19
Georgia Tsintzou and Panagiotis Madesis

3 **Plant Elicitor Peptide Mediated Signalling Cascades During Plant–Pathogen Interaction** . . . 49
Aryadeep Roychoudhury, Sampreet Manna, and Diyasa Banerjee

4 **Inceptin: Exploring Its Role as a Peptide Elicitor in Plant Defense Mechanisms** . . . 99
Sarika Sharma and Shachi Singh

5 **Endogenous Peptides Involved in Plant Growth and Development** . . . 113
Vidushi Yadav

6 **Effector Mediated Defense Mechanisms in Plants against Phytopathogens** . . . 131
Seema Devi, Riddha Dey, Surya Prakash Dube, and Richa Raghuwanshi

7 **Plants Retaliating Defense Strategies against Herbivores** . . . 149
Shweta Verma, Manisha Hariwal, Priya Patel, Priyaka Shah, and Sanjay Kumar

8 **Plant Elicitor Peptides: Mechanism of Action and Its Applications in Agriculture** . . . 171
Data Ram Saini, Pravin Prakash, Savita Jangde, Krishna Kumar, and Ipsita Maiti

9 **Legume Health: Unveiling the Potential of Plant Elicitor Peptides** . . . 199
Krutika S. Abhyankar and Monisha Kottayi

10 **Harnessing Plant Innate Immunity for Improved Biomass Production in Bioenergy Crops** . . . 227
Senthil Nagappan and Dig Vijay Singh

11 **Exogenous Elicitors as Inducers of Environmental Stress Tolerance in Wheat (*Triticum aestivum* L.) Crop** 247
Anjali Yadav and Shachi Singh

12 **Recent Advancement on Peptide Research and their Application in Eco-agriculture** 269
Jyotsna Setty, Pavan Singh, Girish Tantuway, and P. Vijai

13 **Plant Immunity Inducers: Strategies to Identify and Isolate Them to Boost Defense Responses in Plants** 283
Ragiba Makandar

14 **Deep Learning Approaches for Off-targets Prediction in CRISPR-Cas9 Genome Editing to Improve Resistant in Plants** 319
Awadhesh Kumar

15 **Peps, Pathogens, and Pests: Challenges and Opportunities for Usage of Pep Signaling in Sustainable Farming** 335
Alice Kira Zelman and Gerald Alan Berkowitz

Editors and Contributors

About the Editors

Shachi Singh is working as an assistant professor of botany at MMV Banaras Hindu University. She had completed her graduation and post-graduation in botany from Banaras Hindu University and did her master's in biotechnology from BITS, Pilani. She completed her Ph.D. and postdoctoral research work from the Department of Biological Sciences, BITS, Pilani. She was granted UGC NET fellowship and CSIR-JRF and SRF fellowship, as well as young scientist award by DST-SERB. She has also received INSA visiting scientist award. She has published several papers in highly reputed journals. Currently, her research group is working on methods to enhance plant immunity by natural means.

Rajesh Mehrotra graduated from Banaras Hindu University with specialization in genetics and plant breeding. He completed his Ph.D. in biotechnology from Banaras Hindu University while working at National Botanical Research Institute, Lucknow. He did his postdoctoral work at Graduate School of Biostudies, Kyoto University, Japan. His group has published more than 60 publications and has one American and one European patent. He is a Fellow of the Royal Society of Biology, UK. He is a recipient of Prof. V S Rao Foundation, Dr. CR Mitra Best Faculty Award of BITS, Pilani, Dr. Sarvepalli Radhakrishnan Distinguished Professor and Researcher Award, and INSA Bilateral Exchange Award to visit University of Edinburgh, Scotland. He has delivered 30 invited lectures.

Contributors

Krutika S. Abhyankar Division of Biomedical and Life Science, School of Science, Navrachana University, Vadodara, India

Diyasa Banerjee Department of Biotechnology, St. Xavier's College (Autonomous), Kolkata, West Bengal, India

Gerald Alan Berkowitz Department of Plant Science and Landscape Architecture, University of Connecticut, Storrs, CT, USA

Seema Devi Department of Botany, Mahila Mahavidyalaya, Banaras Hindu University, Varanasi, India

Riddha Dey Department of Botany, Mahila Mahavidyalaya, Banaras Hindu University, Varanasi, India

Surya Prakash Dube Department of Botany, Mahila Mahavidyalaya, Banaras Hindu University, Varanasi, India

Georgia Tsintzou Laboratory of Molecular Biology of Plants, Department of Agriculture Crop Production and Rural Environment, School of Agricultural Sciences, University of Thessaly, Volos, Greece

Manisha Hariwal Banaras Hindu University, Varanasi, India

W. P. Thisali Hasara Department of Agricultural Biology, Faculty of Agriculture, University of Ruhuna, Matara and Galle, Sri Lanka

Savita Jangde Institute of Agricultural Sciences, Banaras Hindu University, Varanasi, UP, India

Monisha Kottayi Division of Biomedical and Life Science, School of Science, Navrachana University, Vadodara, India

Sanjay Kumar Banaras Hindu University, Varanasi, India

Krishna Kumar Institute of Agricultural Sciences, Banaras Hindu University, Varanasi, UP, India

Awadhesh Kumar Department of Computer Science, MMV, Banaras Hindu University, Varanasi, India

Ipsita Maiti Institute of Agricultural Sciences, Banaras Hindu University, Varanasi, UP, India

Ragiba Makandar Department of Plant Sciences, School of Life Sciences, University of Hyderabad, Hyderabad, India

Sampreet Manna Department of Biotechnology, St. Xavier's College (Autonomous), Kolkata, West Bengal, India

Senthil Nagappan Sardar Swaran Singh National Institute of Bioenergy, Kapurthala, Punjab, India

Panagiotis Madesis Laboratory of Molecular Biology of Plants, Department of Agriculture Crop Production and Rural Environment, School of Agricultural Sciences, University of Thessaly, Volos, Greece

Institute of Applied Biosciences-INAB, Centre for Research and Technology-CERTH, Thessaloniki, Greece

Priya Patel Banaras Hindu University, Varanasi, India

Pravin Prakash Institute of Agricultural Sciences, Banaras Hindu University, Varanasi, UP, India

Richa Raghuwanshi Department of Botany, Mahila Mahavidyalaya, Banaras Hindu University, Varanasi, India

Nalika P. Ranatnuge Department of Agricultural Biology, Faculty of Agriculture, University of Ruhuna, Matara and Galle, Sri Lanka

Aryadeep Roychoudhury Discipline of Life Sciences, School of Sciences, Indira Gandhi National Open University, New Delhi, India

Data Ram Saini Institute of Agricultural Sciences, Banaras Hindu University, Varanasi, UP, India

Jyotsna Setty Department of Plant Physiology, Rajiv Gandhi South Campus-Banaras Hindu University, Mirzapur, India

Priyaka Shah Banaras Hindu University, Varanasi, India

Sarika Sharma Department of Botany, MMV, Banaras Hindu University, Varanasi, UP, India

Shachi Singh Department of Botany, MMV, Banaras Hindu University, Varanasi, UP, India

Pavan Singh Department of Soil Science and Agricultural Chemistry, JNKVV, College of Agriculture, Tikamgarh, India

Dig Vijay Singh Sardar Swaran Singh National Institute of Bioenergy, Kapurthala, Punjab, India

Girish Tantuway Department of Genetics and Plant Breeding, Rajiv Gandhi South Campus- Banaras Hindu University, Mirzapur, India

Shweta Verma Banaras Hindu University, Varanasi, India

P. Vijai Department of Plant Physiology, Institute of Agricultural Sciences, Banaras Hindu University, Varanasi, India

Vidushi Yadav Department of Applied Science, IIIT, Allahabad, India

Anjali Yadav Department of Botany, MMV, Banaras Hindu University, Varanasi, UP, India

Alice Kira Zelman Department of Plant Science and Landscape Architecture, University of Connecticut, Storrs, CT, USA

1 Plant Elicitor Peptides as Amplifiers of Immune Responses Against Biotic Stressors

Nalika P. Ranatnuge and W. P. Thisali Hasara

Abstract

Plants are continuously challenged by various pathogenic organisms and pests that threaten their existence. However, through millions of years of co-evolution, plants have acquired an array of defense mechanisms that enable them to withstand most biotic stresses effectively. Immunogenic signals perceived at the onset of an infection event trigger a cascade of defense responses within a plant. Some of these immunogenic signals originate externally, such as microbial immunogenic patterns or microbe-associated molecular patterns (MAMPs), which are recognized by pattern recognition receptors (PRR) localized in the plant cell membranes, initiating signaling cascades that induce defense responses. In addition, there are endogenous triggers, which may be constitutively expressed molecules or induced immunogenic factors, that stimulate the induction of immune responses. These induced endogenous immunogenic factors, known as phytocytokines, are detected by membrane-localized receptors (PEPRs) to modulate immune responses. Plant elicitor peptides (Peps) are a group of phytocytokines that amplify downstream signaling mechanisms by binding to receptors containing leucine-rich repeat domains, thereby initiating signaling cascades to reinforce pattern-triggered immunity (PTI), in response to various biotic challenges. Consequently, Peps have emerged as a focal point of research due to their pivotal role in modulating plant defense responses, offering potential applications in agriculture for enhancing plant resistance mechanisms through genetic manipulations and the commercial application of exogenous priming agents. This chapter provides an overview of the known classes of Peps, current insights into their modes of action, and their role in enhancing plant immunity.

N. P. Ranatnuge (✉) · W. P. T. Hasara
Department of Agricultural Biology, Faculty of Agriculture, University of Ruhuna, Mapalana, Kamburupitiya, Sri Lanka
e-mail: nalika@agbio.ruh.ac.lk

S. Singh, R. Mehrotra (eds.), *Plant Elicitor Peptides*,
https://doi.org/10.1007/978-981-97-6374-0_1

Keywords

Biotic stressors · PEPRs · Peps · Plant elicitor peptides · Plant immunity

1.1 Pattern-Triggered Immunity in Plants

Plants are sessile organisms. Due to their immobile nature, plants are equipped with unique mechanisms enabling them to rapidly react to harmful situations such as abiotic and biotic stresses. Since plants are multicellular organisms with compartmentalized anatomy and physiological functions, they must effectively manage such unfavorable circumstances without compromising their well-being. Unlike mammals, plants do not have a circulatory system or mobile defense systems and adaptive immunity. Alternatively, plants display cellular-level innate immunity and systemic signaling mechanisms originating from the site of infection, which alert distant parts of the plant (Dangl and Jones 2001).

Plants possess various passive defense elements such as waxy cuticles, callose layers on cell walls, constitutively produced antimicrobial compounds, and other pre-existing structural barriers to ward off most of the biotic threats. However, when a putative pathogen gets through such barriers or when the plant can "sense" the invasion, a rapid expression of defense responses takes place preventing the disease development (incompatible interaction). The disease develops only when the host is susceptible (compatible interaction) due to the absence of recognition factors or delayed expression of defense responses (Garcia-Brugger et al. 2006).

Apart from the non-host resistance or total immunity in which the plants are not subjected to damage by pathogens due to the absence of recognition factors, "true resistance" is a phenomenon where host–pathogen interactions take place through a complex network of molecular-level communications to suppress an infection. These communications are not solely limited to the molecules present within the plant but also linked to specific molecules produced by pathogens or pests, which plants perceive as "danger signals."

In a plant, resistance is controlled by its resistant (R) genes along with its cognate pathogen avirulent (Avr) genes as described in the gene-for-gene (GFG) model (Flor 1947). In short, R genes in plants and Avr genes in pathogens have co-evolved over millions of years through reciprocal genetic changes such that a "disease" is the outcome of their interplay. Many *R-Avr* gene combinations have been reported (Dangl and Jones 2001; Jones and Dangl 2006; Ngou et al. 2022).

In certain R-*Avr* combinations, the R genes are responsible for producing receptors that can identify pathogen-derived Avr proteins or elicitors (elicitor–receptor model) (Keen et al. 1972), where this elicitor–receptor binding leads to the launching of a signaling cascade that activates cellular defense responses. Resistance induced in this manner is called pattern-triggered immunity or PTI and is considered to be the first layer of active defense against pathogens (Ausubel 2005; Bittel and Robatzek 2007). The receptors that perceive the invaders are known as pattern recognition receptors (PRR) and are localized in the host's cell membranes.

Similar to specific PRRs for Avr proteins of pathogens (pathogen-associated molecular patterns—PAMPs), plants have unique PRRs to detect certain microbes (microbe-associated molecular patterns—MAMPs), herbivores (herbivore-associated molecular patterns—HAMPs), or damages (damage-associated molecular patterns—DAMPs). PRRs are usually either receptor-like kinases (RLKs) or receptor-like proteins (RLPs) which do not contain a protein kinase domain (Ausubel 2005).

Formation of elicitor–receptor complex initiates PTI by activating a multitude of metabolic processes. These include, MAP kinase (mitogen-activated protein kinase) signaling cascade, pathogen-responsive gene expression, phosphorylation of certain proteins, or activation of plasma membrane proteins such as components of receptor molecules, generation of various molecules involved in signaling (such as free calcium, nitric oxide [NO], reactive oxygen species [ROS]), and strengthening the cell wall through the deposition of cellulose or callose layers at sites of infection, all of which contribute to discourage pathogen entry or infection (Nürnberger et al. 2004). Furthermore, various molecules generated during this metabolic boost are engaged in many signaling networks or pathways that amplify and specify the immune responses. Particularly, early expression of genes responsible for phytohormone biosynthesis performs unique amendments in the cellular regulation process depending on the type of parasites (Ton et al. 2002).

Plants have developed these advanced responsive mechanisms as a strategy for their survival against various biotic factors. A single infection event can trigger a multitude of unique metabolic pathways that accommodate the energy expenditure of cells. So inducible defense mechanisms in plants are energy-efficient systems that activate in response to need. In addition, plants require continuous "conditioning" of their defense mechanisms in response to rapidly evolving pathogens and pests; therefore, the response for each type or even each strain of pathogens may need to be unique and effective. Furthermore, many antimicrobial compounds produced during defense responses, such as ROS are cytotoxic (Halliwell and Gutteridge 2007), so their production has to be well controlled. Therefore, the defense systems in a plant are under strict control and continue to be "fine-tuned" along the path of evolution.

As highlighted previously, it is imperative for a plant cell to promptly detect or "sense" the presence of invading pathogens or pests through recognition of their elicitors. While detection mechanisms such as cell membrane receptors and downstream signaling pathways are employed for this purpose, plants appear to utilize additional internal mechanisms as backup support to perceive more robust and expedited immune signals, thereby facilitating the efficient initiation of defensive responses against the invaders. Therefore, in addition to the elicitors originating from attacking organisms, plants also possess a myriad of in-house molecules that emerge as strong alarming signals in response to an injury/infection. These host-derived molecules are known as damage-associated molecular patterns (DAMPs) (Hou et al. 2019; Gigli-Bisceglia et al. 2020).

1.2 Damage-Associated Molecular Patterns (DAMPs) and Their Role in Plant Immunity

DAMPs are endogenous danger signals that secrete extracellularly upon stress or damage, which include ATP, pyridine, glutamate, cell wall fragments, protein fragments, peptide hormones, and reactive oxygen species (Albersheim and Anderson 1971; Chai and Doke 1987; Pearce et al. 1991; McGurl et al. 1992; Pearce et al. 2001; Huffaker et al. 2006; Mousavi et al. 2013; Tanaka et al. 2014; Toyota et al. 2018; Wang et al. 2019).

While PAMPs or MAMPs remain the initial triggers of molecular host defense responses, DAMPs might be considered as the inducers of the second wave of stronger and more prolonged immune responses within the plants (Fig. 1.1). The advantage of DAMPs is that they serve as more general inducers compared to pathogen-specific MAMPs. Thus, DAMPs may be non-specifically expressed in response to microbial pathogens, herbivore attacks, or mechanical damage (Albert 2013).

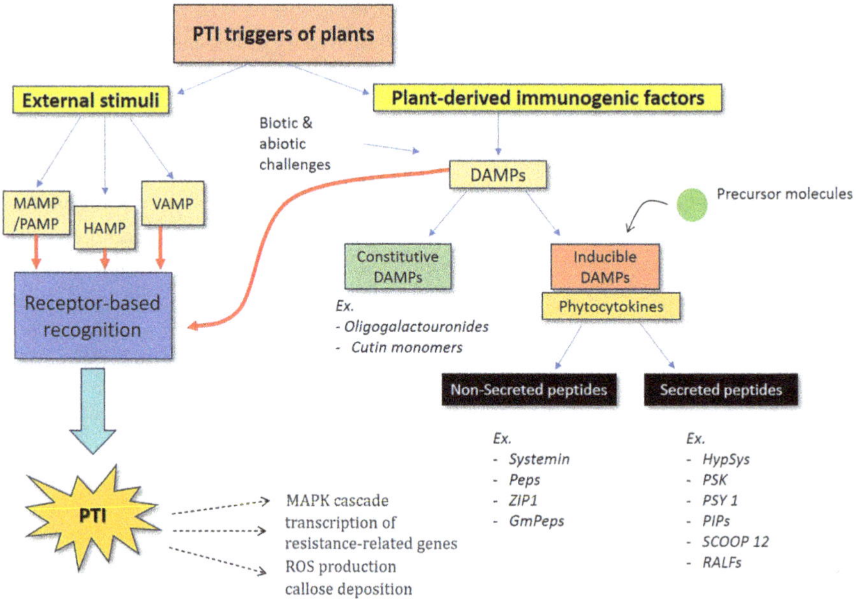

Fig. 1.1 Involvement of endogenous immunogenic factors in the PAMP-triggered immunity (PTI). These plant-derived factors known as damage-associated molecular patterns (DAMPs) amplify the initial stimuli of PTI responses such as microbe-associated molecular patterns (MAMPs)/pathogen-associated molecular patterns (PAMPs), herbivore-associated molecular patterns (HAMPs), and virus-associated molecular patterns (VAMPs). DAMPs initiate molecular signaling cascades similar to PAMPs and others leading to expression of various resistance responses within the host cell

These plant-derived immunogenic factors or DAMPs can be divided into two categories based on their origin. The passive or classical DAMPs originated from constitutively produced essential constituents of a normal, unstressed cell, which usually play an important homeostatic role, for example, cell wall components and ATP. When these molecules are detected freely in the apoplast, they are perceived as signals for immune responses. Two examples of passive DAMPs are oligogalacturonides and cutin monomers. They are passively generated when pathogens release various cell wall degrading enzymes upon their entry. In contrast, inducible DAMPs are peptides generated only during a stressful situation and are considered purely signal molecules produced under the tight control of the host. The production of such inducible compounds may be necessitated by the requirement for additional protection beyond damage-associated detection. These inducible peptidic signal molecules are also known as phytocytokines (Tanaka and Heil 2021; Gust et al. 2017). Phytocytokines are the products of precursor molecules (gene products) that have been subjected to posttranslational modifications. Phytocytokines are synthesized and processed in the cytosol and are released to the apoplasts upon infection or challenge (Luo 2012).

Functionally phytocytokines are similar to animal cytokines, a group of signaling peptides produced in response to various external and internal stimuli such as trauma, inflammation, sepsis, and tumors (Banchereau et al. 2012; Luo 2012). Therefore, phytocytokines are also known as plant peptide hormones that regulate plant immunity (Luo 2012). Furthermore, some peptide hormones originally known to regulate certain physiological functions such as growth and development, reproduction, and abiotic stress tolerance are now known to be involved in plant immunity and vice versa (Hou et al. 2021).

Phytocytokines are broadly classified into two classes: non-secreted peptides and secreted peptides. The former group derives from precursors that lack a signal peptide, while the latter are the products of precursors that contains a signal peptide. Few examples of non-secreted peptides include: Systemin (Pearce et al. 1991; Ryan and Pearce 2003; Wang et al. 2018), Pep1, Pep2, Pep3 (Huffaker et al. 2006, 2011; Liu et al. 2013; Ross et al. 2014; Yamaguchi et al. 2006, 2010), ZIP1 (Ziemann et al. 2018), and GmPep890 and GmPep914 (Yamaguchi and Huffaker 2011). Currently known secreted peptides include: PSK, PSY1 (Amano et al. 2007; Igarashi et al. 2012; Matsubayashi et al. 2002; Matsubayashi and Sakagami 1996; Mosher et al. 2013; Rodiuc et al. 2016; Zhang et al. 2018), PIP1, PIP2 (Hou et al. 2014), IDA, IDL6 (Butenko et al. 2003; Patharkar and Walker 2016; Stenvik et al. 2008; Wang et al. 2017), SCOOP12 (Gully et al. 2019; Hou et al. 2021; Rhodes et al. 2021), RGF7/GLV4 and RGF9/GLV2 (Matsuzaki et al. 2010; Ou et al. 2016; Stegmann et al. 2021; Wang et al. 2021b; Whitford et al. 2012), HypSys (Pearce et al. 2001; Pearce and Ryan 2003), and RALF17, RALF22, and RALFM23 (Haruta et al. 2014; Li et al. 2015; Stegmann et al. 2017; Guo et al. 2018).

The translated precursor gene products are transported via the cytosol to the apoplast and must be modified to serve their functions as phytocytokines. In the case of secreted peptides, precursor proteins consist of three peptide segments, namely amino (N) terminal signal peptide, carboxyl (C)-terminal conserved region, and a middle region called prodomain.

Conversely, non-secreted peptides lack N terminal signal peptide sequence (Matsubayashi 2014; Olsson et al. 2019). Following translation, secreted peptides enter the internal secretory pathways of the endoplasmic reticulum (ER) and Golgi, aided by signal peptides, and are subsequently secreted into the apoplast as mature phytocytokines. During this secretion process, the precursor protein undergoes proteolytic cleavage of the signal peptide and the prodomain, along with other post-translational modifications. Non-secreted peptides, however, do not enter the secretory pathways of the cell, undergo posttranslational modifications within the cytosol (or in the apoplast), and are believed to be released to the apoplast via cellular damage and/or alternative secretory pathways (Ding et al. 2012).

1.3 Plant Elicitor Peptides (Peps)

Among several categories of phytocytokines that have been reported in many higher plants, significant attention has recently been paid to one of its sub-categories known as plant elicitor peptides (Peps) due to the discovery of their vital role in boosting the defense mechanism in plants. Much recent work has been undertaken to understand their cellular synthesis, unique features, different functional roles, downstream signaling mechanisms, and the specific immune responses induced by Peps.

Plant elicitor peptides (Peps) are inducible DAMPs or phytocytokines produced by various plant species (Huffaker et al. 2006). Peps have been identified to induce conserved signals across several plant species, activating immune responses against microbes, nematodes, and herbivores (Huffaker et al. 2011, 2013; Trivilin et al. 2014; Lee et al. 2018; Ruiz et al. 2018; Shinya et al. 2018; Zhang and Gleason 2020).

1.4 Formation and Structure of Peps

Peps represent a significant part of the inducible defense system in plants. They are the end products of PROPEP gene transcripts expressed in response to various stress signals. Peps are short peptides made up of 23–36 amino acids, posttranslationally cleaved from the C termini of their precursor proteins known as PROPEPs (Bartels et al. 2013) by the cysteine protease METACASPASE4 and transferred to the apoplast (Hander et al. 2019; Shen et al. 2019) in response to cellular damage. Peps located in the apoplast are recognized by plasma membrane-located leucine-rich repeat (LRR) receptor-like kinases (RK) (LRR-RKs) referred to as Pep receptors (PEPRs). This recognition event provokes typical PTI responses as in the case of MAMPs or HAMPs (Yamaguchi et al. 2006, 2010).

Current understanding suggests that PROPEPs are localized in the cytoplasm under normal conditions, and in case of damage or infection, the Peps are cleaved off from their precursors and released to the apoplast either passively through the damaged cell membranes or actively in a controlled process (Bartels et al. 2013). Peps liberated to the apoplast are perceived by PEPRs of the same cell or by nearby cells and trigger the immune signaling pathways (Fig. 1.2).

Thus, the cascade of responses initiated by PEP-PEPR-mediated plant defense signaling includes phosphorylation of MAP kinases, generation of cellular signaling mechanisms (such as cyclic GMP, reactive oxygen species, nitric oxide, and various defense-related hormones), induction of Ca^{2+} signals, the release of volatile substances (VOC), boosting resistant gene expression, and callose and lignin deposition on cell walls (Bartels and Boller 2015). In addition, the Pep-PEPR system is also believed to be involved in systemic immune signaling. Studies in *Arabidopsis* revealed that the localized AtPep2 application substantially induces systemic immunity (Ross et al. 2014; Mishina and Zeier 2007). Figure 1.2 summarizes the current knowledge on plant elicitor peptide (Pep)-associated immune signaling mechanisms.

The first discovery of Peps was made in *Arabidopsis*, namely AtPep1, a peptide consisting of 23 amino acids derived from its 92-aa precursor protein known as AtPROPEP1 from an extract of *Arabidopsis* leaves (Huffaker et al. 2006). The AtPep1 is derived from the C terminal of its precursor protein (Fig. 1.3), while the rest of the peptide is enzymatically cleaved off during the posttranslational modifications (Huffaker et al. 2006).

Later, seven other paralogs of the PROPEP1 gene were discovered, namely AtPROPEP2–AtPROPEP8 (Krol et al. 2010; Yamaguchi et al. 2010; Bartels et al. 2013) that give rise to 8 different AtPeps.

The genes that encode precursors of endogenous peptide elicitors are well known to be induced by biotic stressors as well as exogenous application of

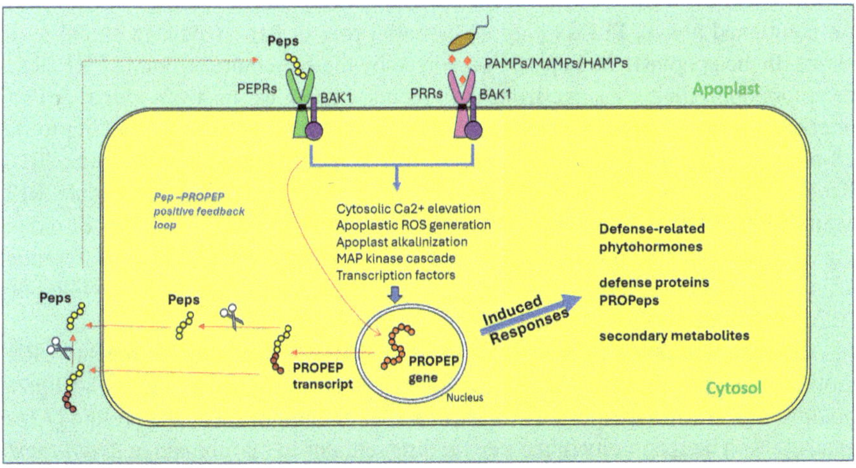

Fig. 1.2 An illustration of plant elicitor peptide (Pep)-associated immune signaling mechanisms. Peps cleaved from PROPEPs are detected by the PEPRs in the apoplast and the downstream immune signals are generated, which are likely to overlap with MAMP-associated signaling pathways. Peps act as inducers of PROPEP genes, creating a positive feedback loop (depicted in red arrows) for the amplification of defense induction signals

1	10	20	30	40	50
MEKSDRRSEE	SHLWIPLQCL	DQTLRAILKC	LGLFHQDSPT	TSSPGTSKQP	

60	70	80	92
KEEKEDVTME	KEEVVVTSR	ATKVKAKQRGKEKVSSGRPGQHN	

Fig. 1.3 The amino acid sequence of the AtPep1 precursor protein PROPEP 1. The AtPep sequence at the C terminus is highlighted. (Adopted from Huffaker et al. 2006)

elicitors (Pearce et al. 1991; Pearce et al. 2001; Huffaker et al. 2006; Huffaker et al. 2011). In *Arabidopsis*, expressions of PROPEP2 and PROPEP3 were upregulated by pathogens, such as Pseudomonas, *Botrytis cinerea*, and *Phytophthora infestans* (Huffaker et al. 2006). In addition, the PROPEP1 gene was activated by wounding, ethylene, methyl jasmonate, flg22, and AtPep1 itself (Yamaguchi and Huffaker 2011; Huffaker et al. 2006). This indicates the role of Peps in innate immunity. Furthermore, the activation of a self-amplifying loop among Peps and their respective PROPEP genes was observed (Huffaker et al. 2006). For instance, the expression of PROPEP was enhanced by AtPep1 in a positive feedback loop that further amplified the immune responses (Gully et al. 2019; Huffaker et al. 2006; Hou et al. 2014; Wang et al. 2021a).

Besides, certain Peps may induce precursor genes of other phytocytokines apart from their own precursor genes. As an example, PIP1 and Pep1 can upscale the expression of the precursor genes of each other (Hou et al. 2014). This indicates that the peptide signaling networks are interconnected and dependent on one another.

1.5 Role of Pep Receptors (PEPRs)

As mentioned above, PEPRs play an essential role in Pep-associated signaling in plants. In the apoplast, the Peps are perceived by these membrane-bound LRR-RKS receptors upon their cleavage from the precursor proteins. In *Arabidopsis*, AtPeps interact with two LRR-RKs, namely AtPEPR1 and AtPEPR2 which belong to the 28-membered family LRR XI of *Arabidopsis* RLKs (Yamaguchi et al. 2006, 2010; Krol et al. 2010; Liu et al. 2010; Postel et al. 2010; Roux et al. 2011; Liu et al. 2013; Majhi et al. 2019). Even though all eight AtPeps display similar functions of inducing the plant defense signaling, AtPEPR1 and AtPEPR2 seem to have a differential affinity for different AtPeps. For example, AtPEPR1 prefers AtPep 1–6, while only AtPep 1–2 are responsive to AtPEPR2 (Yamaguchi et al. 2010).

AtPEPRs consist of three main domains, namely a leucine-rich repeat (LRR) motif, a transmembrane (TM) domain, and a kinase domain. LRR is the ligand binding region which is located in the extracellular component of the protein (Afzal et al. 2008). The specificity of the Pep binding remains in this domain in *Arabidopsis* and is likely to be the same in other plant PEPRs (Zelman and Berkowitz 2023). The kinase domain represents the intracellular segment of the PEPR and contains a putative guanylyl cyclase (GC) domain that is linked to Ca^{2+} elevation in the cell (Ryan 2000). Qi et al. (2010) indirectly proved that the GC domain of AtPEPR1 produced cGMP, the activator of cyclic nucleotide-gated Ca^{2+} conducting channel (CNGC2), which is involved in cytosolic Ca^{2+} elevation.

Upon binding with the LRR-RK receptors, the Peps are assumed to further interact with LRR-RK co-receptors (e.g., BRI1-associated kinase 1—BAK1 and somatic embryo receptor kinase—SERK) to initiate downstream signaling which is similar to MAMP/HAMP signaling pathway (Roux et al. 2011; Hou et al. 2014; Gust et al. 2017). The crystal structure of AtPep bound to the LRR domain of the AtPEPR2 was released in 2015. It confirmed that the heterodimerization of AtPep1 with the PEPR1 and its co-receptor BAK1 is required for its activation (Tang et al. 2015).

It is important to note that BAK1 co-receptor is involved in both MAMP- and AtPep-associated signaling (Heese et al. 2007; Postel et al. 2010; Schulze et al. 2010). This suggests that BAK1 plays a central role in AtPep and MAMP signaling, all of which lead to the expression of a similar set of genes such as MPKinase 3, WRKY transcription factor, PR-1, and PDF 1.2 (Zipfel et al. 2004; Zipfel et al. 2006; Huffaker et al. 2006; Huffaker and Ryan 2007; Yamaguchi et al. 2010).

In addition, another extracellular receptor kinase was found to associate with the LRR motif of AtPEPR1 and AtPEPR2, known as AtAPEX. It works in a ligand-independent manner with the receptor and is reported to be required for the proper functioning of AtPep2-induced responses (Smakowska-Luzan et al. 2018).

1.6 Occurrence of Peps in Plants

Peps have been found in different plant families including Poaceae, Rosaceae, Solanaceae, Fabaceae, and Brassicaceae (Table 1.1) (Huffaker et al. 2013; Huffaker 2015; Lee et al. 2018; Ruiz et al. 2018). A few examples are SmPep1 from *Solanum melongena*, CaPep1 from *Capsicum annuum*, StPep1 from *S. tuberosum*, GmPep1–3 from *Glycine max*, MtPep1 from *Medicago truncatula*, and AhPep1 from *Arachis hypogaea* (peanut) (Huffaker et al. 2013).

Two PROPEPs have been studied in maize plants: ZmPROPEP1, which was induced upon fungal infections, and ZmPROPEP3, induced by the application of oral secretions of *Spodoptera exigua* (Huffaker et al. 2011, 2013). Accordingly, upregulated defense genes were observed when the plants were treated with ZmPep1 and ZmPep3 and displayed enhanced fungal and herbivore resistance (Huffaker et al. 2011, 2013).

In soybean (*Glycine max*), three Peps have been discovered, namely GmPep1, GmPep2, and GmPep3 which demonstrated defense signal induction in response to attacks by chewing insects and nematodes (Huffaker et al. 2013; Lee et al. 2018). OsPep3 was reported in rice (Oryza sativa) that displayed activation of immune responses against bacteria, fungi, and piercing and sucking insect attacks (Shen et al. 2022). SlPep activity against necrotrophic pathogens was reported in tomato (Lori et al. 2015; Yang et al. 2023).

One of the interesting facts is that, despite the similar action of defense induction mediated by PROPEPs from diverse plant species such as *Arabidopsis* and maize, these peptides display a very low homology in amino acid composition (Huffaker

Table 1.1 A list of plant elicitor peptides (Peps) currently described in literature

Family	Plant species	Pep name	Target biotic stress factor	References
Brassicaceae	*Arabidopsis*	AtPep 1–8	Chewing insects, hemibiotrophic pathogens, necrotrophic pathogens	Huffaker et al. (2006), Klauser et al. (2015), and Ma et al. (2013)
Poaceae	*Zea mays*	ZmPep1, ZmPep3	Chewing insects, necrotrophic pathogen	Huffaker et al. (2011, 2013)
	Rice (*Oryza sativa*)	OSPep3	Bacteria, fungi, piercing and sucking insect	Shen et al. (2022)
Solanaceae	Tomato (*Solanum lycopersicum*)	SIPep	Necrotrophic pathogen	Lori et al. (2015) and Yang et al. (2023)
Fabaceae	Soybean (*Glycine max*)	GmPep3	Chewing insects, nematodes	Huffaker et al. (2013) and Lee et al. (2018)
Rosaceae	Peach (*Prunus persica*)	PpPep1, PpPep2	Bacteria; necrotrophic pathogen	Ruiz et al. (2018)

et al. 2011). Not only between different species, according to Yamaguchi et al. (2006), but also among the PROPEPs of Arabidopsis, the total amino acid sequence identity was in the range of 12% and 47%. This dissimilarity suggests that the Peps do not act identically nor have identical structural and functional properties that make them unique even though they initiate conserved downstream signaling mechanisms (Heitz et al. 1997; Huffaker et al. 2006, 2011; Pearce et al. 2007, 2010). Because of this reason, the investigation of Peps in different plant species needs separate in-depth studies. Peps have relatively short peptides so they might tend to evolve faster than other plant-derived complex antimicrobial compounds and this feature might have the evolutionary advantage for the host to rapidly modify its resistant mechanisms against fast-evolving microbial pathogens.

1.7 Pep Signaling in Defense Responses Against Different Pathogens and Herbivores

Peps play an important role in defense signaling against various organisms. In *Arabidopsis*, herbivore attacks induced Pep signaling, hence the defense responses were triggered (Klauser et al. 2015). AtPeps also induced defense against *Pseudomonas syringae* pv. tomato DC3000 and *Pythium irregulare* (Huffaker et al. 2006; Yamaguchi et al. 2010). In maize, ZmPep1 induces defense against pathogens (Huffaker et al. 2011), while ZmPep3 induces anti-herbivore defenses (Huffaker et al. 2013). Pretreatment of maize with ZmPep1 increased resistance to stalk-rot caused by *Colletotrichum graminicola* and leaf blight caused by *Cochliobolus heterostrophus* (Huffaker et al. 2011). Several Peps in *Prunus Persica* (Rosaceae family) displayed defense activation against bacterial infections (Ruiz et al. 2018). Nematicidal responses were observed in soybean Peps (Lee et al. 2018). Several solanaceous crops such as pepper, eggplant, and tomato expressed volatile organic compounds (VOC) in response to Pep application (Huffaker et al. 2013). Trivilin et al. (2014) reported that the resistance of tomatoes against a necrotrophic pathogen was compromised when Pep signaling was disrupted. Among the seven rice Peps characterized so far, OsPep3 displayed defense against both insect pests (brown planthopper) and diseases (*Magnaporthe oryzae* and *Xanthomonas oryzae* pv. oryzae) (Shen et al. 2022).

1.8 Cellular Functions of Peps

AtPeps (Arabidopsis) and ZmPeps (maize) perception have been shown to induce various components of PTI defense responses, such as phosphorylation, generation of ROS (reactive oxygen species), hormone changes, induction of systemic signals, calcium signals, callose and lignin deposition, volatile compound production (VOC), production of a variety of antimicrobial compounds, and the gene expression (Yamaguchi and Huffaker 2011; Huffaker et al. 2013). It is important to note that all the above defense responses are typical plant responses initiated by MAMPs (Zipfel et al. 2004, 2006).

Several studies have highlighted the involvement of different defense-related phytohormones in the functioning of Peps. In maize, increased levels of jasmonic acid (JA) and ethylene (ET) were observed in response to Pep signaling resulting in defense against herbivore attacks (Huffaker et al. 2013). The expression of the two ET-synthesizing genes (ERF-1a and ERF-2b) in *Prunus* was enhanced by Peps (Ruiz et al. 2018). Comparable observations were made concerning the ERF1 gene of tomato (Trivilin et al. 2014). Similarly, the exogenous application of Peps increased ET levels in tomato (Yang et al. 2023).

When the defense mechanism of a cell is switched on, reactive oxygen species (ROS) generation is triggered. ROS are not only involved in destroying pathogens but also serve as signaling molecules in various metabolic pathways such as salicylic acid (SA) signaling and Pep signaling that induce immune responses (Ma et al. 2013; Tintor et al. 2013). Similarly, Peps are known to cause calcium responses (Moyen et al. 1998; Ryan 2000). Transient spikes in calcium ion levels in damaged cells both induce and are induced by ROS generation (Kawano and Bouteau 2013). Therefore, it is evident that Pep signaling, ROS generation, and elevation in cellular calcium levels are interlinked. In *Arabidopsis*, ROS generation was activated by Pep-induced calcium release (Ma et al. 2013). In addition to the involvement of Peps in the early expression of defense actions such as ROS generation and calcium elevation, they are also involved in late defense responses such as callose and lignin depositions on cell walls that structurally protect the plant cells (Zhang et al. 2022).

Furthermore, Peps can modulate defense-related gene expression. Peps are known to promote the defense gene expression against pathogens, for example, PR-1 and PDF1.2 in Arabidopsis (Huffaker et al. 2006). Also in maize, ZmPep3 could induce levels of AOS (allene oxidase synthase), PINs (proteinase inhibitors), and other defense-related transcription (Huffaker et al. 2013). The expression of the defense genes ACS, ERF1, LOXD, PR, and DEF2 in tomato was lower when the SlPROPEP gene was silenced. Similarly, the application of exogenous SlPep enhanced the expression of PR genes, WRKY33A, JA, and ET synthesis genes (Trivilin et al. 2014; Yang et al. 2023). The exact mechanisms and pathways of Pep-induced MPK phosphorylation, cytosolic calcium spikes, and enhancement of ROS are subjects for further study.

1.9 Response of Pathogens to Peps

Certain plant pathogens secrete apoplastic and cytoplasmic (intracellular) effectors to suppress plant defense responses and other cellular functions, thereby making plants vulnerable to infection (Wawra et al. 2012). Some of these effectors block PTI signaling pathways to overcome innate immunity in plants. An effector protein named SePele1, secreted by the sugarcane smut pathogen *Sporisorium scitamineum* was observed to co-express with the Pep receptor ScPEPR in the apoplast. SePele1 has a Pep-like amino acid sequence in the C terminal that could bind to the extracellular LRR domain of the ScPEPR1 to compete with Pep binding, which leads to suppression of ScPep-ScPEPR1 immune responses (Ling et al. 2022). Even though many studies have focused on elucidating the resistant strategies of plants or the virulence mechanisms of pathogens via advanced techniques like metabolomics, proteomics, or DNA/RNA-related omics, mixed transcriptome or proteome data has been seldom available. Approaches similar to weighted gene co-expression network analysis (WGCNA) would be ideal for the study of co-expression of pathogen and host genes during the infection process (Ling et al. 2022).

1.10 Future Directions in Pep Research

Although numerous studies have reported Pep-induced signaling in plants, their occurrence and modes of action across various plant species remain to be fully elucidated. Furthermore, significant knowledge gaps persist regarding Pep synthesis, posttranslational modifications, and the precise mechanisms underlying their downstream signaling. Additionally, a deeper understanding of the interplay between Pep-associated defense induction and other cellular defense mechanisms is needed. Likewise, exploring the feasibility of genetic manipulation of their precursor genes represents a promising avenue for future research.

In terms of their potential applications in agriculture, Peps offer an alternative to conventional pesticides and costly MAMP-based plant protection chemicals, such as harpin (Zelman and Berkowitz 2023). Given that Peps exhibit activity at lower concentrations and possess a simpler peptide structure compared to MAMPs, their commercial synthesis and utilization are likely to be more efficient and cost-effective (Choi et al. 2013). Moreover, integrating signal peptides into crop protection strategies could prime plants to withstand disease without incurring significant crop and yield losses. Furthermore, as plant endogenous compounds, Peps are expected to exhibit reduced persistence in the environment compared to conventional pesticides. Thus, Peps hold promise for playing a significant role in crop protection aspects of commercial agriculture in the future.

References

Afzal AJ, Wood AJ, Lightfoot DA (2008) Plant receptor-like serine threonine kinases: roles in signaling and plant defense. Mol Plant-Microbe Interact 21:507–517

Albersheim P, Anderson AJ (1971) Proteins from plant cell walls inhibit polygalacturonases secreted by plant pathogens. Proc Natl Acad Sci USA 68:1815–1819

Albert M (2013) Peptides as triggers of plant defence. J Exp Bot Dec 64(17):5269–5279. https://doi.org/10.1093/jxb/ert275. Epub 2013 Sep 7. PMID: 24014869

Amano Y, Tsubouchi H, Shinohara H, Ogawa M, Matsubayashi Y (2007) Tyrosine-sulfated glycopeptide involved in cellular proliferation and expansion in Arabidopsis. Proc Natl Acad Sci USA 104(46):18333–18338. https://doi.org/10.1073/pnas.0706403104

Ausubel F (2005) Are innate immune signaling pathways in plants and animals conserved? Nat Immunol 6:973–979. https://doi.org/10.1038/ni1253

Banchereau J, Pascual V, O'Garra A (2012) From IL-2 to IL-37: the expanding the spectrum of anti-inflammatory cytokines. Nat Immunol 13(10):925–931. https://doi.org/10.1038/ni.2406

Bartels S, Boller T (2015) Quo vadis, pep? Plant elicitor peptides at the crossroads of immunity, stress, and development. J Exp Bot 66:5183–5193. https://doi.org/10.1093/jxb/erv180

Bartels S, Lori M, Mbengue M, van Verk M, Klauser D, Hander T, Boni R, Robatzek S, Boller T (2013) The family of peps and their precursors in arabidopsis: differential expression and localization but similar induction of pattern-triggered immune responses. J Exp Bot 64:5309–5321. https://doi.org/10.1093/jxb/ert330

Bittel P, Robatzek S (2007) Microbe-associated molecular patterns (MAMPs) probe plant immunity. Curr Opin Plant Biol 10(4):335–341. https://doi.org/10.1016/j.pbi.2007.04.021

Butenko MA, Patterson SE, Grini PE, Stenvik GE, Amundsen SS, Mandal A, Aalen RB (2003) Inflorescence deficient in abscission controls floral organ abscission in Arabidopsis and identifies a novel family of putative ligands in plants. Plant Cell 15(10):2296–2307. https://doi.org/10.1105/tpc.014365

Chai HB, Doke N (1987) Superoxide anion generation: a response of potato leaves to infection with Phytophthora infestans. Phytopathology 77:645–649

Choi MS, KimW LC, Oh CS (2013) Harpins, multifunctional proteins secreted by gram-negative plant-pathogenic bacteria. Mol Plant-Microbe Interact 26:1115–1122

Dangl JL, Jones JD (2001) Plant pathogens and integrated defence responses to infection. Nature 411:826–833

Ding Y, Wang J, Wang J, Stierhof Y-D, Robinson DG, Jiang L (2012) Unconventional protein secretion. Trends Plant Sci 17:606–615

Flor HH (1947) Inheritance of reaction to rust in flax. J Agric Res 74:241–262

Garcia-Brugger A, Lamotte O, Vandelle E, Bourque S, Lecourieux D, Poinssot B, Wendehenne D, Pugin A (2006) Early signaling events induced by elicitors of plant defenses. Mol Plant-Microbe Interact 19(7):711–724. https://doi.org/10.1094/MPMI-19-0711. PMID: 16838784

Gigli-Bisceglia N, Engelsdorf T, Hamann T (2020) Plant cell wall integrity maintenance in model plants and crop species-relevant cell wall components and underlying guiding principles. Cell Mol Life Sci 77:2049–2077. https://doi.org/10.1007/s00018-019-03388-8

Gully K, Pelletier S, Guillou MC, Ferrand M, Aligon S, Pokotylo I, Perrin A, Vergne E, Fagard M, Ruelland E, Grappin P, Bucher E, Renou JP, Aubourg S (2019) The SCOOP12 peptide regulates defense response and root elongation in Arabidopsis thaliana. J Exp Bot 70:1349–1365. https://doi.org/10.1093/JXB/ERY454

Guo H, Nolan TM, Song G, Liu S, Xie Z, Chen J, Schnable PS, Walley JW, Yin Y (2018) FERONIA receptor kinase contributes to plant immunity by suppressing jasmonic acid signaling in Arabidopsis thaliana. Curr Biol 28:3316–3324

Gust AA, Pruitt R, Nurnberger T (2017) Sensing danger: key to activating plant immunity. Trends Plant Sci 22(9):779–791. https://doi.org/10.1016/j.tplants.2017.07.005

Halliwell B, Gutteridge JMC (2007) Free radicals in biology and medicine, 4th edn. Oxford University Press, New York

Hander T, Fernández-Fernández ÁD, Kumpf RP, Willems P, Schatowitz H, Rombaut D, Staes A, Nolf J, Pottie R, Yao P, Gonçalves A, Pavie B, Boller T, Gevaert K, Van Breusegem F, Bartels S, Stael S (2019) Damage on plants activates Ca2+– dependent metacaspases for release of immunomodulatory peptides. Science 363(6433):eaar7486. https://doi.org/10.1126/science.aar7486

Haruta M, Sabat G, Stecker K, Minkoff BB, Sussman MR (2014) A peptide hormone and its receptor protein kinase regulate plant cell expansion. Science 343(6169):408–411. https://doi.org/10.1126/science.1244454

Heese A, Hann DR, Gimenez-Ibanez S, Jones AM, He K, Li J, Schroeder JI, Peck SC, Rathjen JP (2007) The receptor-like kinase SERK3/BAK1 is a central regulator of innate immunity in plants. Proc Natl Acad Sci USA 104:12217–12222

Heitz T, Bergey DR, Ryan CA (1997) A gene encoding a chloroplast targeted lipoxygenase in tomato leaves is transiently induced by wounding, systemin, and methyl jasmonate. Plant Physiol 114:1085–1093

Hou S, Wang X, Chen D, Yang X, Wang M, Turrà D, Di Pietro A, Zhang W (2014) The secreted peptide PIP1 amplifies immunity through receptor-like kinase 7. PLoS Pathog 10:e1004331

Hou S, Liu Z, Shen H, Wu D (2019) Damage-associated molecular pattern-triggered immunity in plants. Front Plant Sci 10:646. https://doi.org/10.3389/fpls.2019.00646

Hou S, Liu D, Huang S, Luo D, Liu Z, Wang P, Mu R, Han Z, Chai J, Shan L, He P (2021) Immune elicitation by sensing the conserved signature from phytocytokines and microbes via the Arabidopsis MIK2 receptor. bioRxiv preprint. https://doi.org/10.1101/2021.01.28.428652

Huffaker A (2015) Plant elicitor peptides in induced defense against insects. Curr Opin Insect Sci 9:44–50

Huffaker A, Ryan CA (2007) Endogenous peptide defense signals in Arabidopsis differentially amplify signaling for the innate immune response. Proc Natl Acad Sci USA 104:10732–10736

Huffaker A, Pearce G, Ryan CA (2006) An endogenous peptide signal in Arabidopsis activates components of the innate immune response. Proc Natl Acad Sci USA 103:10098–10103. https://doi.org/10.1073/pnas.0603727103

Huffaker A, Dafoe NJ, Schmelz EA (2011) ZmPep1, an ortholog of Arabidopsis elicitor peptide 1, regulates maize innate immunity and enhances disease resistance. Plant Physiol 155(3):1325–1338. https://doi.org/10.1104/pp.110.166710. Epub 2011 Jan 4. PMID: 21205619; PMCID: PMC3046589

Huffaker A, Pearce G, Veyrat N, Erb M, Turlings TC, Sartor R, Shen Z, Briggs SP, Vaughan MM, Alborn HT, Teal PE, Schmelz EA (2013) Plant elicitor peptides are conserved signals regulating direct and indirect antiherbivore defense. Proc Natl Acad Sci USA (14):5707–5712. https://doi.org/10.1073/pnas.1214668110. Epub 2013 Mar 18. PMID: 23509266; PMCID: PMC3619339

Igarashi D, Tsuda K, Katagiri F (2012) The peptide growth factor, phytosulfokine, attenuates pattern-triggered immunity. Plant J 71(2):194–204. https://doi.org/10.1111/j.1365-13X.2012.04950.x

Jones JD, Dangl JL (2006) The plant immune system. Nature 444:323–329

Kawano T, Bouteau F (2013) Crosstalk between intracellular and extracellular salicylic acid signaling events leading to long-distance spread of signals. Plant Cell Rep 32:1125–1138

Keen NT, Partridge J, Zaki AI (1972) Pathogen-produced elicitor of a chemical defense mechanism in soybeans monogenetically resistant to Phytophthora megasperma var. glycinea. (Abstr.). Phytopathology 62:768

Klauser D, Desurmont GA, Glauser G, Vallat A, Flury P, Boller T, Turlings TC, Bartels S (2015) The AtPep-PEPR system is induced by herbivore feeding and contributes to JA-mediated plant defence against herbivory. J Exp Bot 66:5327–5336
Krol E, Mentzel T, Chinchilla D, Boller T, Felix G, Kemmerling B, Postel S, Arents M, Jeworutzki E, Kas A-R et al (2010) Perception of the Arabidopsis danger signal peptide 1 involves the pattern recognition receptor AtPEPR1 and its close homologue AtPEPR2. J BiolChem 285:13471–13479
Lee MW, Huffaker A, Crippen D, Robbins RT, Goggin FL (2018) Plant elicitor peptides promote plant defences against nematodes in soybean. Mol Plant Pathol 19:858–869
Li C, Yeh FL, Cheung AY, Duan Q, Kita D, Liu MC, Maman J, Luu EJ, Wu BW, Gates L, Jalal M, Kwong A, Carpenter H, Wu HM (2015) Glycosylphosphatidylinositol-anchored proteins as chaperones and coreceptors for FERONIA receptor kinase signaling in Arabidopsis. elife 4:e06587. https://doi.org/10.7554/eLife.06587
Ling H, Fu X, Huang N, Zhong Z, Su W, Lin W, Cui H, Que Y (2022) A sugarcane smut fungus effector simulates the host endogenous elicitor peptide to suppress plant immunity. New Phytol 233(2):919–933. https://doi.org/10.1111/nph.17835. Epub 2021 Nov 12. Erratum in: New Phytol 2024 Feb;241(4):1878–1879. PMID: 34716592; PMCID: PMC9298926
Liu YX, Yang XM, Ma J, Wei YM, Zheng YL, Ma XH, Yao JB, Manners JM, Liu CJ (2010) Plant height affects crown rot severity in wheat (Triticum aestivum L.). Phytopathology 100:1276–1281
Liu W, Liu J, Ning Y, Ding B, Wang X, Wang Z et al (2013) Recent progress in understanding PAMP- and effector-triggered immunity against the rice blast fungus Magnaporthe oryzae. Mol Plant 6:605–620. https://doi.org/10.1093/mp/sst015
Lori M, Van Verk MC, Hander T, Schatowitz H, Klauser D, Flury P, Gehring CA, Boller T, Bartels S (2015) Evolutionary divergence of the plant elicitor peptides (peps) and their receptors: interfamily incompatibility of perception but compatibility of downstream signalling. J Exp Bot 66:5315–5325
Luo L (2012) Plant cytokine or phytocytokine. Plant Signal Behav 7(12):1513–1514. https://doi.org/10.4161/psb.22425. Epub 2012 Oct 16. PMID: 23072994; PMCID: PMC3578880
Ma Y, Zhao Y, Walker RK (2013) Berkowitz GA (2013) molecular steps in the immune signaling pathway evoked by plant elicitor peptides: Ca2+-dependent protein kinases, nitric oxide, and reactive oxygen species are downstream from the early Ca2+ signal. Plant Physiol 163:1459–1471
Majhi BB, Sreeramulu S, Sessa G (2019) BRASSINOSTEROID-SIGNALING KINASE5 associates with immune receptors and is required for immune responses. Plant Physiol 180:1166–1184
Matsubayashi Y (2014) Posttranslationally modified small-peptide signals in plants. Annu Rev Plant Biol 65(1):385–413. https://doi.org/10.1146/annurev-arplant-050312-120122
Matsubayashi Y, Sakagami Y (1996) Phytosulfokine, sulfated peptides that induce the proliferation of single mesophyll cells of Asparagus officinalis L. Proc Natl Acad Sci USA 93(15):7623–7627. https://doi.org/10.1073/pnas.93.15.7623
Matsubayashi Y, Ogawa M, Morita A, Sakagami Y (2002) An LRR receptor kinase involved in perception of a peptide plant hormone, phytosulfokine. Science 296(5572):1470–1472. https://doi.org/10.1126/science.1069607
Matsuzaki Y, Ogawa-Ohnishi M, Mori A, Matsubayashi Y (2010) Secreted peptide signals required for maintenance of root stem cell niche in Arabidopsis. Science 329(5995):1065–1067. https://doi.org/10.1126/science.1191132
McGurl B, Pearce G, Orozco-Cardenas M, Ryan CA (1992) Structure, expression, and antisense inhibition of the systemin precursor gene. Science 255(5051):1570–1573. https://doi.org/10.1126/science.1549783
Mishina TE, Zeier J (2007) Pathogen-associated molecular pattern recognition rather than development of tissue necrosis contributes to bacterial induction of systemic acquired resistance in Arabidopsis. Plant J 50:500–513
Mosher S, Seybold H, Rodriguez P, Stahl M, Davies KA, Dayaratne S, Morillo SA, Wierzba M, Favery B, Keller H, Tax FE, Kemmerling B (2013) The tyrosine-sulfated peptide receptors

PSKR1 and PSY1R modify the immunity of Arabidopsis to biotrophic and necrotrophic pathogens in an antagonistic manner. Plant J 73(3):469–482. https://doi.org/10.1111/tpj.12050

Mousavi SA, Chauvin A, Pascaud F, Kellenberger S, Farmer EE (2013) GLUTAMATE RECEPTOR-LIKE genes mediate leaf-to-leaf wound signalling. Nature 500(7463):422–426. https://doi.org/10.1038/nature12478. PMID: 23969459

Moyen C, Hammond-Kosack KE, Jones J, Knight MR, Johannes E (1998) Systemin triggers an increase of cytoplasmic calcium in tomato mesophyll cells: Ca2+ mobilization from intra- and extracellular compartments. Plant Cell Environ 21:1101–1111

Ngou BPM, Ding PT, Jones JDG (2022) Thirty years of resistance: Zig-zag through the plant immune system. Plant Cell 34:1447–1478. https://doi.org/10.1093/plcell/koac041

Nürnberger T, Brunner F, Kemmerling B, Piater L (2004) Innate immunity in plants and animals: striking similarities and obvious differences. Immunol Rev198:249–266

Olsson V, Joos L, Zhu S, Gevaert K, Butenko MA, De Smet I (2019) Look closely, the beautiful may be small: precursor-derived peptides in plants. Annu Rev Plant Biol 70(1):153–186. https://doi.org/10.1146/annurev-arplant-042817-040413

Ou Y, Lu X, Zi Q, Xun Q, Zhang J, Wu Y, Shi H, Wei Z, Zhao B, Zhang X, He K, Gou X, Li C, Li J (2016) RGF1 INSENSITIVE 1 to 5, a group of LRR receptor-like kinases, are essential for the perception of root meristem growth factor 1 in Arabidopsis thaliana. Cell Res 26(6):686–698. https://doi.org/10.1038/cr.2016.63

Patharkar OR, Walker JC (2016) Core mechanisms regulating developmentally timed and environmentally triggered abscission. Plant Physiol 172(1):510–520. https://doi.org/10.1104/pp.16.01004

Pearce G, Ryan CA (2003) Systemic signaling in tomato plants for defense against herbivores: isolation and characterization of three novel defense-signaling glycopeptides hormones coded in a single precursor gene. J Biol Chem 278:30044–30050

Pearce G, Strydom D, Johnson S, Ryan CA (1991) A polypeptide from tomato leaves induces wound-inducible proteinase inhibitor proteins. Science 253:895–898

Pearce G, Moura D, Stratmann J, Ryan CA (2001) Production of multiple plant hormones from a single polyprotein precursor. Nature 411:817–820

Pearce G, Siems WF, Bhattacharya R, Chen YC, Ryan CA (2007) Three hydroxyproline-rich glycopeptides derived from a single petunia polyprotein precursor activate defensin I, a pathogen defense response gene. J Biol Chem 282:17777–17784

Pearce G, Yamaguchi Y, Munske G, Ryan CA (2010) Structure-activity studies of RALF, rapid alkalinization factor, reveal an essential–YISY–motif. Peptides 31:1973–1977. https://doi.org/10.1016/j.peptides.2010.08.01

Postel S, Kufner I, Beuter C, Mazzotta S, Schwedt A, Borlotti A, Halter T, Kemmerling B, Nurnberger T (2010) The multifunctional leucine-rich repeat receptor kinase BAK1 is implicated in Arabidopsis development and immunity. Eur J Cell Biol 89:169–174

Qi Z, Verma R, Gehring C, Yamaguchi Y, Zhao YC, Ryan CA, Berkowitz GA (2010) Ca2+ signaling by plant Arabidopsis thaliana pep peptides depends on AtPepR1, a receptor with guanylyl cyclase activity, and cGMP-activated Ca2+ channels. Proc Natl Acd Sci USA 107:21193–21198

Rhodes J, Yang H, Moussu S, Boutrot F, Santiago J, Zipfel C (2021) Perception of a divergent family of phytocytokines by the Arabidopsis receptor kinase MIK2. Nat Commun 12(1):705. https://doi.org/10.1038/s41467-021-20932-y

Rodiuc N, Barlet X, Hok S, Perfus-Barbeoch L, Allasia V, Engler G, Séassau A, Marteu N, de Almeida-Engler J, Panabières F, Abad P, Kemmerling B, Marco Y, Favery B, Keller H (2016) Evolutionarily distant pathogens require the Arabidopsis phytosulfokine signalling pathway to establish disease. Plant Cell Environ 39(7):1396–1407. https://doi.org/10.1111/pce.12627

Ross A, Yamada K, Hiruma K, Yamashita-Yamada M, Lu X, Takano Y, Tsuda K, Saijo Y (2014) The Arabidopsis PEPR pathway couples local and systemic plant immunity. EMBO J 33:62–75

Roux M, Schwessinger B, Albrecht C, Chinchilla D, Jones A, Holton N, Malinovsky FG, Tor M, de Vries S, Zipfel C (2011) The Arabidopsis leucine-rich repeat receptor-like kinases BAK1/SERK3 and BKK1/SERK4 are required for innate immunity to hemibiotrophic and biotrophic pathogens. Plant Cell 23:2440–2455

Ruiz C, Nadal A, Montesinos E, Pla M (2018) Novel Rosaceae plant elicitor peptides as sustainable tools to control Xanthomonas arboricola pv. Pruni in Prunus spp. Mol Plant Pathol 19:418–431

Ryan CA (2000) The systemin signaling pathway: differential activation of plant defensive genes. Biochim Biophys Acta 1477:112–121

Ryan CA, Pearce G (2003) Systemins: a functionally defined family of peptide signal that regulate defensive genes in Solanaceae species. Proc Natl Acad Sc USA 100:14577–14580

Schulze B, Mentzel T, Jehle A, Mueller K, Beeler S, Boller T, Felix G, Chinchilla D (2010) Rapid heteromerization and phosphorylation of ligand-activated plant transmembrane receptors and their associated kinase BAK1. J Biol Chem 285:9444–9451

Shen W, Liu J, Li JF (2019) Type-II metacaspases mediate the processing of plant elicitor peptides in Arabidopsis. Mol Plant 12(11):1524–1533. https://doi.org/10.1016/j.molp.2019.08.003

Shen W, Zhang X, Liu J, Tao K, Li C, Xiao S, Zhang W, Li JF (2022) Plant elicitor peptide signalling confers rice resistance to piercing-sucking insect herbivores and pathogens. Plant Biotechnol J 20(5):991–1005. https://doi.org/10.1111/pbi.13781. Epub 2022 Feb 18. PMID: 35068048; PMCID: PMC9055822

Shinya T, Yasuda S, Hyodo K, Tani R, Hojo Y, Fujiwara Y, Hiruma K, Ishizaki T, Fujita Y, Saijo Y, Galis I (2018) Integration of danger peptide signals with herbivore-associated molecular pattern signaling amplifies anti-herbivore defense responses in rice. Plant J 94(4):626–637. https://doi.org/10.1111/tpj.13883. Epub 2018 Apr 1. PMID: 29513388

Smakowska-Luzan E, Mott GA, Parys K, Stegmann M, Howton TC, Layeghifard M, Neuhold J, Lehner A, Kong J, Grünwald K, Weinberger N, Satbhai SB, Mayer D, Busch W, Madalinski M, Stolt-Bergner P, Provart NJ, Mukhtar MS, Zipfel C, Desveaux D, Guttman DS, Belkhadir Y (2018) An extracellular network of Arabidopsis leucine-rich repeat receptor kinases. Nature 553(7688):342–346. https://doi.org/10.1038/nature25184. Epub 2018 Jan 10. Erratum in: Nature. 2018 Sep;561(7722):E8. PMID: 29320478; PMCID: PMC6485605

Stegmann M, Monaghan J, Smakowska-Luzan E, Rovenich H, Lehner A, Holton N, Belkhadir Y, Zipfel C (2017) The receptor kinase FER is a RALF-regulated scaffold controlling plant immune signaling. Science 355(6322):287–289. https://doi.org/10.1126/science.aal2541

Stegmann, M, Zecua-Ramirez, P, Ludwig, C, Lee, HS, Peterson, B, Nimchuk, ZL, Belkhadir Y, Hückelhoven, R (2021). RGI-GOLVEN signalling promotes FLS2 abundance to regulate plant immunity. bioRxiv preprint. https://doi.org/10.1101/2021.01.29.428839

Stenvik GE, Tandstad NM, Guo Y, Shi CL, Kristiansen W, Holmgren A, Clark SE, Aalen RB, Butenko MA (2008) The EPIP peptide of INFLORESCENCE DEFICIENT IN ABSCISSION is sufficient to induce abscission in Arabidopsis through the receptor-like kinases HAESA and HAESA-LIKE2. Plant Cell 20(7):1805–1817. https://doi.org/10.1105/tpc.108.059139

Tanaka K, Heil M (2021) Damage-associated molecular patterns (DAMPs) in plant innate immunity: applying the danger model and evolutionary perspectives. Annu Rev Phytopathol 59:53–75

Tanaka K, Choi J, Cao Y, Stacey G (2014) Extracellular ATP acts as a damage-associated molecular pattern (DAMP) signal in plants. Front Plant Sci 5:446. https://doi.org/10.3389/fpls.2014.00446

Tang J, Han Z, Sun Y, Zhang H, Gong X, Chai J (2015) Structural basis for recognition of an endogenous peptide by the plant receptor kinase PEPR1. Cell Res 25:110–120

Tintor N, Ross A, Kanehara K, Yamada K, Fan L, Kemmerling B, Nürnberger T, Tsuda K, Saijo Y (2013) Layered pattern receptor signaling via ethylene and endogenous elicitor peptides during Arabidopsis immunity to bacterial infection. Proc Natl Acad Sci USA 110:6211–6216

Ton J, Van Pelt JA, Van Loon LC, Pieterse CM (2002) Differential effectiveness of salicylate-dependent and jasmonate/ethylene-dependent induced resistance in Arabidopsis. Mol Plant-Microbe Interact 15:27–34

Toyota M, Spencer D, Sawai-Toyota S, Jiaqi W, Zhang T, Koo AJ, Howe GA, Gilroy S (2018) Glutamate triggers long-distance, calcium-based plant defense signaling. Science 361:1112–1115

Trivilin AP, Hartke S, Moraes MG (2014) Components of different signalling pathways regulated by a new orthologue of AtPROPEP1 in tomato following infection by pathogens. Plant Pathol 63:1110–1118

Wang CG, Zhou MQ, Zhang XD, Yao J, Zhang YP, Mou ZL (2017) A lectin receptor kinase as a potential sensor for extracellular nicotinamide adenine dinucleotide in. elife 6:25474. https://doi.org/https://doi.org/10.7554/eLife.25474

Wang L, Einig E, Almeida-Trapp M, Albert M, Fliegmann J, Mithöfer A, Kalbacher H, Felix G (2018) The systemin receptor SYR1 enhances resistance of tomato against herbivorous insects. Nat Plants 4:152–156

Wang J, Hu M, Wang J, Qi J, Han Z, Wang G, Qi Y, Wang HW, Zhou JM, Chai J (2019) Reconstitution and structure of a plant NLR resistosome conferring immunity. Science 364(6435):aav5870. https://doi.org/10.1126/science.aav5870

Wang Y, Li X, Fan B, Zhu C, Chen Z (2021a) Regulation and function of defense-related callose deposition in plants. IJMS 22:2393. https://doi.org/10.3390/ijms22052393

Wang X, Zhang N, Zhang L, He Y, Cai C, Zhou J, Li J, Meng X (2021b) Perception of the pathogen-induced peptide RGF7 by the receptor-like kinases RGI4 and RGI5 triggers innate immunity in Arabidopsis thaliana. New Phytol 230(3):1110–1125. https://doi.org/10.1111/nph.17197

Wawra S, Bain J, Durward E, de Bruijn I, Minor MA, Löbach L, Whisson SC, Bayer P, Porter AJ, Birch PRJ, Secombes CJ, van West P (2012) Host-targeting protein 1 (SpHtp1) from the oomycete Saprolegnia parasitica translocates specifically into fish cells in a tyrosine-O-sulphate–dependent manner. Proc Natl Acad Sci USA 109:2096–2101

Whitford R, Fernandez A, Tejos R, Perez AC, Kleine-Vehn J, Vanneste S, Drozdzecki A, Leitner J, Abas L, Aerts M et al (2012) GOLVEN secretory peptides regulate auxin carrier turnover during plant gravitropic responses. Dev Cell 22(3):678–685. https://doi.org/10.1016/j.devcel.2012.02.002

Yamaguchi Y, Huffaker A (2011) Endogenous peptide elicitors in higher plants. Curr Opin Plant Biol 14(4):351–357

Yamaguchi Y, Pearce G, Ryan CA (2006) The cell surface leucine-rich repeat receptor for AtPep1, an endogenous peptide elicitor in Arabidopsis, is functional in transgenic tobacco cells. Proc Natl Acad Sci USA 103:10104–10109

Yamaguchi Y, Huffaker A, Bryan AC, Tax FE, Ryan CA (2010) PEPR2 is a second receptor for the Pep1 and Pep2 peptides and contributes to defense responses in Arabidopsis. Plant Cell 22:508–522

Yang R, Liu J, Wang Z, Zhao L, Xue T, Meng J, Luan Y (2023) The SlWRKY6-SlPROPEP-SlPep module confers tomato resistance to Phytophthora infestans. Sci Hortic 318:112117

Zelman AK, Berkowitz GA (2023) Plant elicitor peptide (pep) signaling and pathogen defense in tomato. Plan Theory 12:2856

Zhang L, Gleason C (2020) Enhancing potato resistance against root-knot nematodes using a plant-defence elicitor delivered by bacteria. Nat Plants 6:625–629

Zhang H, Hu Z, Lei C, Zheng C, Wang J, Shao S, Li X, Xia X, Cai X, Zhou J, Zhou Y, Yu J, Foyer CH, Shi K (2018) A plant phytosulfokine peptide initiates auxin-dependent immunity through cytosolic ca (2+) signaling in tomato. Plant Cell 30(3):652–667. https://doi.org/10.1105/tpc.17.00537

Zhang J, Li Y, Bao Q, Wang H, Hou S (2022) Plant elicitor peptide 1 fortifies root cell walls and triggers a systemic root-to-shoot immune signaling in Arabidopsis. Plant Signal Behav 17(1):2034270. https://doi.org/10.1080/15592324.2022.2034270

Ziemann S, van der Linde K, Lahrmann U, Acar B, Kaschani F, Colby T, Kaiser M, Ding Y, Schmelz E, Huffaker A, Holton N, Zipfel C, Doehlemann G (2018) An apoplastic peptide activates salicylic acid signalling in maize. Nat Plants 4(3):172–180. https://doi.org/10.1038/s41477-018-0116-y

Zipfel C, Robatzek S, Navarro L, Oakeley EJ, Jones JD, Felix G, Boller T (2004) Bacterial disease resistance in Arabidopsis through flagellin perception. Nature 428:764–767

Zipfel C, Kunze G, Chinchilla D, Caniard A, Jones JDG, Boller T, Felix G (2006) Perception of the bacterial PAMP EF-Tu by the receptor EFR restricts agrobacterium-mediated transformation. Cell 125:746–760

Peptide Elicitors for Defense Against Abiotic Stress

2

Georgia Tsintzou and Panagiotis Madesis

Abstract

One of the major challenges that agriculture is facing is how to meet the world's population's food demands, while improving the sustainability of agricultural practices. Thus, there is a need to develop innovative technologies and products to increase agricultural productivity, despite climate change, technologies that can enhance crop quality and yields while minimizing the environmental impact of farming practices. Plants have a defense mechanism known as pattern-triggered immunity (PTI), enabled upon the detection of both endogenous and external elicitors. Microbe-associated molecular patterns (MAMPs) refer to chemically preserved structures that signal the presence of microorganisms and are used as a description for exogenous elicitors. Endogenous elicitors are produced and identified in various contexts, making it difficult to assess their biological significance. Plant elicitor peptides (PEPs) are a type of endogenous elicitors that help protect plants against bacteria, fungi, and herbivores. Recent research has shown that PEP-triggered signaling pathways are active in response to various stress factors, including abiotic stress. PEPs pose a key role in plant health as they regulate plant development and are involved in PEP-mediated suppression of root growth, among other functions. PEPs are active in angiosperms,

G. Tsintzou
Laboratory of Molecular Biology of Plants, Department of Agriculture Crop Production and Rural Environment, School of Agricultural Sciences, University of Thessaly, Volos, Greece
e-mail: gtsintzou@uth.gr

P. Madesis (✉)
Laboratory of Molecular Biology of Plants, Department of Agriculture Crop Production and Rural Environment, School of Agricultural Sciences, University of Thessaly, Volos, Greece

Institute of Applied Biosciences-INAB, Centre for Research and Technology-CERTH, Thessaloniki, Greece
e-mail: pmadesis@uth.gr

S. Singh, R. Mehrotra (eds.), *Plant Elicitor Peptides*,
https://doi.org/10.1007/978-981-97-6374-0_2

including numerous significant crops, throughout the evolutionary tree. To better understand the effects of PEP application in crop production, comprehensive research in agronomy, physiology, chemistry, biochemistry, and molecular biology must be performed. This will allow the development of precise procedures to aid farmers and agrochemical firms in the production and application of PEPs. This chapter offers a comprehensive up-to-date summary of PEPs as a protective mechanism against abiotic stress.

Keywords

Heat and salinity stress · Oxidative and dehydration stress · Nutrient deficiencies · Microalgae

2.1 Historical Retrospective

Studies on plant cells recognition and response to biological signals through elicitors have been ongoing since the previous century. Proteins called elicitor-binding proteins have been identified and are reported to function as physiological receptors in the signal transduction cascade. Specific high-affinity binding sites have been identified for peptide elicitors. In 1996, Hahn reviewed the properties of these binding sites/proteins, which are consistent with those expected of physiologically important receptors. It is well-known that signal exchange between plant hosts and microbial pathogens leads to the activation of host defenses (Hahn 1996). In 1998, it was reported that the successful defense of plants against pathogens is largely dependent on the use of elicitors, which are substances that can be of either plant or pathogen origin. In the same study, researchers have concluded that fungal peptide elicitors play a key role in mediating pathogen recognition in plants through signaling for defense against fungal pathogens. The term "elicitor" was initially used to describe substances that trigger the accumulation of antimicrobial phytoalexins in plants. Nevertheless, it has since come to encompass any agent that stimulates a defense response in plants (Nürnberger and Nennstiel 1998). The initial peptide to be characterized was AtPep1, which was isolated from the Arabidopsis leaf extract of wounded plants. AtPep1 includes the final 23 amino acids at its C-terminus, which is derived from its precursor protein called PROPEP1. PROPEPs are small proteins that consist of roughly a hundred amino acids, typically encoded by restricted gene clusters. Arabidopsis possesses eight PROPEP genes, while maize has seven genes, five which are believed to be active (Huffaker et al. 2006).

2.2 Introduction

Agricultural sector faces today many challenges, the most pressing is the need to meet the rapidly growing global population's demand for food. This task is particularly daunting when it comes to fostering sustainable agricultural practices. With farmers under pressure to increase crop yields while grappling with the effects of climate change, it is crucial that researchers develop innovative technologies and products that can enhance crop quality and quantity while reducing the environmental impact of agricultural practices (Ronga et al. 2019).

Plants need to balance maximizing their photosynthetic output to create materials for growth, the formation of reproductive organs, and seed production, while simultaneously minimizing the risks of infection, predation, or death due to their sessile nature. These factors are collectively referred to as the so-called growth-defense tradeoff. To achieve sustainable agriculture, it is becoming increasingly important to manage this tradeoff (Zelman and Berkowitz 2023).

Plants possess innate immunity, which is activated by the perception of various chemical compositions of molecules from a range of organisms, including bacteria, fungi, and herbivores. These molecules, which have the ability to trigger an immune response, are commonly referred to as elicitors. Based on their source, they can be classified into different categories, such as microbe-associated molecular patterns (MAMPs), herbivore-associated molecular patterns (HAMPs), and virus-associated molecular patterns (VAMPs). Plants possess the capacity to recognize specific patterns in their environment using pattern recognition receptors (PRRs), which are transmembrane receptors belonging to various classes that activate a similar set of responses, mainly physiological ones. This set of defense responses is known as "PAMP-triggered immunity" or, more accurately, "pattern-triggered immunity" (PTI). The PRRs trigger these responses in plants when they detect certain patterns or molecules (Boller and Felix 2009). Furthermore, a plant host's endogenous pattern can also trigger other pattern-triggered immunity (PTI) even if the host has previously recognized it. These patterns are referred to as damage- and danger-associated molecular patterns (DAMPs). The fact that both damage and danger are used in relation to DAMPs suggests that there are similarities in their formation and function. Fungal enzymes, for example, passively emit oligogalacturonides and cutin monomers when they try to obstruct hyphal entry into the plant body (Ferrari et al. 2013).

Many PROPEPs have been found in different angiosperm species, including important crop plants, despite the minimal sequence similarity among them in the PROPEP gene family of a particular species (Tintor et al. 2013). Pathogen-induced challenges trigger the activation of specific PROPEPs. For example, AtPROPEP1 and ZmPROPEP1 responded to fungal infections, while AtPROPEP3 and ZmPROPEP3 increased in response to herbivore detection. However, in the case of AtPROPEP3 and ZmPROPEP3, only wounding appears to induce the transcription of AtPROPEP5 and AtPROPEP8, and this induction is limited to the midrib of adult leaves. Conversely, AtPROPEP4 and AtPROPEP7 are not induced at all. Furthermore, there was no increase in the transcription of AtPROPEP4, AtPROPEP5, and AtPROPEP6 in response to treatments with jasmonic acid (JA), salicylic acid (SA), or AtPep1 to AtPep6 (Bartels et al. 2013).

The Pep-PEPR system's significance in the realm of plant biology is still needs to be clarified. Despite this interest, a clear understanding of the factors that facilitate and enhance the extracellular PEPs release and activate PEPRs in a molecular level, has not yet been achieved due to the scarcity of experimental data. Currently, two competing models exist. The first model, often referred to as the damage model, posits that PROPEPs and Peps reside in the cytosol and is released upon the loss of cellular integrity resulting from damage. The presence of Peps, detected by neighboring cells, triggers their defense mechanism, creating a barrier that prevents pathogens from entering the plant body via the wounded tissue. This necessitates the constant presence of PROPEPs in most plant cells in order to provide a broad protective effect. However, there is insufficient protein data for PROPEPs, and it is crucial that PROPEPs are rapidly processed into Peps unless they are already as "active" as Peps. Experimental data are required to clarify how and whether PROPEPs are processed. The model of amplification postulates that peptides are discharged into the extracellular space as a response to danger. According to this model, peptides can either prolong the immune response in an active cell via the autocrine pathway or transmit information to nearby cells to elicit their defense response through the paracrine pathway. Notably, PROPEPs are devoid of a conventional signal sequence that facilitates entry into the secretory pathway. Furthermore, PROPEP-YFP fusion proteins failed to localize to the secretory pathway, suggesting that they do not follow the typical route taken by secreted proteins. Therefore, similar to the unconventional exportation of animal interleukin-1β or the yeast mating factor Matα as leaderless secretory proteins (LSPs), PROPEPs or Peps would need to be exported in the same manner as LSPs. In summary, LSPs can either be released through a process in which proteins cross the plasma membrane directly or through the fusion of membrane-bound structures with the plasma membrane (Ding et al. 2012). A considerable number of LSPs is suggested as being released into the apoplast following pathogen attack or treatment with SA. However, the release of PROPEPs has not yet been proven. Depending on the specific PROPEP involved, it is possible that both models are accurate. PROPEPs expression patterns in Arabidopsis vary significantly, whereas overall amino acid sequences show low resemblance. Additionally, they differ in their subcellular localization, suggesting that many of them are expressed upon damage, while the rest are released only after a signaling of danger and released in a tightly regulated way. It is crucial to note that PROPEPs and Peps are assumed to enter the extracellular space to bind to the PEPR-LRR domain and activate PEPRs in both models. In the case that only one single PROPEP is secreted through the secretory pathway, then it can adjunct intracellularly and activate PEPR signaling (Bartels and Boller 2015).

Small peptides, which are the negligible components of the plant proteome, are typically a hundred amino acids or less in length. Apart from serving as antimicrobial peptides, secreted peptides have been established as critical signaling molecules that facilitate both close-range and far-range communication in plants and play a significant role in their growth. Plant peptides are formed by joining precursor proteins or small open reading frames (sORFs) that exist in transcriptomes. These sORFs consist of hundred codons or fewer and can be found in various regions of the transcript. Short protein coding transcripts that are less than a hundred amino acids in length constitute such a region. Another region with this kind of ORFs could be embedded within the primary transcripts of microRNAs, 5′-leader sequence of a messenger RNA. Peptides can be classified based on their unique characteristics, such as those derived from functional or nonfunctional precursor proteins (Gautam et al. 2023). Nonfunctional precursors can incorporate post-translational modifications (PTMs) like tyrosine (Tyr) sulfation and proline (Pro) hydroxylation. They may also have the potential for additional hydrocarbon entities to be substituted to amino acids and may be rich in cysteine (Cys) residues or lack Cys and PTMs. However, in the latter circumstance, they could possibly contain individual amino acids which are critical for their function. An N-terminal sorting sequence that is part of nonfunctional precursor proteins, addresses them into the secretory pathway. In an initial step, the immature peptide is cleaved due to an enzyme belonging to the endoplasmic reticulum (ER) to finally form a mature peptide. Nonetheless, while passing through the ER and Golgi reticulum, the prepropeptide may still possess a prodomain that can undergo post-translational modifications (PTMs) and undergo proteolytic cleavage aiming to produce the mature peptide posing biological activity. Cys-rich peptides with 2–16 Cys residues form disulfide bonds to achieve the correct fold of the active peptide ligand (Datta et al. 2024; Hu et al. 2021). Peptides are defined by their freedom of movement and secretion, which allows them to traverse long or short distances without being hindered by membranes. As ligands, they interact with membrane-bound leucine-rich repeat (LRR) receptor-like kinases (RLKs) that are specifically recognized by these small molecules. In many cases, RLKs work in tandem with co-receptors that enhance the receptors' structural integrity. Plant growth and development are intricately regulated by complicated intracellular signaling pathways that are triggered by the binding of tiny peptide ligands to their appropriate receptor–coreceptor complex. Many tiny peptides have been found to regulate tolerance pathways, allowing plants to adjust to a range of abiotic stressors (Kim et al. 2021).

Plants must find an equilibrium between boosting photosynthesis for growth, the formation of reproductive organs, and seed production, but also reducing the risks of diseases, predation, and death. This intricate equilibrium is known as the growth-defense tradeoff. In light of the growing trend toward sustainable agriculture, effectively managing this tradeoff will become increasingly critical (Figueroa-Macías et al. 2021). Currently, controlling crop diseases involves preventing pathogens from entering fields and greenhouses, minimizing their spread through cultural practices, eliminating plant tissues with a disease, and applying substances to protect plants by targeting the pathogens. However, these methods can have negative impacts on the overall health either for humans or for the ecosystem, as they may be harmful and/or toxic. Furthermore, these strategies do not equip plants to defend themselves. A better management system would merge or adapt current techniques with additional approaches that harness the complex and potent defense mechanisms that plants have developed to activate internal systems that combat stress. Propeptides that have undergone post-translational modifications and do not possess a signal for secretion like prosystemin and PROPEPs seem to be novel in land plants, but not in algae (Bowman 2022).

2.3 Higher Plants' Endogenous Elicitor Peptides

Plants from several species have revealed the presence of endogenous peptide elicitors, which act as key factors for the immunity regulation against pathogens and herbivores. This section will focus on the families of endogenous peptide elicitors, emphasizing their characteristics and the research objectives aimed at deepening our understanding of plant innate immunity.

Defense responses in plants are triggered mainly in two ways. The first one is due to substances synthesized by organisms that are invading the plant and the second one is due to elicitors produced by the host plant in response to injury or infection. These elicitor molecules, which are recognized as danger or alarm signals, refer to protein and oligosaccharide segments and reactive oxygen species (ROS) (Pearce et al. 2001; Yamaguchi and Huffaker 2011). Plants not only detect these molecules locally to initiate defense responses but mostly amplify their defenses by creating identical molecules via specific, regulated activity of enzymes. For instance, the endogenous polygalacturonase function generates cell-wall segment alarms, while endogenous NADPH oxidase function produces ROS alarms (Torres et al. 2002). The genes responsible for the biosynthesis of such enzymes are activated by pathogen or wound-induced activity. Additionally, genes that produce precursors of endogenous peptide elicitors are activated upon stress induced by biotic factors or elicitor activity. Conversely, there is no identification yet for the enzymes responsible for proteolytic processing and releasing the active peptides from their precursors. However, it is possible that they too may be triggered by biotic stress. It is likely that the additional signals released by plants, beyond those generated through cellular damage alone, serve to amplify defense responses for all these damage-associated molecules (McGurl et al. 1992).

Progress in exploring novel groups of plant elicitor peptide and understanding their downstream signaling routes has been made recently. The focus of this chapter is on such elicitor peptides, summarizing their features while discussing the upcoming signaling routes and reactions they control mostly for abiotic stress factors. This chapter categorizes these peptides based on the structure of their precursor proteins and the processing mechanisms involved in releasing active signals. Three types of peptides are discussed. The first one refers to those derived from precursor proteins not containing an N-terminal secretion alarm. The second is for those from precursor proteins with an N-terminal secretion signal. The third one is cryptic peptides derived from proteins with separate primary functions. The first signaling peptide found in plants, tomato systemin, was identified through the search for a host signal that regulates the accumulation of protease inhibitors, which serve as antiherbivory defense proteins. This peptide is extracted from the C-terminus of a 200 amino acid precursor protein referred to as prosystemin. Experimental results from studies on plants treated with systemin or genetically modified plants that either overexpress or suppress prosystemin gene expression have shown that systemin regulates numerous defense responses in tomato plants. In addition, it promotes the protease inhibitors aggregation along with other anti-nutritive proteins. Moreover, systemin enhances the release of plant volatiles, which attract natural enemies of insect herbivores (Corrado et al. 2007; Yamaguchi and Huffaker 2011). The cytosol of vascular phloem parenchyma cells is where prosystemin accumulates. Based on a procedure generated by wound, the activated systemin stimulates jasmonic acid (JA) production in the companion-cell-sieve element complex of the vascular bundle, which in turn leads to systemic protease inhibitor induction. Peptides with a systemin-like structure have been exclusively identified in plants from precursor proteins that possess an N-terminal secretion signal. One example of this is hydroxyproline-rich systemin (HypSys), which shares similarities with anti-nutritive protease inhibitor proteins that are produced in tobacco plants in response to herbivory. Exploring signals that induce the production of protease inhibitors in tobacco led in two 18-amino acid glycopeptides, hydroxyproline-rich systemins I and II (NtHypSysI and NtHypSysII), from tobacco leaves (Narváez-Vásquez and Ryan 2004). Both NtpreproHypSys peptides have in common their precursor protein, which is a common mechanism used by animal and yeast peptide hormones. As for the NtHypSys peptides, their amino acid sequence is comparable to that of systemin, although the polyprolines undergo hydroxylation and subsequent glycosylation with pentose sugar chains after being processed through the secretory system. The transcriptomes of preproHypSys gather in phloem parenchyma cells following injury, and the precursor protein is localized to the cell-wall matrix (Narváez-Vásquez et al. 2005). Isolates of NtHypSys have been extracted from five different solanaceous plants, along with tomato, and one convolvulaceous plant, sweet potato (*Ipomoea batatas*).The use or overexpression of NtHypSys triggers the activation of genes involved in defense responses, including JA biosynthetic enzymes, protease inhibitors, and pathogenesis-related genes. Studies have shown that tobacco plants overexpressing preproHypSys exhibit increased resistance to herbivory by *Helicoverpa armigera* larvae. In the wild tobacco species, NaHypSys plays a crucial role in the biosynthesis process of the 17-hydroxygeranyllinalool diterpene glycosides, which are a group of antiherbivore protective substances. Furthermore, transgenic *N. attenuata* plants with silenced NapreproHypSys genes were found to be more exposed to herbivore catastrophic actions (Heiling et al. 2010; Ren and Lu 2006). Proteins with specific primary functions that generate cryptic peptide signals, known as inceptin, have been found to play an important role in immunoregulation in mammalian systems. The inceptin family of peptides was the first group of cryptides found to regulate plant immunity. These acidic, 11–13 amino acid long, disulfide-bridged peptides were isolated from the oral secretions of fall armyworm (*Spodoptera frugiperda*) larvae. These peptides act as putative elicitors of ethylene (ET) production in cowpea (*Vigna ungiculata*). In addition to triggering ET production, inceptin treatment of cowpea leaves results in the biosynthesis of salicylic acid (SA), jasmonic acid (JA), terpene volatiles, and other metabolites with protective properties. Metabolites, including inceptins, collaborate to restrict the development of fall armyworms. Inceptins are obtained from the controlling segment of the cowpea chloroplastic ATP synthase g-subunit, which has been broken down by proteolytic enzymes present in the armyworm larval gut. Inceptin-like sequences can be found in the g-subunits of plant chloroplastic ATP synthase; however, it is only in the legume genera phaseolus and vigna that these sequences display elicitor activity (Schmelz et al. 2009; Ueki et al. 2007).

Downstream signaling pathways are consistently maintained, as all endogenous peptides collaborate with small molecule defense-related phytohormones in a regulatory loop. In response to peptide-elicitation, accumulation of these phytohormones takes place, and mutant studies have demonstrated that the intact hormone signaling pathway is necessary for downstream peptide-induced defenses. The conserved components of endogenous peptide elicitor signaling include extracellular alkalinization, intracellular Ca^{2+} elevation, ROS production, and MAP kinase activation. Constitutive expression of defense genes in the absence of infection or wounding with enhancement disease resistance is driven by the overexpression of AtPROPEPs. Peps may play a role in maintaining and refining plant reactions to the perception of MAMPs and/or biotic assaults (Qi et al. 2010). The modulation of calcium levels by both endogenous peptides and microbial-associated molecular patterns (MAMPs) has been recently studied. Both flg22 and endogenous peptides enable calcium signaling by binding to their respective receptors, which leads to the activation of a cyclic nucleotide-gated calcium-conducting channel (CNGC2) that is vital for the proper functioning of signal transduction. Despite the dissimilar mechanisms of activation for CNGC2, the intracellular kinase region of AtPEPR1 was revealed to comprise a guanylyl cyclase domain. Even if there is no direct evidence for in vivo activation of CNGC2 due to AtPEPR1-generated cGMP, it appears probable that the cytosolic calcium enhancement induced by AtPep is attributable to the guanylyl cyclase activity of AtPEPR1 (Qi et al. 2010).

The aforementioned signaling mechanisms constitute model responses of plants to MAMPs. Receptor-mediated perception constitutes a common mechanism of signaling between endogenous elicitor peptides and MAMPs. Both PEPRs and MAMP-binding receptors are leucine-rich repeat receptor kinases (LRRRKs) which possess a resemblance in structure with FLS2, EFR, and Xa21, the LRR-RKs that bind the peptide MAMPs. BRI1-associated kinase1 (BAK1) is the sharing coreceptor for complexing with receptors such as the PEPR, EFR, and FLS2 receptors. BAK1 is required for both MAMP and endogenous peptide signaling. The research indicates that BAK1 appears to operate at the junction of AtPep and MAMP signaling pathways. This overlap in signaling was validated in Arabidopsis via experiments for gene expression conducted utilizing MAMPs and AtPep peptides. These experiments showed that both types of stimuli activate a similar cluster of genes related with defense mechanisms, such as MAP kinase 3, WRKY transcription factor genes, and PR-1 and PDF1.2 genes (Schulze et al. 2010; Zipfel et al. 2006).

In the natural environment, the initial number of invading microorganisms likely limits the level of MAMPs and was found to be lower than what is commonly used in laboratory experiments. In such circumstances, the advantage of enhancing the proliferation of MAMP alarms with peptides or endogenous elicitors is clear. Insect basal immunity relies on the binding of Spätzle peptide to the Toll receptor, which mediates the response. Mammalian innate immunity, on the other hand, is activated by peptide cytokines posing inflammatory properties like interleukins which are secreted through an independent pathway after the identification of MAMPs by

Toll-like receptors. PEPs have a conserved function in inducing immune responses in both monocot and dicot species, which diverged approximately 150 million years ago, suggesting that they are functionally analogous to cytokines and Spätzle in plants, serving as peptide mediators of early evolved basal immune responses (Chaw ShuMiaw et al. 2004; Ligoxygakis et al. 2002).

The variety of peptide sequences and precursor protein structures suggests that there are multiple receptor partners and export mechanisms for activating peptide signals. For instance, precursors that have secretion signals are generally exported through the endoplasmic reticulum and Golgi. However, the ones lacking signals might use alternative, leaderless secretion pathways. Cryptides synthesis may result from damage or specific proteolysis. This diversity in signal release helps plants respond to various mechanisms of biotic attack. Although supplied peptides can trigger similar responses, in vivo localization and maturation processes may give rise to distinct defensive roles. Pathogens must develop new strategies to bypass or subvert endogenous peptide-induced defenses that are specific to each species in order to successfully establish an infection (Yamaguchi and Huffaker 2011). Future research should explore several promising avenues to gain a better understanding of plant defenses, such as identifying new receptor-peptide ligand combinations, examining the integration and/or connections between MAMP/HAMP perception and endogenous elicitors, elucidating the secretory and export processes that regulate peptide activation, and determining if genetically modifying endogenous peptide signaling is a feasible approach for raising resilience to herbivores or biotic enemies (Ortiz-Morea and Reyes-Bermudez 2019).

2.4 Abiotic Stress Response

Environmental conditions that are detrimental to plants can have a significant impact on their growth, reproductive potential, and yield, ultimately affecting plant diversity and distribution. Climate change and global warming exacerbate instances of abiotic stress, which are expected to become more frequent and severe in the coming century, presenting a significant challenge for crop production. Abiotic stresses, including salinity, water deficiency, extreme heat/cold, constitute significant obstacles for crop production, leading to substantial yield losses. In order to address challenging environmental conditions, some plants are capable of altering their typical development by diverting resources from reproductive capabilities to stress response mechanisms. These reactions to stress are often specific to particular tissues and influenced by factors like the age of the plant and its tolerance level (Lagiotis et al. 2023).

A number of peptides and signaling pathways are essential for plants to respond to abiotic stressors. Several tiny coding genes are present in the Arabidopsis genome; however, the regulation of plant responses to abiotic stress has only been documented for a small number of such peptides. The function of receptor-like kinases (RLKs) and co-receptors in the pathway for peptide signaling taking place during abiotic stress reactions is one of the many unsolved topics in this field (Kuromori et al. 2022). Furthermore, there are concerns regarding the maturation process of peptide precursors into mature peptides and the long-distance signaling channels that carry these mature peptides to their target regions. In addition, it is not known how peptide-mediated signaling pathways give plants resistance to abiotic stressors or how peptide–receptor interactions result in particular developmental regulations. Two signaling peptides that have been demonstrated to increase drought stress tolerance and cause active ABA accumulation are CLE9 and CLE25, respectively. To improve stress tolerance, the relationship between signaling peptides and conventional phytohormone pathways during abiotic stress could offer new insights. Understanding how plants sense differences in the local concentrations of various nutrients and abiotic stressors to initiate peptide signaling is another topic that needs further investigation. Thus, new strategies for improving crop stress tolerance and boosting agricultural sustainability will become available with the discovery of plant peptides and signaling pathways (Wang et al. 2022).

2.4.1 Oxidative Stress

The majority of environmental stressors, like heat, cold, osmotic stress, dehydration, salinity, injury, and attacks led by pathogens, eventually cause a preliminary and sharp rise in the content of different reactive oxygen species (ROS), also known as the oxidative burst. The IDA peptide is a member of a group of nine genes, of which IDA-Like 1 (IDL1) to IDL5 have been shown to function in plant development (Datta et al. 2024). Two recently discovered genes, IDL6 and IDL7, in the IDA-like peptide family, exhibit expression patterns assuming that they are involved in the regulation processes of abiotic stress. Expression studies indicated that the bacterial elicitor flg22 rapidly induces both IDL7 and similarly acting IDL6, with a maximum expression occurring within 30 min of treatment. Remarkably, co-treatment of Arabidopsis with IDL-7 and flg22 reduces the reactive oxygen species (ROS) burst caused by flg22. The ROS spike caused by flg22 is not lost in idl7 mutants. Taken together, these investigations highlight the possible function of IDA-like peptides, specifically IDL6 and IDL7, as important negative regulators of genes that mediate early stress responses in Arabidopsis seedlings (Datta et al. 2024; Vie et al. 2017). Transcriptomic analysis indicates that plant specimens treated with IDL6 and IDL7 peptides show a substantial reduction in the expression of genes that respond to early stress, such as ZINC FINGER PROTEIN (ZFP) genes, WRKY genes, and genes encoding calcium-dependent proteins. It is noteworthy that H_2O_2 dramatically raises IDL7 expression while the idl7 mutant and the idl6 idl7 double mutants pose a markedly decreased cellular loss level upon exposure to ROS. Furthermore, the phenotype of the IDA mutant and the double knockout mutant for HAE and HSL2 are identical. Interestingly, the overexpression of IDA is unable to restore this phenotype, highlighting the critical function of HAE and HSL2 in mediating the effects of IDA as IDA uses the LRR-RLKs HAE and HSL2 to impact behavior (Stenvik et al. 2008b).

2.4.2 Nutrient Deficiency

Nutrient scarcity or an imbalance can result in stress for plants. In some cases, an excess of one nutrient can lead to a decreased absorption for another. CEPs, CLEs, and RGFs as peptides, are important in nutrient acquisition and can impact plant growth and development.

2.4.2.1 Nitrogen Limitation

Nitrogen (N_2) is a vital element in plants, as it is necessary for the formation of amino acids, DNA/RNA nitrogenous bases, nucleic acids, pigments and chlorophylls, proteins, and useful metabolites that are critical for survival. Its importance focuses on the control of crucial developmental processes, such as root and shoot growth along with flowering. In nature, nitrogen exists in both inorganic forms, namely nitrate (NO_3^-) and ammonium (NH_4^+), and organic forms, including amino acids and urea. Interestingly, the main inorganic form of nitrogen that plants use is nitrate. Plants frequently need inorganic nitrogen fertilizers to make up for the restricted amount of nitrate in their soil. Nevertheless, overuse of these fertilizers is harmful to ecosystems, especially in aquatic situations where eutrophication from excessive nitrogenous fertilizer leaking from fields of agriculture can occur. Small peptides therefore present a viable way to overcome the difficulties brought on by low nitrogen supply and improve its effective use (Vidal et al. 2020). Cell elongation proteins (CEPs) have been extensively studied as small post-translationally modified peptides that act as long-range signaling molecules, playing a vital role in regulating signaling molecular ways in which plants respond to nitrogen levels, nodulation, and root structure. Under circumstances where nitrogen is scarce, CEP1 peptides are synthesized in the roots of Arabidopsis and function as long-distance signaling molecules that travel to the shoot. In the shoot, CEP1 interacts with receptors XIP1/CEPR1 and CEPR2, which initiate a signaling cascade to the roots by upregulating the expression of CEPD polypeptides, specifically CEPD1 and CEPD2. These CEPDs, which belong to the glutaredoxin group, are carried to the roots via the phloem tissue. There, they upregulate the gene that codes for the nitrate transporter, NRT2.1, so encouraging an increased nitrate uptake from the soil. Moreover, CEP3 has been found to regulate how a plant responds to nitrogen deprivation, perhaps improving seedling survival by restricting root growth (Datta et al. 2024; Delay et al. 2019). Closely related to CLE3, CLE2 expresses itself more when there is enough nitrogen available, sugar is scarce, and darkness is present. Remarkably, under these circumstances, mutants lacking CLE2 showed reduced development and yellowing. Surprisingly, the expression of multiple genes in shoots was activated by the activation of CLE2 in roots, especially those involved in regulating glucose metabolism or being stimulated in darkness (Ma et al. 2020). The peptide CLE3 coheres to CLV1-LRRRLK receptor to start a signaling cascade, which in turn inhibits the formation of lateral root primordia under nitrogen shortage. Interestingly, root pericycle cells are able to produce various homologues of CLE peptides, specifically CLE1/3/4/7, under conditions of prolonged nitrogen deficiency. Later, downstream signaling elements create a feedback loop to control the CLE peptide signals' amplitude. This procedure prevents the formation of both lateral root primordia and their separation from the main root system (Araya et al. 2014).

2.4.2.2 Sulfur Deficiency

Sulfur, denoted as S, represents a crucial constituent that is indispensable for the growth and expansion of plants, including the proper activity of their physiological processes. Concerning the vegetal realm, soil-based sulfate is the main source of S, which is absorbed properly by the plant's roots, while the upper sections of the plant additionally utilize gaseous sulfur. Upon absorption by the plant, sulfur undergoes reduction before being assimilated, with cysteine, an amino acid, functioning as the primary precursor for the majority of metabolites rich in S. Sulfur is an integral protein component and serves a vital role in enzymes. These, in turn, contribute to defense mechanisms against biotic and abiotic stress (Datta et al. 2024). Sulfur is vital for the antioxidant system, which facilitates the removal of ROS synthesized upon stress by modulating the sulfur assimilation route. In situations where sulfur is scarce, the CLE peptide affects lateral root development. Thus, the CLV1 receptor kinase interacts to the CLE peptide, initiating downstream signaling. In wild-type plants, lateral root development is hindered when sulfur availability is restricted, but not in clv1 mutant plants. As for the wild types, lateral root development returns when sulfur levels are restored, whereas recovery is less pronounced in clv1 mutant plants. Low lateral root density is the outcome of CLE2 and CLE3 expression declining in a CLV1-independent way in both wild-type and CLV1 mutants during sulfur deprivation (Dong et al. 2019).

2.4.2.3 Phosphorous Deficiency

One important macronutrient that plays a significant role in the growth and development of plants is phosphorus. It is an essential part of proteins, lipids, phosphorylated carbohydrates, and nucleic acids. Phosphate bonds of high-energy are generated with adenine, guanine, and uridine, which are energy carriers for a variety of biological activities. Phosphorus can be found in soil in both organic and inorganic forms while H_2PO_4 is the main inorganic form (Rao et al. 2006). Plant growth and overall productivity in natural ecosystems are severely hampered by the absence of inorganic phosphate (Pi) and HPO_4^{2-} in soil. Through evolution, plants developed many ways to alter their root systems composition in order to deal with the problem of low Pi availability. These changes mainly include shorter primary roots, more root hair development, and improved lateral root production. Additionally, every cell in the root apical meristem (RAM) goes through terminal differentiation, also known as determined growth or root meristem exhaustion (Sánchez-Calderón et al. 2005). As for the wild-type plants, when their roots exposed to low Pi availability bear a significant phenotypic similarity to the RGF1 and RGF2 mutants, suggesting a role for these peptides in controlling root growth as an adaptation to Pi deprivation. Furthermore, plants adjust the structure of their root systems to maximize the uptake of Pi from the soil when it is scarce. When Pi is scarce, RGF2 encourages vertical root development and radial divisions, but RGF1 prevents radial divisions in the root system, which facilitates Pi uptake by defining root hairs (Cederholm and Benfey 2015; Datta et al. 2024). When dividing cells lose their ability to proliferate, a condition known as the exhaustion of the RAM occurs, ultimately leading to the root's differentiation. In this scenario, CLE14 plays a crucial role. Low phosphate root1/low phosphate root2 (LPR1/LPR2) facilitates the iron flow from soil into the RAM in response to limited phosphate (Pi) availability. Consequently, the proximal area of the root meristem experiences an induction in CLE14 expression. CLE14, which is recognized by PEPR2 and CLV2 receptors, efficiently inhibits the growth of main roots by suppressing the PIN/auxin pathway and SCARECROW/SHORT-ROOT. As a result, these processes coordinate the RAM's full differentiation and prevent meristematic cell division. Arabidopsis plants are able to withstand decreased Pi availability in the soil thanks to this adaptation technique (Gutiérrez-Alanís et al. 2017).

2.4.3 Drought Stress

Water deficiency presents a major challenge for plants, as dehydration results in diminished cell turgor, mainly due to a water potential decrease. Abscisic acid (ABA) accumulates as a result of this decrease, setting off a series of physiological and biochemical reactions. Stomata close, osmolytes build up, and most significantly, stress-responsive genes are activated during these reactions. An important class of signaling peptides in plants is the CLAVATA3 (CLV)/embryo-surrounding region related (CLE) peptides. These peptides are typically made up of 12–14 amino acids and are derived from a big precursor protein. A number of CLE peptides are crucial for the upkeep of the root meristem, the development of the shoot and vascular tissue, and the production and function of stomata. Specifically, in *Arabidopsis thaliana*, the dehydration stress tolerance response is mediated by CLE9 and CLE25. A transportable peptide called CLE25 connects dehydration stress tolerance mediated by abscisic acid (ABA) with water deficiency (Takahashi et al. 2018). Water scarcity triggers an increase in the expression of CLE25 in the root vasculature, leading to the migration of peptides to the leaves, which results in stomatal closure. The application of CLE25 peptides to the roots stimulates the expression of the NCED3 gene, which encodes the enzyme Ninecis-Epoxycarotenoid Dioxygenase 3. This enzyme cleaves a precursor molecule of ABA, generating bioactive ABA. The knockout mutants of cle25 are more susceptible to dehydration than wild-type plants. Experiments revealed that CLE25 derived from roots upon drought stress is detected by Barely Any Meristem 1 (BAM1) and BAM2 receptors exclusively in leaves and not roots. The BAM1 BAM2 mutants appeared to become more sensitive to dehydration (Takahashi et al. 2019). All evidences imply that CLE25-BAM modules transmit, from the roots to the leaves, drought signals, facilitating long-distance signaling process that induces accumulation of ABA by upregulating NCED3 expression for stress adaptation and drought resistance. In 2019, Zhang et al. reported that by regulating stomatal closure, CLE9 is involved in elevating drought tolerance. Applying CLE9 peptides or overexpressing CLE9 improves drought tolerance by causing stomatal closure. Mitogen-Activated Protein Kinase (MAPK)3 (MPK3)/6 and two guard cell ABA-signaling components, protein kinase Open Stomata 1 (OST1) and anion channel protein Slow Anion Channel Associated 1 (SLAC1) are required for the stomatal closure which is caused by CLE9, according to genetic research (Kim et al. 2021). Moreover, CLE9 promotes the synthesis of nitric oxide and hydrogen peroxide, which are eliminated in mutants of NADPH oxidase and nitric reductase, respectively. These results imply that CLE9 enhances dehydration tolerance by having in common OST1 and SLAC1 and controlling stomatal closure as feedback to water deficiency. Since guard cells are the only cells in which CLE9 is expressed, stomatal closure is controlled locally by CLE9. Additionally, the bam/clv1 mutants exhibit normal stomatal closure in response to CLE9 peptides, suggesting that BAM receptors are improbable to be part of recognizing CLE9 peptides (Zhang et al. 2019).

Another important peptide is the Inflorescence Deficient in Abscission (IDA) which contributes to the process of floral organ abscission and lateral root emergence by relating with specific receptors, namely LRR-RLKs, HAESA (HAE), and HAESA-Like 2 (HSL2), and the MAPK KINASE4 (MKK4)/5MPK3/6 cascade. In Arabidopsis, the presence of an extended pro-rich motif consisting of 20 amino acids at the C-terminus of IDA is sufficient to trigger floral abscission (Stenvik et al. 2008a). In *Nicotiana benthamiana* leaf tissue, a 12-amino acid peptide with a hydroxylated Pro residue at position 7 was discovered to be the most effective at activating the HSL2 receptor for signaling. When the leaves are subjected to drought conditions, expression of HAE and IDA is triggered in the leaf abscission parts. The results of the mutant analysis revealed that IDA, HAE/HSL2, and MKK4/5 play a crucial role for cauline leaf abscission upon dehydration stress conditions. They also play a role in floral organ abscission and lateral root emergence. IDA is responsible for inducing cell-wall breakdown and cell separation, leading to the aforementioned organ abscission events (Patharkar and Walker 2016).

The PSK peptide, phytosulfokine-α (PSK-α) is a five-amino acid long peptide that has been sulfated at two tyrosine residues. There are seven Arabidopsis genes that are believed to encode precursor proteins for PSK (Kaufmann and Sauter 2019). For cell elongation and primary/lateral root growth, the PSK-PSK receptor (PSKR) and coreceptor Somatic Embryogenesis Receptor Kinase 3 (SERK3) are involved. The production of the PSK pentapeptide from the PSK precursor protein by subtilases (SBTs) and PSK itself is involved in Arabidopsis' ability to withstand drought stress. When exposed to osmotic stressors like mannitol, specific genes responsible for PSK (PSK1, PSK3, PSK4, and PSK5) among with specific SBT genes (SBT1.4, SBT3.7, and SBT3.8) exhibit substantial upregulation (Stührwohldt et al. 2021). Upon osmotic stress, lower root and shoot growth rates are observed in sbt3.8 mutant plants compared to wild type. However, the sensitive phenotype induced by osmotic stress in sbt3.8 mutants can be recovered with PSK peptide treatment. The PSK precursor (proPSK1) is overexpressed in Arabidopsis plants, which exhibit improved root and hypocotyl development as well as resistance to osmotic stress. ProPSK1's processing dependents on the Asp residue that comes right after the cleavage site, which is caused by SBT3.8 at the PSK peptide's C-terminus. The sbt3.8 mutant exhibits impairments in this processing. Moreover, SBT3.8 overexpression in Arabidopsis promotes shoot and root growth and enhances resistance to osmotic stress. These results indicate that SBT3.8 plays a role in drought stress tolerance by mediating PSK peptide processing from the precursor protein proPSK1. Furthermore, tomato blossom drop brought on by dryness is regulated by phytosulfokine signaling. In tomato plants (*Solanum lycopersicum*), overexpression of phytaspase 2 (phyt2), a subtilisin-like protease that cleaves proPSK protein, increases premature flower abscission. In contrast, SlPhyt2 knockdown decreases floral drop in drought-stressed environments, undercovering a function for SlPhyt2 in drought-induced abortion of fruit and flower growth (Kim et al. 2021; Reichardt et al. 2020). The expression of SlPhyt2 is continuously induced in flower pedicels that are close to the abscission zone as a primary result of dehydration. Additionally, SlPSK1 and SlPSK6 are co-induced with SlPhyt2 in response to drought stress. The PSK precursor was discovered to be a possible substrate for SlPhyt2 by a proteomics test using a substrate library. In vitro, mature PSK can be released by SlPhyt2 cleaving peptides containing the PSK pentapeptide of different sizes in an Asp-specific manner. A bioassay using inflorescence revealed that mature PSK induced pedicel abscission. The PSK treatment increased the expression of polygalacturonase genes relevant to tomato abscission and decreased the expression of genes that keep the abscission zone dormant, suggesting that PSK functions as a cue for tomato pedicel abscission. So, these results imply that, in response to drought stress, the subtilase SlPhyt2 which is expressed in the pedicel, generates bioactive PSK that in turn, in the abscission zone, causes an abscission by stimulating cell-wall hydrolases (Kim et al. 2021).

Proteins that consist of an N-terminal secretion alarm, a variable region, one or more CEP domains, and a brief C-terminal region are encoded by the C-terminally encoded peptide (CEP) genes. Precursor proteins of the CEP experience proteolysis and afterwards a pro hydroxylation to yield a CEP peptide of 15-amino acid with bioactive properties (Ogilvie et al. 2014). Along with their respective receptors CEPR1/2, CEPs perform a variety of actions in plant replies to environmental changes. These actions include systemic N-acquisition, the enhancement of lateral root growth in response to sucrose, the suppression of primary root growth, and the regulation of carbon and nitrogen uptake under conditions of nutrient starvation. The diversity of functions that CEPs and CEPR1/2 perform is extensive, demonstrating their importance in plant physiology (Jeon et al. 2021). CEP5 contributes to Arabidopsis's resistance to drought and osmotic stress, as well as negatively regulating the growth of primary and lateral roots. A proteome study indicated that CEP5 modifies a major fraction of the proteins involved in the biological processes "response to stress or to abiotic stimulus," indicating that CEP5 may have a role in the body's defense against abiotic stress. The discovery that CEP5 are able to provide water deficiency resistance was supported by the fact that wild-type seedlings or CEP5-overexpression lines treated with hydroxyprolinated CEP5 peptide showed superior adaptation after dehydration following re-watering. Additionally, CEP5-overexpression lines showed improved resistance to rosette size reduction under osmotic stress and increased transcription factor gene expression induced by osmotic stress. When exposed to osmotic stress, the xylem intermixed with the phloem (xip)/cepr1 cepr2 double mutant did not show rosette size decrease, indicating that CEP5 functions independently of the CEPRs. Significant decreases in DR5-GUS or DR5-LUC activity as well as the expression of auxin-inducible genes Lob Domain-Containing Protein 18 (LBD18), LBD29, and Pin-Formed 1 (PIN1) were seen in the CEP5-overexpression line, which is noteworthy. By influencing proteasome activity, CEP5 stabilizes AUX/IAA proteins, which are negative controllers of auxin adaptors, without changing auxin levels or auxin transport activity. More research is necessary to determine whether and how CEP5's negative regulatory involvement in auxin response through AUX/IAA stabilization is related to osmotic stress tolerance and drought. The results of CEP5 overexpression must also be supported by functional investigation of the cep5 mutants (Kim et al. 2021).

2.4.4 Salinity Stress Response

In 2015, Chien et al. studied AtCAPE1 belonging to the group of CAPE peptides, as a factor for salinity stress regulation in the model plant Arabidopsis. A superfamily of pathogenesis-related proteins (CAP) and specifically the cysteine-rich secretory proteins is the source for CAP-derived peptide 1 (CAPE1) which is constituted from the C-terminus of pathogenesis-related protein1b (PR-1b). CAPE1 was discovered in tomato leaves using a peptidomics approach and induces response against pathogens. This evidence suggests that immune signaling plays a key role in PR-1. Arabidopsis AtCAPE1, comprised of 11 amino acids, has a negative impact on salt stress tolerance. The identification of nine potential CAPEs in Arabidopsis, named PROAtCAPEs, was based on sequence similarity to tomato CAPE1 and the C-terminal conserved motif. Of them, PROAtCAPE1 is mostly downregulated in response to salt stress, and its mutation results in resistance to high-salt environments that restrict growth. On the other hand, in Arabidopsis, the sensitive phenotype of the mutant is restored by applying synthetic AtCAPE1 peptide or overexpressing PROAtCAPE1, suggesting that AtCAPE1 acts as a regulator of salinity resistance in a negative way. Furthermore, salt-inducible genes implicated in detoxification, dehydration response, and osmolyte production are negatively regulated by AtCAPE1 (Chien et al. 2015).

Another group of cysteine-rich peptides (of 5 kDa), which induce rapid alkalinization of plant cells of the extracellular compartments, is called rapid alkalinization factor (RALF) peptides. RALF peptides decrease the proton electrochemical potential that is essential for solute uptake and cause cell development abolition. RALF peptides affect a number of biological activities in plants, such as immunological responses, guard cell migration, root growth, and pollen tube growth and termination during reproductive processes. The LRR-RLK FERONIA (FER) recognizes these peptides, and through interactions with the cell-wall LRR extensions (LRX)3/4/5, the RALF22/23-FER module controls salinity resistance. The LRX proteins are found in the cell wall and are involved in communication between the cell wall and the plasma membrane. They have an N-terminal LRR and a C-terminal extension region. Some of the same symptoms that are seen in the lrx3/4/5 triple mutant, fer mutant, and RALF22 or RALF23 overexpressing crops are lower plant growth and salt hypersensitivity. Physical relationships exist between Ralf peptides and LRX and FER proteins. Salinity leads in separation of mature RALF22 peptides from LRX proteins, which in turn leads in triggering internalization of FER through an endosomal route. RALF22/23-FER and LRX3/4/5 are crucial in regulating plant growth and salinity stress resistance by causing adjustments in the cell wall caused by salinity. Furthermore, these changes may also contribute to FER-mediated defense of roots from bursting due to salinity (Zhao et al. 2018). FERONIA is crucial for maintaining cell-wall integrity, particularly during high salt stress. The LLG1 protein interacts with FER in the endoplasmic reticulum and is essential for FER's localization to the plasma membrane, where it enables RHO GTPase signaling. LLG1, FER and their interactions are critical for preserving the root cell-wall integrity, as shown by the ionic sensitivity and cellular damage observed in llg1 mutant roots. Pectin cross-link fortification can restore growth and cell-wall integrity in FER seedlings upon salinity conditions. In vitro, FER binds with pectin and is present in calcium transients caused by salinity to preserve cell wall integrity after the absence of the stress factor. Moreover, the calcium channel causing FER-mediated calcium transients and the manner in which these transients' control downstream signaling to fix cell-wall damage brought on by salt stress require more research. It appears that RALF1 is not essential for controlling root growth in salt stress because neither the ralf1 mutant nor the aRALF1 RNAi transgenic line significantly inhibited root growth (Feng et al. 2018).

The *Arabidopsis thaliana* plant genome contains eight members of the AtPROPEP group, and among these, AtPep3 is a key factor for the plant's response to salt stress (Bartels et al. 2013). In the AtPROPEP gene cluster, AtPROPEP3 demonstrates the most significant reply to increased salt levels. Salinity resistance could be induced along overexpressing AtPROPEP3 or by administering a synthetic AtPep3 peptide that is generated from the PROPEP3 C-terminal region. In contrast, AtPROPEP3-RNAi lines exhibit heightened sensitivity to salt stress, a condition that can be mitigated with the administration of AtPep3 peptide. Using mass spectrometry, the endogenous AtPep3 peptide was found in plants treated with NaCl. Exogenous AtPep3 peptide treatment is only able to restore plant survival in pepr2 mutants, but high salinity conditions dramatically decrease survival in pepr1, pepr2, and pepr1/2 mutants. This suggests that plants are induced to tolerate salt stress through the recognition of AtPep3 by the PEPR1 receptor. Therefore, it is possible that the AtPEP3–PEPR1 module serves two purposes in terms of immunological responses and salinity stress tolerance (Nakaminami et al. 2018).

2.4.5 Heat Stress Response

The increasing world's temperatures present an extreme obstacle to current agricultural productivity by causing thermal stress, which is followed by a prolonged elevation in temperatures exceeding a crucial point, resulting in irreversible harm to plant growth and development. In particular, the reproductive organs of flowering plants are particularly susceptible to this heat stress, leading to negative consequences on seed yield (De Storme and Geelen 2014). Environmental stressors can have a detrimental impact on the reproductive development of flowering plants, ultimately affecting the yield of seeds. One signaling peptide that plays a role in the high temperature stress response in plants has been identified. The Arabidopsis CLE45, sterility-regulating kinase member (SKM1)/(SMK2) receptor enhances in pollen tube growth. The CLE45 peptide has been shown to extend pollen tube growth in in vitro culture. Moreover, the presence of a double mutation in the SKM1/2 genes, which encode for LRR–RLKs, leads to a pollen tube growth insensitivity to the synthetic CLE45 peptide. Photoaffinity labeling experiments with the SKM1-HaloTag protein demonstrated a physical interaction between SKM1 and CLE45 peptides (Endo et al. 2013). According to GUS reporter expression, it appears that CLE45, which is heat-inducible in pistils, and SKM1, which is present in pollen, operate in the same signaling route. By suppressing CLE45 expression through RNAi or expressing a kinase-dead version of SKM1 as a dominant negative form of SKM1 in the skm1 mutant, it was found that a reduction seed number and size appears when temperatures reach 30 °C but not at room temperature (22 °C). In summary, the CLE45–SKM1/2 pathway supports pollen tube growth even upon heat stress, which finally, leads to optimal seed production (Kim et al. 2021).

2.4.6 Heavy Metal Toxicity

Due to soil and water contamination, heavy metal accumulation—particularly that of arsenic and cadmium (As, Cd)—poses a serious danger to crop quality and productivity worldwide. When plants come into contact with arsenic, they undergo significant physiological and biochemical alterations due to carcinogenic qualities, high toxicity, even at very low levels. Similarly, plants may suffer from heavy metals such as cadmium, which can lead to oxidative stress and dramatically increase ROS levels within cells. Although more research is necessary, we currently know very little about the tiny peptides that control how plants react to stress caused by heavy metals (Singh et al. 2016). Pre-miR408 was reported to encode a small peptide which proved to be important in regulating plant adaptation to arsenic toxicity and sulfur deficiency. Genes targeted by miR408 are downregulated when synthetic miPEP408 is applied, as it increases the expression of the miR408 transcript. Exogenous administration of miPEP408 improved plant sensitivity to a number of stresses. These include conditions with low sulfur levels, arsenic toxicity, and the combined stress of inadequate sulfur and arsenic. Plants overexpressing both miPEP408 and miR408 showed this impact, and lines of miR408 that were CRISPR-edited showed different levels of tolerance to arsenic and low sulfur conditions (Datta et al. 2024). These results provide compelling evidence for the role of miPEP408 in enhancing the physiological and phenotypic effects of miR408. ROS production, including the production of free radicals and nonradical forms, is frequently triggered by abiotic stress factors such as heavy metal toxicity and nutritional deficiencies (Kumar et al. 2023). Elevated levels of reactive oxygen species (ROS) in plants can result in various forms of cellular damage. Research on miPEP408 revealed elevated ROS content in OX lines of miR408 and miPEP408 in comparison with wild-type plants, whereas mutant lines of miR408 exhibited decreased ROS content. Plants that have been genetically engineered to overexpress miPEP408 and miR408 exhibit a diminished utilization of the sulfur reduction pathway, which in turn leads to a decrease in antioxidant molecules such as glutathione and phytochelatins. This reduction in antioxidant capacity is likely to make the plants more susceptible to oxidative stress induced by arsenic and sulfur-deficient conditions. Hence, miPEP408 plays a crucial role in regulating the detoxification of arsenic by modulating the sulfur reduction pathway. The discovery of miPEP408's regulatory function has expanded our understanding of how plants respond to environmental stressors and opened up new possibilities for using miPEP408 in crop development and stress adaptation (Dixit et al. 2016).

2.4.7 Ultra-Violet (UV) Irradiation

The impact of UV light on plant cells has, for the most part, been investigated in terms of the activation of defense-related genes. The relationship between cell cycle-related gene suppression and defense-related gene activation is intriguing. Parsley cell suspension cultures exposed to UV radiation showed a decrease in histone production, indicating a correlation between the expression of a subclass of the histone H3 gene family and cell division. The transcriptional repression of genes encoding a mitotic cyclin and p34cdc2 protein kinase, as well as a subclass of each of the histone H2A, H2B, H3, and H4 gene families, was demonstrated in both fungal elicitor-treated parsley cells and UV-irradiated cells using RNA-blot and run-on transcription assays. The stimulation of flavonoid biosynthesis enzymes coincided with a decrease in histone production, and the inversely linked stimulation of phenylalanine ammonia-lyase suggested that a common shared signaling system is the cause for both cell cycle-related gene suppression and defense-related gene activation. The observed decline in histone H3 mRNA concentration surrounding fungal infection areas in young parsley leaves supports the conclusion that cell division arrest is necessary for the full commitment to transcriptional triggering of mechanisms taking part in UV protection (Logemann et al. 1995).

In 2010, *Rauvolfia verticillata,* its TIA route and the gene for HDR were studied in a molecular base with UV as an eliciting factor. Expression profile investigations demonstrated that exogenous elicitors such as methyl jasmonate, acetyl salicylic acid, abscisic acid, and UV were able to activate RvHDR expression, and that these elicitors also caused an increase in transcription levels when compared to the control. Ultimately, RvHDR was converted into the deadly *E. Coli* HDR mutant strain MG1655, which was able to preserve the mutant's phenotype. This demonstrated that RvHDR possessed the HDR gene's usual function. In addition to providing a candidate gene for metabolic engineering of the TIAs pathway in *R. verticillata*, the cloning, characterization, and functional identification of RvHDR will help to shed light on the biosynthetic mechanism of TIAs precursor and shed further light on the function of HDR at the molecular genetic level (Chen et al. 2010).

In 2015, the photocontrol of PIP-1 elicitor activity was examined to find factors temporally affecting phytoalexin synthesis. Moreover, in tobacco cells, the peptide elicitor PIP-1 can elicit a range of immunological responses. PIP-1 can elicit distinct sorts of reactions based on the length of its stimulation; short-term stimulation results in modest responses, whereas long-term stimulation produces powerful responses, including the formation of the phytoalexin capsidiol. Nevertheless, it is still unknown what essential elements directly control the start of capsidiol production in response to ongoing PIP-1 stimulation. A photocleavable PIP-1 homologue was constructed using 3-amino-3-(2-nitrophenyl) propionic acid as a photocleavable residue. UV radiation can be used to "switch off" the analog's activity without having any negative side effects. After the exposure of tobacco cells to UV radiation for 1 h following treatment, no capsidiol formation was seen, whereas this analog produced a sizable amount of capsidiol when exposed to UV radiation. With the use of this analog, it was discovered that the duration of the PIP-1 stimulation controls the elicitor-inducible 3hydroxy-3-methylglutaryl-CoA reductase activity, which may be linked to the start of capsidiol production (Kim et al. 2015).

2.5 PEPs from Cyanobacteria and Microalgae

Microalgae, which are among the oldest living organisms on Earth, are tiny, single-celled or multi-celled organisms that encompass both prokaryotic cyanobacteria and eukaryotic green algae and diatoms. These autotrophic organisms possess exceptional photosynthetic capabilities and are not influenced by seasonal changes because of the various coping strategies they have evolved in response to different light conditions (Stavridou et al. 2024). The effects of green microalgae *Haematococcus pluvialis* and blue-green microalga *Spirulina platensis* as a source of microalgal elicitors, on the accretion of betalaines and thiophenes in *Beta vulgaris'* and *Tagetes patula*'s hairy root cultures, were investigated, respectively. Results showed that the treatment with extracts of *H. pluvialis* and *S. platensis* increased the biomass of cultured Beta vulgaris hairy roots, which was 165.3 g and 149.4 g fresh wt/L, respectively, compared to the initial inoculum of 1.25 g fresh wt/L. Betalaines accumulation also significantly increased in the hairy root culture treated with *H. pluvialis*, with a 2.28-fold increase on the 15th day compared to the control. Similarly, *S. platensis* extract-treated hairy roots showed a 1.16-fold increase in betalaines on the 25th day. However, *H. pluvialis* extract did not affect the growth or thiophene accumulation in *T. patula* hairy roots, while the accumulation of thiophene increased by 1.2-fold in the cultures treated with *H. pluvialis* extract on the 20th day compared to the untreated control. In summary, the extract from *S. platensis* had an impact on the production of betalaines in the hairy roots of *B. vulgaris*, while the extract from *H. pluvialis* promoted the production of both betalaines and thiophene in the hairy root cultures of *B. vulgaris* and *T. patula*, respectively (Rao et al. 2001).

Cyanobacteria could be important components in agricultural soils. They could enhance soil fertility and crop productivity by providing biological nitrogen fixation, phosphate solubilization, and mineral release. These organisms also generate numerous biologically active molecules, including proteins, vitamins, carbohydrates, amino acids, polysaccharides, and phytohormones, performing an eliciting action to enhance plant growth and assist plants in combating biotic and abiotic stresses. Metabolites from cyanobacteria affect the gene expression of host plants, leading to changes in phytochemical composition. Numerous experiments involving live inoculum or cyanobacterial extracts on various plant species, including rice, wheat, maize, and cotton, showed that they synthesize signaling metabolites (Singh 2014). Plant colonization promotes growth in stressful soil conditions based on releasing bioactive molecules in the rhizosphere that trigger systemic responses in plants to combat. In a study using the cyanobacterium *Scytonema hofmanni*, it was proved to have a positive impact on rice seedlings' growth under salt stress conditions. The extracellular products of *S. hofmanni* partially or completely reversed the growth-inhibiting effects and biochemical alterations caused by NaCl on rice seedlings, including 5-aminolevulinate dehydratase activity, total free porphyrin content, and pigment levels (Rodríguez et al. 2006). Similarly, cyanobacterial extracts can substitute for the loss of phytohormones and promote growth in salt-affected crops. Cyanobacteria have developed various strategies to cope with salt stress conditions. The aforementioned strategies involve the expression of a cyanobacterial flavodoxin in plants, which endows them with the ability to withstand various environmental stresses, such as salt stress (Coba de la Pena et al. 2010). A gene called SsGlc, which encodes a b-1,4-glucanase-like protein (SsGlc), has been analyzed in *Synechocystis* (Tamoi et al. 2007). The SsGlc amino acid sequence revealed a considerable resemblance to the sequences of GH family b-1,4-glucanases (cellulases) derived from multiple sources. This gene was found to have a function in salt stress tolerance and might be employed to improve salt tolerance in plants. When the gene encoding glutaredoxins from Cyanobacterium *Synechocystis* strain (slr1562) was cloned in *Escherichia coli* cells, it was observed to significantly enhance the cells' resistance to drought, oxidative and salt stresses (Gaber and El-Assal 2012). A significant portion of cyanobacterial secondary metabolites consist of peptides or possess peptidic substructures, exhibiting considerable structural diversity. These peptides are primarily believed to be synthesized through NRPS (nonribosomal peptide synthetase) or NRPS/PKS (polyketide synthase) hybrid pathways, and as of now, approximately 600 cyanobacterial peptides have been documented (Kehr et al. 2011). Certain cyanobacterial peptides have been discovered to trigger an antioxidative defense system in plants, which comprises a network of enzymes such as superoxide dismutases, peroxidases, catalases, glutathione S-transferases, and glutathione reductases, as well as a range of low molecular-weight antioxidants including glutathione, ascorbate, and tocopherols. These findings demonstrate the potential of cyanobacterial peptides to promote plant health and growth (Pflugmacher et al. 2007). The use of culture filtrates from cyanobacterial strains like *C. ghosei, H. intricatus,* and *Nostoc* was found to improve germination percentage, radicle length, and coleoptile length of wheat seedlings. This was determined through TLC analyses, which revealed the presence of several amino acids, including histidine (Nanjappan Karthikeyan et al. 2007). Phenylalanine, threonine, glutamate, and glycine were the most abundant free amino acids found in the external media of cyanobacterium *Anabaena siamensis Antarikanonda*, which was isolated from rice paddies in Bangkok, Thailand. This cyanobacterium was grown in media containing N_2, NH_4^+, NO_3^-, and N-starved media (Singh 2014).

Microalgae are known to express robust defense mechanisms in response to pathogen infections or environmental abiotic stress factors. These mechanisms can involve the biosynthesis of antimicrobial substances that could be utilized as pesticides for agriculture (Jena and Subudhi 2019). Microalgae may also be utilized as elicitors of plant defense mechanisms in agriculture. They are used to stimulate the plant's defense responses by releasing chemical compounds and structural components. The direct application of microalgae on crop roots activates the plant's defense responses systemically, as microalgae-plant interactions are reciprocal. This has resulted in the discovery of priming-type responses in various crops, although the specific microalgae and molecules involved remain unknown. For example, root inoculation with *Chlorella vulgaris* in broccoli and guar plants increased the activity of antioxidant enzymes, such as APX, CAT, SOD, and glutathione reductase. Additionally, it led to the accumulation of flavonoid and phenolic compounds (Kusvuran 2021). Industrial applications for this elicitation technique include the production of chemicals of interest by plant cells. Plants employ defensive mechanisms similar to priming to become ready for pathogen attacks. Additionally, by activating SAR and causing a variety of cytological changes, the use of *Chlorella fusca* in cucumber plants through foliar inoculation has been found to decrease the impact of the disease caused by *Colletotrichum orbiculare*, which includes the accumulation of vesicles, the formation of a sheath around penetration hyphae, and the thickness of cell wells adjacent to intracellular hyphae (Kim et al. 2018). Microalgae-polysaccharides make up around 50% of the dry weight in most agricultural applications involving formulations and extracts of microalgae. They are considered a major component and are composed primarily of galactose, xylose, and glucose (Chanda et al. 2019). Microalgae-polysaccharides have been shown to activate plant systemic defensive responses; however, the specific molecules involved remain unclear. Rachidi et al., in 2021, have demonstrated that polysaccharides derived from various *Chlorella, Dunaliella* species belonging to the group of *Chlorophyta*, and *Porphyridium (Rhodophyta)* organisms perform as potent plant elicitors in tomato plants. This leads to an enhanced function of defense-related enzymes and the accumulation of defensive compounds, including polyphenols or steroidal glycoalkaloids (Rachidi et al. 2021). Microalgae are capable of producing and releasing exopolysaccharides that possess eliciting properties, in addition to structural polysaccharides. The exopolysaccharides produced by *Dunaliella salina*, for instance, have been demonstrated to stimulate the activity of CAT, POD, and SOD, as well as the accretion of phenolic substances in tomato leaves (El Arroussi et al. 2018). *Porphyridium sordidum* generates exopolysaccharides that trigger the expression of SA-related genes and PAL activity in A. thaliana leaves, resulting in a reduction of diseases caused by *F. oxysporum* (Drira et al. 2021). It is important to mention that microalgae have the ability to produce a variety of molecules and chemical compounds that act as elicitors for plants. For instance, some types of green microalgae can produce lactic acid, which is a chiral organic acid. It is worth noting that these compounds play a crucial role in plant growth and development (Augustiniene et al. 2022). The specific isoform of D-lactic acid produced by *C. fusca* activates cellular receptors in *A. thaliana*, leading to the expression of genes associated with SA and JA and a reduction in disease induced by *P. syringae* pv. Tomato (Lee et al. 2020). This phenomenon can potentially be utilized on a commercial scale to increase the production of secondary metabolites of interest, such as those found in hairy roots. For instance, the production of glucosamine by *Scenedesmus obliquus* has been shown to enhance the production of spiroketal enol ether diacetylenes in the hairy roots of *Tanacetum parthenium* (Stojakowska et al. 2008). Furthermore, recent research has explored the potential inheritance of plant defense activation by microalgae-elicitors in subsequent generations. In a study, tomato, pepper, and eggplant plants were cultivated in vermicompost along with the microalgae *Ulothrix* spp. (*Chlorophyta*) and *Navicula* spp. (*Ochrophyta*). When the harvested seeds were germinated in the presence of the pathogenic oomycete *Pythium* sp., demonstrated a 90% increase in seedling survival, which suggests inherited resistance (Poveda and Díez-Méndez 2022). Unlike macroalgae, microalgae-biopesticides have been found to be essential against bacteria, fungi, and oomycetes, but the specific responsible molecule is not yet known. Therefore, identifying all the target substances is a future research challenge for scientists. Additionally, most studies conducted with microalgal and macroalgal elicitors reformed in in vitro systems or growth chambers or greenhouses and thus highlighting the requirement for extensive and rigorous tests in the field.

2.6 Conclusion

The need for environmentally friendly alternatives to chemical pesticides and fertilizers, given their detrimental effects on both environmental and human health, already has sparked the search for sustainable solutions in modern agriculture. However, changing traditional farming methods is both a pressing necessity and a challenging task. Given their inability to move, plants have developed intricate defense mechanisms to cope with adverse situations. Climate change-induced abiotic stresses pose a significant threat to crop yields and food security, making it crucial to identify plant peptides and signaling pathways that can improve stress tolerance in crops for agricultural sustainability. Numerous plant peptides and signaling routes have been discovered to be a crucial part in synchronizing crop adaptation to abiotic stressors. The agrochemical market is plagued by a lack of innovative products due to the prolonged time it takes for discoveries to be commercialized. As a result, seven companies control 90% of the market, with their primary income source not derived from producing bioproducts. To apply PEP in crop production, extensive agronomic, physiological, chemical, biochemical, and molecular studies are needed to gain a deeper understanding of the changes induced.

By collecting detailed information on the optimal preparation and utilization of PEPs, precise protocols should be established to assist agrochemical industries and farmers in the production and application of PEPs. Several questions still need to be addressed, such as the complex mechanisms controlling both proteases and maturation of these peptides. Future studies should provide a detailed understanding of the downstream participants, such as receptors or secondary messengers. Playing a critical role, these factors are pivotal in regulating the signaling cascade initiated by peptide-mediated stress responses. Additionally, it is important to investigate the affected tissues of these peptides and investigate the complex interactions between elicitor peptides and phytohormones in the context of abiotic stress regimes. Exploring the mechanisms by which local nutrient and abiotic changes in plants trigger peptide signaling is a daunting task. Climate change intensifies the strain of abiotic stress, posing a considerable risk to crop yields and food security. To mitigate this threat, it is crucial to find ways to bolster stress resilience in crops for sustainable agriculture. One potential solution lies in identifying plant peptides and signaling pathways that could help enhance stress resistance in crops. By focusing on these areas, we may be able to develop new approaches to address the challenges posed by climate change and ensure food security in the near future.

References

Araya T, von Wirén N, Takahashi H (2014) CLE peptides regulate lateral root development in response to nitrogen nutritional status of plants. Plant Signal Behav 9(7):2029–2034. https://doi.org/10.4161/psb.29302

Augustiniene E, Valanciene E, Matulis P, Syrpas M, Jonuskiene I, Malys N (2022) Bioproduction of L-and D-lactic acids: advances and trends in microbial strain application and engineering. Crit Rev Biotechnol 42(3):342–360. https://doi.org/10.1080/07388551.2021.1940088

Bartels S, Boller T (2015) Quo vadis, Pep? Plant elicitor peptides at the crossroads of immunity, stress, and development. J Exp Bot 66(17):5183–5193. https://doi.org/10.1093/jxb/erv180

Bartels S, Lori M, Mbengue M, Van Verk M, Klauser D, Hander T, Boni R, Robatzek S, Boller T (2013) The family of Peps and their precursors in Arabidopsis: differential expression and localization but similar induction of pattern-triggered immune responses. J Exp Bot 64(17):5309–5321. https://doi.org/10.1093/jxb/ert330

Boller T, Felix G (2009) A renaissance of elicitors: perception of microbe-associated molecular patterns and danger signals by pattern-recognition receptors. Annu Rev Plant Biol 60(1):379–406. https://doi.org/10.1146/annurev.arplant.57.032905.105346

Bowman JL (2022) The origin of a land flora. Nat Plants 8(12):1352–1369. https://doi.org/10.1038/s41477-022-01283-y

Cederholm HM, Benfey PN (2015) Distinct sensitivities to phosphate deprivation suggest that RGF peptides play disparate roles in Arabidopsis thaliana root development. New Phytol 207(3):683–691. https://doi.org/10.1111/nph.13405

Chanda M-J, Merghoub N, El Arroussi H (2019) Microalgae polysaccharides: the new sustainable bioactive products for the development of plant bio-stimulants? World J Microbiol Biotechnol 35(11):177. https://doi.org/10.1007/s11274-019-2745-3

Chaw ShuMiaw CS, Chang ChienChang CC, Chen HsinLiang CH, Li W (2004) Dating the monocot-dicot divergence and the origin of core eudicots using whole chloroplast genomes. J Mol Evol 58(4):424–441. https://doi.org/10.1007/s00239-003-2564-9

Chen JW, Liu WH, Chen M, Wang GJ, Peng MF, Chen R, Yang CX, Lan XZ, Ming XJ, Hsieh M, Liao ZH (2010) The HDR gene involved in the TIA pathway from: cloning, characterization and functional identification [Article]. J Med Plants Res 4(10):915–924. https://doi.org/10.1111/nph.16241

Chien P-S, Nam HG, Chen Y-R (2015) A salt-regulated peptide derived from the CAP superfamily protein negatively regulates salt-stress tolerance in Arabidopsis. J Exp Bot 66(17):5301–5313. https://doi.org/10.1093/jxb/erv263

Coba de la Pena T, Redondo FJ, Manrique E, Lucas MM, Pueyo JJ (2010) Nitrogen fixation persists under conditions of salt stress in transgenic Medicago truncatula plants expressing a cyanobacterial flavodoxin. Plant Biotechnol J 8(9):954–965. https://doi.org/10.1111/j.1467-7652.2010.00519.x

Corrado G, Sasso R, Pasquariello M, Iodice L, Carretta A, Cascone P, Ariati L, Digilio MC, Guerrieri E, Rao R (2007) Systemin regulates both systemic and volatile signaling in tomato plants. J Chem Ecol 33(4):669–681. https://doi.org/10.1007/s10886-007-9254-9

Datta T, Kumar RS, Sinha H, Trivedi PK (2024) Small but mighty: peptides regulating abiotic stress responses in plants. Plant Cell Environ 47(4):1207–1223. https://doi.org/10.1111/pce.14792

De Storme N, Geelen D (2014) The impact of environmental stress on male reproductive development in plants: biological processes and molecular mechanisms. Plant Cell Environ 37(1):1–18. https://doi.org/10.1111/pce.12142

Delay C, Chapman K, Taleski M, Wang Y, Tyagi S, Xiong Y, Imin N, Djordjevic MA (2019) CEP3 levels affect starvation-related growth responses of the primary root. J Exp Bot 70(18):4763–4774. https://doi.org/10.1093/jxb/erz270

Ding Y, Wang J, Wang J, Stierhof Y-D, Robinson DG, Jiang L (2012) Unconventional protein secretion. Trends Plant Sci 17(10):606–615. https://doi.org/10.1016/j.tplants.2012.06.004

Dixit G, Singh AP, Kumar A, Mishra S, Dwivedi S, Kumar S, Trivedi PK, Pandey V, Tripathi RD (2016) Reduced arsenic accumulation in rice (Oryza sativa L.) shoot involves sulfur mediated improved thiol metabolism, antioxidant system and altered arsenic transporters. Plant Physiol Biochem 99:86–96. https://doi.org/10.1016/j.plaphy.2015.11.005

Dong W, Wang Y, Takahashi H (2019) CLE-CLAVATA1 signaling pathway modulates lateral root development under sulfur deficiency. Plants 8(4):103. https://doi.org/10.3390/plants8040103

Drira M, Elleuch J, Ben Hlima H, Hentati F, Gardarin C, Rihouey C, Le Cerf D, Michaud P, Abdelkafi S, Fendri I (2021) Optimization of exopolysaccharides production by Porphyridium sordidum and their potential to induce defense responses in Arabidopsis thaliana against Fusarium oxysporum. Biomolecules 11(2):282. https://doi.org/10.3390/biom11020282

El Arroussi H, Benhima R, Elbaouchi A, Sijilmassi B, El Mernissi N, Aafsar A, Meftah-Kadmiri I, Bendaou N, Smouni A (2018) Dunaliella salina exopolysaccharides: a promising biostimulant for salt stress tolerance in tomato (Solanum lycopersicum). J Appl Phycol 30:2929–2941. https://doi.org/10.1007/s10811-017-1382-1

Endo S, Shinohara H, Matsubayashi Y, Fukuda H (2013) A novel pollen-pistil interaction conferring high-temperature tolerance during reproduction via CLE45 signaling. Curr Biol 23(17):1670–1676. https://doi.org/10.1016/j.cub.2013.06.060

Feng W, Kita D, Peaucelle A, Cartwright HN, Doan V, Duan Q, Liu M-C, Maman J, Steinhorst L, Schmitz-Thom I (2018) The FERONIA receptor kinase maintains cell-wall integrity during salt stress through Ca^{2+} signaling. Curr Biol 28(5):666–675.e665. https://doi.org/10.1016/j.cub.2018.01.023

Ferrari S, Savatin D, Sicilia F, Gramegna G, Cervone F, Lorenzo GD (2013) Oligogalacturonides: plant damage-associated molecular patterns and regulators of growth and development. Front Plant Sci 4:49. https://doi.org/10.3389/fpls.2013.00049

Figueroa-Macías JP, García YC, Núñez M, Díaz K, Olea AF, Espinoza L (2021) Plant growth-defense trade-offs: molecular processes leading to physiological changes. Int J Mol Sci 22(2):693. https://doi.org/10.3390/ijms22020693

Gaber A, El-Assal SE (2012) A Cyanobacterium Synechocystis sp. PCC 6803 glutaredoxin gene (slr1562) protects Escherichia coli against abiotic stresses. Am J Agric Biol Sci 7(1):88–96. https://doi.org/10.3844/ajabssp.2012.88.96

Gautam H, Sharma A, Trivedi PK (2023) Plant microProteins and miPEPs: small molecules with much bigger roles. Plant Sci 326:111519

Gutiérrez-Alanís D, Yong-Villalobos L, Jimenez-Sandoval P, Alatorre-Cobos F, Oropeza-Aburto A, Mora-Macías J, Sánchez-Rodríguez F, Cruz-Ramírez A, Herrera-Estrella L (2017) Phosphate starvation-dependent iron mobilization induces CLE14 expression to trigger root meristem differentiation through CLV2/PEPR2 signaling. Dev Cell 41(5):555–570.e553. https://doi.org/10.1016/j.devcel.2017.05.009

Hahn MG (1996) Microbial elicitors and their receptors in plants. Annu Rev Phytopathol 34(1):387–412. https://doi.org/10.1146/annurev.phyto.34.1.387

Heiling S, Schuman MC, Schoettner M, Mukerjee P, Berger B, Schneider B, Jassbi AR, Baldwin IT (2010) Jasmonate and ppHsystemin regulate key malonylation steps in the biosynthesis of 17-hydroxygeranyllinalool diterpene glycosides, an abundant and effective direct defense against herbivores in Nicotiana attenuata. Plant Cell 22(1):273–292. https://doi.org/10.1105/tpc.109.071449

Hu X-L, Lu H, Hassan MM, Zhang J, Yuan G, Abraham PE, Shrestha HK, Villalobos Solis MI, Chen J-G, Tschaplinski TJ (2021) Advances and perspectives in discovery and functional analysis of small secreted proteins in plants. Hortic Res 8(1):130. https://doi.org/10.1038/s41438-021-00570-7

Huffaker A, Pearce G, Ryan CA (2006) An endogenous peptide signal in Arabidopsis activates components of the innate immune response. Proc Natl Acad Sci U S A 103:10098–10103. https://doi.org/10.1073/pnas.0603727103

Jena J, Subudhi E (2019) Microalgae: an untapped resource for natural antimicrobials. In: The role of microalgae in wastewater treatment, pp 99–114. https://doi.org/10.1007/978-981-13-1586-2_8

Jeon BW, Kim M-J, Pandey SK, Oh E, Seo PJ, Kim J (2021) Recent advances in peptide signaling during Arabidopsis root development. J Exp Bot *72*(8):2889–2902. https://doi.org/10.1093/jxb/erab050

Kaufmann C, Sauter M (2019) Sulfated plant peptide hormones. J Exp Bot 70(16):4267–4277. https://doi.org/10.1093/jxb/erz292

Kehr J-C, Picchi DG, Dittmann E (2011) Natural product biosyntheses in cyanobacteria: a treasure trove of unique enzymes. Beilstein J Org Chem 7(1):1622–1635. https://doi.org/10.3762/bjoc.7.191

Kim Y, Miyashita M, Miyagawa H (2015) Photocontrol of elicitor activity of PIP-1 to investigate temporal factors involved in phytoalexin biosynthesis [Article]. J Agric Food Chem 63(25):5894–5901. https://doi.org/10.1021/acs.jafc.5b01910

Kim SJ, Ko EJ, Hong JK, Jeun YC (2018) Ultrastructures of Colletotrichum orbiculare in cucumber leaves expressing systemic acquired resistance mediated by Chlorella fusca. Plant Pathol J 34(2):113. https://doi.org/10.5423/PPJ.OA.09.2017.0204

Kim JS, Jeon BW, Kim J (2021) Signaling peptides regulating abiotic stress responses in plants. Front Plant Sci 12(704):490. https://doi.org/10.3389/fpls.2021.704490

Kumar RS, Sinha H, Datta T, Asif MH, Trivedi PK (2023) microRNA408 and its encoded peptide regulate sulfur assimilation and arsenic stress response in Arabidopsis. Plant Physiol 192(2):837–856. https://doi.org/10.1093/plphys/kiad033

Kuromori T, Fujita M, Takahashi F, Yamaguchi-Shinozaki K, Shinozaki K (2022) Inter-tissue and inter-organ signaling in drought stress response and phenotyping of drought tolerance. Plant J 109(2):342–358. https://doi.org/10.1111/tpj.15619

Kusvuran S (2021) Microalgae (Chlorella vulgaris Beijerinck) alleviates drought stress of broccoli plants by improving nutrient uptake, secondary metabolites, and antioxidative defense system. Hortic Plant J 7(3):221–231. https://doi.org/10.1016/j.hpj.2021.03.007

Lagiotis G, Madesis P, Stavridou E (2023) Echoes of a stressful past: abiotic stress memory in crop plants towards enhanced adaptation. Agriculture 13(11):2090. https://doi.org/10.3390/agriculture13112090

Lee SM, Kim SK, Lee N, Ahn CY, Ryu CM (2020) d-Lactic acid secreted by Chlorella fusca primes pattern-triggered immunity against Pseudomonas syringae in Arabidopsis. Plant J 102(4):761–778. https://doi.org/10.1111/tpj.14661

Ligoxygakis P, Pelte N, Hoffmann JA, Reichhart J-M (2002) Activation of Drosophila Toll during fungal infection by a blood serine protease. Science 297(5578):114–116. https://doi.org/10.1126/science.1072391

Logemann E, Wu SC, Schröder J, Schmelzer E, Somssich IE, Hahlbrock K (1995) Gene activation by UV light, fungal elicitor or fungal infection in Petroselinum crispum is correlated with repression of cell cycle-related genes [Article]. Plant J 8(6):865–876. https://doi.org/10.1046/j.1365-313X.1995.8060865.x

Ma D, Endo S, Betsuyaku S, Shimotohno A, Fukuda H (2020) CLE2 regulates light-dependent carbohydrate metabolism in Arabidopsis shoots. Plant Mol Biol 104(6):561–574. https://doi.org/10.1007/s11103-020-01059-y

McGurl B, Pearce G, Orozco-Cardenas M, Ryan CA (1992) Structure, expression, and antisense inhibition of the systemin precursor gene. Science 255(5051):1570–1573. https://doi.org/10.1126/science.1549783

Nakaminami K, Okamoto M, Higuchi-Takeuchi M, Yoshizumi T, Yamaguchi Y, Fukao Y, Shimizu M, Ohashi C, Tanaka M, Matsui M (2018) AtPep3 is a hormone-like peptide that plays a role in the salinity stress tolerance of plants. Proc Natl Acad Sci U S A 115(22):5810–5815. https://doi.org/10.1073/pnas.1719491115

Nanjappan Karthikeyan NK, Radha Prasanna RP, Lata Nain LN, Kaushik B (2007) Evaluating the potential of plant growth promoting cyanobacteria as inoculants for wheat. Eur J Soil Biol 43(1):23–30. https://doi.org/10.1016/j.ejsobi.2006.11.001

Narváez-Vásquez J, Ryan CA (2004) The cellular localization of prosystemin: a functional role for phloem parenchyma in systemic wound signaling. Planta 218(3):360–369. https://doi.org/10.1007/s00425-003-1115-3

Narváez-Vásquez J, Pearce G, Ryan CA (2005) The plant cell wall matrix harbors a precursor of defense signaling peptides. Proc Natl Acad Sci U S A 102(36):12974–12977. https://doi.org/10.1073/pnas.0505248102

Nürnberger T, Nennstiel D (1998) Fungal peptide elicitors: signals mediating pathogen recognition in plants [Article]. Z Naturforsch C J Biosci 53(3–4):141–150. https://doi.org/10.1515/znc-1998-3-401

Ogilvie HA, Imin N, Djordjevic MA (2014) Diversification of the C-TERMINALLY ENCODED PEPTIDE (CEP) gene family in angiosperms, and evolution of plant-family specific CEP genes. BMC Genomics 15(1):1–15. https://doi.org/10.1186/1471-2164-15-870

Ortiz-Morea FA, Reyes-Bermudez AA (2019) Endogenous peptides: key modulators of plant immunity. In: Bioactive molecules in plant defense. Springer, Cham, pp 159–177. https://doi.org/10.1007/978-3-030-27165-7_10

Patharkar OR, Walker JC (2016) Core mechanisms regulating developmentally timed and environmentally triggered abscission. Plant Physiol 172(1):510–520. https://doi.org/10.1104/pp.16.01004

Pearce G, Moura DS, Stratmann J, Ryan CA (2001) Production of multiple plant hormones from a single polyprotein precursor. Nature 411(6839):817–820. https://doi.org/10.1038/35081107

Pflugmacher S, Aulhorn M, Grimm B (2007) Influence of a cyanobacterial crude extract containing microcystin-LR on the physiology and antioxidative defence systems of different spinach variants. New Phytol 175(3):482–489. https://doi.org/10.1111/j.1469-8137.2007.02144.x

Poveda J, Díez-Méndez A (2022) Use of elicitors from macroalgae and microalgae in the management of pests and diseases in agriculture. Phytoparasitica 51:667–701

Qi Z, Verma R, Gehring C, Yamaguchi Y, Zhao Y, Ryan CA, Berkowitz GA (2010) Ca^{2+} signaling by plant Arabidopsis thaliana Pep peptides depends on AtPepR1, a receptor with guanylyl cyclase activity, and cGMP-activated Ca^{2+} channels. Proc Natl Acad Sci U S A 107(49):21193–21198. https://doi.org/10.1073/pnas.1000191107

Rachidi F, Benhima R, Kasmi Y, Sbabou L, Arroussi HE (2021) Evaluation of microalgae polysaccharides as biostimulants of tomato plant defense using metabolomics and biochemical approaches. Sci Rep 11(1):930. https://doi.org/10.1038/s41598-020-78820-2

Rao SR, Tripathi U, Suresh B, Ravishankar GA (2001) Enhancement of secondary metabolite production in hairy root cultures of beta vulgaris and tagetes patula under the influence of microalgal elicitors. Food Biotechnol 15(1):35–46. https://doi.org/10.1081/fbt-100103893

Rao KM, Raghavendra AS, Reddy KJ (2006) Physiology and molecular biology of stress tolerance in plants. Springer, Dordrecht. https://doi.org/10.1007/1-4020-4225-6

Reichardt S, Piepho H-P, Stintzi A, Schaller A (2020) Peptide signaling for drought-induced tomato flower drop. Science 367(6485):1482–1485. https://doi.org/10.1126/science.aaz5641

Ren F, Lu Y-T (2006) Overexpression of tobacco hydroxyproline-rich glycopeptide systemin precursor A gene in transgenic tobacco enhances resistance against Helicoverpa armigera larvae. Plant Sci 171(2):286–292. https://doi.org/10.1016/j.plantsci.2006.04.001

Rodríguez A, Stella A, Storni M, Zulpa G, Zaccaro M (2006) Effects of cyanobacterial extracellular products and gibberellic acid on salinity tolerance in Oryza sativa L. Saline Syst 2:1–4. https://doi.org/10.1186/1746-1448-2-7

Ronga D, Biazzi E, Parati K, Carminati D, Carminati E, Tava A (2019) Microalgal biostimulants and biofertilisers in crop productions [Review]. Agronomy 9(4):192. https://doi.org/10.3390/agronomy9040192

Sánchez-Calderón L, López-Bucio J, Chacón-López A, Cruz-Ramírez A, Nieto-Jacobo F, Dubrovsky JG, Herrera-Estrella L (2005) Phosphate starvation induces a determinate developmental program in the roots of Arabidopsis thaliana. Plant Cell Physiol 46(1):174–184. https://doi.org/10.1093/pcp/pci011

Schmelz EA, Engelberth J, Alborn HT, Tumlinson JH III, Teal PE (2009) Phytohormone-based activity mapping of insect herbivore-produced elicitors. Proc Natl Acad Sci U S A 106(2):653–657. https://doi.org/10.1073/pnas.0811861106

Schulze B, Mentzel T, Jehle AK, Mueller K, Beeler S, Boller T, Felix G, Chinchilla D (2010) Rapid heteromerization and phosphorylation of ligand-activated plant transmembrane receptors and their associated kinase BAK1. J Biol Chem 285(13):9444–9451. https://doi.org/10.1074/jbc.M109.096842

Singh S (2014) A review on possible elicitor molecules of cyanobacteria: their role in improving plant growth and providing tolerance against biotic or abiotic stress. J Appl Microbiol 117(5):1221–1244. https://doi.org/10.1111/jam.12612

Singh S, Parihar P, Singh R, Singh VP, Prasad SM (2016) Heavy metal tolerance in plants: role of transcriptomics, proteomics, metabolomics, and ionomics. Front Plant Sci 6:165395. https://doi.org/10.3389/fpls.2015.01143

Stavridou E, Karapetsi L, Nteve GM, Tsintzou G, Chatzikonstantinou M, Tsaousi M, Martinez A, Flores P, Merino M, Dobrovic L, Mullor JL, Martens S, Cerasino L, Salmaso N, Osathanunkul M, Labrou NE, Madesis P (2024) Landscape of microalgae omics and metabolic engineering research for strain improvement: an overview. Aquaculture 587:740803. https://doi.org/10.1016/j.aquaculture.2024.740803

Stenvik G-E, Butenko MA, Aalen RB (2008a) Identification of a putative receptor-ligand pair controlling cell separation in plants. Plant Signal Behav 3(12):1109–1110. https://doi.org/10.4161/psb.3.12.7009

Stenvik G-E, Tandstad NM, Guo Y, Shi C-L, Kristiansen W, Holmgren A, Clark SE, Aalen RB, Butenko MA (2008b) The EPIP peptide of INFLORESCENCE DEFICIENT IN ABSCISSION is sufficient to induce abscission in Arabidopsis through the receptor-like kinases HAESA and HAESA-LIKE2. Plant Cell 20(7):1805–1817. https://doi.org/10.1105/tpc.108.059139

Stojakowska A, Burczyk J, Kisiel W, Zych M, Banas A, Duda T (2008) Effects of various elicitors on the accumulation and secretion of spiroketal enol ether diacetylenes in feverfew hairy root culture. Acta Soc Bot Pol 77(1):17–21. https://doi.org/10.5586/asbp.2008.002

Stührwohldt N, Bühler E, Sauter M, Schaller A (2021) Phytosulfokine (PSK) precursor processing by subtilase SBT3. 8 and PSK signaling improve drought stress tolerance in Arabidopsis. J Exp Bot 72(9):3427–3440. https://doi.org/10.1093/jxb/erab017

Takahashi F, Suzuki T, Osakabe Y, Betsuyaku S, Kondo Y, Dohmae N, Fukuda H, Yamaguchi-Shinozaki K, Shinozaki K (2018) A small peptide modulates stomatal control via abscisic

acid in long-distance signalling. Nature 556(7700):235–238. https://doi.org/10.1038/s41586-018-0009-2
Takahashi F, Hanada K, Kondo T, Shinozaki K (2019) Hormone-like peptides and small coding genes in plant stress signaling and development. Curr Opin Plant Biol 51:88–95. https://doi.org/10.1016/j.pbi.2019.05.011
Tamoi M, Kurotaki H, Fukamizo T (2007) β-1, 4-Glucanase-like protein from the cyanobacterium Synechocystis PCC6803 is a β-1, 3-1, 4-glucanase and functions in salt stress tolerance. Biochem J 405(1):139–146. https://doi.org/10.1042/Bj20070171
Tintor N, Ross A, Kanehara K, Yamada K, Fan L, Kemmerling B, Nürnberger T, Tsuda K, Saijo Y (2013) Layered pattern receptor signaling via ethylene and endogenous elicitor peptides during Arabidopsis immunity to bacterial infection [Article]. Proc Natl Acad Sci U S A 110(15):6211–6216. https://doi.org/10.1073/pnas.1216780110
Torres MA, Dangl JL, Jones JD (2002) Arabidopsis gp91phox homologues AtrbohD and AtrbohF are required for accumulation of reactive oxygen intermediates in the plant defense response. Proc Natl Acad Sci U S A 99(1):517–522
Ueki N, Someya K, Matsuo Y, Wakamatsu K, Mukai H (2007) Cryptides: functional cryptic peptides hidden in protein structures. Pept Sci 88(2):190–198. https://doi.org/10.1002/bip.20687
Vidal EA, Alvarez JM, Araus V, Riveras E, Brooks MD, Krouk G, Ruffel S, Lejay L, Crawford NM, Coruzzi GM (2020) Nitrate in 2020: thirty years from transport to signaling networks. Plant Cell 32(7):2094–2119. https://doi.org/10.1105/tpc.19.00748
Vie AK, Najafi J, Winge P, Cattan E, Wrzaczek M, Kangasjärvi J, Miller G, Brembu T, Bones AM (2017) The IDA-LIKE peptides IDL6 and IDL7 are negative modulators of stress responses in Arabidopsis thaliana. J Exp Bot 68(13):3557–3571. https://doi.org/10.1093/jxb/erx168
Wang A, Guo J, Wang S, Zhang Y, Lu F, Duan J, Liu Z, Ji W (2022) BoPEP4, a C-terminally encoded plant elicitor peptide from broccoli, plays a role in salinity stress tolerance. Int J Mol Sci 23(6):3090. https://doi.org/10.3390/ijms23063090
Yamaguchi Y, Huffaker A (2011) Endogenous peptide elicitors in higher plants [Review]. Curr Opin Plant Biol 14(4):351–357. https://doi.org/10.1016/j.pbi.2011.05.001
Zelman AK, Berkowitz GA (2023) Plant elicitor peptide (Pep) signaling and pathogen defense in tomato. Plants 12(15):2856. https://doi.org/10.3390/plants12152856
Zhang L, Shi X, Zhang Y, Wang J, Yang J, Ishida T, Jiang W, Han X, Kang J, Wang X (2019) CLE9 peptide-induced stomatal closure is mediated by abscisic acid, hydrogen peroxide, and nitric oxide in Arabidopsis thaliana. Plant Cell Environ 42(3):1033–1044. https://doi.org/10.1111/pce.13475
Zhao C, Zayed O, Yu Z, Jiang W, Zhu P, Hsu C-C, Zhang L, Tao WA, Lozano-Durán R, Zhu J-K (2018) Leucine-rich repeat extensin proteins regulate plant salt tolerance in Arabidopsis. Proc Natl Acad Sci U S A 115(51):13123–13128. https://doi.org/10.1073/pnas.1816991115
Zipfel C, Kunze G, Chinchilla D, Caniard A, Jones JD, Boller T, Felix G (2006) Perception of the bacterial PAMP EF-Tu by the receptor EFR restricts Agrobacterium-mediated transformation. Cell 125(4):749–760. https://doi.org/10.1016/j.cell.2006.03.037

Plant Elicitor Peptide Mediated Signalling Cascades During Plant–Pathogen Interaction

3

Aryadeep Roychoudhury, Sampreet Manna, and Diyasa Banerjee

Abstract

The constant coevolution between plants and pathogen species causes pathogens to evolve into new strains to evade immune system of plants. Concurrently, plants continue to generate a vast array of different elicitor peptides, small signalling molecules with pivotal roles in orchestrating defence strategies against various biotic stresses, such as pathogen attacks. This chapter delves into elucidating the multifaceted functions and molecular mechanisms involving the regulatory signalling cascades of several peptides, including plant defensins (PDFs), systemin, thionin, hevein-like peptides, knottin-type peptides, α-hairpin family peptides, lipid transfer protein, snakins, Pep2 and 3 (plant elicitor peptides), cyclotides, puroindolines, AtPeps, rapid alkalinization factor 23 (RALF23), PIP1 and 2 (PAMP-induced secreted peptides), serine-rich endogenous peptide 12 (SCOOP12), CLV1 (CLAVATA 1), HYPSYS (hydroxyproline-rich glycopeptide systemins) peptides, PSK (phytosulfokine), and PSY1 (plant peptide containing sulphated tyrosine 1). These peptides ultimately activate the expression of defence-related genes, the production of antimicrobial compounds, and the induction of systemic acquired resistance (SAR) to strengthen plant immunity against pathogen attack (biotic stress). Furthermore, this chapter discusses some of their potential applications in genetic engineering, emphasising the use of these peptides as biocontrol agents as alternatives to harmful chemical pesticides and insecticides. This approach aims to improve crop protection, tolerance, resilience, and resistance from subsequent pathogen infections, leading to sustainable

A. Roychoudhury (✉)
Discipline of Life Sciences, School of Sciences, Indira Gandhi National Open University, New Delhi, India

S. Manna · D. Banerjee
Department of Biotechnology, St. Xavier's College (Autonomous), Kolkata, India

S. Singh, R. Mehrotra (eds.), *Plant Elicitor Peptides*,
https://doi.org/10.1007/978-981-97-6374-0_3

agricultural practises, including improvements in crop quality and yield. Such strategies can be economically advantageous for farmers while simultaneously meeting all criteria for global food security.

Keywords

Elicitor peptides · Defence-related genes · Antimicrobial peptides · Pathogens · Genetic engineering · Global food security

Abbreviations

ROS	Reactive oxygen species
PEP	Plant elicitor peptide
RALF23	Rapid alkalinization factor 23
PIP	PAMP-induced secreted peptides
SCOOP 12	Serine-rich endogenous peptide 12
CLV1	CLAVATA1
HYPSYS	Hydroxyproline-rich glycopeptide systemins
PSK	Phytosulfokine
PSY1	Plant peptide containing sulphated tyrosine1
SAR	Systemic acquired resistance
HR	Hypersensitive response
PDF	Plant defensins
AMP	Antimicrobial peptide
PTI	PAMP-triggered immunity
ETI	Effector-triggered immunity
PAMPs	Pathogen-associated molecular patterns
HAMPs	Herbivore-associated molecular patterns
MAMPs	Microbial-associated molecular patterns
DAMPs	Damage-associated molecular patterns
NBS-LRR	Nucleotide-binding site-leucine rich repeat receptor
TPR1	Topless related protein 1
PRRs	Pattern recognition receptors
CDPKs	Calcium-dependent protein kinases
MAPKs	Mitogen-activated protein kinases
PIP2	Phosphatidylinositol 4,5-bisphosphate
CWI	Cell wall integrity
JA	Jasmonic acid
ACO	1-aminocyclopropane-1-carboxylate oxidase
SYS	Systemin
Pro-SYS	Prosystemin
LAP	Leu aminopeptidase
MeJA	Methyl jasmonate
SERK	somatic embryogenesis receptor kinases

OPDA	Oxo-phytodienoic acid
CWI	Cell wall integrity
LA	Linolenic acid
LOX	Lipoxygenase
AOS	Allene oxide synthase
AOC	Allene oxide cyclase
OPR	12-oxo-phytodienoic acid reductase
JA-Ile	Jasmonoyl-isoleucine
nsLTPs	Non-specific lipid transfer proteins
T3SS	Type III secretion system
FAs	Fatty acids
PCs	Phosphatidylcholines
PIs	Phosphatidylinositols
PGs	Phosphatidylglycerols
LTPs	Lipid transfer proteins
G3P	Glycerol-3-phosphate
RSV	Respiratory syncytial virus
GASA	Gibberellic acid stimulated in Arabidopsis
HTH	helix-turn-helix
BRs	Brassinosteroids
PROPEP	Precursor Peps
LRR-KRs	leucine-rich repeat receptor like kinases
PEPRs	Pep receptors
LRR-RLKs	Leucine-rich repeat kinase receptors
PIN-a/b	Puroindoline-a/b
NILs	Near-isogenic lines

3.1 Introduction

In agricultural practises, the major challenge is protecting plants from pathogen attacks, and in this context, the use of chemical pesticides, herbicides, and insecticides has been a trend for a long time. Still, complete disease resistance in crops is often not fully ensured by using these approaches or even if a crop initially becomes resistant to a particular pathogen, soon the pathogen attains resistance against the defence mechanism of plants and again gains the ability to reinfect the plant, making the plant susceptible to infection by that pathogen. To overcome this problem, earlier breeding strategies between susceptible and resistant parent plant varieties (P1 generation) to produce resistance progeny plants (F1 generation) and nowadays cloning of resistance genes to susceptible plants by genetic engineering has proven effective in this regard. Understanding plant immunity is of foremost importance in this regard. Plant immunity simply means the ability of a plant to prevent or withstand biotic stress rendered by pathogens. This plant immunity is mainly of two types—PAMP-triggered immunity (PTI) and effector-triggered immunity (ETI).

Plant has transmembrane receptors called pattern recognition receptors (PRRs) on their cell surface which are specific for specific components called Pathogen/

Microbial/Herbivore Associated Molecular Patterns (PAMPs/MAMPs/HAMPs) (Abdul Malik et al. 2020) such as bacterial flagellin, fungal chitin present on the pathogen (bacteria, fungi, and herbivore) surface (Dodds and Rathjen 2010). Another class of receptors bind specifically to only the endogenous breakdown products released from the plant cell wall or cell membrane upon pathogen invasion called damage-associated molecular patterns (DAMPs) such as cuticular fragments, cell wall fragments, e.g., oligogalacturonides or cellulose fragments (Ferrari 2013; Nühse 2012). When a pathogen, entering through pores of stomata, hydathodes, or any wound site attaches to the plant surface to invade the plant, PRRs on plant cells bind to their ligands (PAMPs/MAMPs/HAMPs/DAMPs) and initiate defence-responsive signalling in the plant, leading to containment of pathogen and preventing spread of pathogen to other parts of plant body (Jones and Dangl 2006) (Fig. 3.1). However,

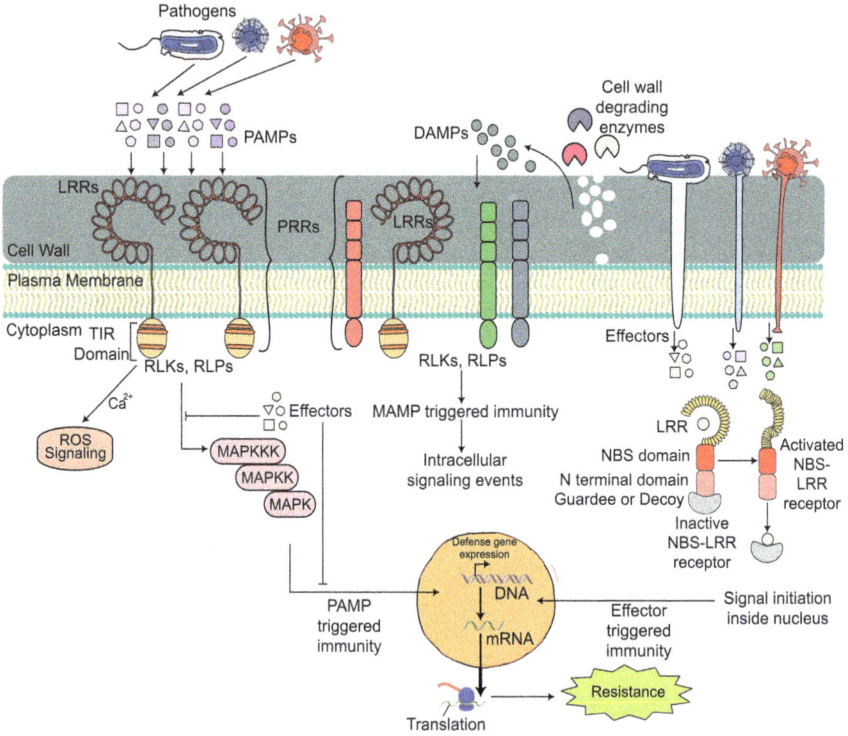

Fig. 3.1 Plant pathogens and integrated immune responses to infection. Pattern recognition receptors detect pathogen-associated molecular patterns (PAMPs) and damage-associated molecular patterns (DAMPs) and initiate plant immune responses. These receptors initiate a series of downstream signalling events, including calcium (Ca^{2+}) and reactive oxygen species (ROS) bursts, activation of calcium-dependent protein kinases (CDPKs) and mitogen-activated protein kinases (MAPKs), and phytohormone synthesis. Furthermore, substantial reprogramming occurs at the transcriptional and metabolic levels. Pathogens develop effectors that circumvent host cell PRRs and engage directly with host proteins. This interaction might result in the destruction or removal of host proteins within the cell. These reactions, known as pattern-triggered immunity (PTI) and effector-triggered immunity (ETI), allow plants to battle infections

this response always creates similar defence molecules in response to different types of pathogen invasion suggesting that PTI which is the plant innate immunity is non-specific and confers a basal level of protection to the plant (Miller et al. 2017).

ETI is more specific in terms of producing defence molecules, i.e., ETI response produces different defence molecules downstream of different types of pathogen attacks (Miller et al. 2017). The concept of this immunity is based on the gene-for-gene model, which means that for each pathogenic gene (*Avr* gene) responsible for disease development in the host plant, there is a corresponding gene (*R* gene) in the host responsible for conferring resistance to the plant (Nguyen et al. 2021). In order to bypass PTI, pathogen sometimes secretes specific virulence molecules called "effectors" encoded by the *Avr* gene in pathogen which can penetrate plant cell wall and cell membrane and can enter the host cytoplasm (Dodds and Rathjen 2010). As a result of coevolution, plants have also evolved resistance mechanisms against these pathogen-secreted effector molecules. Plants have intracellular receptors encoded by *Avr*-specific *R* gene which bind to these effectors and initiate downstream signalling, leading to hypersensitive response (HR) and finally necrotic death of infected cells, restricting pathogen in the infected cells only and limiting further pathogen spread (Dodds and Rathjen 2010). One example of such cytosolic receptor is nucleotide-binding site-leucine rich repeat receptor (NBS-LRR) which contain nucleotide-binding (NB) domains and leucine-rich repeat (LRR) domains (NLRs) (Jones and Dangl 2006) (Fig. 3.1). In the absence of pathogen-secreted effectors, this soluble receptor is kept sequestered and inactive in the cytosol by binding to decoys ("guardees"). In the presence of pathogen-secreted effectors, effector binds to the guardee leading to its dissociation from the receptor, freeing, and activating the receptor (Jones and Dangl 2006). The receptor now enters the nucleus and transcription factors activated by ETI can lead to the expression of various defence-related genes, such as pathogenesis-related proteins. NLRs can interact with transcriptional regulators to influence transcription. For example, TOPLESS-RELATED PROTEIN 1 (TPR1) and the TNL receptor SNC1 interact to decrease negative regulators and enhance immunity. Additionally, SNC1 forms complexes with other transcription factors, including bHLH84, which enhances immune regulation (Garner et al. 2021). Additional instances include how transcriptional factors like WRKY1 and MYB6 interact with barley CNL MLA10 to influence immunological responses (Nguyen et al. 2021). In addition to local responses, the immune response can also lead to SAR, which provides long-distance protection throughout the plant.

3.2 Different Plant Elicitors with Their Defence Actions

Microbes that cause disease are a continual threat of both plants and animals. In this fight, the use of artificial substances like fungicides and antibiotics has been crucial. However, different approaches are required since bacteria can become resistant to these substances. Using antimicrobial proteins and peptides (AMPs) is one such strategy (Morris 2019). During interactions between plants and pathogens, chemicals known as plant elicitors set off defence cascades. Immune receptors in the plasma membrane detect them, triggering defence mechanisms and signalling cascades. Elicitors fall into two categories: general elicitors, which cause defence responses in both host and nonhost plants, and race-specific elicitors, which only cause defence responses in particular host cultivars that result in disease resistance. While race-specific elicitors are linked to *R* gene-mediated signalling, general elicitors alert both host and nonhost plants to the presence of possible pathogens (Mishra et al. 2012). Elicitors are structural elements of pathogens, including chitin, flagellin, glucan, and lipopolysaccharides. They are also implicated in the overall resistance signalling pathways.

Certain general elicitors are selectively active in particular hosts and are only able to identify a restricted range of plants. In order to determine cultivar-specific (gene-for-gene) resistance, race-specific elicitors—such as harpins and *Avr* gene products—play crucial roles as virulence determinants (Mishra et al. 2012). Most plant pathogens produce cell wall-degrading enzymes as their initial elicitor and virulence determinant. These enzymes aid in penetration, supply the pathogen with nutrients, and release pectic fragments that act as endogenous elicitors. We have discussed a few elicitor molecules and their mode of defence responses against pathogenic microbes.

3.2.1 Plant Defensins (PDFs)

Defensins have antifungal and antibacterial properties, can inhibit proteinase and amylase present in insect herbivores thus having a role in plant-microbe interaction (Stotz et al. 2009). They are small (5 kDa in size), globular, 45–54 amino acids long, basic (hence positively charged), and widely present in most of the angiosperms, e.g., *Arabidopsis thaliana* are characterised as "Cysteine-rich peptides" (Nawrot et al. 2013).

3.2.1.1 Structure

Defensins have a Cys-rich domain containing 8-10 Cysteine (Cys) residues which form four to five conserved disulphide bonds (Li et al. 2021). Those plants possessing defensin peptides with eight cysteine residues are termed as "8C" plants, e.g., AlfAFP (present in alfalfa), NaD1 (present in *Nicotiana alata*), Psd1 (present in *Pisum sativum*), VrD1 (present in *Vigna radiata*), Ms-Def1 (present in *Medicago sativa*), ω-hordothionin and Rs-AFPs (present in *Raphanus sativus*), and plants with defensins having ten cysteine residues are termed as "10C" plants, e.g., PhDs (present in *Petunia hybrida*) (Sher Khan et al. 2019). In 8C plants, the disulphide bonds are formed between and arranged in Cys1-Cys8, Cys2-Cys5, Cys3-Cys6, Cys4-Cys7 (indicates number of cysteine residue from N-termini to C-termini) (Sher Khan et al. 2019) and in 10C plants, the disulphide bonds are formed between and arranged in Cys1-Cys10, Cys2-Cys5, Cys3-Cys7, Cys4-Cys8, and Cys6-Cys9 (Lacerda et al. 2014). Other than cysteines, defensin also contains two conserved glycines one at the left flanking site of Cys2 and another at the left flanking site of Cys5 and one glutamic acid at the right flanking site of Cys4 (Fig. 3.2) and one basic

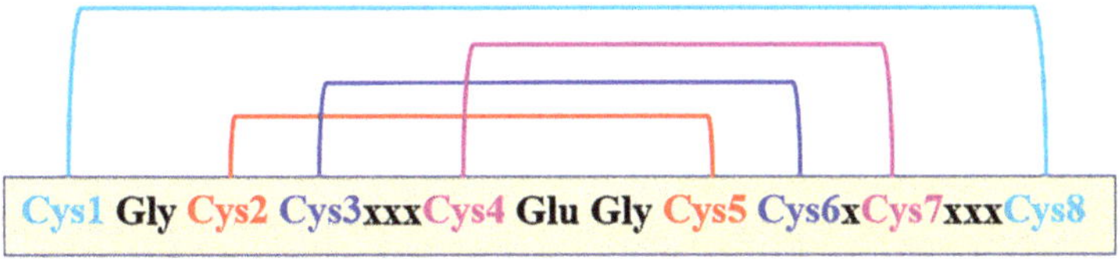

Fig. 3.2 Schematic representation of plant defensin. "Cys" indicates the cysteine amino acid. "x" denotes the conserved amino acid. The light blue, red, dark blue, purple lines denote disulphide bonds between Cys1-Cys8, Cys2-Cys5, Cys3-Cys6, and Cys4-Cys7

amino acid, e.g., arginine or lysine at the N-terminus and an aromatic residue, e.g., phenylalanine or tryptophan (Kovaleva et al. 2020). The cystine knot motif is a characteristic feature found in many defensins which contributes to their structural and functional stability (Molesini et al. 2017). Cystine knot motif is formed by three disulphide bonds between cysteine residues at positions Cys1-Cys4, Cys2-Cys5, and Cys3-Cys6 and their associated connecting peptide segments where two disulphide bonds Cys1-Cys4 and Cys2-Cys5 intertwine with each other to form a loop or ring which is penetrated by Cys3-Cys6 Cys disulphide bond, giving rise to a knotted appearance (Hellinger et al. 2021; Tam et al. 1999). This motif also consists of a triple-stranded antiparallel β-sheet stabilised by these disulphide bonds (Tam et al. 1999). This entire arrangement of cysteine knot confers a highly stable and compact structure of the peptide, which is resistant to proteolytic degradation and other environmental stresses such as high temperature (Molesini et al. 2017). Another essential secondary structural element is the Cys-stabilised αβ (CSαβ) motif of the cysteine knot which is composed of three antiparallel β-sheets (β1, β2, β3) and one α-helix parallel to these β-sheets and arranged in β1-α-β2-β3 pattern connected via three loops (L1, L2, L3) (Kovaleva et al. 2020) (Fig. 3.2). This motif is again stabilised by a total of four disulphide bonds: Cys1-Cys8 disulphide bond connects the N-terminus to the C-terminus. Cys2-Cys5 disulphide bond connects L3 to β2. Cys3-Cys6 and Cys4-Cys7 disulphide bonds connect α to β3 (Kovaleva et al. 2020). The basic nature of defensin is conferred by the gamma core (γ-core) which comprises two antiparallel β-sheets and a β-turn in between (Lacerda et al. 2014). This γ-core contains four basic residues and experiments suggest that this γ-core actually confers the antifungal property to this peptide, as peptides lacking this region are susceptible to fungal diseases (Lacerda et al. 2014).

3.2.1.2 Mechanism of Action

Numerous plant defensins demonstrate noteworthy antifungal efficacy against a diverse range of plant and animal pathogens as well as model species. Notably, certain oomycetes also exhibit susceptibility to the activity of defensins (Carvalho and Gomes 2009). The antifungal activity of defensins is predominantly focused on fungus, with bactericidal activity being uncommon. Defensins that have dual activity against both Gram-positive and Gram-negative bacteria are rare. However, defensins such as Ns-D1 and Ns-D2 from the black seed of the *Nigella sativa* plant have antifungal and antibacterial action against both Gram-positive and

Gram-negative bacterial species (Rogozhin et al. 2011). Two models have been proposed to elucidate the mechanism of action of defensins: the carpet model and the Toroidal pore model (Sher Khan et al. 2019). In both of these models, a commonality lies in the interaction of defensin with the negatively charged membrane. This interaction leads to an augmentation in the permeability of the cell membrane, resulting in its leakage (Sher Khan et al. 2019). Consequently, this disturbance in membrane integrity culminates in cell death.

According to the *Carpet model*, the formation of small pores occurs as a consequence of the inward movement of the hydrophobic sides of antimicrobial peptide (AMP) molecules on the surface of the cell membrane. In this model, the hydrophobic areas of AMPs aggregate and align parallel to the membrane surface, creating a "carpet" effect (Järvå et al. 2018). This aggregation causes the creation of tiny holes, which eventually disrupts the cell membrane integrity and contributes to the antibacterial properties of AMPs. The formation of huge oligomeric complexes, as seen in the NaD1-PA complex structure, supports the concept that defensins adopt a carpet-like arrangement to interact with phospholipids on target membranes (Järvå et al. 2018). NaD1, a class II solanaceous defensin produced in the flowers of *Nicotiana alata*, exhibits antifungal activity against various pathogenic fungi, safeguarding reproductive tissues from potential fungal pathogens (Hayes et al. 2013). NaD1 forms dimers, binding tightly to phosphatidylinositol 4,5-bisphosphate (PI(4,5)P2) on the inner leaflet of the plasma membrane, crucial for its cytotoxic effect on tumour cells. NaD1 enters the cytoplasm of *Fusarium oxysporum*, inducing ROS production, plasma membrane permeabilization, cytoplasmic granulation, and cell death (Poon et al. 2014).

According to the *Toroidal pore model*, defensins interact with the cell membrane, inducing ion permeability (Sher Khan et al. 2019). In this hypothesis, defensins can form toroidal pores in the lipid bilayer of the cell membrane, that increase membrane permeability which allows for the passage of ions, disrupting the cellular homeostasis, as a result of which it leads to cell death (Sher Khan et al. 2019). Defensins target specifically GlcCer (Glucosylceramides) and M(IP)2C (mannosyl-inositolphosphoceramide) (Thevissen et al. 2004). The presence of certain sphingolipids is critical for sensitivity, as evident from resistance in mutants lacking sphingolipid production genes. Defensins may internalise into fungal cells after contact, triggering signalling cascades (e.g., CWI pathway, MAP kinases) and inducing cell death by processes such as ion fluxes and oxidative stress (De Samblanx et al. 1996). Different plant defensins have different modes of responses. Table 3.1 includes specific details such as the identified membrane target, interaction sites, and proposed mechanisms of action for various plant defensins.

3.2.2 Systemin

Systemins are found in plants belonging to the Solanaceae family and are made up of 18 amino acids (Pearce et al. 1991). It was first discovered from the wounded tomato leaves and serves as a protease inhibitor (Pearce et al. 1991) that inhibits the

Table 3.1 The effects of various defensins

Plant defensins	Plant source	Identified membrane target	Proposed mode of action	Key references
RsAFP1	*Raphanus sativus*	Unknown	Recognised as an antifungal defensin	(Terras et al. 1992)
RsAFP2	*Raphanus sativus*	Glucosylceramide (GlcCer); β2-β3 loop, Positions 38 and 39	Interaction with GlcCer, activation of signalling cascades, ion fluxes, apoptosis, cell wall stress, ROS production; fungal cell death	(Aerts et al. 2007)
Psd1	*Pisum sativum*	Glucosylceramide (GlcCer) and/or ergosterol; Psd1 Loop1, Surface Topology	Adsorption on fungal membrane, nuclear translocation, interaction with Cyclin F, cell cycle impairment, fungal cell death	(Cabral et al. 2003)
MsDef1	*Medicago spp.*	Glucosylceramide (GlcCer)	Interaction with fungal membrane, activation of MAP kinase cascades, disruption of Ca^{2+} signalling, delayed fungal cell death, inhibits hyphal growth	(Sagaram et al. 2011)
MtDef4	*Medicago truncatula*	Phosphatidic Acid; RGFRRR motif, C-terminal domains	Recognition of fungal membrane, translocation via RGFRRR motif, interaction with cytosolic PA, disruption of Ca^{2+} signalling, inhibition of germination, inhibition of cell fusion, rapid fungal cell death	(Tetorya et al. 2023)
NaD1	*Nicotiana alata*	PI(4,5)P2; Oligomerization Domain (14-mer)	Interaction with fungal membrane, translocation to cytoplasm, PIP2-mediated oligomerization, membrane permeabilization, intracellular interaction, ROS and NO production, activation of HOG pathway, fungal cell death, granulation of cytoplasm	(Lay et al. 2003)
DmAMP1	*Dahlia merckii*	M(IP)2C	Irreversible binding; triggers K^+ efflux, Ca^{2+} uptake, changes in membrane potential; induces CWI pathway	(Thevissen et al. 2003)
HsAFP1	*Heterotheca subaxillaris*	Cell wall, membrane	Binds to cell wall and cell membrane; induces ROS production; targets mitochondria initiating apoptosis; cytoplasmic penetration	(Struyfs et al. 2020)

enzyme proteinase in the digestive tracts of phytophagous insects and mammalian herbivores (Green and Ryan 1972). Systemin is formed from its precursor prosystemin which is not just a mere precursor; rather this precursor itself is biologically active (Molisso et al. 2022b). Its N-terminal part has been seen to be involved in the expression of resistance genes by induction of octadecanoid signalling pathway to protect plants from a range of insects and fungal pathogens, Jasmonic acid (JA)-responsive genes (*Pin I* and *Pin II*) confer protection to tomato plants by obstructing the digestion process in *Spodoptera littoralis* larva negatively affecting their growth and survival due to lack of nutrients as well as restricting *Botrytis cinerea* growth inside plant organs. *ACO* (1-aminocyclopropane-1-carboxylate oxidase) which is the final enzyme in the ethylene biosynthetic pathway mediates a vast range of defence responses and *PAL* which increases the lignin contents and thickness of cell walls makes the plant surface difficult for parasites to penetrate (Molisso et al. 2022a).

3.2.2.1 Structure

Proteolytic cleavage of the 200 amino acid long precursor prosystemin (Pro-SYS) in the region of amino acids from 179 to 196 releases systemin (SYS) (McGurl et al. 1992). In Pro-SYS, the 179–196 amino acid region has flanking leucine residue towards its N-terminal end with a flanking tetrapeptide fragment (Asn-Asn-Lys-Leu) C-terminal end (McGurl et al. 1992) and it lacks a signal sequence (Zhang et al. 2020). The structure of systemin consists of four to six conserved proline residues and conserved PPKMQTD motif at the C-terminal end (Hu et al. 2018). The ability of prosystemin to modulate several signalling processes is owing to the fact that it is an intrinsically disordered protein (IDP) that possesses long disordered regions. Due to the lack of a stable three-dimensional structure, Pro-SYS is structurally flexible, and therefore can attain a variety of different structural forms and can interact with multiple signalling proteins of different shapes (Zhang et al. 2020). The process of production of SYS from Pro-SYS is explained in Fig. 3.3.

3.2.2.2 Mechanism of Action

Systemin was first isolated in tomato plants and the induction of wounds initiates a systemic reprogramming of leaf cells, leading to the synthesis of more than 20 proteins associated with defence mechanisms (Bergey et al. 1996). The interaction of SYS with its receptor initiates a complex cascade of intracellular events crucial for the activation of a phospholipase A2 (PLA2) and the subsequent release of linolenic acid from membranes in wounded tomato leaves (Sun et al. 2011). This process involves several events, including depolarisation of the plasma membrane, opening of ion channels, an increase in intracellular Ca^{2+} concentration, inactivation of a plasma membrane proton ATPase, activation of a MAP kinase, synthesis of calmodulin, and the activation of PLA2 (Moyen et al. 1998).

The activation of defence response signalling in tomato plants, triggered by wounding and the presence of systemin, involves an increase in the endogenous levels of jasmonate compounds, including jasmonic acid (JA), methyl jasmonate (MeJA), and their metabolic precursor, 12-oxo-phytodienoic acid (12-OPDA) (Sun et al. 2011) (Fig. 3.4). JA is derived from linolenic acid (LA) through the

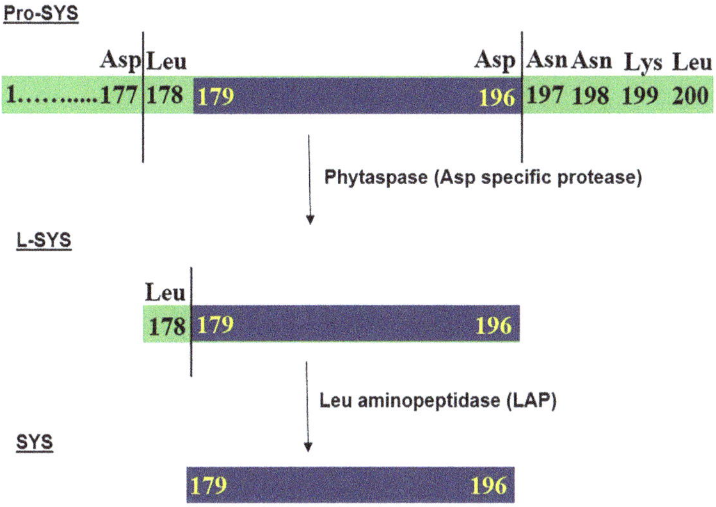

Fig. 3.3 Biosynthesis of SYS from Pro-SYS. In the Pro-SYS, two aspartate (Asp) residues present at positions 177 and 196 (Zhang et al. 2020) are targeted by two Asp specific proteases named phytaspases to cleave Pro-SYS and yield L-SYS (Beloshistov et al. 2018). Then Leu aminopeptidase (LAP) further cleaves the L-SYS at N-terminal leucine residue and yields SYS and this SYS peptide is now finally released in the cytosol followed by its transportation to the extracellular space to mediate downstream signalling responses for defensive purposes (Zhang et al. 2020)

octadecanoid pathway, with key enzymes such as lipoxygenase (LOX), allene oxide synthase (AOS), and allene oxide cyclase (AOC) playing crucial roles in the biosynthesis of 12-OPDA (Ishiguro et al. 2001). These enzymes are localised in the chloroplast, suggesting that the initial steps of JA biosynthesis occur in this organelle. Systemin is implicated in the activation of phospholipase A2 (PLA2), leading to the release of LA from membrane lipids in wounded leaves. Further conversion of OPDA to JA involves 12-oxo-phytodienoic acid reductase (OPR), specifically OPR3, which is located in the peroxisome (Stintzi et al. 2001). The subsequent metabolism of JA results in various products, including MeJA and jasmonoyl-isoleucine (JA-Ile), the latter serving as an active hormone in the defence response (Sun et al. 2011).

The identification of a SYS receptor on the cell surface raises questions about SYS transport throughout plants and its delivery to distal cells. Effects of SYS are likely exerted through an external cell surface receptor. The transport mechanisms may involve the phloem, apoplast, and specific transporters, with scenarios considering both apoplastic and phloem-based transport to activate defence cascades throughout the plant (Ryan 2000). *The modified model* underscores the spatio-temporal regulation of wound and SYS-responsive genes that fall into two distinct classes: *signal transduction pathway genes* and *defensive proteinase inhibitor genes* (Ryan 2000). Signal pathway genes, designated as "early genes," find expression predominantly in vascular bundle cells, whereas proteinase inhibitor genes, referred

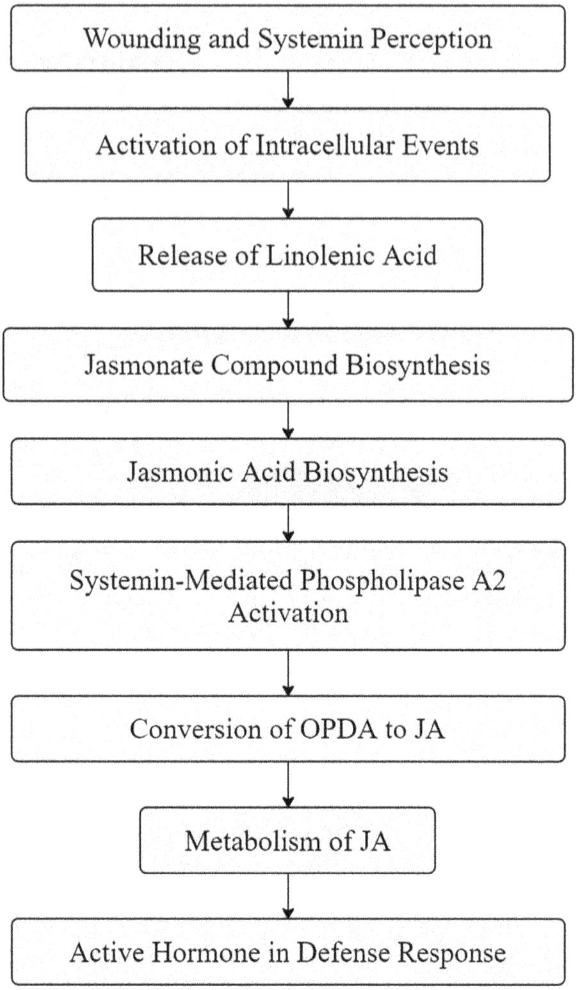

Fig. 3.4 Flowchart of SYS mediated defence response in tomato plants

to as "late genes," are activated in palisade and spongy mesophyll cells (Orozco-Cárdenas et al. 2001).

The model shown in Fig. 3.5 proposes that signal transduction genes act as prompt responders within vascular bundles, initiating the production of second messengers such as OPDA and/or JA. Upon wounding, mRNA transcripts associated with proteinase inhibitors (I and II) [Proteinase inhibitor I is sequestered within central vacuoles of mesophyll cells, while inhibitor II is similarly found within leaf vacuoles (Graham et al. 1986)]; CPI, and CDI (Graham et al. 1986)) exhibit delayed expression, becoming detectable approximately 2 h post-wounding and peak over the subsequent 8 h (Ryan 2000). mRNAs encoding signal pathway components such as LOX, CaM, AOS, and Pro-SYS, on the other hand, appear early, with

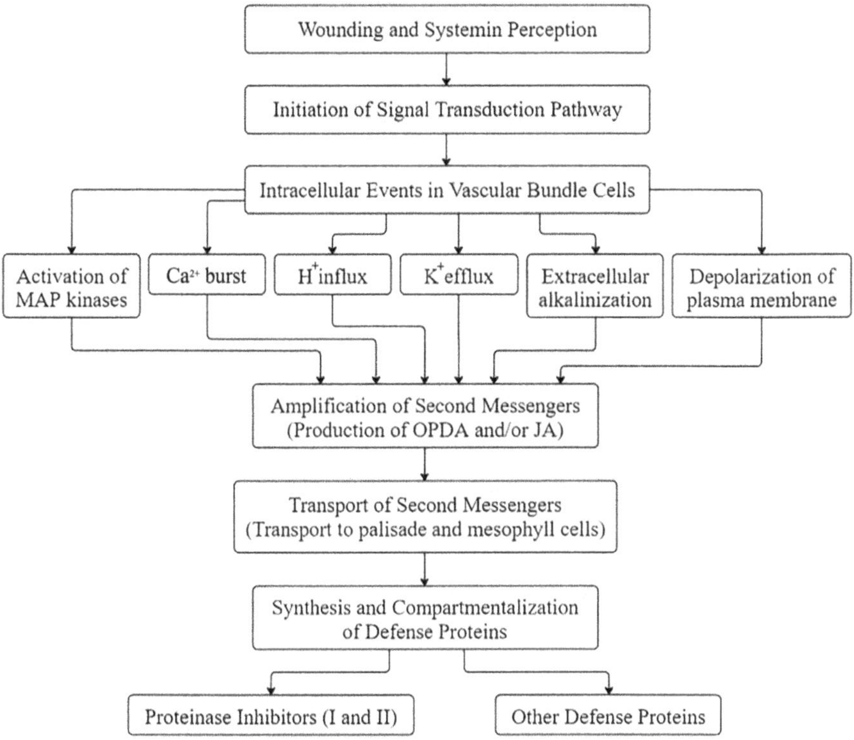

Fig. 3.5 Flowchart of the events in the modified SYS mechanism

detection within 30 min and maximum induction occur 2–3 h after injury. Interestingly, this occurs 5–6 h before the peak expression of proteinase inhibitor mRNAs (Ryan 2000). Conversely, the wound-inducible *Pro-SYS* gene, when fused with the *GUS* reporter gene, induces GUS protein synthesis specifically within the vascular bundles of major and minor veins, as confirmed through colorimetric assays (Corrado et al. 2011).

3.2.3 Thionin

Thionins were initially classified into three categories—α-thionins, β-thionins, and γ-thionins; later on, γ-thionins were considered as plant defensins due to their structural, functional, and sequence similarities with defensins and dissimilarities with α-/β-thionins (Azmi and Hussain 2021). Till now, five classes of thionins have been identified (Table 3.2). α-/β-thionins are considered as plant toxins because of their toxic effects on bacteria, fungi, larvae, and other pathogens (Li et al. 2021). Thionins were first discovered in wheat endosperm as purothionine and later discovered in many monocots and dicots (Odintsova et al. 2018).

Table 3.2 Classification of thionins into five groups

Thionin type	Source	Number of amino acids	Number of cysteine residues	Ionic nature	References
I such as—Purothionin	Seed endosperm in plants belonging to Poaceae family	45	8	highly basic	(Odintsova et al. 2018)
II such as - • α-hordothionin • β-hordothionin	• Leaves and nuts of *Pyrularia pubera* • Leaves of barley *(Hordeum vulgare)*	46–47	8	basic	(Odintsova et al. 2018)
III such as - • Viscotoxicin • Phoratoxicine • Ligatoxicine	Stem and leaves of mistletoe species such as - • *Viscum album* (vascotoxicins) • *Phoradendron tomentosum* (Phoratoxins) • *Phoradendron liga* (Ligatoxin A)	45–46	6	basic	(Azmi and Hussain 2021)
IV such as - Carambine	Seed of Abyssinian cabbage	46	6	neutral	(Azmi and Hussain 2021)
V	Wheat		• 2 disulphide bridges • 1 disulphide loop	acidic or neutral	(Azmi and Hussain 2021)

3.2.3.1 Structure

Thionins are cationic, cysteine-rich peptides composed of 45–48 amino acid residues and contain cysteine-rich domains with 6 or 8 cysteine residues, and 3 or 4 disulphide bonds (Odintsova et al. 2018). Although they can be partially termed as cyclic peptides of ring structure, due to end-to-end joining of N- and C-terminal ends by a disulphide bond (Li et al. 2021), they are not cyclic peptides in true sense because at the two extremities, many amino acids other than cysteines exist (Tam et al. 2015). In thionins containing 8 cysteine residues, this motif is stabilised by four disulphide bonds between Cys1-Cys8, linking β1 to the C-terminal end of coil, Cys2-Cys7 linking the terminal end of β1 to the beginning of β2, Cys3-Cys6 linking α1 to the loop after α2, and Cys4-Cys5 linking α1 and α2. In thionins containing 6 cysteine residues, this stabilisation is maintained by the same set of disulphide bonds except the Cys2-Cys7 disulphide bond (Tam et al. 2015). Their secondary structure has a conserved β1-α1-α2-β2 motif which gives rise to the tertiary structure in the shape of gamma fold where the long vertical arm of the fold is made up of the two antiparallel α-helices (α1-α2 region) composed of hydrophobic residues on its outer surface; the short horizontal arm is made up of the two parallel β-strands

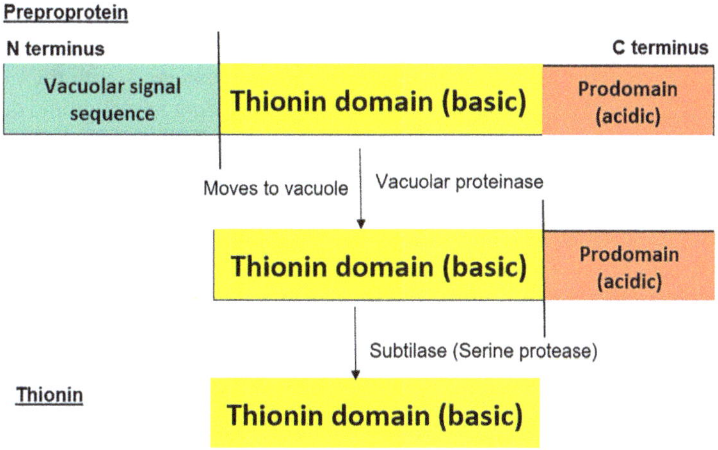

Fig. 3.6 Biosynthesis of thionin from preproprotein. The precursor form of thionin (preproprotein) consisting of a 20 amino acid long signal sequence at the N-terminal end, 45–47 amino acid long basic thionin domain and a 60 amino acid long acidic prodomain at the C-terminal end containing six cysteine residues is formed (Azmi and Hussain 2021; Odintsova et al. 2018). During post-translational processing, the precursor moves to vacuoles as guided by its N-terminal signal sequence during which the C-terminal prodomain keeps the toxic properties of the thionin neutralised till it reaches vacuoles to avoid harm to the plant by its own toxins (Odintsova et al. 2018). Once it reaches vacuole, a proteinase (localised in vacuoles) cleaves the signal peptide of the preproprotein, activating the toxic properties of thionin and a serine protease named subtilase cleaves the remaining C-terminal prodomain, releasing the functionally active thionin (Odintsova et al. 2018)

(β1 and β2-coil) and a short antiparallel β-sheet composed of hydrophilic residues on its outer surface, conferring it the amphipathic nature (Azmi and Hussain 2021; Barashkova et al. 2021). The large groove between these two arms is also composed of hydrophilic residues and contains conserved residue, Tyr13 which is essential for the interaction between thionins and lipids on the plasma membrane of pathogens (Azmi and Hussain 2021). The synthesis of thionins from preproproteins is explained in Fig. 3.6.

3.2.3.2 Mechanism of Action

Thionin has broad selectivity, inducibility, and membrane-interacting properties that makes it effective in protecting plants against pathogens (Taveira et al. 2016). The positively charged portions of thionin interact with negatively charged phospholipids in cell membranes. This interaction is critical for attaching thionins to the membrane surface (Pappas 2011). Thionins also form cation-selective ion channels that may allow the passage of ions through the membrane, affecting membrane integrity and cellular functions (Llanos et al. 2003).

Thionins can damage cell membranes by inducing pore formation or by altering membrane fluidity; as a result, the osmotic balance is lost (Stec et al. 2004). According to the carpet model of membrane permeabilization, thionins accumulate on the membrane surface at high amounts, causing conformational strain on phospholipids and changes in membrane fluidity (Giudici et al. 2003), for example, plant-derived thionin CaThi, against *Candida* species, interacts with the phospholipids in the plasma membrane of *Candida* species, leading to membrane destabilisation and permeabilization, allowing the leakage of cellular contents and ions (Taveira et al. 2016). Thionins also have been reported to induce the production of reactive oxygen species (ROS) in the target microorganisms, including *Candida* species. Thionin-induced membrane permeabilization may produce an increase in ROS levels. ROS can impair cellular activities, causing oxidative damage and contributing to the antibacterial activity of thionins (Hwang et al. 2011).

In vitro studies show that thionins exhibit dose- and time-dependent effects on cell membranes, with membrane destruction occurring within a specific time frame after treatment. The critical dose associated with membrane lysis is around 1 μM, indicating a threshold for toxic effects (Thevissen et al. 1996). CaThi showed a synergistic effect with conventional antifungal agent fluconazole (Taveira et al. 2016). This enhances the antifungal activity against *Candida* species by potentially targeting different pathways or mechanisms within the cells, offering a promising therapeutic strategy.

3.2.4 Hevein-like Peptides

Hevein-like peptides are generally basic in nature, 29–45 amino acid long and mostly serve as antifungal (Santos-Silva et al. 2020) component by binding chitin [polymer of N-acetylglucosamine units linked by beta-(1-4)-glycosidic bonds] on the fungal cell walls. They perform their chitinase activity on chitin by penetrating the fungal hyphae (Ghosh and Roychoudhury 2023), hydrolysing chitin in the cell wall and disrupting the structural integrity of plasma membrane leading to uncontrolled influx–efflux of cellular materials ultimately leading to cell lysis (Santos-Silva et al. 2020).

3.2.4.1 Structure

Their antifungal activity is conferred by the chitin-binding domain with the amino acid sequence SXFGY/SXYGY (where X represents any amino acid) and 6, 8, or 10 cysteines, 7 glycines, and 4 prolines at specific sites in this domain (Kini et al. 2015). Main function of this domain is to bind to chitin (major component of fungal cell wall) (Odintsova et al. 2020) and this ability is determined by the presence of three aromatic residues and one serine in this domain. The secondary structure consists of coil-β1-β2-coil-β3 motif where the antiparallel β-strands and two long coils located on each side are stabilised by 3–5 disulphide bonds (made up of 6-8 Cys residues) form the central β-sheet (Tam et al. 2015). Out of three, two most important disulphide bonds Cys1-Cys4 and Cys2-Cys5 are oriented perpendicular to each other (Odintsova et al. 2020).

3.2.4.2 Mechanism of Action

Hevein-like antimicrobial peptides comprises of 30–45 amino acids, with many stabilising disulphide bridges, and have chitin-binding domains similar to those in class I/IV chitinases (Beliaev et al. 2021). The binding of hevein-like AMPs to chitin and related heptosyl saccharides results in the disruption of fungal cell wall integrity which can lead to the inhibition of growth and ultimately cell lysis. This is likely to result in the disruption of chitin biosynthesis (Beliaev et al. 2021). The chitin-binding site, which contains conserved amino acid residues including aromatic residues and serine, allows hevein-like AMPs to target chitin in fungal cell wall (Beintema 1994). The fungal hyphae can be penetrated by hevein-like peptides, and they can selectively accumulate at particular areas, septa, and hyphal tip. This causes the bursting of the hyphal tip and seepage of hyphal cytoplasm (Wessels 1988). This intracellular targeting may disrupt essential cellular processes and contribute to the fungistatic effect of these peptides. These peptides contain cysteine residues, crucial for forming disulphide bonds which enhance the overall stability (Slavokhotova et al. 2014a). They exhibit target specificity towards fungal cell wall components, disrupting cell wall morphogenesis and interact with fungal proteases like fungalysin, serving as inhibitors (Slavokhotova et al. 2014a).

3.2.5 Knottin-type Peptides

Knottin-type peptides are the smallest in size among different plant antimicrobial peptides and function to inhibit different enzymes, e.g., α-amylase, metalloprotease (carboxypeptidase A), serine-protease (trypsin), and cyclic peptides present in bacteria, fungi, and viruses and also act as insecticides by killing the attacking insects (Le Nguyen et al. 1990).

3.2.5.1 Structure

Knottin-like peptides or cysteine knot peptides are small (around 30–50 amino acid long) and are so named because of the presence of cysteine-rich motifs made up of 39 amino acids and 6 cysteine residues forming 3 intra-disulphide bonds, Cys1-Cys4, Cys2-Cys5, and Cys3-Cys6 (Han et al. 2023), together giving rise to a knot-like structural appearance (Molesini et al. 2017). The knot is basically made up by intertwining several cysteine residues via disulphide bonds, forming a network like arrangement (Li et al. 2021). Although as mentioned earlier, defensins form a knotted arrangement, they differ from knottins in positions of these cysteines (Li et al. 2021). This tightly knotted structure with three disulphide bonds (specially Cys2-Cys5 and Cys3-Cys6 disulphide bonds) confers thermal stability even above 100 °C, chemical and enzymatic (towards endopeptidase, trypsin, and thermolysin) stabilities of Cys-stabilised β-sheet motif (discussed later) (Chiche et al. 2004). They can attain either linear or cyclic conformation with cyclic one being more resistant due to restricted structure unfolding (Heitz et al. 2008). However, in both the cases, the disulphide bridges formed between Cys1-Cys4, Cys2-Cys5, and Cys3-Cys6 creates the ring form. In cyclic conformation, the first two disulphide bonds and the part of the peptide segment containing these bonds cyclise to form cyclic or ring structure; Cys 3rd–6th disulphide bond passes through the space inside the ring, ultimately forming the knot arrangement (Molesini et al. 2017). Even if not in fully cyclic orientation, in pseudocyclic form also, the N-terminus and C-terminus of the peptide contains cysteine residues at terminal or subterminal ends, facilitating looping back followed by disulphide bond formation between those cysteine, conferring enhanced stability to the peptide (Nguyen et al. 2014). Apart from the loop structure, another important feature is the presence of a triple-stranded, antiparallel beta-sheet with at least 10 residues where the 1st and 2nd β-strands are connected by a long loop (Pallaghy et al. 1994). This structural motif functions as a toxin as it blocks the function of larger protein receptors such as ion channels or proteases (Pallaghy et al. 1994).

3.2.5.2 Mechanism of Action

Knottin-like peptides are typically amphipathic molecules containing hydrophobic patches surrounded by hydrophilic residues (Tam et al. 2015). They have numerous qualities, including minimal toxicity, low susceptibility to drug resistance, and a critical function in the innate immunity of insects, providing defence against bacteria, fungi, and parasites (Han et al. 2023). Cyclotide is one of the types of knottin-like peptides, which have interaction with membranes, leading to antimicrobial effects. It usually has weakly positive or neutral charges at physiological pH, reducing electrostatic interactions with membranes. Cyclotides interact with membranes primarily through hydrophobic interactions with lipid tails, facilitated by weak interactions between positively charged residues of cyclotides and the polar head of membranes (Tam et al. 2015). Different cyclotides due to differences in the location of hydrophobic patches within their structures exhibit varied membrane binding modalities. MaK is another knottin-like peptide, comprising of 56 amino acid residues derived from *Monochamus alternatus,* showing antibacterial and nematicidal properties (Han et al. 2023).

3.2.6 α-Hairpinin Family Peptides

The α-hairpinins are short (less than 50 amino acid residues) antimicrobial peptides (AMPs) (Slezina and Odintsova 2023), having role as antifungal, antibacterial, trypsin inhibitors, and ribosome inactivator in insects (Slavokhotova and Rogozhin 2020).

3.2.6.1 Structure

The secondary structure of α-hairpinin peptides such as Sm-AMP-X consists of a helix-loop-helix (α1-turn-α2) motif where the two α-helices (α1 and α2) in antiparallel orientation are joined via a loop and two CXXXC motifs are present in different α-helices (Slavokhotova and Rogozhin 2020). In this unique cysteine motif ($Cys^1X3Cys^2XnCys^3X3Cys^4$) (where X is any amino acid except cysteine), Cys1-Cys4 form one disulphide bond and Cys2-Cys3 form another disulphide bond

(Slavokhotova et al. 2014b). The N- and C-terminal tails are unstructured (Oparin et al. 2012).

3.2.6.2 Mechanism of Action

α-hairpinins are a subset of this peptide family that primarily accumulate in seeds and serve in defence against insect pathogens. The key feature of these inhibitors is a rigid, reactive-site loop that complements the binding site of serine proteinases. The loop between the second and third cysteines contains the functional P1 residue (typically arginine), crucial for binding to the S1 specificity pocket of trypsin (Slavokhotova et al. 2014b). This interaction slows down hydrolysis of the peptide-bond after the critical arginine residue, leading to inhibition. α-hairpinins with ribosome-inactivating activity were discovered by Kimura et al., originating from sponge gourd seeds (Kimura et al. 1997). Initially it was identified as 6.5 k-arginine-glutamate-rich polypeptide (6.5 k-AGRP); half of it existed in full-length form, while the rest was truncated at the C-terminus (Kimura et al. 1997). Another isolated peptide named luffin P1, identical to 6.5 k-AGRP, but lacking two amino acids on both terminal ends (Ng et al. 2011). Luffin P1 exhibited potent inhibitory activity on protein synthesis in cell-free rabbit reticulocyte systems. Luffin P1 formed homotetramers in hydrophobic environment, potentially facilitating its N-glycosidase activity and in hydrophilic conditions, it depolymerised to homodimers (Ng et al. 2011). A group of antifungal α-hairpinins was isolated from barnyard grass seeds, with EcAMP1 being the first plant α-hairpinin demonstrated to have fungistatic activity. It was discovered that EcAMP1 effectively inhibited the germination of fungal spores, especially those of *Fusarium* species (Rogozhin et al. 2018).

3.2.7 Lipid Transfer Protein

Non-specific lipid transfer proteins (nsLTPs) are small cationic peptides with molecular weight, ranging from 6.5 to 10.5 kDa (Liu et al. 2015) and 100 amino acid long (Amador et al. 2021). LTPs have been classified in two ways based on comparison of different parameters between different types of LTPs as shown in Fig. 3.7. Details of Type I nsLTPs and Type II nsLTPs are described in Table 3.3.

3.2.7.1 Structure

Primary structure of non-specific lipid transfer proteins (nsLTPs) contains C-Xn-C-Xn-CC-Xn-CXC-Xn-C-Xn-C motif [also called—8-cysteine conserved motif (8CM)] (Amador et al. 2021). The three-dimensional structure is characterised by the presence of a large tunnel-like internal hydrophobic cavity surrounded by four alpha helices connected by three short flexible loops, an unstructured C-terminal tail and four disulphide bonds formed by eight cysteine residues (Amador et al. 2021). The stability of hydrophobic cavity rendered by disulphide bonds and intramolecular hydrogen bonds confer stability against high temperature, chemicals like denaturing agents, proteases, and digestive enzymes (Lindorff-Larsen and Winther 2001). This hydrophobic cavity or pocket contains the lipid-binding site which

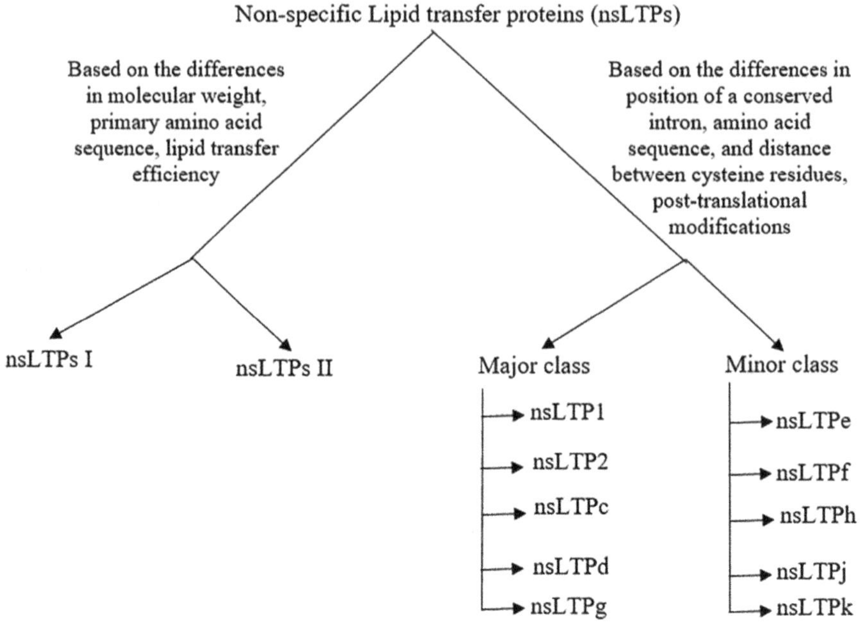

Fig. 3.7 Classification of nsLTPs (Salminen et al. 2016)

reversibly binds to various hydrophobic lipid molecules such as fatty acids, phospholipids, prostaglandin B2, hemolytic derivatives, acyl coenzyme A, sterols, etc. by hydrophobic interactions (Li et al. 2021; Santos-Silva et al. 2020) and transport them between donor and acceptor membranes (Carvalho and Gomes 2007). This peptide is produced from 21–27 amino acid long precursor (Liu et al. 2015) which contains an N-terminal signal peptide, basically required for its translocation to the apoplastic space where the transfer of lipids takes place (Salminen et al. 2016).

Structure of Type II LTP has more flexibility and less specificity in binding to lipid molecules in contrast to Type I, facilitating Type II in ease of binding and accommodating large ligands, e.g., sterols (Amador et al. 2021). Some LTPs have the capability to be integrated to the extracytoplasmic surface on plasma membrane via glycosylphosphatidylinositol lipid anchor and this is possible due to a specific sequence in its C terminal end which allows this post-translational modification of LTP (Santos-Silva et al. 2020).

3.2.7.2 Mechanism of Action

Plant cell wall maintains internal osmotic pressure of the cell and vacuole. This wall limits the rate and direction of cell enlargement. When there are any developmental or physiological cues that trigger cell extension, plant cells enhance wall plasticity through a process called loosening. Previously, a class of proteins known as expansins was identified as mediators of wall loosening. However, in addition to expansins, there exists another group of proteins: the lipid transfer proteins (LTPs) play a

Table 3.3 Difference between Type I nsLTPs and Type II nsLTPs

Topic	Type I nsLTPs	Type II nsLTPs	Reference
Number of alpha helices	Contains four α-helices with second and third helices being longer than Type II.	Contains three extended helices and two single-turn helices.	(Amador et al. 2021; Tam et al. 2015)
Positions of eight cysteines which form four disulphide bonds to form the hydrophobic cavity	Cys1-Cys6, Cys2-Cys3, Cys4-Cys7, and Cys5-Cys8 form the disulphide bridges.	Cys1-Cys5, Cys2-Cys3, Cys4-Cys7, and Cys6-Cys8 form the disulphide bridges.	
Molecular weight	10 kDa	7 kDa	(Melnikova et al. 2022)
Number of amino acids	90–95	70	
Hydrophobic cavity shape	Tunnel-shaped	Conical-shaped	
Lipid-binding residues	• Arg 44 [number of amino acids is as per the LTP1 from *Oryza sativa* (PDB ID 1RZL)] is located near the entrance of the hydrophobic cavity and interacts with hydrophilic head groups of lipids • Lys 35 • Tyr 79 residue forms hydrogen bonds with the polar head group of lipids.	Side chains of • Phe 36 • Tyr 45 • Tyr 48 • Tyr 80 [number of amino acids is as per the LTP2 from *Oryza sativa* (PDB ID 1L6H)]	

crucial role in enhancing cell wall extension. Nieuwland et al. mentioned the function of LTPs as cell wall loosening proteins (Edqvist et al. 2018).

Lipid transfer proteins have also been recognised for their involvement in lipid transfer and deposition, lipid sensing, lipid presentation to other proteins, lipid modification, contributions to disease pathology, cell signalling, stress tolerance, and involvement in apoptosis (Wong et al. 2019). LTPs can directly bind to the proteins secreted by the bacterial type III secretion system (T3SS) to inhibit bacteria to release effector molecules. LTPs can bind and transfer a wide range of ligands, like fatty acids (FAs), phosphatidylcholines (PCs), phosphatidylinositols (PIs), phosphatidylglycerols (PGs), acyl-CoA, prostaglandin B2, sterols, organic solvents, and certain drugs. The presence of a hydrophobic cavity in the structure of LTP molecules facilitates the formation of complexes with ligands (Wong et al. 2019). The exact mechanism of lipid transfer by LTPs remains unclear, but it is hypothesised to involve a shuttle mechanism. In this mechanism, LTP-phospholipid complex interacts with the membrane, facilitating phospholipid exchange between the complex and the membrane (Chiapparino et al. 2016). The LTP first docks to the surface of the donor membrane, where its hydrophobic cavity is used to remove lipid molecules. The LTP diffuses across the cellular environment after undocking from the donor membrane and being loaded with lipids. It then docks with the acceptor membrane and drops its lipid cargo there. The LTP undocks from the acceptor membrane and continues to diffuse after lipid deposition. Lipids can move between membranes with efficiency due to the cyclic mechanism, which supports a number of cellular operations (Wong et al. 2019). The affinity of LTPs for lipid-binding may be impacted by the presence of calmodulin. For example, calmodulin-binding interferes with the lipid-binding site, especially at the Arg 46 residue, in maize Zm-LTP, reducing its lipid-binding effectiveness. On the other hand, the calmodulin-binding site, which is found in the C-terminal region of proteins like Arabidopsis LTP1 and bok choy BP-10, improves the efficacy of lipid binding (Finkina et al. 2016). LTPs involve lipid sensing, where specific domains within LTPs, like the Sec14-like or StARkin domains, may act as lipid sensors by changing the structure of proteins in response to lipid binding. Proteins like OSBP and ORP1L can sense the amounts of cholesterol or sterol and adjust cellular functions accordingly (Wong and Levine 2016). Lipid presentation is aided by LTPs such as microsomal triglyceride transfer protein (MTTP), which facilitates proper CD1 molecule folding and trafficking (Cuchel et al. 2007). Through enzymatic activity inside their hydrophobic chambers, LTPs can alter lipids. For instance, GM2AP hydrolyzes phosphatidylcholine to produce novel lipid species in addition to presenting glycolipids (Sandhoff 2016). Pathologies that involve altered lipid metabolism, in particular, are linked to dysregulation of LTP function, for example, disorders like atherosclerosis and Niemann-Pick type C illness might result from changes in the cholesterol transport routes that are controlled by LTPs (Wong et al. 2019).

Lipid transfer proteins are essential for SAR. Long-distance SAR signal transmission is facilitated by LTPs like DIR1 and AZI1, which help transport mobile SAR signals from the infection site to systemic tissues (Gao et al. 2022). Examples of these signals include glycerol-3-phosphate (G3P) and azelaic acid. They support systemic immunity by facilitating the transfer of lipid signals via plasmodesmata and phloem. High affinity binding of lipids by DIR1 and the formation of complexes with AZI1 facilitate the transmission of SAR signals from chloroplasts to plasmodesmata and systemic locations. These proteins mediate and control the transmission of SAR signals by interactions with other constituents, including plasmodesmata-localising proteins (Gao et al. 2022). Lipid transfer proteins have shown diverse biological activities, including antimicrobial, antiviral, antiproliferative, enzyme inhibitory functions, and fungicidal effects. Some LTPs have antiviral properties that prevent viruses including the influenza A virus and the respiratory syncytial virus (RSV) from replicating (H1N1). Lipid transfer proteins have been shown to inhibit the proliferation of human tumour cells, suggesting potential antiproliferative effects (Finkina et al. 2016). Some LTPs possess enzyme inhibitory activity of proteolytic enzymes and α-amylases, for example, aspartate and serine proteases, as well as cysteine endoproteases, are inhibited by several LTPs (Jones and Marinac 1999). LTPs from certain plant seeds have been found to inhibit the activity of human α-amylase. These properties suggest a role of LTPs in seed germination, seed development, and plant defence against insects and herbivores (Finkina et al. 2016). The mechanism of LTP action is represented schematically in Fig. 3.8.

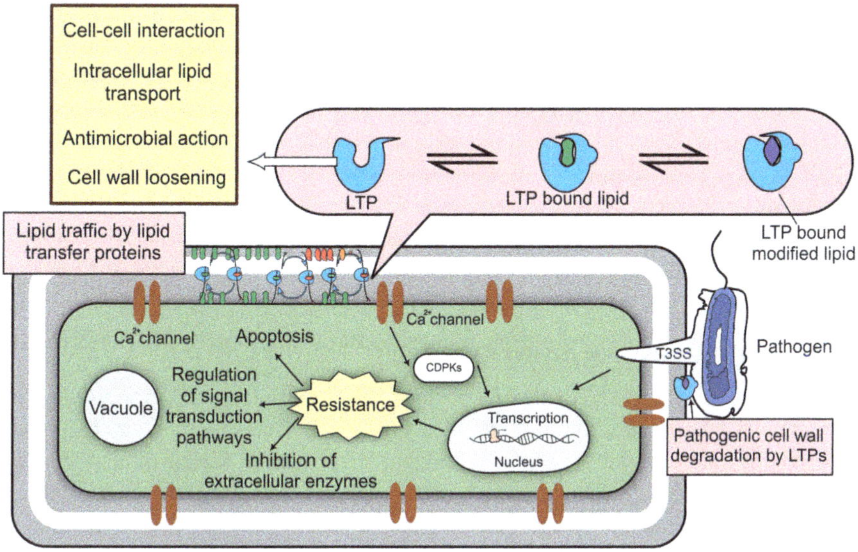

Fig. 3.8 Mechanism of LTP action. LTPs play a crucial role in defending against pathogens by disrupting bacterial cell membranes. LTPs may also interact directly with the key proteins of the bacterial type III secretion system, thereby inhibiting bacterial activity and preventing the secretion of harmful effectors into host cells. LTPs can act as sensors by detecting changes in lipid conformation upon binding and transmitting that information to proteins. This interaction between a signalling protein and lipid-bound LTP may lead to lipid-dependent signalling pathways. LTPs can modify lipids, resulting in conformational changes that impact lipid function

3.2.8 Snakins

Snakin, another cysteine-rich peptide, was first discovered in potato and pepper plants and has roles in response to a variety of biotic agents such as bacteria, fungi, and nematodes (Oliveira-Lima et al. 2017). In potato, snakins (StSN1, StSN2) have been reported to be generally expressed in axillary buds, petals, floral buds, stem, tubers, petals, carpels, and stamens except sepals, roots, and stolons (Segura et al. 1999; Berrocal-Lobo et al. 2002). Apart from antimicrobial activities, snakins also have roles in fruit development (Ando and Grumet 2010), lateral root development (Zimmermann et al. 2010), growth and cell division as it is expressed in young actively dividing plant cells of shoot apex, apical bud, vascular stem, and root tissues. Their expression level reduces as the plant ages towards maturity (Iqbal and Khan 2023). Bacterial pathogens (e.g., *Erwinia chrysanthemi* and *Ralstonia solanacearum*) were shown to downregulate the expression of *snakin* genes for successful infection in potato (Berrocal-Lobo et al. 2002).

3.2.8.1 Structure

Snakins are small in size (~7 kDa), 63 amino acid long, positively charged and share several similarities to the proteins of GASA (gibberellic acid stimulated in

Arabidopsis) family found in Arabidopsis (Su et al. 2020). Snakins belonging to Snakin/GASA family contain 18–29 residue long signal peptide at its long N-terminus, a variable site whose sequence and composition vary greatly among different classes of snakins and 60 amino acid long GASA domain at its C-terminus which contains 12 conserved cysteine residues forming six disulphide bonds (Cys1-Cys9, Cys2-Cys7, Cys3-Cys4, Cys5-Cys11, Cys6-Cys12, and Cys8-Cys10) that confer stability to the structure (Porto and Franco 2013). Snakin has a helix-turn-helix (HTH) motif which along with disulphide bonded structure facilitates binding of snakin to target sites (negatively charged cell membrane, protein or DNA) on pathogens (Iqbal and Khan 2023).

3.2.8.2 Mechanism of Action

Snakin/GASA play crucial roles in promoting or inhibiting cell elongation, cell division, root formation, stem growth, flowering time, fruit ripening, and responses to biotic or abiotic stresses such as pathogen defence, heat, salt or oxidative stress (Nahirñak et al. 2012). Snakin/GASA peptides suppress a variety of bacterial and fungal growth at incredibly low concentrations, functioning in plant innate immunity (Nahirñak et al. 2012). These peptides exhibit antimicrobial properties in vitro and in vivo through their interactions with pathogens, including fungi, viruses, and Gram-positive and Gram-negative bacteria.

One hypothesis suggests that the high cationic nature of Snakin/GASA peptides enables them to interact with the negatively charged components of their target sites, leading to a non-specific pore-forming effect in the membrane of bacteria, fungi, and plant protoplasts and destabilising or interacting with the target (Rodríguez-Decuadro et al. 2018). The presence of a DNA-binding motif in snakin/GASA peptides suggests that it may also target DNA by deregulating microbial gene expression. Another hypothesis suggests that snakin/GASA peptides function through the biosynthesis and transduction processes of phytohormones such as gibberellins (GA), abscisic acid (ABA), and brassinosteroids (BR) acting as phytohormonal signalling transducers and integrators (Nahirñak et al. 2012).

The presence of putative redox-active sites in snakin/GASA peptides suggests that it may be involved in redox and hormonal signalling pathways, as they contain conserved cysteine residues which potentially form disulphide bridges regulating the ascorbate levels (Wigoda et al. 2006). This modulates the reactive oxygen species (ROS) status in plants. Certain snakin/GASA proteins, such as GASA4 and GASA5, have been implicated in regulating floral meristem identity, seed size and weight, hypocotyl elongation, and responses to heat stress (Rubinovich and Weiss 2010).

3.2.9 Plant Elicitor Peptides (Pep)

Plant elicitor peptides (Peps) are molecules produced by plants endogenously to trigger pattern-triggered immunity (PTI) against bacterial and fungal pathogens and other herbivores (Boller and Felix 2009). PTI is triggered by elicitors released by pathogens or present on pathogen surfaces termed as PAMP (pathogen-associated molecular pattern) [e.g., bacterial-derived peptide flg22 (Zipfel et al. 2004), fungal-derived chitin (Iizasa et al. 2010)] or HAMP (herbivore associated molecular pattern) or molecules which are the breakdown products of wounding or infection termed as DAMP (damage-associated molecular pattern) [e.g., cutin monomers, cell wall fragments such as oligogalacturonides or cellulose fragments (Ferrari 2013; Nühse 2012), and plant elicitor peptides (Peps) (Albert 2013)]. Pep (20–23 amino acid long) is produced by metacaspase-mediated cleavage (Shen et al. 2019) from the C-terminal part of PROPEP (Precursor Pep) and is then exported to extracellular space. On pathogen invasion, from the disrupted cell, Pep is released and binds to its transmembrane leucine-rich repeat kinase receptors (LRR-KRs) known as Pep receptors (PEPRs) of adjacent cells to mediate downstream signalling for upregulating the expression of defence genes (Ruiz et al. 2018).

The Pep-PEPR system has been extensively studied in model plants like Arabidopsis and maize, and has been shown to activate multiple defence pathways, making it a potential strategy to increase plant resistance to pathogen attack (Lori et al. 2015). Peps are perceived by leucine-rich repeat kinase receptors (LRR-RLKs) known as Pep receptors (PEPRs) and somatic embryogenesis receptor kinases (SERKs). When Peps bind to these receptors, a series of downstream signalling events are triggered, including phosphorylation events, Ca^{2+}-dependent signalling, activation of mitogen-activated protein kinases (MAPKs), production of nitric oxide and ROS, coactivation of salicylic acid (SA), jasmonic acid (JA), and ethylene (ET) hormone pathways, and production of antimicrobial pathogenesis-related (PR) proteins and phenylpropanoids (Hu et al. 2018).

3.2.10 Puroindolines

Puroindolines (PINs) are small (13 kDa in size), basic, cysteine-rich protein, found in wheat endosperm and have two isoforms, viz., puroindoline-a (PIN-a) which is the major component conferring more antimicrobial properties and puroindoline-b (PIN-b) which is the minor component conferring more antimicrobial properties (Bottier et al. 2008). Puroindoline-A (PinA) and puroindoline-B (PinB) share 60% sequence identity and are folded by five disulphide bonds (Talukdar et al. 2021).

3.2.10.1 Structure

Puroindoline-a, which is a 13 amino acid long tryptophan-rich domain (FPVTWRWWKWWKG-NH2) (Phillips et al. 2011), serves as bactericidal component against both Gram-positive and Gram-negative bacteria (Haney et al. 2013). This positively charged domain binds to the negatively charged head groups of phospholipids on the bacterial membrane and destabilises it not by forming pores but by imposing curvature to the microbial membrane (Jing et al. 2003). During this interaction with the microbial membrane, it forms a partially helical amphipathic structure with all the positive residues (lysine and arginine) orienting themselves close to the planar side of the aromatic indole rings of tryptophan and becoming surrounded by the indole rings, resulting in the formation of a stable structure (by cation-π interactions) and better disintegrity of microbial membranes (Jing et al. 2003). Puroindoline-b which is a 12 amino acid long tryptophan-rich domain (FPVTWPTKWWKG-NH2) (Alfred et al. 2013) also has antimicrobial activities, but of lower efficacy simply due to the presence of less positively charged and tryptophan residues in this sequence.

3.2.10.2 Mechanism of Action

Puroindolines are amphipathic proteins found in wheat (*Triticum aestivum*) endosperm, which protect plants from bacterial, fungal, and viral attacks. They are small (13 kDa) proteins with a unique dual role in determining wheat quality and enhancing crop resistance to various pathogens (Talukdar et al. 2021). The primary mechanism by which the puroindolines function is by creating ion channels in biological membranes, which is essential to their antibacterial activity. The tryptophan-rich domain (TRD) of puroindolines is what gives them their antibacterial qualities. Three tryptophan residues in PinB (WPTKWWK) and five tryptophan residues in PinA (WRWWKWWK) make up the TRD (Haney et al. 2013). Puroindolines play a significant role in determining grain texture, particularly in wheat. They are proteins that are associated with the softness or hardness of grains. When both puroindoline a and b are functional and expressed in the grain, the texture is soft. Near-isogenic lines (NILs) have been used to study the relative effects of the puroindolines on grain texture (Hogg et al. 2003). In vitro studies have shown that PinA and PinB full proteins or shorter peptides exhibit antimicrobial activity against various microorganisms, including *Escherichia coli, Staphylococcus aureus, Staphylococcus epidermidis, Bacillus subtilis, Listeria monocytogenes*, and fungi like *Candida albicans* and *Aspergillus flavus* (Haney et al. 2013). Puroindolines are thought to be essential to natural defence mechanism of wheat seeds during dormancy, defending them from fungus and other diseases. One way by which puroindolines exhibit antimicrobial activity is by defending wheat seeds against dry seed decay caused by *Penicillium spp*. Seeds that contain puroindolines have better germination rates than seeds that do not when infected with *Penicillium spp.* spores. Fungal diseases including rice blast (*Magnaporthe grisea*) and sheath blight (*Rhizoctonia solani*) are less common in rice grown in experiments using wheat puroindoline genes (Giroux et al. 2003). Puroindoline-transformed wheat (such as the cultivar HiLine, which was transformed using the maize ubiquitin promoter to produce puroindoline genes) showed improved resistance to a variety of fungal pathogens, including *Fusarium culmorum* (scab), *Gaeumannomyces graminis* (take-all disease), and other fungal diseases (Giroux et al. 2003) (Table 3.4). Table 3.4 summarizes these ten major plant elicitor peptides, along with additional compounds, highlighting their modes of action and antimicrobial properties in defense against foreign pathogens.

Table 3.4 Summary table of different plant antimicrobial peptides with their mode of action

Antimicrobial peptides	Target pathogen	Source plant	Tissue of expression	Mode of action	References
Plant defensins (PDFs)	Fungi: • *Fusarium* spp. • *Botrytis cinerea* • *Verticillium dahliae* • *Penicillium expansum* • *Alternaria alternata* Bacteria: • *Pectobacterium carotovorum* • *Clavibacter michiganensis* ssp. *michiganensis*	• *Arabidopsis thaliana* • *Erythrina crista-galli*	Leaves, pods, tubers, fruit, roots, bark, floral organs	Plant defensins primarily interact with membrane lipids, bind to specific membrane receptor targets, and can permeabilize the plasma membrane of certain bacterial pathogens, leading to inhibition of DNA synthesis or transcription, while also exhibiting antifungal activity against human pathogens and inhibiting mammalian voltage-gated potassium channels.	(Hu et al. 2018; Santos-Silva et al. 2020; Stotz et al. 2009)
Systemin	Fungi: • *Botrytis cinerea* (necrotrophic) • *Glomus clarum*	Tomato	Vascular bundles	The plant systemin acts as a signalling peptide that triggers the production of jasmonic acid, initiating defence responses against pathogens by activating defence genes and inducing the synthesis of defence compounds like protease inhibitors, contributing to plant resistance	(Hu et al. 2018; Zhang and Hu 2017)

(continued)

Table 3.4 (continued)

Antimicrobial peptides	Target pathogen	Source plant	Tissue of expression	Mode of action	References
Thionin	Bacteria • *Pseudomonas solanacearum* • *Xanthomonas phaseoli* • *Xanthomonas campestris* • *Erwinia amylovora* • *Corynebacterium fascians* Fungi • *Fusarium solani* • *Sclerotinia sclerotiorum* • *Phytophthora infestans*	Families - • Poaceae • Brassicaceae • Papaveraceae	Seeds, stems, roots, leaves and flowers	Plant thionins, a class of antimicrobial peptides, contribute to the first line of defence against pathogens by inhibiting their growth and development, providing rapid innate immunity against a wide range of microbial pathogens, and enhancing disease resistance when overexpressed	(Taveira et al. 2016; Hao et al. 2016; De Caleya et al. 1972)
Hevein-like peptides	Fungi: • *Cercospora beticola* • *Fusarium culmorum* • *Phytophthora infestans* • *Ascochyta lycopersici* • *Verticillium dahliae* • *Giberella zeae* • *Alternaria nicotianae* • *Fusarium moniliforme* Bacteria • *Pseudomonas syringae*	• Radish • Barley • Maize • Arabidopsis • Spinach • Grapevine • Wheat • Onion • Shepherd's purse • *Hevea brasiliensis*	Seed, leaves, latex.	Plant hevein-like peptides act by inhibiting fungal metalloproteases, thereby impeding pathogen growth, and they exhibit antifungal activity, contributing to plant defence against a variety of pathogens.	(Slavokhotova et al. 2014a; Huang et al. 2002)

Knottin-type peptides	Fungi • *Alternaria alternata* • *Cercospora personata* • *Aspergillus niger*	• Maize • Potato • Wheat	Seeds	Plant knottin-type peptides act as antimicrobial peptides with diverse biological functions, including promoting resistance to biotic and abiotic stresses, enhancing symbiotic interactions, and exhibiting antimicrobial, cytotoxic, insecticidal, and anti-HIV activities, by binding to molecular targets and forming conserved disulphide bonds, contributing to plant defence against bacteria, fungi and viruses.	(Han et al. 2023; Attah et al. 2022; Yang et al. 2007)
α-hairpinin family peptides	Phytopathogenic fungi and bacteria	• Maize • Barnyard grass • Wheat	Seeds	α-hairpinins inhibit insect and fungal pathogens, through their rigid, reactive-site loop that complements the binding site of serine proteases. They slow down the hydrolysis of peptide bonds and inhibit protein synthesis, while also interacting with fungal cell wall components to induce apoptosis and programmed cell death.	(Ng et al. 2011; Slavokhotova and Rogozhin 2020)
Lipid Transfer Protein	Fungi: • *Fusarium solani* • *Colletotrichum lindemuthianum* • *Candida tropicalis*	All land plants including • Barley • Pepper • Rice • Sunflower • Wheat	Tapetal cytoplasm of anther tissues, Extracellular space or microspore	Lipid transfer proteins (LTPs) enhance plant cell wall extension through a process of wall loosening by interacting with the plant cell membrane to facilitate the exchange of phospholipids, while also exhibiting diverse biological activities such as antimicrobial, antiviral, and enzyme inhibitory functions that contribute to plant defence and cellular stress responses	(Wong et al. 2019; Fang et al. 2023; Zottich et al. 2011)

(continued)

Table 3.4 (continued)

Antimicrobial peptides	Target pathogen	Source plant	Tissue of expression	Mode of action	References
Snakins	Fungi: • *Botrytis cinerea* • *Fusarium solani* • *Fusarium culmorum* • *Fusarium oxysporum* • *f. sp conglutinans* • *Fusarium oxysporum* • *f. sp lycopersici* • *Plectosphaerella cucumerina* • *Colletotrichum graminicola* • *Colletotrichum lagenarium* • *Bipolaris maydis* • *Aspergillus flavus* Bacteria: • *Clavibacter michiganensis* • *Ralstonia solanacearum* • *Erwinia chrysanthemi* • *Rhizobium meliloti*	• *Arabidopsis thaliana* • *Solanum lycopersicum*	Leaf, Stem, Roots	Snakin/GASA peptides exhibit antimicrobial properties and interact with negatively charged components of bacterial and fungal cell membranes to form pores and destabilise targets, while also potentially targeting DNA, influencing phytohormone biosynthesis and signalling, and regulating redox and hormonal pathways for plant growth and stress responses.	(Nahirñak et al. 2012; Berrocal-Lobo et al. 2002)
Plant elicitor peptides (Pep)	Bacteria and fungi, as well as herbivores	• *Arabidopsis thaliana* • Maize	Roots	Plant Elicitor Peptides (PEPs) defend against pathogens by activating pattern recognition receptors (PRRs) on plant cells, initiating signalling cascades that lead to the production of antimicrobial compounds, reinforcement of cell walls and induction of SAR, thereby enhancing the overall immunity of the plants.	(Lori et al. 2015; Zhang et al. 2022a)

Puroindolines	Fungi: • *Alternaria brassicola* • *Ascophyta pisi* • *Botrytis cinerea* • *Verticillium dahliae* • *Fusarium culmorum* Bacteria: • *Staphylococcus aureus* • *Micrococcus luteus* • *Klebsiella* sp. • *Bacillus cereus*	In the seeds of plants from the Triticeae tribe, which includes • Wheat • Barley • Rye	Specifically, they are located in the starchy endosperm of wheat grains and are crucial for determining grain hardness	Puroindolines defend against pathogens by forming ion channels in biological membranes, disrupting microbial cell integrity and causing cell death, while also enhancing the natural resistance of plants to bacterial, fungal and viral attacks through antimicrobial properties of tryptophan-rich domain.	(Marion et al. 2007; Chugh et al. 2015)
Cyclotides	Bacteria • *Oldenlandia affinis* • *Chassalia parviflora* • *Viola odorata*	Families - • Rubiaceae • Violaceae	Flowers, Leaves, Stems, Roots	Cyclotides, a class of plant-derived cyclic peptides, serve as defence molecules in plants against various pests and pathogens, exhibiting insecticidal, nematicidal, molluscicidal, antimicrobial, antitumor, and protease inhibitory activities, attributed to their unique cyclic cysteine knot structure that confers exceptional stability and membrane-disrupting properties.	(Craik 2012; Weidmann and Craik 2016)

(continued)

Table 3.4 (continued)

Antimicrobial peptides	Target pathogen	Source plant	Tissue of expression	Mode of action	References
AtPeps (*Arabidopsis thaliana* Pep)	• Bacteria: *Pseudomonas syringae* • Oomycete: *Pythium irregulare*	• *Arabidopsis thaliana* • Tomato (*Solanum lycopersicon*)	The apoplast, including the endodermis, xylem and xylem pole pericycle cells of the meristematic zone, roots	Defence against pathogens by modulating the activity of the plasma membrane H^+-ATPases, a proton pump responsible for maintaining pH and ion homeostasis in plants. AtPeps enhance the activity of the H^+-ATPases, leading to increased proton efflux and membrane depolarisation, which in turn triggers the activation of downstream defence responses, such as the production of ROS and activation of mitogen-activated protein kinases.	(Seidel 2022; Ortiz-Morea et al. 2016; Zelman and Berkowitz 2023)
Rapid alkalinization factor 23 (RALF23)	Fungi: *Sclerotinia sclerotiorum*	• *Arabidopsis thaliana* • *Brassica napus*	Vegetative tissues	The plant Rapid alkalinization factor 23 (AtRALF23) functions as a defence response against pathogens by inhibiting plant resistance, likely through modulating the activity of cell wall-associated receptor-like kinases, such as FERONIA, which is crucial for triggering female-male signal communication and regulating cell growth and development. AtRALF23 has been shown to inhibit root growth in *Arabidopsis thaliana*, and its signalling pathway involves calmodulin-like protein 38 (CML38) essential for root growth inhibition, but not for the alkalinization response.	(Campos et al. 2018; Gonneau et al. 2018; He et al. 2022)

PIP1 and 2 (PAMP-induced secreted peptides)	• Bacteria: *Pseudomonas syringae* • Fungi: *Fusarium oxysporum*	*Arabidopsis thaliana*	Roots	The plant PIP1 and PIP2 (PAMP-induced secreted peptides) function as defence response against pathogens by inducing stomatal closure and activating immune responses, including cytosolic Ca^{2+} elevation, ROS burst, MAP kinase activation, and transcriptional reprogramming. PIP1 and PIP2 are perceived by plasma membrane receptors and induce the expression of defence genes, suggesting a role in stomatal immunity and eliciting defence responses. AtPIP1 binds to the RLK7 receptor with a more stable free energy and less binding area than AtPIP2, indicating a more specific and stable interaction between AtPIP1 and its receptor.	(Yu et al. 2023; Leng et al. 2021; Najafi et al. 2022)
Serine-rich endogenous peptide 12 (SCOOP12)	Fungi: *Fusarium oxysporum*	• Brassicaceae family • Solanaceae family • *Arabidopsis thaliana*	Roots	The plant serine-rich endogenous peptide 12 (PROSCOOP12) regulates defence response and root elongation in *Arabidopsis thaliana* by acting as a negative regulator of defence against necrotrophic pathogens, such as the bacteria *Erwinia amylovora*, through the phospholipid pathway and ROS regulation. This peptide is involved in biotic and oxidative stress responses and has properties of phytocytokines, activating the phospholipid signalling pathway and inducing downstream defence responses.	(Gully et al. 2019; Guillou et al. 2022; Zhang et al. 2022)

(continued)

Table 3.4 (continued)

Antimicrobial peptides	Target pathogen	Source plant	Tissue of expression	Mode of action	References
CLV1 (CLAVATA 1)	Bacteria: *Ralstonia solanacearum*	• *Arabidopsis thaliana* • Tobacco (*Nicotiana tabacum*)	Maintenance of central region of meristem and flower meristem, Anther development, Vascular tissues, Root system architecture	Plant CLAVATA1 (CLV1) functions in defence responses against pathogens by regulating key developmental processes, such as meristem maintenance, anther development, vascular tissue formation, and root system architecture, through interactions with small signalling peptides and receptor-like kinases.	(Hazak and Hardtke 2016; Zhang et al. 2019)
PSK (Phytosulfokine)	Bacteria: • *Botrytis cinerea* • *Pseudomonas syringae*	• *Arabidopsis thaliana* • *Zea mays* (corn) • *Asparagus officinalis* (asparagus) • *Oryza sativa* (rice) • Tomato	Root nodules of leguminous plants	Phytosulfokine (PSK) is a plant peptide hormone that optimises plant growth and defence via glutamine synthetase GS2 phosphorylation in tomato. PSK interacts with the calcium-dependent protein kinase CPK28, which in turn phosphorylates the key enzyme of nitrogen assimilation, viz., glutamine synthetase GS2 at two sites (Serine 334 and Serine 360). GS2 phosphorylation at S334 specifically regulates plant defence, whereas S360 regulates growth, uncoupling the PSK-induced effects on defence responses and growth regulation.	(Ding et al. 2023; Li et al. 2024; Hu et al. 2018)

PSY1 (Plant peptide containing sulphated tyrosine1)	Bacteria • *Erwinia carotovora* • *Agrobacterium radiobacter* • *Agrobacterium rhizogenes* • *Clavibacter michiganensis* • *Curtobacterium flaccumfaciens*	*Arabidopsis thaliana*	Root elongation zone, shoot apical meristem	PSYs are a class of post-translationally modified peptides that play crucial roles in plant growth and development, including primary root growth regulation. PSYs, such as PSY1, are tyrosine-sulphated peptides that interact with specific receptors to trigger downstream signalling cascades, leading to the regulation of various physiological processes.	(Matsubayashi and Sakagami 2006; Amano et al. 2007; Pelegrini et al. 2011)
Macadamia (β-barrelins)	Bacteria • *Clavibacter michiganensis* • *Chalara elegans* • *Ralstonia solanacearum* • *Phytophthora cryptogea* • *Sclerotinia sclerotiorum* • *Fusarium oxysporum* • *Cercospora nicotianae*	• *Macadamia integrifolia* • *Capsicum chinense* • *Phaseolus vulgaris*	Seeds	Macadamia β barrelins activate defence responses in *Macadamia integrifolia* by exhibiting antifungal properties, serving as a first line of defence against fungal pathogens through a unique Greek key β-barrel fold, which is distinct among plant antimicrobial proteins, potentially reflecting a similar mode of action to yeast killer toxins.	(McManus et al. 1999; Marcus et al. 1999)
Impatiens-like (Ib-AMPs)	Bacteria • *Staphylococcus aureus*	*Impatiens balsamina*	Leaves	*Impatiens* sp. exhibit increased leaf palatability with rising temperature, potentially due to defence chemicals rather than nutrient changes, impacting herbivore preferences.	(Thevissen et al. 2005)

(continued)

Table 3.4 (continued)

Antimicrobial peptides	Target pathogen	Source plant	Tissue of expression	Mode of action	References
Thaumatin-like protein (TLP)	Fungi • *Rhizoctonia solani* • *Botrytis cinerea* • *Fusarium oxysporum* • *Mycosphaerella arachidicola* • *Trichoderma viride*	• Rice • Barley • Maize • Moss • Black Cottonwood	Roots, bulbs, stems, leaves, buds, flowers, and sprouts	The antifungal activity of TLPs (classified as the PR-5 or Pathogenesis-Related-5 protein family) may be attributed to the inhibition of xylanase, α-amylase, and trypsin activities, as well as the ability to rupture fungal membranes by pore formation and inducing programmed cell death in fungi following binding to the plasma membrane receptor proteins. TLPs might help plants mitigate various abiotic stresses and are recognised as osmotin-like proteins based on their sequence homology with the osmotin that accumulates in response to osmotic stress. The expression of TLPs is influenced by phytohormones such as methyl jasmonate, ethylene, salicylic acid and abscisic acid, which mediate plant defence responses towards attack by fungal pathogens.	(Misra et al. 2016; Anisimova et al. 2022; Singh et al. 2013)

3.3 Applications of Plant Antimicrobial Peptides and Future Prospects

The plant antimicrobial peptides have a huge potential and application for advancing agricultural science and food security. Researchers can use this pathogen elicitor–receptor interaction and its specificity for developing crops with heightened resistance to diseases and pests.

1. For crop protection:
   (a) Isolating and applying AMPs to plants by spraying might serve to protect plants (Ghosh and Roychoudhury 2023).
   (b) The worldwide problem of severe loss of agricultural crops can be addressed by employing genetic engineering approaches utilising these antimicrobial plant peptides (AMPs) as discussed in this section.

   For example,

(a) Incorporating alfalfa antifungal peptide *defensin* gene into potato plants has been shown to be effective in resisting infection caused by *Verticillium dahilae* (Tam et al. 2015).
(b) Incorporating *snakin-1* gene in potato plants has been shown to be effective in resisting infection caused by *Rhizoctonia solani* and *Erwinia carotovora* (Ghosh and Roychoudhury 2023).

"Elicitation" is a biotechnological tool to increase the production of secondary metabolites and biomass in plant cell cultures and organ cultures. Plant defence methods could become more efficient and eco-friendly with the discovery of new elicitors (natural, chemical, and recombinant elicitors) and a greater comprehension of their mechanisms of action (Kaur et al. 2022).

2. Usage as food preservative in food industries:

Usage of artificial and chemical preservatives has been in use for so long to prevent food waste. These preservatives harm the consumer body in several adverse ways. Instead, usage of natural products like AMPs would help to counter this problem. Being broad-spectrum antimicrobial component, they have the potential to serve as food preservative.

However, there are some drawbacks due to the following reasons (Baindara and Mandal 2022):

- AMPs sometimes impart cytotoxic effects. The food being preserved with AMPs when consumed by humans, it will lead to cytotoxicity to humans which is undesirable.
- AMPs have low stability and degrade quite easily.
- Being larger in size, AMPs often act as immunogen and trigger immune response when the AMP-preserved food is consumed by humans.

Due to these above-mentioned reasons, natural AMPs cannot be used; instead, research on synthetic AMP generation has been started and this approach of generating AMPs synthetically, generally includes extreme ranges of temperature, pH levels, and salt concentration to ensure their stability and retain optimal bioactivity in food industries (Baindara et al. 2016). Not only generation, the improvement of antimicrobial capabilities of AMP can also be done by genetic engineering and proteomics approach. However, the artificial synthesis of these peptides is quite expensive as compared to cheap chemical harmful preservatives, e.g., sodium benzoate (Baindara and Mandal 2022).

3. Strategies for plant disease management:
   (a) Chemically synthesised pesticides, fertilisers, and agrochemicals are a huge problem for the environment and also compromises health. Agrochemicals using AMPs that boost plant immune system hold immense potential for solving the global food security and safety amidst the challenges posed by climate change (Jakupi and Demirbas 2022). This can be used to enhance crop resilience against various stressors triggered by climate change, making highly efficient elite cultivars which can survive drastic environmental conditions like extreme temperature, drought, and shifting precipitation patterns (Iriti and Vitalini 2020).
   (b) Use of AMPs can compensate for the yield losses caused by plant diseases, like 16% annually due to pathogen attack (Iriti and Vitalini 2020). The use of plant protection products also stimulate the plant immune system to elevate the levels of bioactive phytochemicals in plant foods, such as polyphenols, thus enhancing the nutritional quality of food and promoting consumer health (Mishra et al. 2012).

By reducing the dependency on traditional agrochemicals like pesticides, which pose risks to human health, the environment, and non-target organisms, these modified strategies promote safer agricultural practises and contribute to environmental sustainability. More innovations and researches in this area are needed that could result in more sustainable and effective plant protection solutions that support resilience and general health of plants in addition to counter diseases (Mishra et al. 2012).

4. Understanding the complex signalling pathways:
   (a) Plant defence responses makes it easier to distinguish between different elicitor–receptor interactions, such as those associated with induced systemic resistance (ISR) and SAR, which opens the door to focused approaches to improve plant resilience (Abdul Malik et al. 2020). Research into finding compounds that can mimic the effects of well-known inducers such as benzothiadiazole (BTH) or acibenzolar-S-methyl (ASM) is ongoing and could lead to the creation of new plant defence strategies that could revolutionise crop protection against pests and diseases (Kouzai et al. 2018).

An incompatible pathogen attack can trigger both systemic and localised reactions that trigger the synthesis of secondary metabolites like phenolics, terpenoids, alkaloids, and phytoalexins (Ruiz-García and Gómez-Plaza 2013). These secondary metabolites then help form physical barriers against pathogen invasion and its replication in addition to having strong antimicrobial properties. With the goal of using elicitors to boost or induce the biosynthesis of particular compounds in plants, the concept of "*Elicitor engineering*" can lead to increasing crop productivity and sustainability (Abdul Malik et al. 2020).

(b) When a pathogen invades a plant, pattern recognition receptors (PRRs) on the plants detect pathogen-associated molecular patterns (PAMPs) of the extracellular pathogens, which triggers subsequent downstream signalling cascades with the activation of mitogen-activated protein kinases (MAPKs), that causes the nucleus to start transcribing defence-responsive genes. These molecules usually indicate a basal level of resistance. Plants also utilise a more powerful defence mechanism called the hypersensitive response (HR) (Abdul Malik et al. 2020). Pathogens attempting to evade recognition by PRRs often secrete effector molecules into the plant cell cytosol bypassing the cell wall. These effectors can inhibit the MAPK pathways and prevent the activation of pathogen-triggered immunity (PTI). However, upon entry into the plant cell, these effectors may be recognised by specific *R* genes, leading to the induction of HR. To harness this HR-mediated defence mechanism, transgenic approach may be undertaken. The strategies involve cloning an effector gene from the pathogen under a constitutive promoter, for which a corresponding *R* gene is present in the plant. Any PAMP should activate this promoter to guarantee that the effector gene is continuously expressed in response to pathogen infection. As a result, when effectors are produced, the corresponding *R* gene is activated, activating HR and conferring resistance (Ruiz-García and Gómez-Plaza 2013).

With this method, researchers can elicit HR-mediated resistance against a wide range of pathogen infections, including ones for which the precise effector is unclear. Through the utilisation of the innate recognition abilities of *R* genes, the transgenic plant may efficiently establish a strong defence mechanism, guaranteeing increased resistance against pathogenic assaults (Benhamou 1996).

5. Conversion of unstable proteins into stable form:

Presence of multiple cysteine residues forming disulphide bonds in AMPs confer compact structure, stability to high temperature, prevention from degradation in presence of denaturing chemical agents or degrading enzymes. Incorporation of cysteine residues makes the cysteine-scaffold and modify the structures of biologically active compounds (which are unstable and easily get degraded by digestive enzymes) to a stable form, making them resistant to digestive enzymes with no deteriorating effect on their bioactivity (Tam et al. 2015).

3.4 Conclusion

Plants, despite being immotile organisms and not having the abilities to escape danger, have unique abilities to resist and deal with several adverse conditions such as biotic stress posed by different bacteria, fungi, and other parasites. A number of different peptides having roles as bactericidal, fungicidal, herbicidal agents in plants via different signalling pathways identify and kill or prevent the spread of the pathogens, thus successfully mounting biotic stress response in plants. Due to their innate broad-spectrum antimicrobial activities, they can be used in healthcare as an alternative to antibiotics and agriculture. Unfortunately, the cytotoxicity imparted by them, low stability and high expense of their synthetic production have limited their usage in the healthcare system and therefore, they have not yet received any clinical approval for their use as drug except cationic AMPs which have been approved to be used in aerosol sprays for patients with cystic fibrosis (Nawrot et al. 2013). However, research is being continuously done in this field to identify and isolate novel peptides and apply them for the improvement of human health and stability for disease resistance in agricultural crops by either spraying them or by genetic engineering strategies. Future research needs to focus on developing peptides with reduced toxicity, greater stability with lower costs via improved biotechnological approaches.

Overall, this chapter encompasses structures and roles of different plant peptides with antimicrobial activities, utilising their ability to generate bacteria/fungi/pest-resistant transgenic crops to address the issues of global food security in the current scenario of economic crisis and population growth. This chapter also emphasises their potential contributions in the healthcare system with impetus on further research in this field.

References

Abdul Malik NA, Kumar IS, Nadarajah K (2020, January 31) Elicitor and receptor molecules: orchestrators of plant defense and immunity. Int J Mol Sci 21(3):963. https://doi.org/10.3390/ijms21030963

Aerts AM, François IE, Meert EM, Li QT, Cammue BP, Thevissen K (2007) The antifungal activity of RsAFP2, a plant defensin from *Raphanus sativus*, involves the induction of reactive oxygen species in Candida albicans. J Mol Microbiol Biotechnol 13(4):243–247. https://doi.org/10.1159/000104753

Albert M (2013, September 7) Peptides as triggers of plant defence. J Exp Bot 64(17):5269–5279. https://doi.org/10.1093/jxb/ert275

Alfred RL, Palombo EA, Panozzo JF, Bariana H, Bhave M (2013, March 1) Stability of puroindoline peptides and effects on wheat rust. World J Microbiol Biotechnol 29(8):1409–1419. https://doi.org/10.1007/s11274-013-1304-6

Amador VC, Santos-Silva CAD, Vilela LMB, Oliveira-Lima M, De Santana Rêgo M, Roldan-Filho RS, De Oliveira Silva RL, Lemos AB, De Oliveira WD, Ferreira-Neto JRC, Crovella S, Benko-Iseppon AM (2021, October 21). Lipid Transfer Proteins (LTPs)—Structure, Diversity and Roles beyond Antimicrobial Activity. Antibiotics, 10(11), 1281. https://doi.org/10.3390/antibiotics10111281

Amano Y, Tsubouchi H, Shinohara H, Ogawa M, Matsubayashi Y (2007, November 13) Tyrosine-sulfated glycopeptide involved in cellular proliferation and expansion in Arabidopsis. Proc Natl Acad Sci 104(46):18333–18338. https://doi.org/10.1073/pnas.0706403104

Ando K, Grumet R (2010, July) Transcriptional profiling of rapidly growing cucumber fruit by 454-pyrosequencing analysis. J Am Soc Hortic Sci 135(4):291–302. https://doi.org/10.21273/jashs.135.4.291

Anisimova OK, Kochieva EZ, Shchennikova AV, Filyushin MA (2022, March 11) Thaumatin-like protein (TLP) genes in garlic (*Allium sativum* L.): genome-wide identification, characterization, and expression in response to *Fusarium proliferatum* infection. Plants 11(6):748. https://doi.org/10.3390/plants11060748

Attah AF, Lawal BA, Yusuf AB, Adedeji OJ, Folahan J, Akhigbe KO, Roy T, Lawal AA, Ogah NB, Olorundare OE, Chamcheu JC (2022, November 28) Nutritional and pharmaceutical applications of under-explored knottin peptide-rich phytomedicines. Plants. https://doi.org/10.3390/plants11233271

Azmi S, Hussain MK (2021, January 14) Analysis of structures, functions, and transgenicity of phytopeptides defensin and thionin: a review. Beni-Seuf Univ J Basic Appl Sci (Print). https://doi.org/10.1186/s43088-020-00093-5

Baindara P, Mandal SM (2022, August 11) Plant-derived antimicrobial peptides: novel preservatives for the food industry. Foods 11(16):2415. https://doi.org/10.3390/foods11162415

Baindara P, Singh N, Ranjan M, Nallabelli N, Chaudhry V, Pathania GL, Sharma N, Kumar A, Patil PB, Korpole S (2016) Laterosporulin10: a novel defensin like Class IId bacteriocin from *Brevibacillus* sp. strain SKDU10 with inhibitory activity against microbial pathogens. Microbiology 162(8):1286–1299. https://doi.org/10.1099/mic.0.000316

Barashkova AS, Sadykova VS, Salo VA, Zavriev SK, Rogozhin EA (2021) Nigellothionins from black cumin (*Nigella sativa* L.) seeds demonstrate strong antifungal and cytotoxic activity. Antibiotics (Basel, Switzerland) 10(2):166. https://doi.org/10.3390/antibiotics10020166

Beintema JJ (1994, August 22) Structural features of plant chitinases and chitin-binding proteins. FEBS Lett 350(2–3):159–163. https://doi.org/10.1016/0014-5793(94)00753-5

Beliaev DV, Yuorieva NO, Tereshonok DV, Tashlieva II, Derevyagina MK, Meleshin AA, Rogozhin EA, Kozlov SA (2021) High Resistance of potato to early blight is achieved by expression of the Pro-SmAMP1 gene for hevein-like antimicrobial peptides from common chickweed (*Stellaria media*). Plants (Basel, Switzerland) 10(7):1395. https://doi.org/10.3390/plants1007139

Beloshistov RE, Dreizler K, Galiullina RA, Tuzhikov AI, Serebryakova MV, Reichardt S, Shaw J, Taliansky ME, Pfannstiel J, Chichkova NV, Stintzi A, Schaller A, Vartapetian AB (2018) Phytaspase-mediated precursor processing and maturation of the wound hormone systemin. New Phytol 218(3):1167–1178. https://doi.org/10.1111/nph.14568

Benhamou N (1996, July) Elicitor-induced plant defence pathways. Trends Plant Sci 1(7):233–240. https://doi.org/10.1016/1360-1385(96)86901-9

Bergey DR, Howe GA, Ryan CA (1996) Polypeptide signaling for plant defensive genes exhibits analogies to defense signaling in animals. Proc Natl Acad Sci USA 93(22):12053–12058. https://doi.org/10.1073/pnas.93.22.12053

Berrocal-Lobo M, Segura A, Moreno M, López G, García-Olmedo F, Molina A (2002, March 1) Snakin-2, an antimicrobial peptide from potato whose gene is locally induced by wounding and responds to pathogen infection. Plant Physiol 128(3):951–961. https://doi.org/10.1104/pp.010685

Boller T, Felix G (2009) A renaissance of elicitors: perception of microbe-associated molecular patterns and danger signals by pattern-recognition receptors. Annu Rev Plant Biol 60:379–406. https://doi.org/10.1146/annurev.arplant.57.032905.105346

Bottier C, Géan J, Desbat B, Renault A, Marion D (2008, August 28) Structure and orientation of puroindolines into wheat galactolipid monolayers. Langmuir 24(19):10901–10909. https://doi.org/10.1021/la800697s

Cabral KM, Almeida MS, Valente AP, Almeida FC, Kurtenbach E (2003) Production of the active antifungal *Pisum sativum* defensin 1 (Psd1) in *Pichia pastoris*: overcoming the inefficiency of the STE13 protease. Protein Expr Purif 31(1):115–122. https://doi.org/10.1016/s1046-5928(03)00136-0

Campos WF, Dressano K, Ceciliato PH, Guerrero-Abad JC, Silva AL, Fiori CS, Morato do Canto A, Bergonci T, Claus LA, Silva-Filho MC, Moura DS (2018, February) *Arabidopsis thaliana* rapid alkalinization factor 1–mediated root growth inhibition is dependent on calmodulin-like protein 38. J Biol Chem 293(6):2159–2171. https://doi.org/10.1074/jbc.m117.808881

Carvalho ADO, Gomes VM (2007) Role of plant lipid transfer proteins in plant cell physiology-a concise review. Peptides 28(5):1144–1153. https://doi.org/10.1016/j.peptides.2007.03.004

Carvalho ADO, Gomes VM (2009, May) Plant defensins—Prospects for the biological functions and biotechnological properties. Peptides 30(5):1007–1020. https://doi.org/10.1016/j.peptides.2009.01.018

Chiapparino A, Maeda K, Turei D, Saez-Rodriguez J, Gavin AC (2016) The orchestra of lipid-transfer proteins at the crossroads between metabolism and signaling. Prog Lipid Res 61:30–39. https://doi.org/10.1016/j.plipres.2015.10.004

Chiche L, Heitz A, Gelly JC, Gracy J, Chau PT, Ha PT, Hernandez JF, Le-Nguyen D (2004) Squash inhibitors: from structural motifs to macrocyclic knottins. Curr Protein Pept Sci 5(5):341–349. https://doi.org/10.2174/1389203043379477

Chugh V, Kaur K, Singh D, Kumar V, Kaur H, Dhaliwal HS (2015) Molecular characterization of diverse wheat germplasm for puroindolineproteins and their antimicrobial activity. Turk J Biol 39:359–369. https://doi.org/10.3906/biy-1405-30

Corrado G, Agrelli D, Rocco M, Basile B, Marra M, Rao R (2011, June 1) Systemin-inducible defence against pests is costly in tomato. Biol Plant 55(2):305–311. https://doi.org/10.1007/s10535-011-0043-5

Craik DJ (2012, February 15) Host-defense activities of cyclotides. Toxins. https://doi.org/10.3390/toxins4020139

Cuchel M, Bloedon LT, Szapary PO, Kolansky DM, Wolfe ML, Sarkis A, Millar JS, Ikewaki K, Siegelman ES, Gregg RE, Rader DJ (2007, January 11) Inhibition of microsomal triglyceride transfer protein in familial hypercholesterolemia. N Engl J Med 356(2):148–156. https://doi.org/10.1056/nejmoa061189

De Caleya RF, Gonzalez-Pascual B, García-Olmedo F, Carbonero P (1972, May) Susceptibility of phytopathogenic bacteria to wheat purothionins in vitro. Appl Microbiol 23(5):998–1000. https://doi.org/10.1128/am.23.5.998-1000.1972

De Samblanx GW, Fernandez del Carmen A, Sijtsma L, Plasman HH, Schaaper WM, Posthuma GA, Fant F, Meloen RH, Broekaert WF, van Amerongen A (1996) Antifungal activity of synthetic 15-mer peptides based on the Rs-AFP2 (*Raphanus sativus* antifungal protein 2) sequence. Pept Res 9(6):262–268

Ding S, Lv J, Hu Z, Wang J, Wang P, Yu J, Foyer CH, Shi K (2023) Phytosulfokine peptide optimizes plant growth and defense via glutamine synthetase GS2 phosphorylation in tomato. EMBO J 42(6):e111858. https://doi.org/10.15252/embj.2022111858

Dodds PN, Rathjen JP (2010, June 29) Plant immunity: towards an integrated view of plant–pathogen interactions. Nat Rev Genet 11(8):539–548. https://doi.org/10.1038/nrg2812

Edqvist J, Blomqvist K, Nieuwland J, Salminen TA (2018, August) Plant lipid transfer proteins: are we finally closing in on the roles of these enigmatic proteins? J Lipid Res 59(8):1374–1382. https://doi.org/10.1194/jlr.r083139

Fang C, Wu S, Li Z, Pan S, Wu Y, An X, Long Y, Wei X, Wan X (2023) A systematic investigation of lipid transfer proteins involved in male fertility and other biological processes in maize. Int J Mol Sci 24(2):1660. https://doi.org/10.3390/ijms24021660

Ferrari S (2013) Oligogalacturonides: plant damage-associated molecular patterns and regulators of growth and development. Front Plant Sci 4. https://doi.org/10.3389/fpls.2013.00049

Finkina EI, Melnikova DN, Bogdanov IV, Ovchinnikova TV (2016, June 15) Lipid transfer proteins as components of the plant innate immune system: structure, functions, and applications. Acta Nat 8(2):47–61. https://doi.org/10.32607/20758251-2016-8-2-47-61

Gao H, Ma K, Ji G, Pan L, Zhou Q (2022, September 2) Lipid transfer proteins involved in plant–pathogen interactions and their molecular mechanisms. Mol Plant Pathol 23(12):1815–1829. https://doi.org/10.1111/mpp.13264

Garner CM, Spears BJ, Su J, Cseke LJ, Smith SN, Rogan CJ, Gassmann W (2021, February 23) Opposing functions of the plant TOPLESS gene family during SNC1-mediated autoimmunity. PLoS Genet 17(2):e1009026. https://doi.org/10.1371/journal.pgen.1009026
Ghosh P, Roychoudhury A (2023, December 23) Plant peptides involved in abiotic and biotic stress responses and Reactive Oxygen Species (ROS) signaling. J Plant Growth Regul. https://doi.org/10.1007/s00344-023-11194-7
Giroux MJ, Sripo T, Gerhardt S, Sherwood J (2003, December) Puroindolines: their role in grain hardness and plant defence. Biotechnol Genet Eng Rev 20(1):277–290. https://doi.org/10.1080/02648725.2003.10648047
Giudici M, Pascual R, de la Canal L, Pfüller K, Pfüller U, Villalaín J (2003, August) Interaction of viscotoxins A3 and B with membrane model systems: implications to their mechanism of action. Biophys J 85(2):971–981. https://doi.org/10.1016/s0006-3495(03)74536-6
Gonneau M, Desprez T, Martin M, Doblas VG, Bacete L, Miart F, Sormani R, Hématy K, Renou J, Landrein B, Murphy E, Van De Cotte B, Vernhettes S, De Smet I, Höfte H (2018) Receptor kinase THESEUS1 is a rapid alkalinization factor 34 receptor in Arabidopsis. Curr Biol 28(15):2452–2458.e4. https://doi.org/10.1016/j.cub.2018.05.075
Graham JS, Hall G, Pearce G, Ryan CA (1986) Regulation of synthesis of proteinase inhibitors I and II mRNAs in leaves of wounded tomato plants. Planta 169(3):399–405. https://doi.org/10.1007/BF00392137
Green TR, Ryan CA (1972) Wound-induced proteinase inhibitor in plant leaves: a possible defense mechanism against insects. Science (New York, NY) 175(4023):776–777. https://doi.org/10.1126/science.175.4023.776
Guillou MC, Balliau T, Vergne E, Canut H, Chourré J, Herrera-León C, Ramos-Martín F, Ahmadi-Afzadi M, D'Amelio N, Ruelland R, Zivy M, Renou JP, Jamet L, Aubourg S (2022, December 16) The PROSCOOP10 gene encodes two extracellular hydroxylated peptides and impacts flowering time in Arabidopsis. Plants. https://doi.org/10.3390/plants11243554
Gully K, Pelletier S, Guillou MC, Ferrand M, Aligon S, Pokotylo I, Perrin A, Vergne E, Fagard M, Ruelland E, Grappin P, Bucher E, Renou JP, Aubourg S (2019) The SCOOP12 peptide regulates defense response and root elongation in *Arabidopsis thaliana*. J Exp Bot 70(4):1349–1365. https://doi.org/10.1093/jxb/ery454
Han X, Zhou T, Hu XY, Zhu Y, Shi Z, Chen S, Liu Y, Weng X, Zhang F, Wu S (2023, December 17) Discovery and characterization of MaK: a novel knottin antimicrobial peptide from *Monochamus alternatus*. Int J Mol Sci. https://doi.org/10.3390/ijms242417565
Haney EF, Petersen AP, Lau CK, Jing W, Storey DG, Vogel HJ (2013) Mechanism of action of puroindoline derived tryptophan-rich antimicrobial peptides. Biochim Biophys Acta 1828(8):1802–1813. https://doi.org/10.1016/j.bbamem.2013.03.023
Hao G, Stover E, Gupta G (2016, July 22) Overexpression of a modified plant thionin enhances disease resistance to citrus canker and Huanglongbing (HLB). Front Plant Sci. https://doi.org/10.3389/fpls.2016.01078
Hayes BM, Bleackley MR, Wiltshire JL, Anderson MA, Traven A, van der Weerden NL (2013) Identification and mechanism of action of the plant defensin NaD1 as a new member of the antifungal drug arsenal against *Candida albicans*. Antimicrob Agents Chemother 57(8):3667–3675. https://doi.org/10.1128/AAC.00365-13
Hazak O, Hardtke CS (2016, June 23) CLAVATA 1-type receptors in plant development. J Exp Bot 67(16):4827–4833. https://doi.org/10.1093/jxb/erw247
He YH, Zhang ZR, Xu YP, Chen SY, Cai XZ (2022) Genome-wide identification of rapid alkalinization factor family in *Brassica napus* and functional analysis of BnRALF10 in immunity to *Sclerotinia sclerotiorum*. Front Plant Sci 13:877404. https://doi.org/10.3389/fpls.2022.877404
Heitz A, Avrutina O, Le-Nguyen D, Diederichsen U, Hernandez JF, Gracy J, Kolmar H, Chiche L (2008) Knottin cyclization: impact on structure and dynamics. BMC Struct Biol 8:54. https://doi.org/10.1186/1472-6807-8-54
Hellinger R, Muratspahić E, Devi S, Koehbach J, Vasileva M, Harvey PJ, Craik DJ, Gründemann C, Gruber CW (2021) Importance of the cyclic cystine knot structural motif for immunosuppressive effects of cyclotides. ACS Chem Biol 16(11):2373–2386. https://doi.org/10.1021/acschembio.1c00524

Hogg AC, Sripo T, Beecher B, Martin JM, Giroux MJ (2003, November 27) Wheat puroindolines interact to form friabilin and control wheat grain hardness. Theor Appl Genet 108(6):1089–1097. https://doi.org/10.1007/s00122-003-1518-3

Hu Z, Zhang H, Shi K (2018) Plant peptides in plant defense responses. Plant Signal Behav 13(8):e1475175. https://doi.org/10.1080/15592324.2018.1475175

Huang RH, Xiang Y, Liu XZ, Zhang Y, Hu Z, Wang DC (2002, May 22) Two novel antifungal peptides distinct with a five-disulfide motif from the bark of *Eucommia ulmoides* Oliv. FEBS Lett 521(1–3):87–90. https://doi.org/10.1016/s0014-5793(02)02829-6

Hwang B, Hwang JS, Lee J, Lee DG (2011, February) The antimicrobial peptide, psacotheasin induces reactive oxygen species and triggers apoptosis in *Candida albicans*. Biochem Biophys Res Commun 405(2):267–271. https://doi.org/10.1016/j.bbrc.2011.01.026

Iizasa E, Mitsutomi M, Nagano Y (2010, January) Direct binding of a plant LysM receptor-like kinase, LysM RLK1/CERK1, to chitin in vitro. J Biol Chem 285(5):2996–3004. https://doi.org/10.1074/jbc.m109.027540

Iqbal A, Khan RS (2023) Snakins: antimicrobial potential and prospects of genetic engineering for enhanced disease resistance in plants. Mol Biol Rep 50(10):8683–8690. https://doi.org/10.1007/s11033-023-08734-5

Iriti M, Vitalini S (2020, January 24) Sustainable crop protection, global climate change, food security and safety—plant immunity at the crossroads. Vaccines 8(1):42. https://doi.org/10.3390/vaccines8010042

Ishiguro S, Kawai-Oda A, Ueda J, Nishida I, Okada K (2001) The DEFECTIVE IN ANTHER DEHISCIENCE gene encodes a novel phospholipase A1 catalyzing the initial step of jasmonic acid biosynthesis, which synchronizes pollen maturation, anther dehiscence, and flower opening in Arabidopsis. Plant Cell 13(10):2191–2209. https://doi.org/10.1105/tpc.010192

Jakupi M, Demirbas S (2022, December 31) Production of secondary metabolites through elicitors: their application in agriculture. Tekirdag Namik Kemal Univ Inst Nat Appl Sci. https://doi.org/10.55848/jbst.2022.23

Järvå M, Lay FT, Phan TK, Humble C, Poon IKH, Bleackley MR, Anderson MA, Hulett MD, Kvansakul M (2018, May 17) X-ray structure of a carpet-like antimicrobial defensin–phospholipid membrane disruption complex. Nat Commun 9(1). https://doi.org/10.1038/s41467-018-04434-y

Jing W, Demcoe AR, Vogel HJ (2003) Conformation of a bactericidal domain of puroindoline a: structure and mechanism of action of a 13-residue antimicrobial peptide. J Bacteriol 185(16):4938–4947. https://doi.org/10.1128/JB.185.16.4938-4947.2003

Jones JDG, Dangl JL (2006, November) The plant immune system. Nature 444(7117):323–329. https://doi.org/10.1038/nature05286

Jones BL, Marinac LA (1999, December 31) Purification and partial characterization of a second cysteine proteinase inhibitor from ungerminated barley (*Hordeum vulgare* L.). J Agric Food Chem 48(2):257–264. https://doi.org/10.1021/jf9903556

Kaur S, Samota MK, Choudhary M. et al. (2022) How do plants defend themselves against pathogens-Biochemical mechanisms and genetic interventions. Physiol Mol Biol Plants 28, 485–504. https://doi.org/10.1007/s12298-022-01146-y

Kimura M, Park SS, Sakai R, Yamasaki N, Funatsu G (1997, January) Primary structure of 6.5k-arginine/glutamate-rich polypeptide from the seeds of sponge gourd (*Luffa cylindrica*). Biosci Biotechnol Biochem 61(6):984–988. https://doi.org/10.1271/bbb.61.984

Kini SG, Nguyen PQ, Weissbach S, Mallagaray A, Shin J, Yoon HS, Tam JP (2015) Studies on the chitin binding property of novel cysteine-rich peptides from *Alternanthera sessilis*. Biochemistry 54(43):6639–6649. https://doi.org/10.1021/acs.biochem.5b00872

Kouzai Y, Noutoshi Y, Inoue K, Shimizu M, Onda Y, Mochida K (2018) Benzothiadiazole, a plant defense inducer, negatively regulates sheath blight resistance in *Brachypodium distachyon*. Sci Rep 8(1):17358. https://doi.org/10.1038/s41598-018-35790-w

Kovaleva V, Bukhteeva I, Kit OY, Nesmelova IV (2020) Plant defensins from a structural perspective. Int J Mol Sci 21(15):5307. https://doi.org/10.3390/ijms21155307

Lacerda AF, De Vasconcelos RAR, Pelegrini PB, Grossi-de-Sá MF (2014, April 2) Antifungal defensins and their role in plant defense. Front Microbiol. https://doi.org/10.3389/fmicb.2014.00116

Lay FT, Schirra HJ, Scanlon MJ, Anderson MA, Craik DJ (2003) The three-dimensional solution structure of NaD1, a new floral defensin from *Nicotiana alata* and its application to a homology model of the crop defense protein alfAFP. J Mol Biol 325(1):175–188. https://doi.org/10.1016/s0022-2836(02)01103-8

Le Nguyen D, Heitz A, Chiche L, Castro B, Boigegrain RA, Favel A, Coletti-Previero MA (1990) Molecular recognition between serine proteases and new bioactive microproteins with a knotted structure. Biochimie 72(6–7):431–435. https://doi.org/10.1016/0300-9084(90)90067-q

Leng H, Jiang C, Song X, Lu M, Wan X (2021) Poplar aquaporin PIP1;1 promotes Arabidopsis growth and development. BMC Plant Biol 21(1):253. https://doi.org/10.1186/s12870-021-03017-2

Li J, Hu S, Jian W, Xie C, Yang X (2021) Plant antimicrobial peptides: structures, functions, and applications. Bot Stud 62(1):5. https://doi.org/10.1186/s40529-021-00312-x

Li Y, Di Q, Luo L, Yu L (2024, January 5) Phytosulfokine peptides, their receptors, and functions. Front Plant Sci 14. https://doi.org/10.3389/fpls.2023.1326964

Lindorff-Larsen K, Winther JR (2001) Surprisingly high stability of barley lipid transfer protein, LTP1, towards denaturant, heat and proteases. FEBS Lett 488(3):145–148. https://doi.org/10.1016/s0014-5793(00)02424-8

Liu F, Zhang X, Lu C, Zeng X, Li Y, Fu D, Wu G (2015, July 2) Non-specific lipid transfer proteins in plants: presenting new advances and an integrated functional analysis. J Exp Bot. https://doi.org/10.1093/jxb/erv313

Llanos P, Henriquez M, Minic J, Elmorjani K, Marion D, Riquelme G, Molg J, Benoit E (2003, October 15) Neuronal and muscular alterations caused by two wheat endosperm proteins, puroindoline-a and alpha1-purothionin, are due to ion pore formation. Eur Biophys J 33(3). https://doi.org/10.1007/s00249-003-0353-4

Lori M, van Verk MC, Hander T, Schatowitz H, Klauser D, Flury P, Gehring CA, Boller T, Bartels S (2015, May 22) Evolutionary divergence of the plant elicitor peptides (Peps) and their receptors: interfamily incompatibility of perception but compatibility of downstream signalling. J Exp Bot 66(17):5315–5325. https://doi.org/10.1093/jxb/erv236

Marcus JP, Green JL, Goulter KC, Manners JM (1999, September) A family of antimicrobial peptides is produced by processing of a 7S globulin protein in *Macadamia integrifolia* kernels. Plant J 19(6):699–710. https://doi.org/10.1046/j.1365-313x.1999.00569.x

Marion D, Bakan B, Elmorjani K (2007, March) Plant lipid binding proteins: properties and applications. Biotechnol Adv 25(2):195–197. https://doi.org/10.1016/j.biotechadv.2006.11.003

Matsubayashi Y, Sakagami Y (2006) Phytosulfokine. In: Handbook of biologically active peptides. Academic Press, pp 29–32. https://doi.org/10.1016/b978-012369442-3/50009-x

McGurl B, Pearce G, Orozco-Cardenas M, Ryan CA (1992) Structure, expression, and antisense inhibition of the systemin precursor gene. Science (New York, NY) 255(5051):1570–1573. https://doi.org/10.1126/science.1549783

McManus AM, Nielsen KJ, Marcus JP, Harrison SJ, Green JL, Manners JM, Craik DJ (1999, October) MiAMP1, a novel protein from *Macadamia integrifolia* adopts a greek key β-barrel fold unique amongst plant antimicrobial proteins. J Mol Biol 293(3):629–638. https://doi.org/10.1006/jmbi.1999.3163

Melnikova DN, Finkina EI, Bogdanov IV, Tagaev AA, Ovchinnikova TV (2022, December 20) Features and possible applications of plant lipid-binding and transfer proteins. Membranes (Basel). https://doi.org/10.3390/membranes13010002

Miller RN, Costa Alves GS, Van Sluys MA (2017) Plant immunity: unravelling the complexity of plant responses to biotic stresses. Ann Bot 119(5):681–687. https://doi.org/10.1093/aob/mcw284

Mishra AK, Sharma K, Misra RS (2012, June) Elicitor recognition, signal transduction and induced resistance in plants. J Plant Interact 7(2):95–120. https://doi.org/10.1080/17429145.2011.597517

Misra RC, Sandeep KM, Kumar S, Ghosh S (2016, May 6) A thaumatin-like protein of *Ocimum basilicum* confers tolerance to fungal pathogen and abiotic stress in transgenic Arabidopsis. Sci Rep 6(1). https://doi.org/10.1038/srep25340

Molesini B, Treggiari D, Dalbeni A, Minuz P, Pandolfini T (2017) Plant cystine-knot peptides: pharmacological perspectives. Br J Clin Pharmacol 83(1):63–70. https://doi.org/10.1111/bcp.12932

Molisso D, Coppola M, Buonanno M, Di Lelio I, Aprile AM, Langella E, Rigano MM, Francesca S, Chiaiese P, Palmieri G, Tatè R, Sinno M, Barra E, Becchimanzi A, Monti SM, Pennacchio F, Rao R (2022a, May 24) Not only systemin: prosystemin harbors other active regions able to protect tomato plants. Front Plant Sci. https://doi.org/10.3389/fpls.2022.887674

Molisso D, Coppola M, Buonanno M, Di Lelio I, Monti SM, Melchiorre C, Amoresano A, Corrado G, Delano-Frier JP, Becchimanzi A, Pennacchio F, Rao R (2022b) Tomato prosystemin is much more than a simple systemin precursor. Biology 11(1):124. https://doi.org/10.3390/biology11010124

Morris CF (2019, May 27) The antimicrobial properties of the puroindolines, a review. World J Microbiol Biotechnol 35(6). https://doi.org/10.1007/s11274-019-2655-4

Moyen C, Hammond-Kosack KE, Jones J, Knight MR, Johannes E (1998, November) Systemin triggers an increase of cytoplasmic calcium in tomato mesophyll cells: Ca^{2+} mobilization from intra- and extracellular compartments. Plant Cell Environ 21(11):1101–1111. https://doi.org/10.1046/j.1365-3040.1998.00378.x

Nahirñak V, Almasia NI, Hopp HE, Vazquez-Rovere C (2012) Snakin/GASA proteins: involvement in hormone crosstalk and redox homeostasis. Plant Signal Behav 7(8):1004–1008. https://doi.org/10.4161/psb.20813

Najafi J, Gjennestad RS, Kissen R, Brembu T, Bartosova Z, Winge P, Bones AM (2022, November 30) PAMP-induced secreted peptide-like 6 (PIPL6) functions as an amplifier of plant immune response through RLK7 and WRKY33 module. https://doi.org/10.1101/2022.11.30.518506

Nawrot R, Barylski J, Nowicki G, Broniarczyk J, Buchwald W, Goździcka-Józefiak A (2013, October 4) Plant antimicrobial peptides. Folia Microbiol. https://doi.org/10.1007/s12223-013-0280-4

Ng YM, Yang Y, Sze KH, Zhang X, Zheng YT, Shaw PC (2011, April) Structural characterization and anti-HIV-1 activities of arginine/glutamate-rich polypeptide Luffin P1 from the seeds of sponge gourd (*Luffa cylindrica*). J Struct Biol 174(1):164–172. https://doi.org/10.1016/j.jsb.2010.12.007

Nguyen PQ, Wang S, Kumar A, Yap LJ, Luu TT, Lescar J, Tam JP (2014) Discovery and characterization of pseudocyclic cystine-knot α-amylase inhibitors with high resistance to heat and proteolytic degradation. FEBS J 281(19):4351–4366. https://doi.org/10.1111/febs.12939

Nguyen QM, Iswanto ABB, Son GH, Kim SH (2021) Recent advances in effector-triggered immunity in plants: new pieces in the puzzle create a different paradigm. Int J Mol Sci 22(9):4709. https://doi.org/10.3390/ijms22094709

Nühse TS (2012) Cell wall integrity signaling and innate immunity in plants. Front Plant Sci 3. https://doi.org/10.3389/fpls.2012.00280

Odintsova TI, Slezina MP, Istomina EA (2018, September 27) Plant thionins: structure, biological functions and potential use in biotechnology. Vavilov J Genet Breed 22(6):667–675. https://doi.org/10.18699/vj18.409

Odintsova T, Shcherbakova L, Slezina M, Pasechnik T, Kartabaeva B, Istomina E, Dzhavakhiya V (2020) Hevein-like antimicrobial peptides wamps: structure-function relationship in antifungal activity and sensitization of plant pathogenic fungi to tebuconazole by WAMP-2-derived peptides. Int J Mol Sci 21(21):7912. https://doi.org/10.3390/ijms21217912

Oliveira-Lima M, Benko-Iseppon AM, Neto JRCF, Rodriguez-Decuadro S, Kido EA, Crovella S, Pandolfi V (2017) Snakin: structure, roles and applications of a plant antimicrobial peptide. Curr Protein Pept Sci 18(4):368–374. https://doi.org/10.2174/1389203717666160619183140

Oparin PB, Mineev KS, Dunaevsky YE, Arseniev AS, Belozersky MA, Grishin EV, Egorov TA, Vassilevski AA (2012) Buckwheat trypsin inhibitor with helical hairpin structure belongs to a new family of plant defence peptides. Biochem J 446(1):69–77. https://doi.org/10.1042/BJ20120548

Orozco-Cárdenas ML, Narváez-Vásquez J, Ryan CA (2001) Hydrogen peroxide acts as a second messenger for the induction of defense genes in tomato plants in response to wounding, systemin, and methyl jasmonate. Plant Cell 13(1):179–191

Ortiz-Morea FA, Savatin DV, Dejonghe W, Kumar R, Luo Y, Adamowski M, Van Den Begin J, Dressano K, De Oliveira GP, Zhao X, Lü Q, Madder A, Friml J, Moura DS, Russinova E (2016, September 20) Danger-associated peptide signaling in Arabidopsis requires clathrin. Proc Natl Acad Sci USA. https://doi.org/10.1073/pnas.1605588113

Pallaghy PK, Nielsen KJ, Craik DJ, Norton RS (1994) A common structural motif incorporating a cystine knot and a triple-stranded beta-sheet in toxic and inhibitory polypeptides. Protein Sci: A Publication of the Protein Society 3(10):1833–1839. https://doi.org/10.1002/pro.5560031022

Pappas PG (2011, August) The role of azoles in the treatment of invasive mycoses: review of the Infectious Diseases Society of America guidelines. Curr Opin Infect Dis 24:S1–S13. https://doi.org/10.1097/01.qco.0000399602.83515.ac

Pearce G, Strydom D, Johnson S, Ryan CA (1991) A polypeptide from tomato leaves induces wound-inducible proteinase inhibitor proteins. Science (New York, NY) 253(5022):895–897. https://doi.org/10.1126/science.253.5022.895

Pelegrini PB, Del Sarto RP, Silva ON, Franco OL, Grossi-de-Sá MF (2011, January 1) Antibacterial peptides from plants: what they are and how they probably work. Biochem Res Int. https://doi.org/10.1155/2011/250349

Phillips RL, Palombo EA, Panozzo JF, Bhave M (2011, January) Puroindolines, Pin alleles, hordoindolines and grain softness proteins are sources of bactericidal and fungicidal peptides. J Cereal Sci 53(1):112–117. https://doi.org/10.1016/j.jcs.2010.10.005

Poon IK, Baxter AA, Lay FT, Mills GD, Adda CG, Payne JA, Phan TK, Ryan GF, White JA, Veneer PK, van der Weerden NL, Anderson MA, Kvansakul M, Hulett MD (2014, April 1) Phosphoinositide-mediated oligomerization of a defensin induces cell lysis. elife 3. https://doi.org/10.7554/elife.01808

Porto WF, Franco OL (2013, June) Theoretical structural insights into the snakin/GASA family. Peptides 44:163–167. https://doi.org/10.1016/j.peptides.2013.03.014

Rodríguez-Decuadro S, Barraco-Vega M, Dans PD, Pandolfi V, Benko-Iseppon AM, Cecchetto G (2018, June 8) Antimicrobial and structural insights of a new snakin-like peptide isolated from *Peltophorum dubium* (Fabaceae). Amino Acids 50(9):1245–1259. https://doi.org/10.1007/s00726-018-2598-3

Rogozhin EA, Oshchepkova YI, Odintsova TI, Khadeeva NV, Veshkurova ON, Egorov TA, Grishin EV, Salikhov SI (2011, February) Novel antifungal defensins from Nigella sativa L. seeds. Plant Physiol Biochem 49(2):131–137. https://doi.org/10.1016/j.plaphy.2010.10.008

Rogozhin E, Zalevsky A, Mikov A, Smirnov A, Egorov T (2018, November 2) Characterization of hydroxyproline-containing hairpin-like antimicrobial peptide EcAMP1-Hyp from barnyard grass (*Echinochloa crusgalli* L.) seeds: structural identification and comparative analysis of antifungal activity. Int J Mol Sci 19(11):3449. https://doi.org/10.3390/ijms19113449

Rubinovich L, Weiss D (2010, November 4) The Arabidopsis cysteine-rich protein GASA4 promotes GA responses and exhibits redox activity in bacteria and in planta. Plant J 64(6):1018–1027. https://doi.org/10.1111/j.1365-313x.2010.04390.x

Ruiz C, Nadal A, Foix L, Montesinos L, Montesinos E, Pla M (2018, January 23) Diversity of plant defense elicitor peptides within the Rosaceae. BMC Genom Data. https://doi.org/10.1186/s12863-017-0593-4

Ruiz-García Y, Gómez-Plaza E (2013, January 25) Elicitors: a tool for improving fruit phenolic content. Agriculture 3(1):33–52. https://doi.org/10.3390/agriculture3010033

Ryan CA (2000) The systemin signaling pathway: differential activation of plant defensive genes. Biochim Biophys Acta 1477(1–2):112–121. https://doi.org/10.1016/s0167-4838(99)00269-1

Sagaram US, Pandurangi R, Kaur J, Smith TJ, Shah DM (2011) Structure-activity determinants in antifungal plant defensins MsDef1 and MtDef4 with different modes of action against *Fusarium graminearum*. PLoS One 6(4):e18550. https://doi.org/10.1371/journal.pone.0018550

Salminen TA, Blomqvist K, Edqvist J (2016) Lipid transfer proteins: classification, nomenclature, structure, and function. Planta 244(5):971–997. https://doi.org/10.1007/s00425-016-2585-4

Sandhoff K (2016, November) Neuronal sphingolipidoses: membrane lipids and sphingolipid activator proteins regulate lysosomal sphingolipid catabolism. Biochimie 130:146–151. https://doi.org/10.1016/j.biochi.2016.05.004

Santos-Silva CAD, Zupin L, Oliveira-Lima M, Vilela LMB, Bezerra-Neto JP, Ferreira-Neto JRC, Ferreira J, De Oliveira Silva RL, De Jesús Pires C, Aburjaile FF, De Oliveira MF, Kido DA, Crovella S, Benko-Iseppon AM (2020, January 1) Plant antimicrobial peptides: state of the art, in silico prediction and perspectives in the omics era. Bioinform Biol Insig. https://doi.org/10.1177/1177932220952739

Segura A, Moreno M, Madueño F, Molina A, García-Olmedo F (1999) Snakin-1, a peptide from potato that is active against plant pathogens. Mol Plant Microbe Interact: MPMI 12(1):16–23. https://doi.org/10.1094/MPMI.1999.12.1.16

Seidel T (2022, June 30) The plant V-ATPase. Front Plant Sci 13. https://doi.org/10.3389/fpls.2022.931777

Shen W, Liu J, Li J (2019, November 1) Type-II metacaspases mediate the processing of plant elicitor peptides in Arabidopsis. Mol Plant (Print). https://doi.org/10.1016/j.molp.2019.08.003

Sher Khan R, Iqbal A, Malak R, Shehryar K, Attia S, Ahmed T, Ali Khan M, Arif M, Mii M (2019) Plant defensins: types, mechanism of action and prospects of genetic engineering for enhanced disease resistance in plants. 3 Biotech 9(5):192. https://doi.org/10.1007/s13205-019-1725-5

Singh NK, Kumar KR, Kumar D, Shukla P, Kirti PB (2013) Characterization of a pathogen induced thaumatin-like protein gene AdTLP from *Arachis diogoi*, a wild peanut. PLoS One 8(12):e83963. https://doi.org/10.1371/journal.pone.0083963

Slavokhotova AA, Rogozhin EA (2020) Defense peptides from the α-Hairpinin family are components of plant innate immunity. Front Plant Sci 11:465. https://doi.org/10.3389/fpls.2020.00465

Slavokhotova AA, Naumann TA, Price NPJ, Rogozhin EA, Andreev YA, Vassilevski AA, Odintsova TI (2014a, September 24) Novel mode of action of plant defense peptides – hevein-like antimicrobial peptides from wheat inhibit fungal metalloproteases. FEBS J 281(20):4754–4764. https://doi.org/10.1111/febs.13015

Slavokhotova AA, Rogozhin EA, Musolyamov AK et al (2014b) Novel antifungal α-hairpinin peptide from *Stellaria media* seeds: structure, biosynthesis, gene structure and evolution. Plant Mol Biol 84:189–202. https://doi.org/10.1007/s11103-013-0127-z

Slezina MP, Odintsova TI (2023) Plant antimicrobial peptides: insights into structure-function relationships for practical applications. Curr Issues Mol Biol 45(4):3674–3704. https://doi.org/10.3390/cimb45040239

Stec B, Markman O, Rao U, Heffron G, Henderson S, Vernon L, Brumfeld V, Teeter M (2004, December) Proposal for molecular mechanism of thionins deduced from physico-chemical studies of plant toxins. J Pept Res 64(6):210–224. https://doi.org/10.1111/j.1399-3011.2004.00187.x

Stintzi A, Weber H, Reymond P, Browse J, Farmer EE (2001) Plant defense in the absence of jasmonic acid: the role of cyclopentenones. Proc Natl Acad Sci USA 98(22):12837–12842. https://doi.org/10.1073/pnas.211311098

Stotz HU, Thomson JG, Wang Y (2009) Plant defensins: defense, development and application. Plant Signal Behav 4(11):1010–1012. https://doi.org/10.4161/psb.4.11.9755

Struyfs C, Cools TL, De Cremer K, Sampaio-Marques B, Ludovico P, Wasko BM, Kaeberlein M, Cammue BPA, Thevissen K (2020) The antifungal plant defensin HsAFP1 induces autophagy, vacuolar dysfunction and cell cycle impairment in yeast. Biochim Biophys Acta Biomembr 1862(8):183255. https://doi.org/10.1016/j.bbamem.2020.183255

Su T, Han M, Cao D, Xu M (2020) Molecular and biological properties of snakins: the foremost cysteine-rich plant host defense peptides. J Fungi (Basel, Switzerland) 6(4):220. https://doi.org/10.3390/jof6040220

Sun JQ, Jiang HL, Li CY (2011) Systemin/Jasmonate-mediated systemic defense signaling in tomato. Mol Plant 4(4):607–615. https://doi.org/10.1093/mp/ssr008

Talukdar PK, Turner KL, Crockett TM, Lu X, Morris CF, Konkel ME (2021) Inhibitory effect of puroindoline peptides on Campylobacter jejuni growth and biofilm formation. Front Microbiol 12:702762. https://doi.org/10.3389/fmicb.2021.702762

Tam JP, Lu YA, Yang JL, Chiu KW (1999) An unusual structural motif of antimicrobial peptides containing end-to-end macrocycle and cystine-knot disulfides. Proc Natl Acad Sci USA 96(16):8913–8918. https://doi.org/10.1073/pnas.96.16.8913

Tam JP, Wang S, Wong KH, Tan W (2015, November 16) Antimicrobial peptides from plants. Pharmaceuticals 8(4):711–757. https://doi.org/10.3390/ph8040711

Taveira GB, Carvalho AO, Rodrigues R, Trindade FG, Da Cunha M, Gomes VM (2016) Thionin-like peptide from *Capsicum annuum* fruits: mechanism of action and synergism with fluconazole against Candida species. BMC Microbiol 16:12. https://doi.org/10.1186/s12866-016-0626-6

Terras FR, Schoofs HM, De Bolle MF, Van Leuven F, Rees SB, Vanderleyden J, Cammue BP, Broekaert WF (1992) Analysis of two novel classes of plant antifungal proteins from radish (*Raphanus sativus* L.) seeds. J Biol Chem 267(22):15301–15309

Tetorya M, Li H, Djami-Tchatchou AT, Buchko GW, Czymmek KJ, Shah DM (2023) Plant defensin MtDef4-derived antifungal peptide with multiple modes of action and potential as a bio-inspired fungicide. Mol Plant Pathol 24(8):896–913. https://doi.org/10.1111/mpp.13336

Thevissen K, Ghazi A, De Samblanx GW, Brownlee C, Osborn RW, Broekaert WF (1996, June) Fungal Membrane responses induced by plant defensins and thionins. J Biol Chem 271(25):15018–15025. https://doi.org/10.1074/jbc.271.25.15018

Thevissen K, François IE, Takemoto JY, Ferket KK, Meert EM, Cammue BP (2003) DmAMP1, an antifungal plant defensin from dahlia (Dahlia merckii), interacts with sphingolipids from *Saccharomyces cerevisiae*. FEMS Microbiol Lett 226(1):169–173. https://doi.org/10.1016/S0378-1097(03)00590-1

Thevissen K, Warnecke DC, François IE, Leipelt M, Heinz E, Ott C, Zähringer U, Thomma BP, Ferket KK, Cammue BP (2004) Defensins from insects and plants interact with fungal glucosylceramides. J Biol Chem 279(6):3900–3905. https://doi.org/10.1074/jbc.M311165200

Thevissen K, François IE, Sijtsma L, Amerongen AV, Schaaper WM, Meloen R, Posthuma-Trumpie T, Broekaert WF, Cammue BP (2005, July) Antifungal activity of synthetic peptides derived from *Impatiens balsamina* antimicrobial peptides Ib-AMP1 and Ib-AMP4. Peptides 26(7):1113–1119. https://doi.org/10.1016/j.peptides.2005.01.008

Weidmann J, Craik DJ (2016) Discovery, structure, function, and applications of cyclotides: circular proteins from plants. J Exp Bot 67(16):4801–4812. https://doi.org/10.1093/jxb/erw210

Wessels JGH (1988, March) A steady-state model for apical wall growth in fungi. Acta Bot Neerl 37(1):3–16. https://doi.org/10.1111/j.1438-8677.1988.tb01576.x

Wigoda N, Ben-Nissan G, Granot D, Schwartz A, Weiss D (2006, October 31) The gibberellin-induced, cysteine-rich protein GIP2 from *Petunia hybrida* exhibits in planta antioxidant activity. Plant J 48(5):796–805. https://doi.org/10.1111/j.1365-313x.2006.02917.x

Wong L, Levine T (2016, April 11) Lipid transfer proteins do their thing anchored at membrane contact sites … but what is their thing? Biochem Soc Trans 44(2):517–527. https://doi.org/10.1042/bst20150275

Wong LH, Gatta AT, Levine TP (2019) Lipid transfer proteins: the lipid commute via shuttles, bridges and tubes. Nat Rev Mol Cell Biol 20(2):85–101. https://doi.org/10.1038/s41580-018-0071-5

Yang X, Xiao Y, Wang X, Pei Y (2007) Expression of a novel small antimicrobial protein from the seeds of motherwort (*Leonurus japonicus*) confers disease resistance in tobacco. Appl Environ Microbiol 73(3):939–946. https://doi.org/10.1128/AEM.02016-06

Yu XS, Wang HR, Lei FF, Li RQ, Yao HP, Shen JB, Ain NU, Cai Y (2023, November 24) Structure and functional divergence of PIP peptide family revealed by functional studies on PIP1 and PIP2 in *Arabidopsis thaliana*. Front Plant Sci 14. https://doi.org/10.3389/fpls.2023.1208549

Zelman AK, Berkowitz GA (2023, August 3) Plant Elicitor Peptide (Pep) signaling and pathogen defense in tomato. Plants 12(15):2856. https://doi.org/10.3390/plants12152856

Zhang H, Hu Y (2017, November 6) Long-distance transport of prosystemin messenger RNA in tomato. Front Plant Sci. https://doi.org/10.3389/fpls.2017.01894
Zhang C, Chen H, Zhuang RR, Chen YT, Deng Y, Cai TC, Wang SY, Liu QZ, Tang RH, Shan SH, Pan RL, Chen LS, Zhuang WJ (2019) Overexpression of the peanut CLAVATA1-like leucine-rich repeat receptor-like kinase AhRLK1 confers increased resistance to bacterial wilt in tobacco. J Exp Bot 70(19):5407–5421. https://doi.org/10.1093/jxb/erz274
Zhang H, Zhang H, Lin J (2020, March 27) Systemin-mediated long-distance systemic defense responses. New Phytol (Print). https://doi.org/10.1111/nph.16495
Zhang J, Li Y, Bao Q, Wang H, Hou S (2022a) Plant elicitor peptide 1 fortifies root cell walls and triggers a systemic root-to-shoot immune signaling in Arabidopsis. Plant Signal Behav 17(1):2034270. https://doi.org/10.1080/15592324.2022.2034270
Zhang J, Zhao J, Yang Y, Bao Q, Li Y, Wang H, Hou S (2022b, May 12) EWR1 as a SCOOP peptide activates MIK2-dependent immunity in Arabidopsis. J Plant Interact 17(1):562–568. https://doi.org/10.1080/17429145.2022.2070292
Zimmermann R, Sakai H, Hochholdinger F (2010) The gibberellic acid stimulated-like gene family in maize and its role in lateral root development. Plant Physiol 152(1):356–365. https://doi.org/10.1104/pp.109.149054
Zipfel C, Robatzek S, Navarro L, Oakeley EJ, Jones JD, Felix G, Boller T (2004) Bacterial disease resistance in Arabidopsis through flagellin perception. Nature 428(6984):764–767. https://doi.org/10.1038/nature02485
Zottich U, Da Cunha M, Carvalho AO, Dias GB, Silva NC, Santos IS, do Nacimento VV, Miguel EC, Machado OL, Gomes VM (2011) Purification, biochemical characterization and antifungal activity of a new lipid transfer protein (LTP) from *Coffea canephora* seeds with α-amylase inhibitor properties. Biochim Biophys Acta 1810(4):375–383. https://doi.org/10.1016/j.bbagen.2010.12.002

4 Inceptin: Exploring Its Role as a Peptide Elicitor in Plant Defense Mechanisms

Sarika Sharma and Shachi Singh

Abstract

Plants have developed an intrinsic system to perceive and react to pathogenic microorganisms. A crucial aspect of plant immunity involves pattern recognition receptors (PRRs) that identify conserved microbial- or herbivore-associated molecular patterns (MAMPs or HAMPs, respectively). Inceptin has been identified as a potent elicitor, which elicits a wide range of defense responses in plants, including the production of reactive oxygen species (ROS), induction of defense-related genes, release of volatile organic compounds, and the accumulation of phytoalexins and other antimicrobial compounds. Inceptin recognition is mediated by specific receptor kinases present on the cell surface, initiating a complex signaling cascades which ultimately leads to the activation of defense responses. Recent research on inceptin has made it an attractive candidate for developing sustainable and environment-friendly crop protection strategy. This chapter summarizes the key features of inceptin based on previous investigations of receptor-mediated HAMP perception and signaling cascades. Furthermore, we also discuss the role of this peptide in underlying mechanisms in agriculture for enhancing plant defense responses and improving crop productivity.

Keywords

Plant immunity · MAMPS · HAMPS · Inceptin · Phytoalexins

S. Sharma · S. Singh (✉)
Department of Botany, MMV, Banaras Hindu University, Varanasi, India

S. Singh, R. Mehrotra (eds.), *Plant Elicitor Peptides*,
https://doi.org/10.1007/978-981-97-6374-0_4

4.1 Introduction

Plants have evolved a sophisticated system to detect and respond to a variety of pathogenic microorganisms. Plants possess a well-coordinated network of biochemical reactions that can be triggered in response to signals, leading to the activation of defense-related enzymes (Palani et al. 2016). A key component of plant immunity is the recognition of conserved microbial- or herbivore-associated molecular patterns (MAMPs or HAMPs, respectively) by pattern recognition receptors (PRRs) (Boutrot and Zipfel 2017). This recognition process must then initiate and coordinate defensive reprogramming such as the production of antimicrobial compounds, strengthening of cell walls, the hypersensitive response, and emission of volatile compounds (Singh 2014).

Recognition of pathogen-derived compounds and subsequent signaling that influence plant defenses through activating such responses are termed elicitation. It is widely acknowledged that a plant's response to herbivory attack often differ from that of even careful simulations of the mechanical wounding which mimics the damage caused by herbivore feeding (Mithöfer and Boland 2008). The insect specific elicitors, found in insect oral secretions (OS) or oviposition fluids, are frequently responsible for the specificity of the defense responses. These elicitors, originating from insects, encompass fatty acid derivatives, enzymes, as well as some other proteins and peptides (Bonaventure et al. 2011).

Volicitin was the first identified fatty amino acid-conjugate (FAC) elicitor isolated from OS of *Spodoptera exigua* larvae (Truitt et al. 2004). Volicitin has the ability to attract predators by inducing the release of volatile organic compounds in maize (Alborn et al. 1997). Besides FACs, sulfooxy fatty acids elicitor, Caeliferin A is found in the OS of grasshopper (Truitt et al. 2004). The oral secretion of insects also possesses proteins and peptides (as elicitor), which are subsequently injected into plant tissues and serve a diverse array of functions (Aljbory and Chen 2018). In lepidopteran OS, several types of elicitors have been identified such as glucose oxidase (GOX), isolated from *Helicoverpa zea* OS (Eichenseer et al. 1999; Musser et al. 2005), β-Glucosidase from the OS of white cabbage butterfly larvae elicits the production of volatiles from cabbage plants, these volatiles attract parasite wasps to feeding larvae (Mattiacci et al. 1995; Diezel et al. 2009). Additional types of enzyme elicitors include lipase and phospholipase C in OS of insects. Lipase enzyme found in the OS of the desert locust, *Schistocerca gregaria*, has been shown to enhance the production of oxylipins (Oxygenated fatty acid derivatives) in *Arabidopsis*. Similarly, the phospholipase C enzyme derived from *Spodoptera frugiperda* elicits the production of proteinase inhibitors in corn (*Zea mays*) plant (Acevedo et al. 2018).

In addition to these, one well studied potent peptide elicitor in plant defense responses is inceptin. Inceptin is a small peptide molecule cross-linked by disulfide bonds and found in the fall armyworm's OS (Schmelz et al. 2006). The origin of inceptin can be attributed to the early 2000s, during which researchers investigating plant–herbivore interaction, detected a newly recognized peptide derived from the insect-related peptide (IRP). This endogenous plant compound is released upon damage to plant tissues, typically during insect herbivory or pathogen attack and emerges a cascade of immune signaling pathways. Inceptin peptide was found to be highly effective elicitor of defense responses in various plant species, such as legumes, grasses and solanaceous crops (Schmelz et al. 2006; Huffaker et al. 2013).

Schmelz et al. (2007) specifically focused on the inceptin in two legumes: cowpea (*Vigna unguiculata*) and common bean (*Phaseolus vulgaris*). Their findings revealed that inceptin is a peptide elicitor derived from the chloroplastic ATP synthase γ-subunit of insect *Spodoptera frugiperda* (fall armyworm). Additionally, Schmelz et al. (2012) reported alongside *S. frugiperda*, the three highest accumulators of active inceptin related peptides were reported in *S. Exigua, Ostrinia nubilalis,* and *Heliothis zea*. Inceptins are transformed into plant elicitors through the action of digestive protease into the gut of insects (Shinya et al. 2016; Schmelz et al. 2006, 2007). Thereafter, it is delivered back into the plant tissues as a component of oral secretions. Moreover, inceptin can be recognized through HAMP using cell surface receptor, known as inceptin receptor (INR), unique to particular legume species that confers binding, signaling, and defense outputs during biotic stress (Steinbrenner et al. 2020). INR is leucine-rich repeat receptor like protein (LRR-RLP) that is specifically expressed in legumes. LRR-RLPs are a class of cell surface receptors that lack an intracellular kinase domain but have ability to perceiving signals. Thus, previous studies support that INR initiates downstream signaling cascades through the interaction of inceptin.

Recognition of inceptin by receptor INR through HAMP can lead to activation of plant defense responses. Previous studies have provided evidence that the scratch-wounding in *Vigna unguiculata* (cowpea) leaves did not activate the production of ethylene (ET), jasmonic acid (JA), salicylic acid (SA), as well as (E)-48-dimethyl-1,3,7-nonatriene (DMNT). In contrast, the interaction of inceptin led strong induction of ET, JA, and SA in plants, which lead activation of proteinase inhibitor (Schmelz et al. 2006, 2007). Along with inceptin, other elicitors such as volicitin, caeliferin A also elicit phytohormones. However, activities are idiosyncratically distributed across angiosperm diversity (Schmelz et al. 2009). Research indicates that inceptin induces volatile production in plants, thereby contributing to direct and indirect plant defenses against herbivores (Takemoto and Takabayashi 2012). Therefore, inceptin plays a crucial role as a herbivore-associated molecular pattern (HAMP) elicitor, triggering potent anti-herbivore defense responses in plants. In contrast, mere mechanical wounding or damage alone elicits comparatively weaker defense responses. This highlights the specificity of the plant's recognition system for herbivore-derived elicitors like inceptin. According to transcriptomic studies, INR signaling may amplify wound induced phosphorylation cascades. Alternatively, INR activation may initiate separate signaling pathways that converge on common transcriptional targets involved in defense gene expressions. Furthermore, the activation of INR by inceptin induce many cell surface receptors and therefore potentially recognition of damage associated molecular pattern (DAMP) (Steinbrenner et al. 2022). This leads to systemic immune response in plants against plant herbivory.

4.2 Structure and Perception of Inceptin

Inceptin is a short 11 amino acid peptide regulator of defence responses in *Vigna* and *Phaseolus* species of legume plant. The structure consisting the sequence +ICDINGVCVDA- (Fig. 4.1a) (Schmelz et al. 2007) derived from *Spodoptera frugiperda* in cowpea and common beans. The disulfide-bridge containing peptide is synthesized after the cleavage of the chloroplast ATP synthase γ-subunit inside insect gut during herbivory. At this point, sequence integrity is necessary for the regulatory action of the γ-subunit and cATPase activation arises once a disulfide bond between regulatory subunit and inhibitory Ɛ subunit. Inceptins peptide contains two cysteine residues, one which forms a disulfide bridge with the Ɛ-subunit of the chloroplastic ATP synthase complex.

Inceptins belong to a category of signaling molecules known as cryptides, which are peptide signals derived from proteins with unrelated functions, rather than originating from dedicated precursor propeptides. Inceptins represent the first identified cryptides in plants and are synthesized due to the gradual proteolysis from cATPase by digestive enzyme in insect gut and its effectiveness is recognized as an elicitor in OS. Consequently, while feeding, OS containing inceptins activate anti-herbivore defenses (Schmelz et al. 2007). Variation in the active inceptin peptide may arise due to the differences in cleavage sites and/or variation among plant species of cATPase γ-subunit. Insect larvae that feed exclusively on root tissues do not elicit inceptin mediated defense responses in plants. This observation can be attributed to the absence of cATPase in root tissues, which serves as the precursor protein from which inceptins are proteolytically derived (Schmelz et al. 2006).

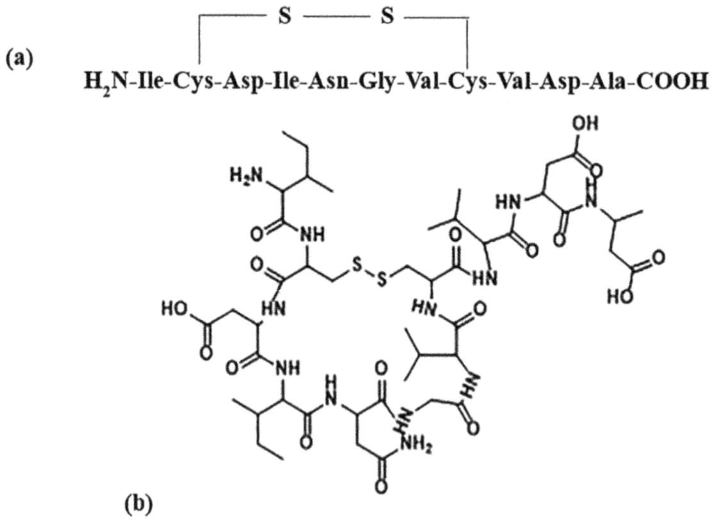

Fig. 4.1 Cowpea derived inceptin structure, (**a**) amino acid sequences of inceptin, (**b**) Chemical structure of inceptin (Jones et al. 2022)

Several forms of inceptins can be investigated in the larval oral secretion probably due to non-specific proteolysis. Nevertheless, the C-terminal alanine is indispensable for its activity. For instance, the substitution of lysine for the C-terminal alanine in the spinach, inceptin peptide led to its cleavage and consequent loss its ability to defense (Schmelz et al. 2007). Moreover, substituting alanine in the 11mer inceptin peptide reveals that Asp10 is essential for elicitor activity while substitution at Asp3 and Cys8 significantly diminish bioactivity. Interestingly, replacing Cys2, which bonds with Cys8, through disulfide bridge has minimal impact on elicitor activity. Therefore, cyclization results only for stability, not for signaling function. Hence, studies proved that the only substitutions, additions, or truncations abolished induced ET activity were within the last four amino acids of C terminus, which was insignificant in N-terminus (Schmelz et al. 2006, 2007). The oral secretion of fall armyworm (FAW) contained wild-type inceptin, *Vu-In* while other four inceptin related peptides have been identified, consisting of predominantly *Vu-In* and lesser amount of *Vn-* $^{E+}$*In* ($^{+}$EICDINGVCVDA^{-}), *Vu*-$^{GE+}$In ($^{+}$GEICDINGVCVDA^{-}), and *Vu*-In^{-A} ($^{+}$ICDINGVCVD^{-}). The C-terminal truncated *Vu*-In^{-A} lacked ET, SA, JA and DMNT inducing activity in contrast to other inceptin related peptides (Schmelz et al. 2007). In the study of Schmelz et al. (2012), the C-terminal truncated and inactive *Vu*-In^{-A} was found predominantly in the velvet bean caterpillar (VBC) OS in comparison to active *Vu-In*.

Structural studies have revealed that inceptin adopts a β-hairpin conformation, which is essential for its recognition. The recognition of inceptin by plant cell is mediated by specific cell surface receptor known as pattern recognition receptors (PRRs). In the model plant *Arabidopsis thaliana*, the PRR for inceptin has been identified as the leucine-rich repeat receptor kinase (LRR-RLK), Flagellin-Sensing 2 (FLS2) (Kouzai et al. 2014). Interestingly, FLS2 also function as the receptor for the bacterial PAMP, flagellin, suggesting that plants have evolved to recognize multiple PAMPs using a single receptor. Steinbrenner et al. (2020) elucidated the inceptin receptor, designated as INR, via forward genetic mapping. This receptor exclusive to legume species mediates the inceptin-induced response and enhanced defense response in tobacco. By identifying the INR in the *Phaseolinae* subtribe, which include important crop species like common bean and cowpea, Steinbrenner et al. (2020) have paved the way to exploit this receptor-mediated defense mechanisms in other related crop plants through genetic engineering or breeding approaches. The researcher approached to identify the physical and functional differences of INR in different crop plants based on (1) previous investigations of receptor-mediated HAMP perception and (2) research on genetic variations in HAMP sensitivity among accessions of given plant species (Wan et al. 2019). To characterize the genetic map underlying INR function, a series of cowpea (*Vigna unguiculata*) accessions was subjected to treatment with variant forms of the inceptin peptide. These variant forms of inceptin exhibited attenuated biological activity as compared to the wild-type molecule. One of the variant of inceptin is C-terminal truncated inceptin (+ICDINGVCVD−) or and *Vu*-In^{-A}, found in the OS of a legume specialist herbivore (*Anticarsia gemmatalis*), lacked ET, SA, JA, and DMNT inducing activity in contrast to other inceptin related peptides (Schmelz et al. 2007). These changes uncovered remarkable variations in defense response within cowpea germplasm, which was unexpectedly not observable upon treatment with the wild-type inceptin peptide. A cross between inceptin sensitive and insensitive cowpea lines was subsequently performed to develop a recombinant inbred line (RIL) population, which enabled quantitative trait loci (QTL) mapping to delineate the genomic regions associated with differential inceptin sensitivity. Consequently, these approaches led to the discovery of a single genetic locus associated with inceptin sensitivity. Thus, the researchers identified the leucine-rich repeat (LRR)-receptor like protein (RLP) INR, encoded by *INR* gene and successfully conferred the inceptin sensitivity to *Nicotiana benthamiana*.

In the recent studies, the recognition of inceptin by the receptor INR, which belongs to the LRR-RK family (Ngou et al. 2021), exhibits broad distribution across monocot and dicot plants as elucidated by Huffaker et al. 2006. Additionally, LRR-RLP inceptin receptor (INR) demonstrates specificity toward the legume tribe Phaseoleae (Schmelz et al. 2006, 2007, 2012; Snoeck et al. 2022). Expression of INR in non-native species such as *Nicotiana benthamiana* (tobacco) permits inceptin-induced association with co-receptors SERK/BAK1 and SOBIR1. Upon binding of inceptin, INR undergoes homo-dimerization and phosphorylation, leading to the activation of downstream signaling cascades. This includes the production of reactive oxygen species, calcium signaling, activation of mitogen-activated protein kinase (MAPKs), and transcriptional reprogramming to induce defense-related genes and results in reduced insect growth rate (Yan et al. 2012; Boller and Felix 2009).

4.3 Mechanism of Signal Transduction

In order to detect insects attack, plants must possess the ability to sense chemical signals produced by the attackers. This involves the recognition of herbivores associated elicitors, generated by insect, followed by the activation of signaling mechanisms upon their perception. The understanding of the molecular constituents involved in how plants recognize elicitors and translate this recognition into downstream signaling events, ultimately activating specific physiological processes, remains limited. Some studies revealed that elicitors bind to its receptor found in plasma membrane (Truitt et al. 2004), while others have shown that plant responses to insect herbivory is due to changes in ion fluxes across plasma membrane such as calcium ions, activation of protein kinases, and generation of reactive oxygen species (ROS) (Maffei et al. 2004). Initially, in the model plant *Arabidopsis thaliana*, the receptor for peptide signaling was based on Flagellin-Sensing 2 (FLS2) which is activated by the peptide flagellin (flg22) (Kouzai et al. 2014). Interestingly, the first identified endogenous peptide signal was the *Arabidopsis thaliana* Pep1 (AtPep1) to regulate plant antimicrobial defense and communicates with either one of two LRR-RLKs, AtPEPR1 and AtPEPR2 (Yamaguchi et al. 2006, 2010). The peptides (Peps) systemin and inceptin signal transduction in *Arabidopsis* has similar signaling patterns as AtPep1 and flg22 signaling (Chen and Mao 2020). Accordingly, upon interacting with their receptors, AtPeps function as signal inducer similar to systemin and triggers the production of salicylic acid (SA) along with jasmonic acid (JA) and ethylene (ET) and other secondary messengers such as ROS, calcium influx, and nitric oxide (NO) (Huffaker et al. 2006; Bartels et al. 2013) Moreover, these signals trigger defense genes and production of secondary metabolites (Huffaker et al. 2006; Bartels et al. 2013; Yamaguchi and Huffaker 2011).

Defense eliciting HAMPs, inceptins, present during caterpillar herbivory on cowpea (*Vigna unguiculata*) is termed as *Vu*-In. When chloroplastic ATP synthase γ-subunits are ingested into gut of insect, the proteins are converted into inceptin through proteolysis. The oral secretion of insect is delivered back into plant tissue containing inceptin along with glucose oxidase (GOX). Released inceptin interacts with the cell surface localized leucine-rich repeat (LRR-RLP) receptor, recognized as inceptin receptor (INR). This receptor differs from LRR-RLKs in lack of an intracellular kinase domain. Because of this kinase deficiency, LRR type surface receptors strongly associated with somatic embryogenesis receptor kinase (SERK) co-receptors. Additionally, RLPs constitutively associate with the adapter RK suppressor of BIR1 (SOBIR1) for signal transduction (Fig. 4.2) (Steinbrenner et al. 2020). The extracellular domain of INR and SERK have been crystallized with inceptin indicate that binding of inceptin stabilizes the interaction of INR with SERK like AtPep1 (Huffaker 2015). The binding of inceptin to receptor complex activates several intracellular signaling pathways that converge to elicit plant defense responses.

One of the earliest responses is the production of reactive oxygen species (ROS), such as superoxide and hydrogen peroxide (H_2O_2) after phosphorylation of co-receptor, which can directly inhibit pathogen growth and act as signaling molecules

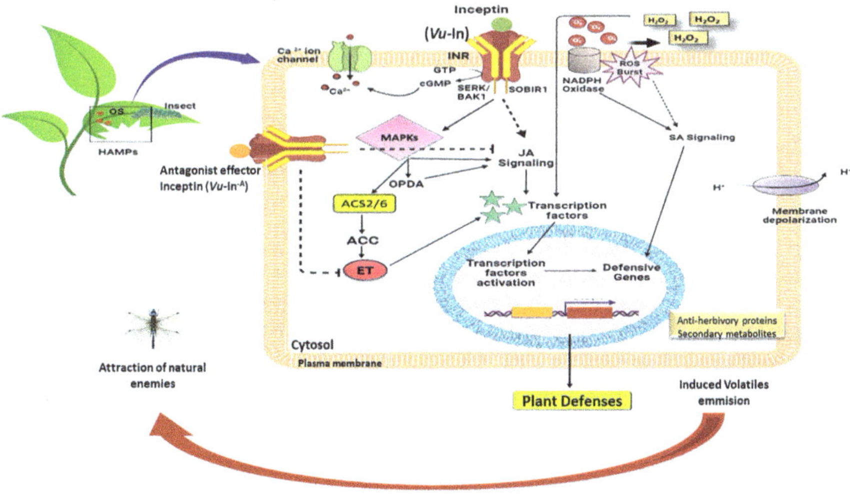

Fig. 4.2 Conceptual schematic signaling pathway of inceptin peptide and role of oral secretions (OS) in the early activation of plant defenses

to amplify the defense response. The ROS burst is mediated upon the phosphorylation of plasma membrane localized NADPH oxidases which is effectively induced by either inceptin or GOX (Kadota et al. 2014). The production of H_2O_2 triggers the SA signaling pathway directly. On the other side, INR contain a cytosolic guanylyl cyclase domain, capable of producing cGMP. It has been nevertheless been proposed that guanylyl cyclase activity of receptor may produce enough cGMP to activate the plasma membrane located cyclic nucleotide-gated Ca^{2+} channel CNGC2 and promote the influx of Ca^{2+} in cytosol (Qi et al. 2010). Moreover, inceptin perception also triggers the activation of mitogen-activated protein kinase (MAPK) cascades, which induce the various pathways such as SA, ET, and JA (Fig. 4.2) (Wasternack and Hause 2013). Consequently, this lead to the phosphorylation and activation of numerous transcription factors that regulate the expression of defense-related genes (Meng and Zhang 2013). These defense genes encode a variety of antimicrobial proteins, such as pathogenesis-related (PR) proteins, as well as enzymes involved in the biosynthesis of secondary metabolites with anti-herbivory properties. In addition to transcriptional changes, inceptin signaling also leads to the reinforcement of plant cell walls through the deposition of callose, a β-1,3-glucan polymer (Huffaker and Ryan 2007). This physical barrier can prevent or slow the penetration of herbivores into plant tissues. Moreover, the defense responses elicited by inceptin perception are not limited to the initially challenged cells but can also spread to neighboring cells and tissues, a process known as systemic acquired resistance. This allows the plant to mount a comprehensive defense against the invading pathogen.

As shown in Fig. 4.2, inceptin (*Vu*-In) and inceptin related peptide (*Vu*-In^{-A}) occurring in many lepidoptera larvae, interact with putative HAMPs receptors, inceptin receptor (INR). INR located on the plant cell surface and recognize the small peptide (*Vu*-In or *Vu*-In^{-A}). Upon binding of peptide to INR, peptide stabilizes interactions with co-receptors SERK/BAK1 and SOBIR1 which results in phosphorylation of co-receptors. Thus, in turn phosphorylation of NADPH Oxidase to activate superoxide formation leading to oxidative burst of reactive oxygen species (ROS). ROS further contributes to H_2O_2 production and elevated salicylic acid (SA) signaling. INR are also hypothesized to possess guanylyl cyclase activity, potentially producing cGMP that could activate calcium channel, leading to an influx of calcium ions into the cytosol. Binding of *Vu*-In positively regulate the mitogen-activated protein kinases (MAPKs) signaling and depolarized the membrane potential. MAPKs act as positive regulator of jasmonate (JA) and ethylene (ET) accumulation and signaling. JA can be synthesized direct or indirect (through precursor 12-oxophytodienoic acid, OPD) MAPKs pathway. Together these signals events, SA, JA, and ET signaling, facilitate transcriptional reprogramming, leading to the accumulation of defense genes. These defensive genes are responsible for the production of anti-herbivory proteins, secondary metabolites, and volatile compounds that can facilitate attraction of natural enemies of herbivory. In contrast, *Vu*-In^{-A} has antagonistic function and suppresses the phytohormone productions.

4.4 Contribution to Local and Systemic Immunity

Plants and insects have evolved intricate mechanism for adaptation. Plants can recognize specific peptide fragments or elicitors through HAMPs or MAMPs and induce the defense mechanism against insect herbivory. Depolarization of the plasma transmembrane potential, alterations in cytosolic Ca^{+2} levels, reactive oxygen species (ROS) burst, and activation of mitogen-activated protein kinase (MAPK) are some of the early defense responses that arise upon plant perception (Chen and Mao 2020). Recent studies reveal that these reactions are able to induce JA, SA, and ET pathways (Wasternack and Hause 2013). As a result of the activation of signaling pathways, various genes, enzymes, and pathogenesis-related (PR) proteins are expressed, leading to the production and accumulation of defensive compounds. These compounds include proteinase inhibitors, toxic secondary metabolites (such as alkaloids, terpenoids, etc.) as well as volatiles that attract natural enemies of herbivores (Schmelz et al. 2009).

One study revealed that the comparison of foliar volatile profile with those emitted in response to herbivory in maize after interaction of ZmPep3, which could recapitulate all observed volatiles except green leafy volatiles. The application of ZmPep3 damage little bit but does mimic the volatile organic compounds (VOC) inducing activity of OS (Huffaker et al. 2013). Thus, ZmPep3 was a potent peptide elicitor which exhibited direct or indirect anti-herbivore defenses such as regulating phytohormone biosynthesis, accumulation of transcripts, and metabolites production. Furthermore, the application of ZmPep3 to maize leaves was adequate to inhibit the growth of herbivore, *S. exigua* larvae. Additionally, the treated leaves attracted the naive parasitoid wasp *Cotesia marginiventris*, a natural enemy that parasitizes Lepidopteran larvae (Huffaker et al. 2013).

The binding of inceptin to the receptor INR triggers a vigorous defense responses in plants, characterized by the production of phytoalexins, reinforcement of cell wall, and the hypersensitive responses, effectively limiting the proliferation of pathogen and herbivory (Kaku et al. 2006; Wu and Baldwin 2010). Studies have shown that synthetic inceptin can induce the production of ethylene (ET), JA, SA, volatile emission, cinnamic acid, and transcripts of cystatin (a protease inhibitor) in cowpea plants (Schmelz et al. 2006, 2009), as well as increase and pools of predominant volatile (*E*)-4,8-dimethyl-1,3,7-nonatriene (DMNT) (Schmelz et al. 2006, 2007, 2012; Carroll et al. 2008). Despite the fact that fatty acid-amide conjugates (FACs) from *S. frugiperda* OS induce anti-herbivore defense responses in certain plant species, these compounds are ineffective at triggering such defenses in cowpea plants.

Inceptins were identified as elicitor responsible for induced ethylene production and indicating that inceptin induce both direct and indirect defenses. As candidate for direct defenses, inceptin also strongly influenced the accumulation of specific Peroxidase (POX) transcript, upregulated gene expression of a cystatin protease inhibitor and stimulated the production of cinnamic acid, which is a precursor of phenylpropanoid defense metabolites in plants (Schmelz et al. 2006; Steinbrenner et al. 2022). In addition to the production of metabolites, inceptin triggers indirect defense responses against herbivory including the production of volatile organic compounds such as terpenes as well as indole and methyl salicylate (Schmelz et al. 2006). These plant volatiles released in response to herbivory serve to recruit natural enemies of *S. frugiperda*, facilitating an indirect defense against pest.

Certainly, the slower growth rate and reduced biomass of larvae after consuming on inceptin pre-treated cowpea leaves showed that inceptin play a pivotal role at onset of these signaling cascades to initiate diverse responses. On the other hand, the induced volatile secretion facilitates the attraction of natural enemies of insect herbivory, also regulates the defense response in plant (Schmelz et al. 2006).

4.5 Conclusion

Inceptin is a potent peptide elicitor derived from insect herbivores that triggers robust defense responses in plants upon recognition by INR receptor. As a potent elicitor of defense responses in plants, inceptin has been shown to confer broad spectrum protection against a wide range of biotic and abiotic stresses, including pathogens, pests, and herbivores. Elucidating the detailed structural features, its perception mechanisms by INR receptor and the associated downstream signaling has provided valuable insights into plant innate immunity pathways. Notably, the identification of genetic determinants and regulatory networks involved in inceptin responsiveness has paved the way for targeted breeding strategies and genetic engineering approaches to new avenues for agricultural applications such as to develop crop varieties with enhanced stress tolerance and resilience. In summary, inceptin represents an effective peptide elicitor, originating from herbivor-induced plant response. Through the molecular understanding of structure of inceptin and its receptor INR, and mechanism of signaling unveiled new avenue for crop protection, yield enhancement and development of environmentally friendly agricultural practices. Inceptin research represents an exciting frontier with significant implications for sustainable agriculture.

4.6 Future Perspectives

Recent research has shed light on the peptide elicitor inceptin and the mechanisms through which plants may recognize herbivores and subsequently activate defense responses. Potential areas for future efforts that could lead to major breakthroughs are related to elucidating the molecular mechanisms underlying inceptin perception and signal transduction in different crop species. The characterized and identified variant forms of inceptin opens avenues for identifying of compatible binding proteins and plant genes encoding their HAMP receptors. The future research challenge in this field will demonstrate that these binding proteins are indeed genuine receptors capable of initiating signal transduction pathways and subsequent regulation of the defense mechanisms upon activation. Additionally, there is a need to discern potential constituents of signal transduction pathways through biochemical and genetic approaches, examining specific signal perception mechanisms.

Furthermore, the potential of inceptin as a biocompatible and biodegradable alternative to conventional chemical pesticides warrants exploration. Inceptin-based formulations could be developed for foliar application or seed treatment, providing a more sustainable approach to crop protection. Continued research efforts should also focus on optimizing inceptin production and delivery methods, ensuring cost effectiveness, and scalability for commercial applications. Collaboration between academia, industry, and regulatory bodies will be crucial in navigating the regulatory framework and facilitating the commercialization of inceptin-based products.

References

Acevedo FE, Peiffer M, Ray S, Meagher R, Luthe DS, Felton GW (2018) Intraspecific differences in plant defense induction by fall armyworm strains. New Phytol 218(1):310–321. https://doi.org/10.1111/nph.14981

Alborn HT, Turlings TCJ, Jones TH et al (1997) An elicitor of plant volatiles from beet armyworm oral secretion. Science 276(5314):945–949. https://doi.org/10.1126/science.276.5314.945

Aljbory Z, Chen MS (2018) Indirect plant defense against insect herbivores: a review. Insect Sci 25(1):2–23. https://doi.org/10.1111/1744-7917.12436

Bartels S, Lori M, Mbengue M, van Verk M, Klauser D, Hander T, Böni R, Robatzek S, Boller T (2013) The family of Peps and their precursors in Arabidopsis: differential expression and localization but similar induction of pattern-triggered immune responses. J Exp Bot 64(17):5309–5321. https://doi.org/10.1093/jxb/ert330

Boller T, Felix G (2009) A renaissance of elicitors: perception of microbe-associated molecular patterns and danger signals by pattern-recognition receptors. Annu Rev Plant Biol 60:379–406. https://doi.org/10.1146/annurev.arplant.57.032905.105346

Bonaventure G, VanDoorn A, Baldwin IT (2011) Herbivore-associated elicitors: FAC signaling and metabolism. Trends Plant Sci 16(6):294–299. https://doi.org/10.1016/j.tplants.2011.01.006

Boutrot F, Zipfel C (2017) Function, discovery, and exploitation of plant pattern recognition receptors for broad-spectrum disease resistance. Annu Rev Phytopathol 55:257–286. https://doi.org/10.1146/annurev-phyto-080614-120106

Carroll MJ, Schmelz EA, Teal PE (2008) The attraction of Spodoptera frugiperda neonates to cowpea seedlings is mediated by volatiles induced by conspecific herbivory and the elicitor inceptin. J Chem Ecol 34:291–300. https://doi.org/10.1007/s10886-007-9414-y

Chen CY, Mao YB (2020) Research advances in plant-insect molecular interaction. F1000Res 9:F1000 Faculty Rev-198. https://doi.org/10.12688/f1000research.21502.1

Diezel C, von Dahl CC, Gaquerel E, Baldwin IT (2009) Different lepidopteran elicitors account for cross-talk in herbivoryinduced phytohormone signaling. Plant Physiol 150(3):1576–1586. https://doi.org/10.1104/pp.109.139550

Eichenseer H, Mathews MC, Bi JL, Murphy JB, Felton GW (1999) Salivary glucose oxidase: multifunctional roles for helicoverpa zea? Arch Insect Biochem Physiol 42(1):99–109. https://doi.org/10.1002/(SICI)1520-6327(199909)42:1<99::AID-ARCH10>3.0.CO;2-B

Huffaker A (2015) Plant elicitor peptides in induced defense against insects. Curr Opin Insect Sci 9:44–50. https://doi.org/10.1016/j.cois.2015.06.003

Huffaker A, Ryan CA (2007) Endogenous peptide defense signals in Arabidopsis differentially amplify signaling for the innate immune response. Proc Natl Acad Sci USA 104(25):10732–10736. https://doi.org/10.1073/pnas.0703343104

Huffaker A, Pearce G, Ryan CA (2006) An endogenous peptide signal in Arabidopsis activates components of the innate immune response. Proc Natl Acad Sci USA 103(26):10098–100103. https://doi.org/10.1073/pnas.0603727103

Huffaker A, Pearce G, Veyrat N, Erb M, Turlings TC, Sartor R, Shen Z, Briggs SP, Vaughan MM, Huffaker A, Pearce G, Veyrat N, Erb M, Turlings TC, Sartor R, Shen Z, Briggs SP, Vaughan MM, Alborn HT, Teal PE, Schmelz EA (2013) Plant elicitor peptides are conserved signals regulating direct and indirect antiherbivore defense. Proc Natl Acad Sci USA 110(14):5707–5712. https://doi.org/10.1073/pnas.1214668110

Jones AC, Felton GW, Tumlinson JH (2022) The dual function of elicitors and effectors from insects: reviewing the 'arms race' against plant defenses. Plant Mol Biol 109:427–445. https://doi.org/10.1007/s11103-021-01203-2

Kadota Y, Sklenar J, Derbyshire P, Stransfeld L, Asai S, Ntoukakis V, Jones JD, Shirasu K, Menke F, Jones A, Zipfel C (2014) Direct regulation of the NADPH oxidase RBOHD by the PRR-associated kinase BIK1 during plant immunity. Mol Cell 54(1):43–55. https://doi.org/10.1016/j.molcel.2014.02.021

Kaku H, Nishizawa Y, Ishii-Minami N, Akimoto-Tomiyama C, Dohmae N, Takio K, Minami E, Shibuya N (2006) Plant cells recognize chitin fragments for defense signaling through a plasma membrane receptor. Proc Natl Acad Sci USA 103(29):11086–11091. https://doi.org/10.1073/pnas.0508882103

Kouzai Y, Mochizuki S, Nakajima K, Desaki Y, Hayafune M, Miyazaki H, Yokotani N, Ozawa K, Minami E, Kaku H, Shibuya N, Nishizawa Y (2014) Targeted gene disruption of OsCERK1 reveals its indispensable role in chitin perception and involvement in the peptidoglycan response and immunity in rice. Mol Plant Microbe Interact 27(9):975–982. https://doi.org/10.1094/MPMI-03-14-0068-R

Maffei M, Bossi S, Spiteller D, Mithöfer A, Boland W (2004) Effects of feeding Spodoptera littoralis on lima bean leaves. I. Membrane potentials, intracellular calcium variations, oral secretions, and regurgitate components. Plant Physiol 134(4):1752–1762. https://doi.org/10.1104/pp.103.034165

Mattiacci L, Dicke M, Posthumus MA (1995) beta-Glucosidase: an elicitor of herbivore-induced plant odor that attracts host-searching parasitic wasps. Proc Natl Acad Sci USA 92(6):2036–2040. https://doi.org/10.1073/pnas.92.6.2036

Meng X, Zhang S (2013) MAPK cascades in plant disease resistance signaling. Annu Rev Phytopathol 51:245–266. https://doi.org/10.1146/annurev-phyto-082712-102314

Mithöfer A, Boland W (2008) Recognition of herbivory-associated molecular patterns. Plant Physiol 146(3):825–831. https://doi.org/10.1104/pp.107.113118

Musser RO, Cipollini DF, Hum-Musser SM, Williams SA, Brown JK, Felton GW (2005) Evidence that the caterpillar salivary enzyme glucose oxidase provides herbivore offense in solanaceous plants. Arch Insect Biochem Physiol 58(2):128–137. https://doi.org/10.1002/arch.20039

Ngou BPM, Ahn HK, Ding P, Jones JDG (2021) Mutual potentiation of plant immunity by cell-surface and intracellular receptors. Nature 592(7852):110–115. https://doi.org/10.1038/s41586-021-03315-7

Palani NA, Seethapathy PA, Jeyaraman RA, Kathaperumal AR, Shanmugam VA (2016) Systemic elicitation of defense related enzymes suppressing Fusarium wilt in mulberry (Morus spp.). Afr J Microbiol Res 10(22):813–819. https://doi.org/10.5897/AJMR2015.7900

Qi Z, Verma R, Gehring C, Yamaguchi Y, Zhao Y, Ryan CA, Berkowitz GA (2010) Ca2+ signaling by plant Arabidopsis thaliana Pep peptides depends on AtPepR1, a receptor with guanylyl cyclase activity, and cGMP-activated Ca2+ channels. Proc Natl Acad Sci USA 107(49):21193–21198. https://doi.org/10.1073/pnas.1000191107

Schmelz EA, Carroll MJ, LeClere S, Phipps SM, Meredith J, Chourey PS, Alborn HT, Teal PE (2006) Fragments of ATP synthase mediate plant perception of insect attack. Proc Natl Acad Sci USA 103(23):8894–8899. https://doi.org/10.1073/pnas.0602328103

Schmelz EA, LeClere S, Carroll MJ, Alborn HT, Teal PE (2007) Cowpea chloroplastic ATP synthase is the source of multiple plant defense elicitors during insect herbivory. Plant Physiol 144(2):793–805. https://doi.org/10.1104/pp.107.097154

Schmelz EA, Engelberth J, Alborn HT, Tumlinson JH III, Teal PE (2009) Phytohormone-based activity mapping of insect herbivore-produced elicitors. Proc Natl Acad Sci 106(2):653–657. https://doi.org/10.1073/pnas.0811861106

Schmelz EA, Huffaker A, Carroll MJ, Alborn HT, Ali JG, Teal PE (2012) An amino acid substitution inhibits specialist herbivore production of an antagonist effector and recovers insect-induced plant defenses. Plant Physiol 160(3):1468–1478. https://doi.org/10.1104/pp.112.201061

Shinya T, Hojo Y, Desaki Y, Christeller JT, Okada K, Shibuya N, Galis I (2016) Modulation of plant defense responses to herbivores by simultaneous recognition of different herbivore-associated elicitors in rice. Sci Rep 6:32537. https://doi.org/10.1038/srep32537

Singh S (2014) A review on possible elicitor molecules of cyanobacteria: their role in improving plant growth and providing tolerance against biotic or abiotic stress. J Appl Microbiol 117(5):1221–1244. https://doi.org/10.1111/jam.12612

Snoeck S, Abramson BW, Garcia AGK, Egan AN, Michael TP, Steinbrenner AD (2022) Evolutionary gain and loss of a plant pattern-recognition receptor for HAMP recognition. Elife 11:e81050. https://doi.org/10.7554/eLife.81050

Steinbrenner AD, Muñoz-Amatriaín M, Chaparro AF, Aguilar-Venegas JM, Lo S, Okuda S, Glauser G, Dongiovanni J, Shi D, Hall M, Crubaugh D, Holton N, Zipfel C, Abagyan R, Turlings TCJ, Close TJ, Huffaker A, Schmelz EA (2020) A receptor-like protein mediates plant immune responses to herbivore-associated molecular patterns. Proc Natl Acad Sci USA 117(49):31510–31518. https://doi.org/10.1073/pnas.2018415117

Steinbrenner AD, Saldivar E, Hodges N, Guayazán-Palacios N, Chaparro AF, Schmelz EA (2022) Signatures of plant defense response specificity mediated by herbivore-associated molecular patterns in legumes. Plant J 110(5):1255–1270. https://doi.org/10.1111/tpj.15732

Takemoto H, Takabayashi J (2012) Exogenous application of liquid diet, previously fed upon by pea aphids Acyrthosiphon pisum (Harris), to broad bean leaves induces volatiles attractive to the specialist parasitic wasp Aphidius ervi (Haliday). J Plant Interact 7(1):78–83. https://doi.org/10.1080/17429145.2011.625475

Truitt CL, Wei HX, Paré PW (2004) A plasma membrane protein from Zea mays binds with the herbivore elicitor volicitin. Plant Cell 16(2):523–532. https://doi.org/10.1105/tpc.017723

Wan WL, Fröhlich K, Pruitt RN, Nürnberger T, Zhang L (2019) Plant cell surface immune receptor complex signaling. Curr Opin Plant Biol 50:18–28. https://doi.org/10.1016/j.pbi.2019.02.001

Wasternack C, Hause B (2013) Jasmonates: biosynthesis, perception, signal transduction and action in plant stress response, growth and development. An update to the 2007 review in Annals of Botany. Ann Bot 111(6):1021–1058. https://doi.org/10.1093/aob/mct067

Wu J, Baldwin IT (2010) New insights into plant responses to the attack from insect herbivores. Annu Rev Genet 44:1–24. https://doi.org/10.1146/annurev-genet-102209-163500

Yamaguchi Y, Huffaker A (2011) Endogenous peptide elicitors in higher plants. Curr Opin Plant Biol 14(4):351–357. https://doi.org/10.1016/j.pbi.2011.05.001

Yamaguchi Y, Pearce G, Ryan CA (2006) The cell surface leucine-rich repeat receptor for AtPep1, an endogenous peptide elicitor in Arabidopsis, is functional in transgenic tobacco cells. Proc Natl Acad Sci USA 103(26):10104–10109. https://doi.org/10.1073/pnas.0603729103

Yamaguchi Y, Huffaker A, Bryan AC, Tax FE, Ryan CA (2010) PEPR2 is a second receptor for the Pep1 and Pep2 peptides and contributes to defense responses in Arabidopsis. Plant Cell 22(2):508–522. https://doi.org/10.1105/tpc.109.068874

Yan L, Ma Y, Liu D, Wei X, Sun Y, Chen X, Zhao H, Zhou J, Wang Z, Shui W, Lou Z (2012) Structural basis for the impact of phosphorylation on the activation of plant receptor-like kinase BAK1. Cell Res 22(8):1304–1308. https://doi.org/10.1038/cr.2012.74

Endogenous Peptides Involved in Plant Growth and Development

5

Vidushi Yadav

Abstract

Many aspects of a plant's growth, development, and response to external stimuli are regulated by plant peptides. This chapter provides an overview of a number of important plant peptide families, such as CEP (C-terminally encoded peptides), IDA (inflorescence deficient in abscission), and CIF (casparian strip integrity factor). PAMP-induced secreted peptides or PIPs PSY PEP peptides are involved in the defensive response of plants, including PSY1-plant (peptide containing sulfated tyrosine 1), PSKs (phytosulfokine receptor 1), RGF/GLV/CLEL (root growth factor, golven, Cle-Like), and others. PEP peptides also promote the manufacturing of antimicrobial chemicals and enhance resistance to diseases. Plant immune response peptides are produced by PAMPS (pathogen-associated molecular patterns) and are crucial for controlling nutrient uptake and root development. Peptide hormones have a role in several root development processes in plants, including cell division, the formation of lateral roots, the production of root hair, and the maintenance of the root meristem, etc. Despite a growing number of research on the role of signal peptides in plant root development, the specific functions of many of these peptides are still unclear though this study may help to understand the role of plant peptide and their function for develop understanding and future research. Additionally, this work may contribute to understanding of the function and relevance of plant peptides in future research and development.

Keywords

Plant peptides · Plant growth · Nutrient uptake · Root development · Cell division

V. Yadav (✉)
Department of Applied Science, IIIT, Allahabad, India

S. Singh, R. Mehrotra (eds.), *Plant Elicitor Peptides*,
https://doi.org/10.1007/978-981-97-6374-0_5

5.1 Introduction

Short sequences of amino acids called peptides are essential to many biological activities in plants. These peptides serve as signal molecules and initiate cell-to-cell signaling during various biochemical and physiological processes. Certain peptides govern defense responses at the time of plant–microbe interactions, whereas others control plant growth. Since peptides are important regulators of plant development and responses to environmental stimuli, there has been a growing interest in learning more about their roles and modes of action in recent years.

Plant peptides' roles in regulating and responding to abiotic and biotic stresses were long underestimated, but over the last 20 years, they have become more recognized as a class of hormone molecules, capable of acting locally and systemically at extremely low concentrations (nM to pM) to regulate plant development and stress responses (Ryan 2000). The majority of functional, bioactive peptides are generated by proteolytic processing of a precursor protein of more than 100 amino acids, or a short open reading frame (sORF) of 50–100 amino acids. The proteolytic reaction that produces the peptide can happen in minutes, producing urgent signals and preparing the organism for the impending stress. Since their primary sequences vary and post-translational modification events like hydroxylation, sulfation, and proteolytic cleavage effectively control their production and activity, peptides serve a broad range of roles in inducing nearly all major physiological responses.

Over 1000 genes generating putative secretory peptides were predicted in the genome of *Arabidopsis thaliana*, making plants ten times more likely than mammals to have peptide receptors or transporters. In 1991, the first peptide hormone recovered from plants was discovered to be connected to stress (Pearce et al. 1991). The majority of the known peptide hormones are connected to development; however, the total is still far less than expected. Only three peptide families were known to trigger a stress response prior to 2008, and they were all found via screening techniques led by bioassays. Over the last 5 years, a large number of peptides discovered by in silico screening have been demonstrated to control stress reactions, and a number of them have also been proven to be triggered by stress at the peptide level (Ryan 2000). Peptides can have lengths as short as 10 kDa. They fall into two groups: degraded peptides, which are the end products of proteolytic enzyme activity during protein turnover, and bioactive peptides, which are produced by the specific action of peptidases on bigger precursor proteins. Although both types of peptides are by-products of proteolysis, their intracellular behaviors differ, these peptides often exhibit unique sequence characteristics (Ryan 2000). A list of endogenous peptides involved in plant growth and development is presented in Table 5.1 and their effect on different plant tissues is shown in Fig. 5.1.

Table 5.1 Peptide families, their major receptors and sequences

	Peptide family	Major Receptor(s) (LRR-RLK subfamily)	Amino Acid Sequence
1	CEP-C-TERMINALLY ENCODED PEPTIDEs	CEPR-CEP RECEPTOR (XI)	CEP1: DFROTNPGNSOGVGH
2	CIF-CASPARIAN STRIP INTEGRITY FACTORs	GSO1/SGN3, GSO2, GASSHO/ SCENGEN (XI)	CIF1: DYGNNSOSORLERPPFKLIPN
3	CLE-CLAVATA3/ EMBRYO SURROUNDING REGION- related	CLV1, BAM-CLAVATA1, BARELY ANY MERISTEM	CLV3:
4	(CLV3/R-type)	(XI)	RTVOSGODPLHH
5	CLE	PXY/TDR-PXY, TDIF RECEPTOR	CLE41:
6	(TDIF/H-type)	(XI)	HEVOSGONPISN
7	IDA + IDL-INFLORESCENCE DEFICIENT IN ABSCISSION, IDA-LIKE	HAE, HSL-HAESA HAE-LIKE (XI)	IDA: PIPPSAOSKRHN
8	PIP + PIPL-PAMP-INDUCED SECRETED PEPTIDE, PIP-LIKE	RLK7-RECEPTOR-LIKE KINASE 7 (XI)	TOLS2/PIPL3: SGOSRRGAGH
9	PSK-PHYTOSULFOKIN	PSKR-PSK RECEPTOR (X)	PSK1: YIYTQ
10	PSY1-PLANT PEPTIDE CONTAINING SULFATED TYROSINE1	PSYR-PSY1 RECEPTOR (X)	PSY1:DYGDPSANPKHDPGVoOS
11	RGF/GLV/CLEL -ROOT GROWTH FACTOR, GOLVEN, CLE-LIKE	RGFR/RGI-RGF RECEPTOR, ROOT GROWTH FACTOR INSENSITIV	REG(FX1I)/GLV11/CLEL8: DYSNPGHHPORHN

5.2 CEP-C-Terminally Encoded Peptides

A 15-amino acid post-translational peptide called a C-terminally encoded peptide (CEP) was discovered in *Arabidopsis* and is essential for nodulation, lateral root development, and long-distance root-to-shoot signaling under nitrogen starvation conditions. Although nitrogen deficiency affects the expression of CEP gene members in *Arabidopsis*, little is known about their function under other abiotic stress situations (Tabata et al. 2014). CEPs are known to regulate the growth and architecture of roots, impacting root hair development, lateral root production, and primary root growth (Delay et al. 2013). In leguminous plants, certain CEPs play a role in the development of root nodules, where nitrogen-fixing bacteria help to convert

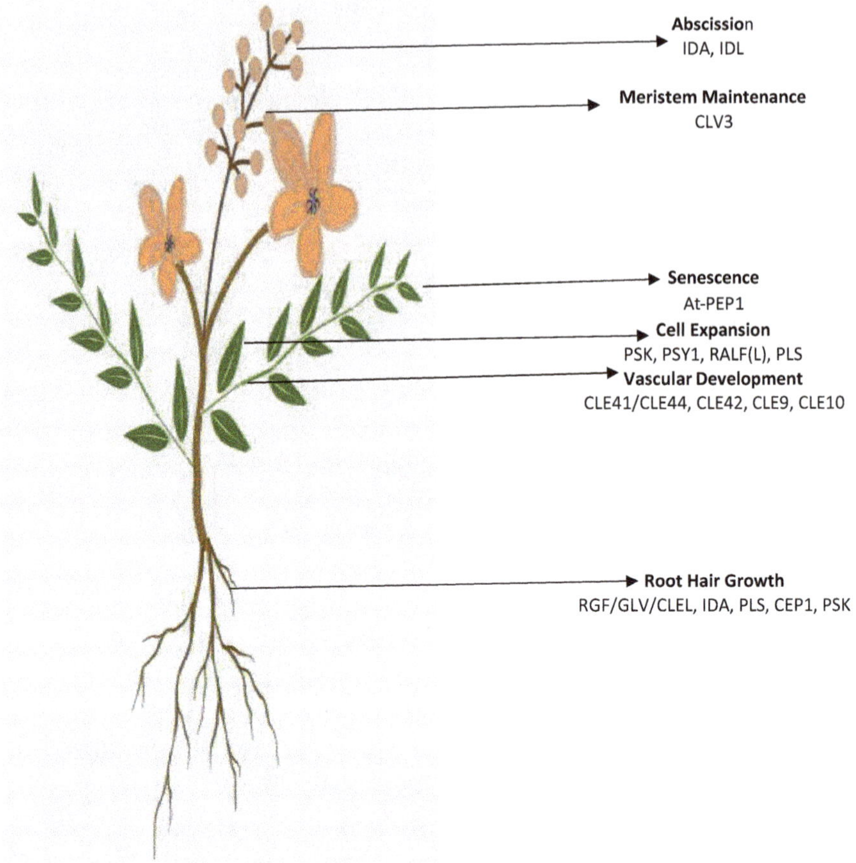

Fig. 5.1 Figure representing the role of endogenous peptides on plant growth and development

atmospheric nitrogen into a usable form for plant growth. Soil nitrate levels can affect CEP expression, linking them to nitrogen metabolism and plant responses to nutrient availability (Mohd-Radzman et al. 2016). In *Medicago*, MtCEP1, expressed in primary root tips, vascular tissues, and lateral organ primordia, affects nodule formation and lateral root development. Exogenous synthetic MtCEP1 peptide overexpression enhances nodule formation and inhibits lateral root development through interaction with the CRA2 receptor. CRA2 knockout plants exhibit a highly branching root phenotype, indicating interaction with MtCEP1. Overexpression of AtCEP5 in the phloem pole pericycle shows a similar phenotype to MtCEP1. Under different abiotic stress conditions, the AtCEP3 mutant displays accelerated development of both main and lateral roots (Roberts et al. 2016). In addition to root architecture roles, CEP genes are involved in nitrate uptake during nitrogen starvation. CEP(s) act as long-distance root-to-shoot signaling molecules, upregulating gene for nitrate transporter (NRT2.1) under nitrogen starvation. CEP interaction with

CEPR, also known as XIP1, in shoots leads to the expression of CEPD1 and CEPD2, which are necessary for upregulating NRT2.1 (Tabata et al. 2014). Plant peptide hormones, including CEPs, initiate signaling cascades by binding to receptor kinases in response to internal or external stimuli. CEP genes, encoding peptides with an NSS, variable domain, one or more CEP domains, and a short C-terminal extension, play significant roles in root growth under abiotic stress. They are involved in lateral root and nodule production, root and stem growth regulation, and responses to nitrogen-deficient conditions, linking them to abiotic stress responses (Delay et al. 2013; De Coninck and De Smet 2016). CEPs' existence in a wide range of plant species suggests evolutionary conservation, though their number and types vary, indicating diverse roles (Tabata et al. 2014). CEPs are recognized by specific receptors on the surface of plant cells, triggering subsequent signaling processes. These receptors are related to the leucine-rich repeat receptor kinase (LRR-RK) protein family (Roberts et al. 2016). CEPs generally range from 15 to 50 amino acids, distinguished by a conserved C-terminal motif crucial for their biological action (Mohd-Radzman et al. 2016). Before post-translational modification, the CEP polypeptide includes a variable domain, one or more CEP domains, a short C-terminal extension, and an N-terminal secretion signal (NSS). The peptides undergo post-translational modification, with proline residues being chemically modified to hydroxyproline or arabinosylated hydroxyproline, affecting protein flexibility and receptor interaction (Delay et al. 2013).

5.3 CIF-Casparian Strip Integrity Factor

Root endodermal cell walls are impregnated with hydrophobic lignin-based structures that serve as a physical barrier in extracellular space (Naseer et al. 2012). These structures, known as Casparian strips (CSs), were first reported in 1869 by Robert Caspary, inspired the original moniker of Casparian Strip Integrity Factor (CIF) peptides (Hose et al. 2001). The CIF family peptides can be identified by five genes, CIF1–4 and TWS1 converted in mature forms consisting of 21–24 amino acids and are tyrosine (Tyr)-sulfated (Geldner 2013). Through nanoliquid studies of the supernatant obtained through hydroponics grown Arabidopsis plants which were inducing the At2g16385 gene, the chemical structure of the encoded CIF family peptide, CIF1 has been determined (Shiono et al. 2014). The identification of the mature CIF1 peptide, which is a tyrosine-sulfated peptides of 21 amino acids containing two hydroxyprolines, exhibits several common characteristics within the CIF peptide family. Similar to other plant tyrosine-sulfated peptides like root growth factors (RGFs) and plant peptides containing sulfated tyrosine (PSYs), the amino acid aspartic acid (Asp) and tyrosine (DY) form a sulfation motif at the N-terminus of mature CIF peptides (Matsubayashi 2014). The DY motif is also present in immature phytosulfokines (PSKs), although asp is removed throughout maturity (Sauter 2015). Additionally, the "Lord of the Rings (LOTR) screen," a further advanced genetic screen, reidentified the sgn1 and sgn3 mutants in addition to additional intriguing CS-deficient mutants (Pfister et al. 2014). According to a recent study, the SGN pathway is a receptor/peptide signaling system that is comprised of SGN factors. Although the receptor complex that interacts with sgn3/gso1 and Somatic Embryogenesis Receptor Kinase proteins detects the CIF1 and CIF2 peptides that have been sulfated by TPST/SGN2, it is unclear which members of the SERK family are involved in this pathway. (Okuda et al. 2020). Following this, the receptor-peptide complex signal is passed to RBOHF/SGN4, a plasma membrane-anchored kinase protein, which then produces hydrogen peroxide and superoxide dismutases in extracellular spaces (Dubreuil et al. 2018). As a result, specific peroxidases using endodermis-derived monolignol catalyze radical-driven lignin polymerization in CSs (Fujita et al. 2020). The SGN pathway not only activates signals on plasma membranes but also regulates the transcription of genes that are essential for creating fully closed CASP domains (Roppolo et al. 2014). Additionally, CIF peptides induce surplus lignin/suberine deposition or an explosion of reactive oxygen species (ROS) in endodermal cells, among other previously reported phenomena (Geldner 2013). These events are explained by the fact that CIF peptides enhance suberine biosynthesis, which is lignin polymerization, and the development of genes sensitive to oxidative stress in addition to controlling the creation of functional CSs. (Roppolo et al. 2014). It is interesting to note that CIF2 therapy dramatically altered the transcription of several genes related to metabolites, hormone transporters, water channels, and nutrients (Shiono et al. 2014). The modified endodermal cell wall state, which has a significant impact on small molecule movement in extracellular spaces, may be the cause of these transcriptional alterations. Mitogen-activated protein kinase (MAPK) activation is thought to have been responsible for at least some of the CIF-induced transcriptional alterations, as observed in pattern-triggered immune responses (Matsubayashi 2014). It has been proposed that the CSs in plant roots serve as extracellular barriers. However, due to the difficulty in evaluating the impact of CS deficiencies and the unavailability of appropriate mutations, no direct experimental data has been presented. There is an excessive lignin accumulation in Enhanced Suberin 1 (ESB1), MYB36, CASP1, CASP3, LOTR1, and LOTR2 that is not associated with the Casparian Strip, as well as several mutants with CS abnormalities.suberin to compensate for CS imperfections, which could change how tiny molecules move through the SGN pathway. (Andersen et al. 2018). This restriction was addressed with the discovery of the sgn3/gso1 and cif1 cif2 mutants, which, despite CS deficiencies, do not cause excessive suberization or lignification (Lee et al. 2013). The cif1 cif2 and sgn3/gso1 mutants had recurrent discontinuous CS dots (Pfister et al. 2014). As previously reported, the phenotype of the SGN3/GSO1 receptors mutant is consistent with the CIF1 CIF2 double mutants, which show clear disruptions in the Casparian Strip (Dubreuil et al. 2018). Similar to the "gassho" (Japanese for "two palms joined") characteristic, the SGN3/GSO1 GSO2 double mutant displays a fused-cotyledon phenotype., in addition to the CS deficiency (Okuda et al. 2020). On agar plates, however, cif1 cif2 double mutant seeds develop without any problems, and the developing seedlings show no visible defects, like ectopic cotyledon adhesion or deformed embryonic cuticles. (Fujita et al. 2020). These differences in phenotype between the sgn3/gso1 gso2 and cif1 cif2 double mutants suggest that there are additional ligands for SGN3/GSO1 and GSO2.

5.4 CLE-Clavata3/Embryo Surrounding Region Related

It was initially discovered that Clavata3 (CLV3) in *Arabidopsis* encodes an extremely tiny extracellular protein that regulates the quantity of cells within the shoot apical meristem (SAM) (Clark et al. 1995). In 2006, it was found that the mature structure of the CLV3 peptide is a dodecapeptide (Kondo et al. 2006). Moreover, it was discovered that CLV3 and other CLE peptides function as arabinosylated glycopeptides (Shinohara and Matsubayashi 2015). Thirty-two CLE genes are present in *Arabidopsis* (Jun et al. 2010), of them, WUSCHEL (WUS) and WUSCHEL-related homeobox 5 (WOX5) are homeodomain transcription factors that are expressed in the root apical meristem (RAM) and SAM, respectively (Deyhle et al. 2007). CLV3 and CLE40 are in charge of maintaining these structures, respectively (Hobe et al. 2003). Stem cell homeostasis in the SAM and RAM is preserved via the interconnected, self-correcting feedback loop that is formed by the CLE-WOX signaling partners (Stahl et al. 2009). The C-terminus's 14-aa CLE domain and a potential N-terminal secretory signal peptide are two common structural elements of the small (about 100 amino acid) proteins encoded by the CLE gene family (Fletcher et al. 1999). The amino acid sequences of CLE proteins are significantly divergent, with the exception of the sequences found inside the CLE domain (Wang and Fiers 2010). CLV3, domain-swap as well as deletion examinations show that the CLE domain and signal peptide are essential with to the transcribed protein to function, implying the CLE domain is mainly responsible for CLV3 action in plants and that CLE peptides need to be secreted (Ni and Clark 2006). In real terms, biochemical studies have discovered peptides with the CLE domain that contain 12 or 13 amino acids (Kondo et al. 2006). Furthermore, CLV3 overexpression resembled the behavior of *Arabidopsis* terminating SAM and RAM when exposed to chemically produced CLV3 peptides corresponding to the CLE domain (Ren et al. 2006). The smallest unit known to demonstrate CLV3 action is a 12-amino acid peptide known as RTVPSGPDPLHH or MCLV3 (Kondo et al. 2006). CLV3 functions as a particular peptide ligand released that emerges via the CLE domain in plants (Clark et al. 1995). It has been determined that the perception of CLE peptides requires a number of cell surface receptors (Matsubayashi 2014). Transmission of the signal requires three primary clv1, the clv2-suppressor of llp1–2 (sol2)/coryne (crn) complex, and the newly discovered receptor-like protein kinase 2 (rpk2)/toadstool 2 (toad2) are examples of receptor kinase complexes (Muller et al. 2008). Therefore, it appears that the CLE-receptor kinase-WOX module is preserved in the vascular cambium, RAM, and SAM meristems (Matsubayashi 2014). Additionally, it appears that this WOX-CLE-receptor kinase module, which controls meristem activity, is conserved in maize, rice, and grasses (Suzaki et al. 2008). In addition to its role in meristem maintenance, CLE activity is essential for specific plant–microbe interactions (Betsuyaku et al. 2011). The quantity of freshly formed nodules in leguminous plants has been shown to be adversely regulated by the CLV1 homologues hyper-nodulation aberrant root (har) 1, sym29, nitrate tolerant symbiotic (nts)-1, and super numeric nodules (SUNN) (Schnabel et al. 2005). It is hypothesized that this inhibitory signal originates in the growing nodules in the roots and is then sent through the shoot to control the development of nodules in the roots (Okamoto et al. 2009). The possibility that CLE peptides could be used by plants to transmit signals both short range and long range is highlighted by the recent identification of the CLEs involved in this autoregulation of the nodule as candidate mobile signals (Betsuyaku et al. 2011). The green alga *Chlamydomonas reinhardtii* is one of the plants that have the CLE gene family, albeit there is considerable conservation in the Chlamydomonas CLE precursor's amino acid sequences (Hanada et al. 2011). The only organisms with CLE genes were multicellular plants and phytoparasitic nematodes, with the exception of *Chlamydomonas reinhardtii* (Oelkers et al. 2008). The moss *Physcomitrella patens* CLE precursor proteins appear to act as peptide hormones because they contain a signal peptide sequence that points them in the direction of the extracellular area (Whitewoods et al., 2018). The SAM and RAM's development is controlled by CLE peptides (Fletcher et al. 1999). CLEs are involved in the function of procambial cells in the vascular meristem, another form of stem cell tissue (Hirakawa et al. 2008). The distribution of water, nutrients, and other materials required for plant development and defense, as well as mechanical support, are all made possible by the vascular tissues (De Rybel et al. 2016). The vascular tissues of plants are composed of two types of distinct conducting tissues, xylem and phloem, and one type of meristematic tissue, known as procambial tissue (Lucas et al. 2013). These tissues create bundles that connect all parts of the plant. Procambial cells can develop into either tissue and are found sporadically between xylem and phloem. These procambial cells create xylem and phloem cells on opposite sides of the vascular development (Hirakawa et al. 2008).

5.5 IDA- Inflorescence Deficient in Abscission

In *Arabidopsis*, a mutant inflorescence lacking in abscission (ida) gene was found sensitive to ethylene. In this mutant, floral organs stay connected to the flower even after ripe seeds have shed (Butenko et al. 2003). IDA is primarily produced in the abscission zone and encodes a 77-amino acid preproprotein with an N-terminal secretion signal (Stenvik et al. 2006). Five IDA-like genes, known as IDL1 through IDL5, have been discovered in the *Arabidopsis* genome (Stenvik et al. 2008). C-terminus of IDA, IDLs has a conserved signature sequence, called PIP peptide (pv/iPpSa/gPSk/rk/rHN) (Butenko et al. 2003). The IDA mutant phenotype may be restored by treating it with artificially synthesized PIP peptide, suggesting that IDA functions as a peptide hormone to control organ abscission (Stenvik et al. 2006). The inflorescence deficient in abscission (AtIDA) peptide ligand was found to be essential for the regulation of floral organ abscission in Arabidopsis, since the ida mutant was unable to abscise its floral organs (Butenko et al. 2003). The abscission process occurs in cell files that make up the abscission zone (AZ) in wild type (wt) plants at the border between the organ to be shed and the main plant body (González-Carranza et al. 2002). Early on, in conjunction with the apical meristem growth of the lateral organs, an AZ is created. AZ cells differ from their neighbors in that they are small, lack vacuoles, and have thick cytoplasm (Patterson 2001). Similar biological mechanisms exist in many plant organs and stages of development are regulated by the same signaling module, which includes plasma membrane-bound receptors and *Arabidopsis* peptide ligands (Sundaresan and Alandete-Saez 2010). The finding that IDA and HAE (receptor-like protein kinase) suggest that IDA-HAE/HSL2 signaling may regulate cell separation processes outside of AZ cells. It is expressed in dehiscence regions of matured siliques, which go through process of cell separation to facilitate seed shedding (Cho et al. 2008). According to earlier gene expression research, IDA and the IDL genes may be involved in controlling several cell-to-cell separation processes that occur throughout the development of plants (Stenvik et al. 2008). Investigation of the roles of IDA, HAE, and HSL2 in lateral root emergence was inspired by the significant pIDA: GUS activity in the cells encircling the lateral root systems (LRs) (Cho et al. 2008). Subcellular localization investigations have demonstrated that the impacted gene, IDA, produces a brief protein with a peptide signaling from the N-terminus that instructs the protein to extracellular environment (Stenvik et al. 2006). Five *Arabidopsis* IDA-LIKE (IDL) transcripts were found using in silico methods and gene expression analysis (Stenvik et al. 2008). The conserved 20 amino acid, the EPIP motif is identified by its proline-rich domain at the C-terminus, was discovered by A comparison of amino acid sequences between members of the IDA family and IDL family (Stenvik et al. 2008). The EPIP domain is composed of two domains, one with an N-terminal secretion signal and another with a variable region that is short and rich in proline molecules of the IDA and IDL proteins mirrored those of CLAVATA3 (CLV3), suggesting that IDA served as a tiny signaling peptide (Stenvik et al. 2006). IDL genes, which encode tiny secreted proteins, control organ shedding in a variety of plants, including citrus, *Litchi chinensis*, and *Arabidopsis thaliana*. These proteins have conserved EPIP motifs in them (Stenvik et al. 2008). Floral organ separation is not observed in the *ida* mutant, although early abscission is caused by IDA overexpression (Butenko et al. 2003). The IDA-HAE/HSL2 complex has been shown through experimentation to cause a MITOGEN-ACTIVATED PROTEIN KINASE (MAP) cascade in the AZ cell's cytoplasm (Cho et al. 2008). This pathway inhibits the transcription factor KNAT1, also known as KNOTTED-LIKE HOMEOBOX-like (KNOX-like), which permits KNAT2 and KNAT6 to cause organ abscission (Cho et al. 2008). IDA is involved in the ET-response pathway in *Populus*, *L. chinensis*, and palm oil (Stenvik et al. 2008). On the other hand, it has been suggested that IDA from *A. thaliana* is controlled in a way that is independent of ET (Cho et al. 2008). The genetic mechanism leading to organ separation appears to include comparable components. Nevertheless, different plant hormones and other variables regulate their action in certain plant species (Butenko et al. 2003). One possible explanation for this phenomenon is evolutionary adaptability to various environments and lifestyles (Patterson 2001). Thus, it is crucial for production control to comprehend the molecular regulation of abscission, particularly in crop species (González-Carranza et al. 2002). Every outcome is extremely important, not just to researchers but also to breeders who operate in this area. This knowledge facilitates the characterization and selection of new cultivars and species with enhanced characteristics, as well as aids in the explanation of abscission (Sundaresan and Alandete-Saez 2010).

5.6 PSY PSY1-Plant Peptide Containing Sulfated Tyrosine1

18-amino acid sulfated and glycosylated peptide containing sulfated tyrosine 1 (PSY1) stimulates cell division and proliferation in the elongation zone of roots, To promote cell division and proliferation, the PSY1 RECEPTOR (PSY1R) interacts with and phosphorylates H+-ATPases (AHAs) in the plasma membrane (Amano et al. 2007a). Xanthomonas oryzae pv. oryzae (Xoo), a biotrophic pathogen in rice produces RaxX, a PSY1-like sulfated peptide that binds to the host PSY1R to activate its downstream signaling process (Pruitt et al. 2015). By attaching itself directly to the immune receptor XA21, RaxX initiates the immunological response (Pruitt et al. 2015). It is still unclear what PSY signaling's basic functions are during the growth and development of plants. According to a recent study, three orphan LRR-RKs mitigate the exchange between a plant and development as well as the stress reaction by acting as the PSY peptides' direct ligand receptors. (Fukuda and Ito 2021). Peptides PSY8, PSY6, and PSY5 demonstrating PSY1 sequence similarities were identified by nanoliquid chromatography-coupled tandem mass spectrometry (LC-MS/MS) analysis (Fukuda and Ito 2021). PSY5 and PSY6 interact with three similar LRR-RKs named PSYR1, PSYR2, PSYR3, which are closely related as detected through photoaffinity labeling tests (Fukuda and Ito 2021). Moreover, synthesized sulfated RaxX peptides RaxX16 and RaxX21 bind to PSYR2 and PSYR3. These findings demonstrated that PSYRs are able to identify both RaxX and PSY family peptides (Pruitt et al. 2015). The poly-mutant of sulfated peptide hormones is replaced by the tyrosyl protein mutation of sulfotransferase tpst-1, that is deficient in producing all tyrosine-sulfated peptides (Fukuda and Ito 2021). The receptor variants of the triple mutant psyr1,2,3, as well as Tpst-1 abnormalities, were examined in order to comprehend the molecular mechanism behind PSY–PSYR signaling (Amano et al. 2007a). The PSY5 peptide was chosen as a representative of the PSY family because it was successful in correcting the root development problems in tpst-1 (Amano et al. 2007b). On the other hand, psyr1,2,3 mutants have increased root growth (Fukuda and Ito 2021), while the PSY5 therapy had no impact on Psyr1,2,3 root growth (Fukuda and Ito 2021). A quadruple mutant known as tpst-1/psyr1,2,3 was produced in order to validate the receptor mutant phenotype (Fukuda and Ito 2021) and the investigations led to the conclusion the repetitive detrimental regulators of plant development PSYR1, PSYR2, PSYR3 (Fukuda and Ito 2021). As ligand receptors for PSY family peptides, LRR-RKs, PSYR1, PSYR2, and PSYR3, mediate the switching between two opposing pathways: stress response and plant development (Fukuda and Ito 2021), these research findings revealed a hitherto unidentified mechanism for cell-to-cell communication that allows plants to withstand environmental stress. Adjacent to the injured tissue areas, viable cell layers undergo preventive stress responses induced by ligand-deprivation-dependent activation mechanisms (Fukuda and Ito 2021). This strikes a balance between stress tolerance and energetic costs, enabling optimal plant growth under adverse conditions. The molecular processes behind this pathway, however, are still unknown. For instance, nothing is known about the PSY–PSYR signaling pathway, particularly with regard to the trade-off between growth and stress response (Fukuda and Ito 2021). Mass spectrometric analysis and coimmunoprecipitation revealed the co-receptors that had previously been identified were not bound by PSYR3-GFP (Fukuda and Ito 2021). Using gain-of-function techniques such as linking receptor-like kinase3 chimera and dimerizing designed receptor–co-receptor pairs induced by rapamycin, it is possible to determine whether or not plants' associations with SERK co-receptors activate PSYR signaling, which in turn induces receptor characteristics and subsequent cellular reactions. Receptor-like kinase3 chimera and rapamycin-induced dimerization of specific receptor–co-receptor pairs are two examples of gain-of-function approaches that could be used to address the question of whether or not plants' associations with SERK co-receptors activate PSYR signaling, induces receptor characteristics, and consequent cellular reactions. It's interesting to note that 26 stress-related transcription factor genes are expressed much more in tpst-1 than in the wild type. This suggests that the PSY–PSYR pathway regulates these multiple stress-related transcription factors, which in turn control the balance between optimum plant growth and stress response (Amano et al. 2007a). Identifying targeted genes for these transcription factors that are involved in PSY-PSYR signaling downstream circuit will be interesting (Fukuda and Ito 2021). It has been reported that crosstalk between different phytohormones is facilitated by PSY1R signaling (Pruitt et al. 2015).

5.7 PSKs-Phytosulfokine Receptor 1

A class of sulfated plant peptides known as phytosulfokine (PSK) has bioactivity at nanomolar doses. However, as we comprehend currently, their functions considerably outweigh growth regulation and encompass reproductive, biotic, and abiotic stress responses. The 18-amino acid plant peptide PSY1 and the disulfated pentapeptide PSK, which have tri-L-arabinose side chains connected to a sulfated Tyr (Y2SO3H) residue and a hydroxyproline repeat (hP16tri-L-ara), are two examples of growth factors present in Arabidopsis thaliana (Amano et al. 2007b). Both the disulfated pentapeptide PSK and the 18-amino acid plant peptide PSY1 are growth factors present in Arabidopsis thaliana. These factors consist of tri-L-arabinose side chains connected to a hydroxyproline repeat (hP16tri-L-ara) and a sulfated Tyr (Y2SO3H) residue. The addition of media from an earlier cell culture can counteract this population impact; this priming is referred to as the feeder effect. A reliable bioassay was developed to quantify the activity of the chemical component in suspension-cultured *Asparagus officinalis* mesophyll cells that causes this feeder effect. This bioassay enabled the identification of two distinct peptides by mass spectrometry-based Edman degradation. Two different activities in asparagus have been identified: the disulfated peptide PSK-α, which includes five amino acids, and the C-terminally shortened disulfated tetrapeptide PSK-β (Tyr(SO3H)-Ile-Tyr(SO3H)-Thr) Isoleucine-Sulphotyrosine (SI)glutamine-threonine-sulphotyrosine (Matsubayashi et al. 1996). In *Arabidopsis*, a single gene encodes the tyrosylprotein sulfotransferase (TPST), which sulfates the PSK precursor. Root elongation is greatly promoted at 0.3 nM and greater doses of PSK in the tpst-1 background without endogenous active PSK, signifying that PSK binds its specific target with high affinity. The disulfated pentapeptide's specificity was validated by analysis of 12 PSK analogues, which also disclosed a fundamental need of Tyrosine(sulfate)-Isoleucine-Tyrosine(sulfate) which is N-terminal peptide. The unsulfated pentapeptide exhibits poor root growth promotion at 1 μM and is inactive below 100 nM in terms of encouraging cell division at low density. As a result, PSK's biological activity is increased by about 1000 times upon sulfonylation of its two tyrosyl side chains (Matsubayashi et al. 1996). Small gene families encode 80–120 amino acid prepeptides, which are used to synthesize PSK. The modified pentapeptide YIYSQ was produced by mutating the cDNA of rice (Oryza sativa) OsPSK. This pentapeptide was then extracted from the transformed cells' culture medium, demonstrating the transcript will transform into the fully formed peptide. The Presence of the C-terminal in the growth of tetrapeptide YIYS medium indicates that PSK-β is a secondary product resulting from the breakdown of PSK-α substrate. The Propeptide shows minimal primary sequence conservation, with a single instance of a unique pattern that includes the YIYTQ peptide in the C-terminal half (Matsubayashi et al. 1996). Transmembrane proteins called PSK receptors have an extracellular domain and an intracellular kinase domain at the C-terminus. They are a part of the receptor-like kinase (RLK) family, which has over 1100 members in rice and over 600 members in *Arabidopsis*. An N-terminal signal peptide directs the PSK receptor proteins toward the secretory route, and a single transmembrane helix anchors them to the plasma membrane. PSK was shown to bind to a membrane fraction enriched in rice suspension cell plasma membranes as well as to plasma membrane preparations of tobacco, tomato, asparagus, and maize using photoaffinity labeling and radiolabeled PSK. Low-density cell cultures of rice and asparagus experience growth inhibition; this is alleviated when nanomolar doses of PSK were added to the culture medium. Nevertheless, PSK does not stimulate cell proliferation in the absence of auxin and cytokinin, indicating that in undifferentiated cultured cells, PSK functions as a quorum signal that makes cells receptive to cytokinin and auxin-induced cell division (Matsubayashi et al. 1996).

5.8 Root Growth Factor (RGF), Golven (GLV), and Cle-like (CLEL)

Root growth factor (RGF), Golven (GLV), and Cle-Like (CLEL) peptides are a family of secreted signal molecules that undergo post-translational modifications. Three distinct research groups used various approaches to identify them in *Arabidopsis*. The chemical production of mature RGF peptides restored the tertiary structure of tyrosine within the context of PSK and PSY1 (Ito et al. 2006). Additionally, they were given the term GOLVENs because, when overexpressed, they may result in irregular root waving, presumably as a result of a decrease in gravitropism (Whitford et al. 2008). Eleven RGFs related to various aspects of plant development, including maintaining the roots and root hair, formation, root and their hypocotyl gravitropism, as well as lateral formation of roots, have been discovered in *Arabidopsis* (Fisher and Turner 2007). RGF polypeptides at maturity are broken down by cleaves of their preproproteins by proteases, just like other secreted signal peptides. A precursor protein typically consists of 79–182 amino acids, with a conserved RGF domain at the C-terminal and an peptide signaling N-terminal that directs it into the secretion of pathways (Whitford et al. 2012). The 13–16 amino acid residues that make up mature peptides typically need to undergo post-translational changes, such as proline hydroxylation and tyrosine sulfation (Ohyama et al. 2009). Auxin has the ability to upregulate TPST expression, according to earlier research and TPST is capable of sulfating AtRGF1 (Ito et al. 2006). Apical meristem growth is mediated by RGIs/RGFRs, a class of leucine-rich repeat receptor-like kinases, which can recognize correctly modified RGFs and upregulate the two downstream transcription factor genes, PLATHOLA1 (PLT1) and PLT2 (Fisher and Turner 2007). Three independent study groups used various techniques to identify RGIs/RGFRs. Research revealed that the rgi1,2,3,4,5 quintuple mutant completely lacks sensitivity to exogenously administered AtRGF1 and displays a severe short root phenotype with reduced meristem size (Whitford et al. 2012). Protein interaction study and dot-blotting demonstrated that AtRGI1's extracellular domain can directly bind AtRGF1. AtRGF1 can cause its receptors, including AtRGI1 and AtRGI2, to become auto phosphorylated and ubiquitinated (Ito et al. 2006). RGFs have a variety of roles in plant evolution; many of them are produced in roots, although others are also expressed in particular tissues or types of cells, shoots, and reproductive organs. Entire functional investigations of RGF peptides in the model Arabidopsis, a eudicot plant, have been conducted. However, little is known about the evolution and roles of RGF peptides in other major clades, especially in earliest terrestrial plants that diverged. Ohyama et al. (2009) conducted a number of investigations of the 24 different species of plants, representing several plants generations, have the growth factor receptor gene family. RGFs and RGIs in liverworts raise the possibility that RGFs developed after plant roots were initiated. Previous research revealed that the majority of plants have experienced several rounds of whole-genome duplications (WGDs). For instance, *Arabidopsis* had at least two rounds of tetraploidization, while soybeans (Glycine max) underwent numerous rounds of WGDs (Fisher and Turner 2007).

5.9 PIP-PAMP-Induced Secreted Peptides/PIP-LIKE

The identification of AtPIP1 and AtPIP2 in *Arabidopsis* was achieved by the activation of receptor-like kinase 7 (RLK7), which is triggered by pathogen-associated molecular patterns (PAMPs) (Hou et al. 2014). Protopeptides with an N-terminal signal peptide (SP) is used to create PIPs, much like other small secreted peptides (SSPs). Once the SP is eliminated, the precursor peptide proceeds via the secretory route and is converted into a mature peptide consisting of 15–25 amino acids. Under the influence of exogenous PAMPs like flagellin and chitin, the expression of the precursor genes of AtPIP1 and AtPIP2 was markedly elevated, which in turn stimulated PAMP-triggered immunity (PTI) responses, such as bursts of reactive oxygen species and the control of the expression of defense genes (Hou et al. 2014). Additionally, it has been demonstrated that AtPIP3 controls plant immunity by modifying the interaction between the signaling pathways for salicylate and jasmonate (Hou et al. 2014). PIP1 identified in potatoes (*Solanum tuberosum*) is thought to contribute in plant resistance against infection by the potato virus Y (PVY) (Hou et al. 2014). Furthermore, TOLS2/PIP-Like3 peptide, via its interaction with RLK7 has shown to suppress the growth of lateral roots in plant species (Toyokura et al. 2019). In fact, the presence of homologous prePIP proteins in several dicotyledonous and monocotyledonous plant species, including rice, soybean, grape, and maize, suggests that PIP is widely distributed and functional in a wide range of taxa (Hou et al. 2014).

Less is known about the PIP family of tiny secreted peptides, only around eleven members of the *Arabidopsis* PIP family are known. Each member of the prePIP family possesses the signal peptide, a highly conserved SGP-rich C-terminus motif, and a variable region in between, which are features of post-translational modifications of secretory peptide precursors (Hou et al. 2014). Proline in the SGP motif is a possible target for hydroxylation, which is a significant characteristic of this family. Although AtPIP2 has two SGPs and AtPIP1 has only one, it is unclear how the two vary in terms of activity and functionality (Hou et al. 2014).

5.9.1 Future Challenges

Peptide hormones are involved in a variety of processes related to the growth of plant; however, the precise roles of many of these peptides remain mostly unknown. Peptide hormones and membrane-localized receptors mediate cell-to-cell communication, which is one of the fundamental mechanisms that control the growth and development of multicellular organisms. When these peptide hormones connect to the extracellular domains of receptors, physicochemical interactions are transformed into physiological outputs that trigger downstream signaling, which modifies the receptors' conformational modifications to affect cellular activities and destinies. One of the main concerns of contemporary biological research is the identification of hormone receptor couples because membrane-localized receptors serve as master switches of intricate intracellular signaling cascades.

Understanding of short peptide signals has grown significantly, not just in terms of their detection by in silico predictions or biochemical tests, but also in terms of their functional characterization. Both the genetic and the bioassay techniques, however, have limitations. For example, the genetic approach may be hampered by the functional redundancy of many signaling peptides, while bioassay methods may not be able to detect peptide hormones in tissues due to their normally low physiological concentrations. Still, a greater emphasis on producing knockout mutants, for instance, using TILLING will be necessary to clarify the precise function(s) of peptide signals in developmental processes. Since wide overexpression and the exogenous administration of peptides can produce nonspecific or unrelated symptoms, loss-of-function data are particularly crucial.

Almost all of the enzymes engaged in proteolytic processing and many of the enzymes involved in post-translational modification of peptide signals are yet unknown. The identification of the genes and/or gene families that encode these enzymes could prove significant for a more thorough understanding of known peptide signals as well as providing a helpful starting point for the discovery of new peptide signals through genetic methods, given their importance for the correct activation of many known small peptide signals. The vast number of potential peptide ligand–receptor (kinase) pairs of which only a very small number have been demonstrated to form pairs represents a significant knowledge gap. Even though there are several ways to link peptide signals to their appropriate receptors, physical contact inevitably requires a biochemical technique. In order to do this, techniques for larger-scale research on these interactions must be used, and instruments for studying the dynamics of peptide ligand–receptor and receptor–receptor interactions in vivo must be created. Research has demonstrated that, when produced in non-native expression patterns, some receptors may functionally substitute for one another. Comprehending the reasons behind and mechanisms underlying the interchangeability of these proteins will aid in the clarification of additional peptide signaling pathways that lack recognized ligand or receptor pairings. The findings that there are several potential connections rather than just one peptide-receptor combination will probably be further supported by such an approach. Determining the peptide ligand–receptor (kinase) pairings will also shed light on the various peptides' (differential) functions within overlapping expression regions.

References

Amano Y, Tsubouchi H, Matsubayashi Y (2007a) Another homologue of the arabidopsis phytosulfokine receptor gene PSKR1 is a positive regulator of plant growth. Plant Physiol 143(3):1346–1356

Amano Y, Tsubouchi H, Shinohara H, Ogawa M, Matsubayashi Y (2007b) Tyrosine- sulfated glycopeptide involved in cellular proliferation and expansion in Arabidopsis. Proc Natl Acad Sci 104(43):18333–18338

Andersen TG, Naseer S, Ursache R, Wybouw B, Smet W, De Rybel B et al (2018) Diffusible repression of cytokinin signalling produces endodermal symmetry and passage cells. Nature 555(7697):529–533

Betsuyaku S, Sawa S, Yamada M (2011) The function of the CLE peptides in plant development and plant-microbe interactions. The Arabidopsis Book/American Society of Plant Biologists 9

Butenko MA, Patterson SE, Grini PE, Stenvik GE, Amundsen SS, Mandal A, Aalen RB (2003) Inflorescence deficient in abscission controls floral organ abscission in Arabidopsis and identifies a novel family of putative ligands in plants. Plant Cell 15(10):2296–2307

Cho SK, Larue CT, Chevalier D, Wang H, Jinn TL, Zhang S, Walker JC (2008) Regulation of floral organ abscission in Arabidopsis thaliana. Proc Natl Acad Sci 105(40):15629–15634

Clark SE, Running MP, Meyerowitz EM (1995) CLAVATA3 is a specific regulator of shoot and floral meristem development affecting the same processes as CLAVATA1. Development 121(7):2057–2067

De Coninck B, De Smet I (2016) Plant peptides–taking them to the next level. J Exp Bot 67:4791–4795

De Rybel B, Mähönen AP, Helariutta Y, Weijers D (2016) Plant vascular development: from early specification to differentiation. Nat Rev Mol Cell Biol 17(1):30–40

Delay C, Imin N, Djordjevic MA (2013) CEP genes regulate root and shoot development in response to environmental cues and plant hormones. Plant Signal Behav 8(5):e24723

Dubreuil C, Roppolo D, Geldner N (2018) Control of Casparian strip development and function in roots. Curr Opin Plant Biol 41:1–7

Deyhle F, Sarkar AK, Tucker EJ, Laux T (2007) WUSCHEL regulates cell differentiation during anther development. Dev Biol 302(1):154–159

Fisher K, Turner S (2007) PXY, a receptor-like kinase essential for maintaining polarity during plant vascular-tissue development. Curr Biol 17(12):1061–1066

Fletcher JC, Brand U, Running MP, Simon R, Meyerowitz EM (1999) Signaling of cell fate decisions by CLAVATA3 in Arabidopsis shoot meristems. Science 283(5409):1911–1914

Fujita S, De Bellis D, Edel KH, Köster P, Andersen TG, Schmid-Siegert E (2020) Lignin-based barrier restricts pathogen invasion in the roots of Arabidopsis. Nat Commun 11:3522

Fukuda K, Ito T (2021) Orphan LRR-RKs regulate plant development and stress response through PSY signaling. J Plant Res 134(1):123–134

Geldner N (2013) The endodermis: development and differentiation of the plant's inner skin. Protoplasma 250(5):3–20

González-Carranza ZH, Elliott KA, Roberts JA (2002) Expression of polygalacturonases and evidence to support their role during cell separation processes in Arabidopsis thaliana. J Exp Bot 53(377):1405–1415

Hanada K, Zou C, Lehti-Shiu MD, Shinozaki K, Shiu SH (2011) Importance of lineage-specific expansion of plant tandem duplicates in the adaptive response to environmental stimuli. Plant Physiol 157(2):732–745

Hirakawa Y, Kondo Y, Fukuda H (2008) Arabidopsis BES1 and BZR1 homologs cooperate with phloem intercalated with xylem to regulate vascular cell differentiation. Development 135(13):2411–2419

Hobe M, Müller R, Grünewald M, Brand U, Simon R (2003) Loss of CLE40, a protein functionally equivalent to the stem cell restricting signal CLV3, enhances root waving in Arabidopsis. Dev Genes Evol 213(8):371–381

Hose E, Clarkson DT, Steudle E, Schreiber L, Hartung W (2001) The exodermis: a variable apoplastic barrier. J Exp Bot 52(365):2245–2264

Hou S, Liu D, Huang Y, Yu S (2014) The Arabidopsis thaliana secretome: current status and future directions. Front Plant Sci 5:296

Ito S, Matsukawa T, Matsubayashi Y (2006) Sulfated peptides as potent and selective inhibitors of human plasma kallikrein. J Med Chem 49(13):3960–3963

Jun J, Fiume E, Roeder AH, Meng L, Sharma VK, Osmont KS et al (2010) Comprehensive analysis of CLE polypeptide signaling gene expression and overexpression activity in Arabidopsis. Plant Physiol 154(4):1721–1736

Lee Y, Rubio MC, Alassimone J, Geldner N (2013) A mechanism for localized lignin deposition in the endodermis. Cell 153(2):402–412

Lucas WJ, Groover A, Lichtenberger R, Furuta K, Yadav SR, Helariutta Y et al (2013) The plant vascular system: evolution, development and functions. J Integr Plant Biol 55(4):294–388

Matsubayashi Y (2014) Posttranslationally modified small-peptide signals in plants. Annu Rev Plant Biol 65:385–413

Matsubayashi Y, Ogawa M, Kihara H, Niwa M, Sakagami Y, Haga K (1996) The endogenous sulfated pentapeptide phytosulfokine-α stimulates tracheary element differentiation of isolated mesophyll cells of zinnia. Plant Physiol 111(2):433–438

Mohd-Radzman NA, Djordjevic MA, Imin N (2016) Nitrogen modulation of legume root architecture signaling pathways involves dynamic regulation of CEP genes. Front Plant Sci 7:1260

Muller R, Bleckmann A, Simon R (2008) The receptor kinase CORYNE of Arabidopsis transmits the stem cell–limiting signal CLAVATA3 independently of CLAVATA1. Plant Cell 20(4):934–946

Naseer S, Lee Y, Lapierre C, Franke R, Nawrath C, Geldner N (2012) Casparian strip diffusion barrier in Arabidopsis is made of a lignin polymer without suberin. Proc Natl Acad Sci 109(24):10101–10106

Ni J, Clark SE (2006) Evidence for functional conservation, sufficiency, and proteolytic processing of the CLAVATA3 CLE domain. Plant Physiol 140(2):726–733

Kondo T, Sawa S, Kinoshita A, Mizuno S, Kakimoto T, Fukuda H, Sakagami Y (2006) A plant peptide encoded by CLV3 identified by in situ MALDI-TOF MS analysis. Science 313(5788):845–848

Oelkers K, Goffard N, Weiller GF, Gresshoff PM, Mathesius U, Frickey T (2008) Bioinformatic analysis of the CLE signaling peptide family. BMC Plant Biol 8:1–15

Ohyama K, Shinohara H, Ogawa-Ohnishi M, Matsubayashi Y (2009) A glycopeptide regulating stem cell fate in Arabidopsis thaliana. Nat Chem Biol 5(8):578–580

Okamoto S, Ohnishi E, Sato S, Takahashi H, Nakazono M, Tabata S, Kawaguchi M (2009) Nod factor/nitrate-induced CLE genes that drive HAR1-mediated systemic regulation of nodulation. Plant Cell Physiol 50(1):67–77

Okuda S, Fujita S, Moretti A, Hohmann U, Doblas V.G, Ma Y, Hothorn M. (2020). Molecular mechanism for the recognition of sequence-divergent CIF peptides by the plant receptor kinases GSO1/SGN3 and GSO2. Proceedings of the National Academy of Sciences, 117(5), 2693-2703.

Patterson SE (2001) Cutting loose: abscission and dehiscence in Arabidopsis. Plant Physiol 126(2):494–500

Pearce G, Strydom D, Johnson S, Ryan CA (1991) A polypeptide from tomato leaves induces wound-inducible proteinase inhibitor proteins. Science 253(5022):895–897

Pfister B, Lu KJ, Eicke S, Feil R, Lunn JE, Streb S, Zeeman SC (2014) Genetic evidence that chain length and branch point distributions are linked determinants of starch granule formation in Arabidopsis. Plant Physiol 165(4):1457–1474

Pruitt RN, Locci F, Wanke F, Zhang L, Saile SC, Joe A et al (2015) The rice XA21 binding protein 3 (XB3) is a ubiquitin E3 ligase required for full Xa21-mediated disease resistance. Plant J 83(4):779–793

Ren XY, Vorst O, Fiers MW, Stiekema WJ, Nap JP (2006) In plants, highly expressed genes are the least compact. Trends Genet 22(10):528–532

Roberts I et al (2016) CEP5 and XIP1/CEPR1 regulate lateral root initiation in Arabidopsis. J Exp Bot 67(16):4889–4899

Roppolo D, Boeckmann B, Pfister A, Boutet E, Rubio MC, Dénervaud-Tendon V et al (2014) Functional and evolutionary analysis of the CASPARIAN STRIP MEMBRANE DOMAIN PROTEIN family. Plant Physiol 165(4):1709–1722

Ryan, C. A. (2000). The systemin signaling pathway: differential activation of plant defensive genes. Biochim Biophys Acta Protein Struct Mol Enzymol, 1477(1–2), 112–121

Sauter M (2015) Phytosulfokine peptide signalling. J Exp Bot 66(17):5161–5169

Schnabel E, Journet EP, de Carvalho-Niebel F, Duc G, Frugoli J (2005) The Medicago truncatula SUNN gene encodes a CLV1-like leucine-rich repeat receptor kinase that regulates nodule number and root length. Plant Mol Biol 58:809–822

Shinohara H, Matsubayashi Y (2015) Reevaluation of the CLV 3-receptor interaction in the shoot apical meristem: dissection of the CLV 3 signaling pathway from a direct ligand-binding point of view. Plant J 82(2):328–336

Shiono K, Ando M, Nishiuchi S, Takahashi H, Watanabe K, Nakamura M et al (2014) RCN1/OsABCG5, an ATP-binding cassette (ABC) transporter, is required for hypodermal suberization of roots in rice (Oryza sativa). Plant J 80(1):40–51

Stenvik GE, Butenko MA, Urbanowicz BR, Rose JK, Aalen RB (2006) Overexpression of INFLORESCENCE DEFICIENT IN ABSCISSION like genes induce organ abscission in Arabidopsis. Gene 442(1–2):129–136

Stenvik GE, Tandstad NM, Guo Y, Shi CL, Kristiansen W, Holmgren A, Clark SE, Aalen RB, Butenko MA (2008) The EPIP peptide of INFLORESCENCE DEFICIENT IN ABSCISSION is sufficient to induce abscission in Arabidopsis through the receptor-like kinases HAESA and HAESA-LIKE2. Plant Cell 20(7):1805–1817

Stahl Y, Wink RH, Ingram GC, Simon R (2009) A signaling module controlling the stem cell niche in Arabidopsis root meristems. Curr Biol 19(11):909–914

Sundaresan V, Alandete-Saez M (2010) Pattern formation in miniature: the female gametophyte of Arabidopsis. Cold Spring Harb Perspect Biol 2(7):a001560

Suzaki T, Yoshida A, Hirano HY (2008) Functional diversification of CLAVATA3-related CLE proteins in meristem maintenance in rice. Plant Cell 20(8):2049–2058

Tabata R, Sumida K, Yoshii T, Ohyama K, Shinohara H, Matsubayashi Y (2014) Perception of root-derived peptides by shoot LRR-RKs mediates systemic N-demand signaling. Science 346(6207):343–346

Toyokura K, Watanabe K, Nakashima T, Sugiyama S, Yamamoto T, Tian L (2019) Lateral root formation is blocked by a TOLS2-PIPL3 peptide in Arabidopsis. Nat Commun 10(1):2138

Wang G, Fiers M (2010) CLE peptide signaling during plant development. Protoplasma 240:33–43

Whitewoods CD, Cammarata J, Venza ZN, Sang S, Crook AD, Aoyama T et al (2018) CLAVATA was a genetic novelty for the morphological innovation of 3D growth in land plants. Curr Biol 28(15):2365–2376

Whitford R, Fernandez A, De Groodt R (2008) Identification of peptides controlling root meristem growth in Arabidopsis thaliana. J Biol Chem 283(41):28579–28588

Whitford R, Fernandez A, Tejos R, Perez AC (2012) GOLVEN secretory peptides regulate auxin carrier turnover during plant gravitropic responses. Dev Cell 22(3):678–685

Yamaguchi Y, Pearce G, Ryan CA (2011) The cell surface leucine-rich repeat receptor for AtPep1, an endogenous peptide elicitor in Arabidopsis, is functional in transgenic tobacco cells. Proc Natl Acad Sci 103(26):10104–10109

Effector Mediated Defense Mechanisms in Plants against Phytopathogens

6

Seema Devi, Riddha Dey, Surya Prakash Dube, and Richa Raghuwanshi

Abstract

Plants in environment are often exposed to pathogenic attack to which they respond through numerous morphological, biochemical, and molecular mechanisms. Plant parasites while forming association with host delivers proteins and suites of effector molecules needed for infestation. The host system in turn triggers innate immunity through the interactive signaling pathways involving the membrane-bound or intracellular receptors that perceive different elicitors such as pathogen-associated molecular patterns (PAMPs/MAMPs), damage-associated molecular patterns (DAMPs) released by microbial enzymes or effectors, thereby causing a PAMP-triggered immunity (PTI) or an effector-triggered immunity (ETI). PTI results into basal immunity, whereas ETI confers durable immunity that finally results in hypersensitive responses (HR). While hypersensitive responses and systemic acquired resistance (SAR) are triggered in plants on exposure to different compatible pathogens, the induced systemic resistance (ISR) are elicited by plant growth-promoting rhizobacteria (PGPR), categorized as incompatible pathogens. Phytopathogenic bacteria, oomycetes, and fungi inject effectors into host plant body thereby damaging the host metabolism causing infection. These effectors many times trigger resistance in host plant as well. Effectors of non-pathogenic endophytic microbes establish a different response by founding a mutualistic relationship with host plant, positively regulating the plant health. The present review therefore discusses effectoromics undermining plant expressions during infection and host immune responses.

S. Devi · R. Dey · S. P. Dube · R. Raghuwanshi (✉)
Department of Botany, Mahila Mahavidyalaya, Banaras Hindu University, Varanasi, India

S. Singh, R. Mehrotra (eds.), *Plant Elicitor Peptides*,
https://doi.org/10.1007/978-981-97-6374-0_6

Keywords

Biotic stress · Elicitors · Pathogen-associated molecular patterns (PAMPs) · Damage-associated molecular patterns (DAMPs) · Hypersensitive response (HR) · Systemic acquired resistance (SAR) and induced systemic resistance (ISR)

6.1 Introduction

Natural selection results in the survival of organisms in their niche and the genome evolution helped the microbes to successfully complete their life cycle under the changing environmental conditions. Among the genome were the set of genes that encoded molecules called effectors which facilitated infection and inducing defense response in the host. Effectors are significant in establishing network with host and are mostly beneficial to the organisms producing them. They are also produced by symbiotic and beneficial microbes like rhizobacteria or mycorrhizae. The effectors can be categorized under proteins, secondary metabolites, small RNAs, majority being proteinaceous in nature. Effectors help in microbial penetration in host, its proliferation, host immune suppression, and nutrient acquisition from host (Zhang et al. 2022). Investigation and understanding effectors are largely owed to plant–pathogen interactions. Phytopathogenic interactions are of three different types, those which eventually kill the host by secreting some cell wall degrading enzymes or toxins categorized as necrotrophs, those which depend on living host throughout their life as biotrophs and those that initially depends on living host but they kill their host at later stages of infection as hemi-biotrophs. Effectors of biotrophs help them in entering the host stealthily and avoid recognition by the host. Thus, in response to the diverse pathogenic nature, plants too have developed a multilayer defense system which includes the constitutive and the induced defense systems. Constitutive defense while inherently present in the plants includes the host cell structure which prevents the entry of pathogens. On the other hand, inducible defenses triggered in response to pathogen attack requires specific detection systems comprising signal transductions that can detect the presence of a pathogen or herbivore thereby activating gene expression and metabolism accordingly. Inducible plant defense response includes pathogen-associated molecular pattern (PAMP) that trigger PTI. This pathogen-associated molecular pattern (MAMPs/DAMPs) are recognized by pattern recognition receptor (PRRs) that result in the activation of Ca^{2+} and mitogen activated protein (MAP) kinase signaling cascades (Felix and Boller, 2003). As plants have co-evolved with pathogens, the pathogens acquired effectors as virulence factor, and plants evolved new PRRs (R) protein which recognized effectors to induce a defense response called effector-triggered immunity (ETI). Any signal that is received by the plant cells and activates its defense pathway is considered as elicitor. PTI involves the recognition of some general elicitors (MAMPs, PAMPs, and DAMPs) that confers low level basal immunity (Fig. 6.1A), whereas ETI recognizes specific effectors or elicitors that confers durable resistance (Fig. 6.1B). Therefore, PTI and ETI are considered as primary and secondary innate immunity, respectively. ETI is an amplified PTI response that results in

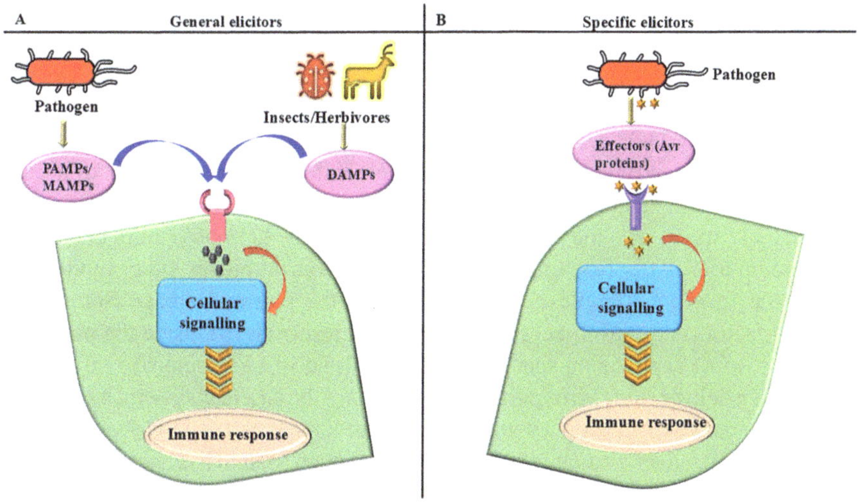

Fig. 6.1 Categorization of elicitors (**A**) General (non-specific) that confers primary innate immunity (basal resistance); (**B**) Specific elicitors that confers secondary innate immunity (durable resistance)

hypersensitive response (HR) at the infection site and systemic acquired resistance (SAR). Therefore, ETI is considered to be more efficient than PTI in reducing plant disease.

On the other hand, during mutualist interactions, the MAMPs secreted by the symbiont set off the alarm and plant defense responses. Mycorrhizae and PGPR produce effectors to downregulate the plant defense during colonization of plant roots. These organisms trigger a weak (MAMP) triggered immunity (MTI) in plants as compared to true pathogens. The plants later get benefited by the induced systemic resistance (ISR) once the symbiosis is established. Thus, undermining effectors of plant pathogens or endophytes and their role in plant immunity can be helpful in developing novel strategies in disease management, plant growth and development.

6.2 Pathogen Invasion and Plant Defense Mechanism

Entry of pathogens into the plant tissue is a crucial step in causing disease. Some pathogens directly enter into the host tissue by penetrating the cuticle and cell wall with the help of some secreted lytic enzymes, which digest plants mechanical barriers. Bacteria move toward the host through chemotaxis process and then enter into the host tissue through stomata, lenticels, hydathodes, trichomes, floral parts, wounds, and lateral roots. After entering into the host, pathogenic bacteria inhabit the apoplast.

Phytopathogenic fungi infect the plant tissue by using modified forms of their hyphae called appressoria as infection structure on the plant surface followed by its penetration into the plant tissue. The fungal spore adheres to the host tissue surface and directs the formation of germ tube that finally forms appressoria. This appressoria forms a peg like structure that weakens the host cell through enzymatic release. After entering into leaf epidermal cells, hyphae get divided into two types of hyphae called bulbous and lobed infectious hyphae, which later grow inter and intracellularly. *Cladosporium fulvum*, a biotrophic fungi, instead of forming haustoria grow in apoplast (Thomma et al. 2005).

Entry of viruses into the host cell takes place by the help of vectors or through physical injuries. After reaching inside the plant, virus mobilizes either in virion or non-virion form via symplastic pathway across the plasmodesmata. Some virus encoded factors and movement proteins assist their movement inside the plant system. Also, some sap feeding species of arthropods acts like a vector for viral infection and can deliver some virus particles directly into the vascular system. Nematodes and aphids feed by inserting a stylet directly into a plant cell. These diverse pathogen classes all deliver virulence factors inside the plant system to enhance microbial fitness. Entry of pathogens and activation of inducible defense response is primarily restricted by some physical barriers like cell wall, cuticle, and periderm. The second line of defense involves biochemical mechanism such as antimicrobial enzymes or secondary metabolites production. Pathogens that overcome these host defense responses produce certain signal molecules that further induce host defense response by two interconnected mechanisms called microbial (or pathogen) associated molecular pattern (MAMP/PAMP) triggered immunity (MTI/PTI) and effector-triggered immunity (ETI) as depicted in Fig. 6.2. The perception of pathogen-associated pattern leads to the activation of some defensive plant hormones signaling such as ethylene, jasmonic acid, and salicylic acid (Meng and Zhang 2013). During

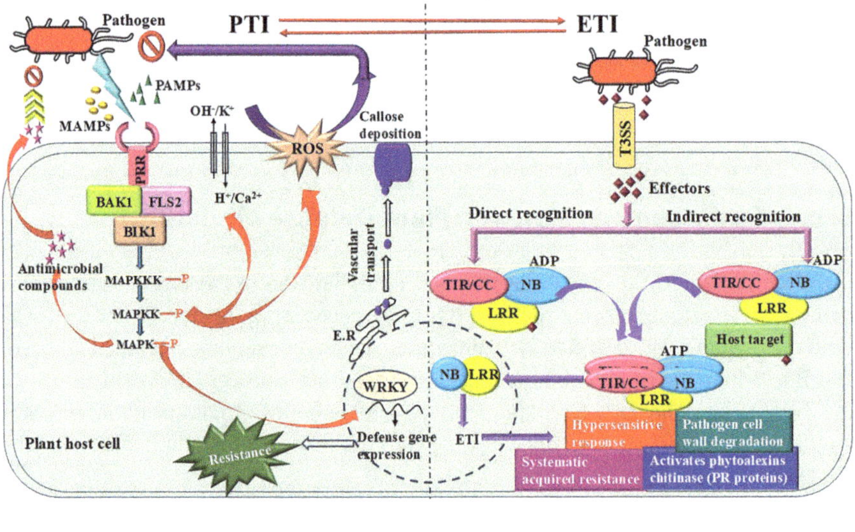

Fig. 6.2 Schematic representation of modes of plant immunity (**A**) Pattern-triggered immunity (PTI); (**B**) Effector-triggered immunity (ETI)

hypersensitive response, quick and rapid localized cell death takes place within hours and prevents the spreading of infection by programmed cell death of infected tissue. The local pathogen challenges that implements resistance in host plants to subsequent attack known as systemic acquired resistance, develops over a period of time. SAR works by forming defense related compounds like salicylic acid, pathogenesis related (PR) proteins like β-1,3-glucanases, chitinase, and hydrolytic enzymes (Park et al. 2007). Initially after pathogen attack level of SA rises in the zone of infection, and it establish SAR in distal parts of the plant. Methyl salicylate is a mobile form of SA transported via the phloem or through plasmodesmata to rest of the plant parts (Kiefer and Slusarenko, 2003). SAR is triggered by compatible pathogens, whereas ISR is induced by incompatible pathogens. Non-pathogenic bacteria like the plant growth-promoting rhizobacteria (PGPR) can trigger ISR in plants by activation of various cellular defense responses, which are subsequently gets induced upon pathogen attack. These responses include the oxidative burst, cell wall reinforcement, formation of defense enzymes, and production of secondary metabolites.

6.2.1 Microbial/Pathogen-Associated Molecular Pattern (MAMP/PAMP) Triggered Immunity (MTI/PTI)

MAMPs are microbial components or molecules produced by microbes and are general (non-specific) elicitors. These include ergosterol, cold shock proteins, xylanase, peptidoglycan, pili, flagella, bacterial lipopolysaccharide, and fungal chitin. Some particular MAMPs trigger immune response in very species-specific manner such as EF-Tu is recognized only by the members of family Brassicaceae (Zipfel et al. 2006). Other MAMPs like chitin, lipopolysaccharide, and flagellin elicitates defense response, even if there is less degree of specificity with host species (Kunze et al. 2004). Molecular alarm signal that arises from the host tissue itself either due to damage caused by microbes or by chewing insects are collectively known as damage-associated molecular pattern (DAMPs). Extracellular MAMPs produced by microbes and DAMPs released in response to microbial enzymes bind to pattern recognition receptor (PRRs) present on the outer surface of cell, carrying an external ligand binding domain having leucine rich repeats (LRR), a hydrophobic transmembrane domain and an internal catalytic domain inside the cell with serine/threonine kinase (Shiu and Bleecker, 2001). The PRR family includes some receptor like kinases (RLK) and receptor like proteins (RLP) a few listed in Table 6.1. Similarly, RLP is structurally similar but they lack intracellular cytosolic domain. Instead of catalytic domain they require adaptor proteins for signal transduction. Some of the best studied MAMP-PRR combinations are EF-Tu; EFR, Flg22; FLS2 and Ax21; Xa21. Flg22 is a peptide derived from flagellin acts as a ligand for FLS2 (Chinchilla et al. 2006; Gómez-Gómez and Boller 2000). Small peptide elf18 from the N-terminus of EF-Tu act as a ligand for receptor EFR (Chinchilla et al. 2007), and Ax21 a peptide derived from *Xanthomonas oryzae* is recognized by receptorXa21 (Lee et al. 2009). Binding of MAMP to the ligand binding side of PRR

Table 6.1 MAMPs/DAMPs under different pathogenic categories and their associated receptors in host

Pathogen	MAMPs/DAMPs	PRR	References
Bacteria	Flagellin	FLS2	Gómez-Gómez et al. (2001)
	Ax21	Xa21	Lee et al. (2009)
	Ef-Tu	EFR	Kunze et al. (2004)
	Peptidoglycan hairpin	Unidentified	Erbs et al. (2008)
	Xoo derived	LRR-RLK	Lee et al. (2009)
Fungi	Ergosterol	Unidentified	Granado et al. (1995)
	Chitin	CERK1	Shimizu et al. (2010)
	Xylanase	LeEIX2	Ron and Avni (2004)
	β-glucan	GEBP	Kishimoto et al. (2011)
	Cerebrosides	Unidentified	Umemura et al. (2002)
Oomycetes	Lipid-transfer protein	Unidentified	Osman et al. (2001)
	Transglutaminase	Unidentified	Brunner et al. (2002)
	β-glucan	GnGBP	Klarzynski et al. (2000)
	Cellulose-binding elicitor lectin (CBEL)	Unidentified	Gaulin et al. (2006)
DAMPs	Prosystemin	LRR-RLK	Scheer and Ryan (2002)
	PEPR1	LRR-RLK	Krol et al. (2010)
	Homogalacturonan	EGF-RLK	Brutus et al. (2010)

causes the receptor to heterodimerize, bringing the two cytosolic kinase domains together and thereby promoting their activation by phosphorylation. This phosphorylation creates high affinity docking sites for intracellular signaling proteins. On binding with Flg22, kinase domain of FLS2 gets phosphorylated and dimerizes with brassinosteroids intensitive 1 (BRI1)-associated kinase (BAK1) and botrytis-induced kinase (BIK1) (Kunze et al. 2004). Mutual trans autophosphorylation of the kinase domain of BIK1 and FLS2/BAK1 results in the release of BIK1 to activate downstream signaling (Laluk et al. 2011). In *Arabidopsis* discharged BIK1 activates two downstream MAPK signaling cascades which lead to the activation of WRKY family of transcription factors (Wang 2012). Inside the nucleus, these WRKYs bind with the promoter region of defense genes and activate their expression (Navarro et al. 2004). The initial defense response shown by host plants against these MAMPs are influx of K^+, H^+, Ca^{2+}, and Cl_2 ions through the plasma membrane. Increase in cytoplasmic calcium level is the main step in PTI response. PRR phosphorylation and the G protein molecular switch activate the Ca^{2+} conducting channels present in plasma membrane. This increase in Ca^{2+} level is sensed by some calcium binding proteins like calcium-dependent protein kinases (CDPKs) and calmodulin (Reddy and Reddy, 2004), which activates some defense responses such as salicylic acid (SA) formation, inhibition of reactive oxygen species (ROS) production, and stomatal closure. ROS are more toxic form of hydrogen peroxide and superoxide anion. They were also shown to have antimicrobial activities through in-direct mechanisms like cell wall strengthening via oxidative cross linking of glycoproteins, activation of SA synthesis pathway or SAR that associates with systemic propagation of the oxidative burst and activation of MAPK signaling cascade. Defense responses shown by host plants against MAMPs/PAMPs/DAMPs produced by pathogens includes closing of stomata, callose (β-1,3-glucan polymer) deposition between the plasma membrane and cell wall along with strengthening of cell wall by the crosslinking of glycoproteins. The elevation of cytoplasmic Ca^{2+} level by CAS signaling pathway finally results in closure of stomata. High extracellular Ca^{2+} level induces the accumulation of H_2O_2 and NO inside the guard cells that trigger the movement of Ca^{2+}into guard cells that finally leads to stomatal closure (Pandey and Somssich, 2009).

6.2.2 Effector-Triggered Immunity (ETI)

As the plants and pathogens started co-evolving, the pathogens produced harmful and specific elicitors to manifest disease. Some effectors, like AvrPto, AvrPtoB, HopF2, and HopAI1, suppressed the microbial triggered immunity by targeting sites in signaling cascades (Jones and Dangl 2006). The initial step in ETI signal transduction is the recognition of Avr and R proteins. Effectors produced by microbes are recognized by new resistance (R) protein which reside intracellularly and belong to the nucleotide binding leucine rich repeat (NB-LRR) protein family as listed in Table 6.2. NB-LRR comprises of a N terminal effector binding domain, a C terminal LRR domain, and a central NB region (Zhang et al. 2007). Based on the presence of variable N terminal NB-LRRs are of two types belonging to coiled coil and toll/interleukin receptor families (Wu et al. 2014). The coiled coil NB-LRR (CNL) are found in both monocots and dicots whereas, toll/interleukin NB-LRR (TNL) are found only in dicots (Zhang and Zhou, 2010). Effectors are produced by avirulence (Avr) gene and they enter into the cytoplasm of plant cell by type III secretion system (T3SS) (Grant et al. 2006). Inside the plant cell, these bacterial, fungal, and oomycetes effectors mainly target the ubiquitin-proteasome system, autophagy components, immune receptors, reactive oxygen species (ROS), phytohormones (Khan et al. 2018; Lo presti et al. 2015), chromatin configuration and also acts as a factor that increases the nutrients availability for pathogen survival (Feng and Zhou 2012). After entering into the host, they may also affect the apoplastic region and cytoplasm where these effectors target the intracellular organelles (Robin et al. 2018). Some effectors like AvrPtoB, AvrPto, HopF2, and HopAI1 downregulate the MTI by targeting few sites in signaling cascades. Receptors like NB-LRR recognize effectors either by directly interacting with the effector molecule or by indirect interaction between the receptor and effector mediated by some accessory proteins. When effectors are absent, NB-LLRs are present in their inactive ADP binding state (Zhang et al. 2007). Direct interaction between effectors and receptors induces conformational changes in receptor that result in ADP/ATP exchange and activates downstream signaling (Takken and Tameling, 2009). Indirect interaction of effectors with NB-LRR involves the interaction of plant protein RIN4 as an accessory protein. Effector AvrRpt2 (RPS2) of *Pseudomonas syringae* targets the

Table 6.2 Effectors and their R genes, reported in different pathogens

Pathogen type	Pathogen	Effector/Avr proteins	R Genes	References
Bacteria	*Pseudomonas syringae*	AvrB	Rpg1-b, Rpg2	Chisholm et al. (2006)
	Erwinia amylovora	AvrPtoB	Rps4 and Prf	White et al. (2000)
	Xanthomonas axonopodis	AvrBs1	Bs1, Bs3, Bs4, Rxv, and Xv3	Simonich and Innes, 1995
	Xanthomonas oryzae	AvrXa10	Xa3	Leach et al. (1993)
	Xanthomonas campestris	Hax3, Hax4	Bs4	Zou et al. (2010)
	Xanthomonas campestris	AvrB6, PthN, PthN2	B1	Yang (1996)
	Xanthomonas axonopodis	AvrBsT	BsT	Chisholm et al. (2006)
	Ralstonia solanacearum	PopP2	Rrs1-R	Lahaye (2004)
Fungi	*Cladosporium fulvum*	Avr2	Cf-2	Dixon et al. (1996)
	Leptosphaeria maculans	AvrLm1	Rlm1	Gout et al. (2006)
	Magnaporthe oryzae	Avr-Pita	Pi-ta	Orbach et al. (2000)
	Rhynchosporium secalis	Nip1	Rrs-1	Rohe et al. (1995)
	Blumeria graminis	Avra10	Mla10	Ridout et al. (2006)
	Phytophthora sojae	Avr3a	R3a	Armstrong et al. (2005)
	Phytophthora sojae	Avr1b-1	Rps1b	Chisholm et al. (2006)
	Phytophthora infestans	Avr4	R4	Van poppel et al. (2008)
	Hyaloperonospora parasitica	Atr13	RPP13	Rentel et al. (2008)
	Melampsora lini	AvrL567	L5, L6, and L7	Dodds et al. (2004)
	Rhynchosporium sechalis	Nip1, Nip2, and Nip3	Rrs-1	Rohe et al. (1995)
	Blumeria graminis	AvrK1	Mlk1	Ridout et al. (2006)

Arabidopsis RPM1 interacting protein 4 (RIN4) for degradation to trigger ETI (Kim et al. 2005). Three different effectors recognize and target the same host protein; therefore, manipulating RIN4 in such cases is more important for pathogenesis (Wilton et al. 2010). In case of *Arabidopsis*, the R protein (RRS1-R) interacts with the effector PopP2 (TIR NB-LRR) released by *Ralstonia solanacearum*. The

RRS1-R-PopP2 complex is then translocated into the nucleus for further downstream signaling. Downstream signaling initiated by both pattern recognition receptors and NLRs results in the production of reactive oxygen species, callose deposition, and in the activation of defense-related genes. Many times, the ETI and PTI responses overlap with each other in downstream signaling including the activation of both MAPK signaling cascade and WRKY transcription factors. This involves the formation of various antimicrobial compounds (saponin, triterpenes), polymers (Lignin, callose), hydrolytic enzymes (glucanases, chitinases), and phytoalexin (isoflavonoids, sesquiterpenes). This subsequently induces the activation of few (PR) genes surrounding the infected host tissue for the synthesis of ethylene, jasmonic acid, and salicylic acid (Nomura et al. 2012). Salicylic acid concentration rises sharply in the infected part of the host plant and binds to receptor NPR3 (nonexpressor PR Genes 3), which facilitates the turnover of NPR1 (cell death suppressor) via ubiquitination (Fu et al. 2012), that result in cell death, the hypersensitive response (HR). HR responses are quick and within hours it inhibits infection in other areas by programmed cell death (PCD) of infected tissue. SAR is a form of inducible resistance that develops in distal or healthy tissue after local infection in plant against subsequent attack. SAR is developed by the accumulation of salicylic acid and its derivatives SA-glucoside (SAG). In *Arabidopsis*, mutation in ics1(isochorismate synthase1), an enzyme required in SA synthesis, decreases the SAR response (Wildermuth et al. 2001). SA rises in the zone of infection, and it establishes in other parts of plant by a long-distance communication through phloem tissue (Tuzun and Kuć 1985). Furthermore, in tobacco plants, it was experimentally proved that long-distance transmission of SAR signals was disrupted by removing the phloem tissue above the pathogen inoculated site. DIR1(defective in induced resistance1), a lipid-transfer protein, which is present in the phloem sieve elements and companion cells, shows structural similarities to the LTP2 family of lipid-transfer proteins and is required for long-distance signaling in SAR (Champigny et al. 2011). Similarly, SAR response is also abolished in transgenic *Arabidopsis* and tobacco plants containing SA degrading salicylate hydroxylase (SH) encoded by nahG gene. It was found that methyl salicylate may be the mobile signal (transported via the vascular system) for SAR and it is not degraded by salicylate hydroxylase (SH).

6.2.2.1 Bacterial Effectors

Bacterial effector molecules are mostly excreted through the type III, IV, and VI secretory system of which type III effectors (T3Es) are best studied in plant defense. Bacterial T3Es are mainly associated with host defense suppression exhibited through interference with transcription, signaling pathways, secretory pathways, nutrient acquisition, and bacterial colonization. Transcription activator like effectors (TALEs) of *Xanthomonas* sp. is the main class of T3SS effectors that acts like a transcription factor on binding near the promoter region of host genes in nucleus targeting host nutrient transporters and various factors involved in promoting disease susceptibility (Schwartz et al. 2017). While in bacterial system, the defense-related avirulence effectors induces ETI in the presence of resistant proteins like RipB and Rip, in few bacterias like *Xanthomonas* T3E suppress ETI and PTI related cell death. Effectors AvrPtoB and AvrPto produced by *Pseudomonas syringae* trigger resistance in tomato plant having Prf as NB-LRR protein, whereas some effectors such as AvrB, AvrRpm1, and AvrRpt2 specify immunity in *Arabidopsis* plants containing NB-LRR proteins such as RPS2, RPS5, and RPM1 (Dangl and Jones, 2001). Genetic and molecular studies have observed that these NB-LRR receptors do not show direct interaction with their corresponding effector molecules, but they show indirect interaction with effector protein by associating with some other host proteins (Day et al. 2005). Bacterial effector AvrPphB interacts with host protein PBS1 and latter gets associated with the NB-LRR protein RPS5. AvrPphB possesses cysteine protease properties that cleave the protein PBS1of *Arabidopsis*. Cleavage of PBS1 protein induces a conformation change in RPS5 that changes it from inactive GDP-binding state to active GTP-binding state that finally triggers resistance. Effectors AvrRpm1 and AvrB both trigger resistance in host plants having the resistance gene RPM1. Their recognition is carried out by a membrane localized protein called RPM1 interacting protein (RIN4). Both AvrRpm1 and AvrB phosphorylates the RIN4 present in host plants (Day et al. 2005). Even the biochemical process induced by both AvrB and AvrRpm1 are not known but they bind ADP and show some structural features similar to protein kinases (Mackey et al. 2002). Effector AvrRpt2 acts like a cysteine protease and gets activated by interacting with a cyclophilin protein present in the host cell (Mackey et al. 2003). On binding with AvrB, RIN4 also interacts with another R protein RPS2 (Coaker et al. 2005). RIN4 on cleaving with AvrRpt2 releases RPS2 and triggers resistance. It also inhibits the RPM1 activation by interacting with AvrB or AvrRpm (Desveaux et al. 2007). Thus, the ability of RPS2 to guard RIN4 and AvrRpt2 to inhibit RIN4-dependent RPM1 resistance appears to be an arm race in the host pathogen co-evolution for existence. Another example of evolutionary arm race can be seen during the interaction between *Pseudomonas syringae* effectors AvrPto and AvrPtoB from *Pseudomonas syringae* with their corresponding R protein complex. Effector AvrPto induces resistance by interacting with a protein kinase Pto associated with Prf in tomato plants (Xing et al. 2007). Few structural and biochemical studies have shown that effector AvrPto acts like a protine kinase inhibitor on interacting with Pto (Tang et al. 1996). Such interaction may become suitable for the activation of Prf protein. Like Pto, Fen also interacts with another effector AvrPtoB to induce resistance (Dong et al. 2009). N-terminus of both AvrPto and AvrPtoB attaches with the P + 1 loop of Pto that is mainly necessary for the substrate binding and confirms that AvrPtoB inhibits the Pto kinase (Rosebrock et al. 2007). The C-terminus of AvrPtoB acts like an ubiquitin E ligase and it degrades Fen via the E3 ubiquitin ligase pathway (Abramovitch et al. 2006). A recent report indicated that Pto is not ubiquitinated by AvrPtoB but it can phosphorylate the C-terminus of AvrPtoB to inhibit its E3 ubiquitin ligase activity. The interactions of both effectors AvrPto and AvrPtoB with their corresponding Prf-Pto and Prf-Fen complexes once again confirm the R gene mediated resistance and host–pathogen co-evolution theory.

6.2.2.2 Fungal Effectors

Among eukaryotes, fungal and oomycete are the major disease-causing pathogens. Fungal effectors are mostly proteins, protein toxins and other metabolites that interfere with various plant processes. Effectors either promote the harmful effects of fungal pathogens or allow them to colonize host plants as a symbiont. These effectors molecules can easily get attached at cell wall, can reside in apoplast and cytoplasm, and can also mobilize to other compartments of a cell (Kubicek et al. 2014). Mutation and deletion of effector gene result in less severe disease symptoms. Pep1 effector secreted from *Ustilago maydis* accumulates in the apoplast after penetrating the fungal hyphae from epidermal cells of maize. Deletion or mutation of Pep1 induces strong defense responses and blocks all the adverse effects of fungal intracellular hyphae in plants. Pep1 suppresses the plant defense response by inhibiting the activity of POX12, a peroxidase enzyme formed in maize plant (Hemetsberger et al. 2012). Effector Pit2 secreted from *Ustilago maydis* inhibits the activity of cysteine proteases present in apoplastic region of maize, whose function is to promote the salicylic acid triggered defense response in host plant (Doehlemann et al. 2011). Recent findings have suggested that effectors molecules produced by pathogens can also inhibit secondary metabolites pathway. Tin2 effector of *Ustilago maydis* interacts with the cytoplasmic protein kinase ZmTTK1 of maize and induces the biosynthesis of anthocyanin which negatively regulates the lignin production by reducing the level of its precursor p-coumaric acid. Lignin acts like a physical barrier to pathogens spread (Tanaka et al. 2014). Effectors of necrotrophic fungi include secondary metabolites, peptide toxins, necrosis, and ethylene inducing peptide1 (Nep1) that induces plant cell death (Manning et al. 2007). In case of necrotrophs, host protein should have sensitivity for their compatible effectors. For establishing a compatible interaction, necrotrophs transfer effectors toward their corresponding R proteins of host and destroy the mechanism that triggers resistance to biotrophic pathogens (Oliver et al. 2012). The effector ToxA of *Stagonospora nodorum* R protein Tsn1 in wheat after reaching the chloroplast suppresses the activity of thylakoid forming protein ToxABP1. This interaction interrupts photosynthesis, finally resulting into cell death (Qutob et al. 2006). As discussed earlier, fungal effectors promote host plant colonization by targeting their signaling components and metabolic pathways. Mainly all fungal and oomycetes effectors have protease inhibitor activity. Effector Pit2 from *Ustilago maydis* and EPIC1, EPIC1B from *Phytophthora infestans* inhibits apoplastic PIP1 and Rcr3 cysteine protease in tomato. Inhibition of these apoplastic proteases is a main step for establishing compatibility (Mueller et al. 2013). Avr2 effector from *Cladosporium fulvum* interacts with Rcr3 a Cysteine protease and activates Cf-2-mediated immunity. Avr2 binding induces some conformational changes in Rcr3 that is monitored by the Cf-2-protein resulting in hypersensitive response and resistance produced by wild type Avr2 against *Cladosporium fulvum* (Angot et al. 2006). The Avr-Pita gene of *Magnaporthe oryzae* encodes metalloprotease which acts like an avirulence factor and is recognized on the cell surface by a receptor-like protein Pita in rice and it leads to a hypersensitive response in the infected part of plant when they are infected with *Magnaporthe oryzae* (Jia et al. 2000). Some effectors like Foa3 produced by *Fusarium oxysporum* and Rip1 by *Ustilago maydis* suppress MTI in host plant. Effector BAS2 of *Colletotrichum gloeosporioides* aid in the formation of pathogen reproductive structures and effector Lep1 produced by *Ustilago maydis* help in attachment of hyphae and its proliferation inside the host plant (Fukada et al. 2021). Necrotrophic wheat pathogens such as *Parastagnospora nodorum* and *Pyrenophora tritici repentis* binds with host protein encoded by susceptibility (S) genes. This interaction results in susceptibility of disease instead of providing resistance to host. So, this interaction is known as inverse gene for gene interaction. Like (R) protein and Avr interaction, this interaction also leads to programmed cell death (PCD) triggered by an oxidative burst. Effector SnTox1 of *Parastagnospora nodorum* and victorin of *Cochliobolus victoriae* is an example of necrotrophic effector that interacts with S gene product in their host. Some effectors also interfere with ubiquitin-proteasome system of host to promote disease. The effector AvrPiz-t of *Magnaporthe oryzae* inhibits the proteasomal activity of RING E3 ubiquitin ligase APIP6. GALA effector secreted by *Ralstonia solanacearum* inhibits the SKP1-like protein of the SCF-type E3 ubiquitin ligase system (Van Esse et al. 2008).

6.3 Effectors in Plant Beneficial Microbe Interaction

Mutualistic pathogens reprogram the host–plant immune responses to facilitate colonization. The mutualistic microbes trigger MTI and the plants later get benefited by the systemic resistance developed against diverse pathogens through ISR (Pieterse et al. 2014). Well-reported ISR inducers among bacteria and fungi include *Pseudomonas* spp., *Rhizobium* spp., *Trichoderma* sp., and *Serendipita indica* (Jaiswal et al. 2021). Effectors play an important role in shaping microbial communities, their compatibility with host and other microbes also determining their life style. Effectors secreted by various beneficial microbes target some hormone signaling pathways in host plants. Effector, SP7 produced by *Glomus intraradices,* alters the ethylene production by interacting with ERF19 a transcription factor involved in ethylene signaling (kloppholz et al. 2011). Effector MiSSP7 of *Laccaria bicolor* prevents the degradation of plant repressor proteins such as PtJAZ5 and PtJAZ6, which would otherwise result in the activation of jasmonic acid forming genes that plays a role in plant's defense (Porras-Alfaro and Bayman 2011). Effectors also induce mutualistic interaction between endophytes and plants. Effector RiCRN1 of *Rhizophagus irregularis* develops a fungal interactive structure called arbuscules (Wawra et al. 2016).

All mycorrhizae are not considered as endophytes as they have originated from different phylogenetic groups and all root endophytes also do not actively participate in nutrients transfer as mycorrhizae do (Plett et al. 2014). Effectors produced by endophytic microbes also manipulate MTI to establish a symbiotic relationship between endophyte and host. Effector produced by endophytic fungus *Pestalotiopsis* sp. hydrolyzes the chitin oligomers for inhibiting the chitin-triggered immunity in rice host plants through its chitin deacetylase activity (Cord-Landwehr et al. 2016). Effector FGBI secreted from endophytic fungus *Piriformospora indica* prevents

MTI triggered by β-glucan by interacting with β-glucan in the cell wall of fungus. (Wawra et al. 2016). Bacteria that belong to the genera *Variovorax* and *Acidovorax* residing in the rhizospheric part of *Arabidopsis thaliana* protect the *Arabidopsis thaliana* from various disease-causing fungi, bacteria, and oomycetes by inducting systemic resistance against these pathogens (Durán et al. 2018). Effector Hyde1 from endophytic bacteria *Acidovorax* show antibacterial activity against *Escherichia coli* and other bacterial isolates as well.

6.4 Conclusion

Effectors repertoires play an integral role in plant–microbe interactions, which help the pathogens and the endophytes to enter and proliferate in the host through suppressing the host recognition system, interfering with phytohormone synthesis, regulating host gene expression and protein translocation. Better understanding of the underlying mechanism of interaction between effector molecules and their host plants is crucial in disease management. Effector-assisted selection of germplasm with resistance genes or plant lacking susceptible genes to certain effectors can be helpful in resistance breeding and finding novel resistance genes. Advancements in effector biology can also be achieved through gene editing tools, R gene pyramiding or manipulating the S gene. These molecules involved in beneficial and non-beneficial plant interactions are proven mines in understanding plant growth and disease management.

References

Abramovitch RB, Janjusevic R, Stebbins CE, Martin GB et al (2006) Type III effector AvrPtoB requires intrinsic E3 ubiquitin ligase activity to suppress plant cell death and immunity. PNAS 103(8):2851–2856

Angot A, Peeters N, Lechner E, Vailleau F, Baud C et al (2006) *Ralstonia solanacearum* requires F-box-like domain-containing type III effectors to promote disease on several host plants. PNAS 103:14620–14625

Armstrong MR, Whisson SC, Pritchard L, Bos JI, Venter E, Avrova AO et al (2005) An ancestral oomycete locus contains late blight avirulence gene Avr3a, encoding a protein that is recognized in the host cytoplasm. PNAS 102(21):7766–7771

Brunner F, Rosahl S, Lee J, Rudd JJ, Geiler C, Kauppinen S et al (2002) Pep-13, a plant defense-inducing pathogen-associated pattern from *Phytophthora transglutaminases*. EMBO J 16; 21(24):6681–6688

Brutus A, Sicilia F, Macone A, Cervone F, De Lorenzo G et al (2010) A domain swap approach reveals a role of the plant wall-associated kinase 1 (WAK1) as a receptor of oligogalacturonides. PNAS107 20:9452–9457

Champigny MJ, Shearer H, Mohammad A, Haines K, Neumann M, Thilmony R et al (2011) Localization of DIR1 at the tissue, cellular and subcellular levels during systemic acquired resistance in *Arabidopsis* using DIR1: GUS and DIR1: EGFP reporters. BMC Plant Boil 11:1–16

Chinchilla D, Bauer Z, Regenass M, Boller T, Felix G et al (2006) The *Arabidopsis* receptor kinase FLS2 binds flg22 and determines the specificity of flagellin perception. Plant Cell 18(2):465–476

Chinchilla D, Zipfel C, Robatzek S, Kemmerling B, Nürnberger T, Jones JD, Boller T et al (2007) A flagellin-induced complex of the receptor FLS2 and BAK1 initiates plant defence. Nature 448(7152):497–500

Chisholm ST, Coaker G, Day B, Staskawicz BJ et al (2006) Host-microbe interactions: shaping the evolution of the plant immune response. Cell J 124(4):803–814

Cord-Landwehr S, Melcher RL, Kolkenbrock S, Moerschbacher BM et al (2016) A chitin deacetylase from the endophytic fungus *Pestalotiopsis* sp. efficiently inactivates the elicitor activity of chitin oligomers in rice cells. Sci Rep 6(1):38018

Dangl JL, Jones JD (2001) Plant pathogens and integrated defence responses to infection. Nature 411(6839):826–833

Day B, Dahlbeck D, Huang J, Chisholm ST, Li D, Staskawicz BJ et al (2005) Molecular basis for the RIN4 negative regulation of RPS2 disease resistance. Plant Cell 17(4):1292–1305

Desveaux D, Singer AU, Wu AJ, McNulty BC, Musselwhite L, Nimchuk Z et al (2007) Type III effector activation via nucleotide binding, phosphorylation, and host target interaction. PLoS Pathog 3(3):e48

Dixon MS, Jones DA, Keddie JS, Thomas CM, Harrison K, Jones JD et al (1996) The tomato Cf-2 disease resistance locus comprises two functional genes encoding leucine-rich repeat proteins. Cell J 84(3):451–459

Dodds PN, Lawrence GJ, Catanzariti AM, Ayliffe MA, Ellis JG et al (2004) The *Melampsora lini* AvrL567 avirulence genes are expressed in haustoria and their products are recognized inside plant cells. Plant Cell 16(3):755–768

Doehlemann G, Reissmann S, Aßmann D, Fleckenstein M, Kahmann R et al (2011) Two linked genes encoding a secreted effector and a membrane protein are essential for *Ustilago maydis*-induced tumour formation. Mol Microbiol 81(3):751–766

Dong J, Xiao F, Fan F, Gu L, Cang H, Martin GB, Chai J et al (2009) Crystal structure of the complex between *pseudomonas* effector AvrPtoB and the tomato Pto kinase reveals both a shared and a unique interface compared with AvrPto-Pto. Plant Cell 21(6):1846–1859

Durán P, Thiergart T, Garrido-Oter R, Agler M, Kemen E, Schulze-Lefert P, Hacquard S et al (2018) Microbial interkingdom interactions in roots promote *Arabidopsis* survival. Cell 175(4):973–983

Erbs G, Silipo A, Aslam S, De Castro C, Liparoti V, Flagiello A et al (2008) Peptidoglycan and muropeptides from pathogens *agrobacterium* and *Xanthomonas* elicit plant innate immunity: structure and activity. Chem Biol 15(5):438–448

Felix G, Boller T (2003) Molecular sensing of bacteria in plants: the highly conserved RNA-binding motif RNP-1 of bacterial cold shock proteins is recognized as an elicitor signal in tobacco. J Biol Chem 278(8):6201–6208

Feng F, Zhou JM (2012) Plant-bacterial pathogen interactions mediated by type III effectors. Curr. Opin. Plant Biol., 15, 469–476

Fu ZQ, Yan S, Saleh A, Wang W, Ruble J, Oka N et al (2012) NPR3 and NPR4 are receptors for the immune signal salicylic acid in plants. Nature 486(7402):228–232

Fukada F, Rössel N, Münch K, Glatter T, Kahmann R et al (2021) A small *Ustilago maydis* effector acts as a novel adhesin for hyphal aggregation in plant tumors. New Phytol 231(1):416–431

Gaulin E, Drame N, Lafitte C, Torto-Alalibo T, Martinez Y, Ameline-Torregrosa C et al (2006) Cellulose binding domains of a *Phytophthora* cell wall protein are novel pathogen-associated molecular patterns. Plant Cell 18(7):1766–1777

Gómez-Gómez L, Boller T (2000) FLS2: an LRR receptor–like kinase involved in the perception of the bacterial elicitor flagellin in *Arabidopsis*. Mol Cell 5(6):1003–1011

Gómez-Gómez L, Bauer Z, Boller T et al (2001) Both the extracellular leucine-rich repeat domain and the kinase activity of FLS2 are required for flagellin binding and signalling in *Arabidopsis*. Plant Cell 13(5):1155–1163

Gout L, Fudal I, Kuhn ML, Blaise F, Eckert M, Cattolico L et al (2006) Lost in the middle of nowhere: the AvrLm1 avirulence gene of the Dothideomycete *Leptosphaeria maculans*. Mol Microbiol 60(1):67–80

Granado J, Felix G, Boller T et al (1995) Perception of fungal sterols in plants (subnanomolar concentrations of ergosterol elicit extracellular alkalinization in tomato cells). Plant Physiol 107(2):485–490

Grant SR, Fisher EJ, Chang JH, Mole BM, Dangl JL et al (2006) Subterfuge and manipulation: type III effector proteins of phytopathogenic bacteria. Ann Rev Microbiol 60:425–449

Hemetsberger C, Herrberger C, Zechmann B, Hillmer M, Doehlemann G et al (2012) The Ustilago maydis effector Pep1 suppresses plant immunity by inhibition of host peroxidase activity. PLoS Pathog 8(5):e1002684

Jaiswal SK, Mohammed M, Ibny FY, Dakora FD et al (2021) Rhizobia as a source of plant growth-promoting molecules: potential applications and possible operational mechanisms. Front Sustain Food Syst 4:619676

Jia Y, McAdams SA, Bryan GT, Hershey HP, Valent B et al (2000) Direct interaction of resistance gene and avirulence gene products confers rice blast resistance. EMBO J 9(15):4004–4014

Jones JD, Dangl JL (2006) The plant immune system. Nature 444(7117):323–329

Khan M, Seto D, Subramaniam R, Desveaux D (2018) Oh, the places they'll go! A survey of phytopathogen effectors and their host targets. Plant J 93(4):651–663

Kiefer IW, Slusarenko AJ (2003) The pattern of systemic acquired resistance induction within the *Arabidopsis* rosette in relation to the pattern of translocation. Plant Physiol 132(2):840–847

Kim MG, Da Cunha L, McFall AJ, Belkhadir Y, DebRoy S, Dangl JL, Mackey D et al (2005) Two pseudomonas syringae type III effectors inhibit RIN4-regulated basal defense in *Arabidopsis*. Cell 121(5):749–759

Kishimoto K, Kouzai Y, Kaku H, Shibuya N, Minami E, Nishizawa Y et al (2011) Enhancement of MAMP signaling by chimeric receptors improves disease resistance in plants. Plant Signal Behav 6(3):449–451

Klarzynski O, Plesse B, Joubert JM, Yvin JC, Kopp M, Kloareg B, Fritig B et al (2000) Linear β-1, 3 glucans are elicitors of defense responses in tobacco. Plant Physiol 124(3):1027–1038

Kloppholz S, Kuhn H, Requena N et al (2011) A secreted fungal effector of *glomus intraradices* promotes symbiotic biotrophy. Curr Biol 21(14):1204–1209

Krol E, Mentzel T, Chinchilla D, Boller T, Felix G, Kemmerling B et al (2010) Perception of the Arabidopsis danger signal peptide 1 involves the pattern recognition receptor AtPEPR1 and its close homologue AtPEPR2. J Biol Chem 285(18):13471–13479

Kubicek CP, Starr TL, Glass NL et al (2014) Plant cell wall–degrading enzymes and their secretion in plant-pathogenic fungi. Annu Rev Phytopathol 52:427–451

Kunze G, Zipfel C, Robatzek S, Niehaus K, Boller T, Felix G et al (2004) The N terminus of bacterial elongation factor Tu elicits innate immunity in *Arabidopsis* plants. Plant Cell 16(12):3496–3507

Lahaye T (2004) Illuminating the molecular basis of gene-for-gene resistance; *Arabidopsis thaliana* RRS1-R and its interaction with *Ralstonia solanacearum* popP2. Trends Plant Sci 9(1):1–4

Laluk K, Luo H, Chai M, Dhawan R, Lai Z, Mengiste T et al (2011) Biochemical and genetic requirements for function of the immune response regulator BOTRYTIS-INDUCED KINASE1 in plant growth, ethylene signalling, and PAMP-triggered immunity in *Arabidopsis*. Plant Cell 23(8):2831–2849

Leach JE, Hopkins C, Guo A, Choi SH, Mazzola M, Ryba-White M, White FF et al (1993) A family of avirulence genes from *Xanthomonas oryzae* pv. *oryzae* is involved in resistant interactions in rice. In *Proceedings of the 6th international symposium on molecular plant-microbe interactions, Seattle, Washington, USA, July 1992*. In *Advances in molecular genetics of plant-microbe interactions, Vol 2*. Springer Dordrecht, pp. 221–230

Lee SW, Han SW, Sririyanum M, Park CJ, Seo YS, Ronald PC et al (2009) A type I–secreted, sulfated peptide triggers XA21-mediated innate immunity. Science 326(5954):850–853

Lo Presti L, Lanver D, Schweizer G, Tanaka S, Liang L, Tollot M et al (2015) Fungal effectors and plant susceptibility. Annu Rev Plant Biol 66:513–545

Mackey D, Holt BF, Wiig A, Dangl JL et al (2002) RIN4 interacts with *pseudomonas syringae* type III effector molecules and is required for RPM1-mediated resistance in *Arabidopsis*. Cell 108(6):743–754

Mackey D, Belkhadir Y, Alonso JM, Ecker JR, Dangl JL et al (2003) *Arabidopsis* RIN4 is a target of the type III virulence effector AvrRpt2 and modulates RPS2-mediated resistance. Cell 112:379–389

Manning VA, Hardison LK, Ciuffetti LM et al (2007) Ptr ToxA interacts with a chloroplast-localized protein. Mol Plant-Microbe Interact 20(2):168–177

Meng X, Zhang S (2013) MAPK cascades in plant disease resistance signaling. Annu Rev Phytopathol 51:245–266

Mueller AN, Ziemann S, Treitschke S, Aßmann D, Doehlemann G (2013) Compatibility in the *Ustilago maydis*–maize interaction requires inhibition of host cysteine proteases by the fungal effector Pit2. PLoS Patho 9(2):e1003177

Navarro L, Zipfel C, Rowland O, Keller I, Robatzek S, Boller T, Jones JD et al (2004) The transcriptional innate immune response to flg22. Interplay and overlap with Avr gene-dependent defense responses and bacterial pathogenesis. Plant Physiol 135(2):1113–1128

Nomura H, Komori T, Uemura S, Kanda Y, Shimotani K, Nakai K et al (2012) Chloroplast-mediated activation of plant immune signalling in *Arabidopsis*. Nat Commun 3(1):926

Oliver RP, Friesen TL, Faris JD, Solomon PS et al (2012) *Stagonospora nodorum*: from pathology to genomics and host resistance. Annu Rev Phytopathol 50:23–43

Orbach MJ, Farrall L, Sweigard JA, Chumley FG, Valent B et al (2000) A telomeric avirulence gene determines efficacy for the rice blast resistance gene pi-ta. Plant Cell 12(11):2019–2032

Osman H, Vauthrin S, Mikes V, Milat ML, Panabières F, Marais A et al (2001) Mediation of elicitin activity on tobacco is assumed by elicitin-sterol complexes. Mol Biol Cell 12(9):2825–2834

Pandey SP, Somssich IE (2009) The role of WRKY transcription factors in plant immunity. Plant Physiol 150(4):1648–1655

Park SW, Kaimoyo E, Kumar D, Mosher S, Klessig DF et al (2007) Methyl salicylate is a critical mobile signal for plant systemic acquired resistance. Science 318(5847):113–116

Pieterse CM, Zamioudis C, Berendsen RL, Weller DM, Van Wees SC, Bakker PA et al (2014) Induced systemic resistance by beneficial microbes. Annu Rev Phytopathol 52:347–375

Plett JM, Daguerre Y, Wittulsky S, Vayssières A, Deveau A, Melton SJ et al (2014) Effector MiSSP7 of the mutualistic fungus *Laccaria bicolor* stabilizes the *Populus* JAZ6 protein and represses jasmonic acid (JA) responsive genes. PNAS 111(22):8299–8304

Porras-Alfaro A, Bayman P (2011) Hidden fungi, emergent properties: endophytes and microbiomes. Annu Rev Phytopathol 49:291–315

Qutob D, Kemmerling B, Brunner F, Kufner I, Engelhardt S, Gust AA et al (2006) Phytotoxicity and innate immune responses induced by Nep1-like proteins. Plant Cell 18(12):3721–3744

Reddy VS, Reddy AS (2004) Proteomics of calcium-signalling components in plants. Phytochemistry 65(12):1745–1776

Rentel MC, Leonelli L, Dahlbeck D, Zhao B, Staskawicz BJ et al (2008) Recognition of the *Hyaloperonospora parasitica* effector ATR13 triggers resistance against oomycete, bacterial, and viral pathogens. PNAS 105(3):1091–1096

Ridout CJ, Skamnioti P, Porritt O, Sacristan S, Jones JD, Brown JK et al (2006) Multiple avirulence paralogues in cereal powdery mildew fungi may contribute to parasite fitness and defeat of plant resistance. Plant Cell 18(9):2402–2414

Robin GP, Kleemann J, Neumann U, Cabre L, Dallery JF, Lapalu N, O'Connell RJ et al (2018) Subcellular localization screening of *Colletotrichum higginsianum* effector candidates identifies fungal proteins targeted to plant peroxisomes, golgi bodies, and microtubules. Front Plant Sci 9:353255

Rohe M, Gierlich A, Hermann H, Hahn M, Schmidt B, Rosahl S, Knogge W et al (1995) The race-specific elicitor, NIP1, from the barley pathogen, *Rhynchosporium secalis*, determines avirulence on host plants of the Rrs1 resistance genotype. EMBO J 14(17):4168–4177

Ron M, Avni A (2004) The receptor for the fungal elicitor ethylene-inducing xylanase is a member of a resistance-like gene family in tomato. Plant Cell 16(6):1604–1615

Rosebrock TR, Zeng L, Brady JJ, Abramovitch RB, Xiao F, Martin GB et al (2007) A bacterial E3 ubiquitin ligase targets a host protein kinase to disrupt plant immunity. Nature 448(7151):370–374

Scheer JM, Ryan CA Jr (2002) The systemin receptor SR160 from *Lycopersicon peruvianum* is a member of the LRR receptor kinase family. PNAS 99(14):9585–9590

Schwartz AR, Morbitzer R, Lahaye T, Staskawicz BJ et al (2017) TALE-induced bHLH transcription factors that activate a pectate lyase contribute to water soaking in bacterial spot of tomato. PNAS 114(5):E897–E903

Shimizu T, Nakano T, Takamizawa D, Desaki Y, Ishii-Minami N, Nishizawa Y et al (2010) Two LysM receptor molecules, CEBiP and OsCERK1, cooperatively regulate chitin elicitor signalling in rice. Plant J 64(2):204–214

Shiu SH, Bleecker AB (2001) Receptor-like kinases from *Arabidopsis* form a monophyletic gene family related to animal receptor kinases. PNAS 98(19):10763–10768

Simonich MT, Innes RW (1995) A disease resistance gene in *Arabidopsis* with specificity for the avrPph3 gene of *pseudomonas syringae* pv. *Phaseolicola*. Mol Plant-Microbe Interact 8(4):637–640

Takken FLW, Tameling WIL (2009) To nibble at plant resistance proteins. Science 324(5928):744–746

Tanaka S, Brefort T, Neidig N, Djamei A, Kahnt J, Vermerris W et al (2014) A secreted *Ustilago maydis* effector promotes virulence by targeting anthocyanin biosynthesis in maize. elife 3:e01355

Tang X, Frederick RD, Zhou J, Halterman DA, Jia Y, Martin GB et al (1996) Initiation of plant disease resistance by physical interaction of AvrPto and Pto kinase. Science 274:2060–2063

Thomma BP, Van Esse HP, Crous PW, De Wit PJ et al (2005) *Cladosporium fulvum* (syn. *Passalora fulva*), a highly specialized plant pathogen as a model for functional studies on plant pathogenic Mycosphaerellaceae. Mol. Plant Pathol 6(4):379–393

Tuzun S, Kuć J (1985) Movement of a factor in tobacco infected with *Peronospora tabacina Adam* which systemically protects against blue mold. Physiol Plant Pathol 26(3):321–330

Umemura K, Ogawa N, Koga J, Iwata M, Usami H et al (2002) Elicitor activity of cerebroside, a sphingolipid elicitor, in cell suspension cultures of rice. Plant Cell Physiol 43(7):778–784

Van Esse HP, Van't Klooster JW, Bolton MD, Yadeta KA, van Baarlen P, Boeren S et al (2008) The *Cladosporium fulvum* virulence protein Avr2 inhibits host proteases required for basal defense. Plant Cell 20(7):1948–1963

Van Poppel PM, Guo J, van de Vondervoort PJ, Jung MW, Birch PR, Whisson SC, Govers F et al (2008) The *Phytophthora infestans* avirulence gene Avr4 encodes an RXLR-dEER effector. Mol Plant-Microbe Interact 21(11):1460–1470

Wang ZY (2012) Brassinosteroids modulate plant immunity at multiple levels. Proc Nat Acad Sci 109(1):7–8

Wawra S, Fesel P, Widmer H, Timm M, Seibel J, Leson L et al (2016) The fungal-specific β-glucan-binding lectin FGB1 alters cell-wall composition and suppresses glucan-triggered immunity in plants. Nat Commun 7(1):13188

White FF, Yang B, Johnson LB (2000) Prospects for understanding avirulence gene function. Curr Opin Plant Biol 3(4):291–298

Wildermuth MC, Dewdney J, Wu G, Ausubel FM et al (2001) Isochorismate synthase is required to synthesize salicylic acid for plant defence. Nature 414(6863):562–565

Wilton M, Subramaniam R, Elmore J, Felsensteiner C, Coaker G, Desveaux D et al (2010) The type III effector HopF2 Pto targets *Arabidopsis* RIN4 protein to promote *pseudomonas syringae* virulence. PNAS 107(5):2349–2354

Wu L, Chen H, Curtis C, Fu ZQ (2014) Go in for the kill: how plants deploy effector-triggered immunity to combat pathogens. Virulence 5(7):710–721

Xing W, Zou Y, Liu Q, Liu J, Luo X, Huang Q et al (2007) The structural basis for activation of plant immunity by bacterial effector protein AvrPto. Nature 449(7159):243–247

Yang Y, Yuan Q, Gabriel DW (1996) Water soaking function (s) of XcmH1005 are redundantly encoded by members of the *Xanthomonas* avr/pth gene family. MPMI-Mol Plant Microbe Interact 9(2):105–113

Zhang J, Zhou JM (2010) Plant immunity triggered by microbial molecular signatures. Mol Plant 3(5):783–793

Zhang J, Shao F, Li Y, Cui H, Chen L, Li H et al (2007) A *pseudomonas syringae* effector inactivates MAPKs to suppress PAMP-induced immunity in plants. Cell Host Microbe 1(3):175–185

Zhang S, Li C, Si J, Han Z, Chen D et al (2022) Action mechanisms of effectors in plant-pathogen interaction. Int J Mol Sci 23(12):6758

Zipfel C, Kunze G, Chinchilla D, Caniard A, Jones JD, Boller T, Felix G et al (2006) Perception of the bacterial PAMP EF-Tu by the receptor EFR restricts *agrobacterium*-mediated transformation. Cell 125(4):749–760

Zou H, Zhao W, Zhang X, Han Y, Zou L, Chen G et al (2010) Identification of an avirulence gene, a vrxa5, from the rice pathogen *Xanthomonas oryzae* pv. *Oryzae*. Sci China Life Sci 53:1440–1449

Plants Retaliating Defense Strategies against Herbivores

7

Shweta Verma, Manisha Hariwal, Priya Patel, Priyaka Shah, and Sanjay Kumar

Abstract

Every organism including plants possesses their own shield of protection to survive on this earth. They perform every effort against pathogens resistance, tolerance, and even multiple secretions to defend themselves. Plants exploit physical defense as primitive line of defense where plant structural parts play significant role such as thorns, prickles, and trichomes. However, the acute, active, and spontaneous defense mechanism initiated by the plants in response to attack of external pathogen is chemical defense. Within this, various plant as well as pathogen derived elicitor's, employ their duty as signal recognizer, receiver and as transferor to the plant defense system. Elicitor's triggers synthesis and secretion of secondary metabolites which causes the extensive ditch toward herbivory. A plethora of secondary metabolites can directly affect the herbivores or indirectly encounter the pathogen by attracting predators of them belonging to above trophic levels, which unintentionally rescues the plant. These natural chemical compounds such as terpenoids, lipases, glucosinolates, cyanogenic glycosides, polypeptides, and fatty acid conjugates can be anti-nutritive or toxic to herbivore pathogens which minimize their fecundity, growth, and even survival.

Keywords

Physical defense · Chemical defense · Elicitor · Herbivory · Anti-nutritive · Secondary metabolites

---

S. Verma · M. Hariwal · P. Patel · P. Shah · S. Kumar (✉)
Banaras Hindu University, Varanasi, India
e-mail: skumar.bot@bhu.ac.in

S. Singh, R. Mehrotra (eds.), *Plant Elicitor Peptides*,
https://doi.org/10.1007/978-981-97-6374-0_7

7.1 Introduction

Plants demand various fundamental elements in certain amount for survival within their evident life cycle phases of its growth, development, and their metabolisms. Simultaneously, they also deal with several abiotic and biotic environmental stresses in this voyage. Various ecological factors, which are non-living and impact negatively to the plants, are considered as abiotic stresses (Verma et al. 2013). Even fundamental elements such as temperature, light, water, etc. can also be regarded as abiotic stress causing agents. When their extreme fluctuations harm the plants they cause stress, it can be very high or too low temperature, water excessive or deficient conditions, salinity challenges, ultraviolet radiations, as well as load of heavy metal stress. Environmental factors responsible for abiotic stresses can be majorly compensated by acclimatization, adaptation, and repair mechanisms of plants. Whereas, the living organisms that cause stress to other living beings are considered as biotic stresses. These stresses are imposed via heterotrophic organisms, specifically herbivores, which are the major damaging entities, e.g., bacteria, virus, fungi, parasites, insects, and herbivores (Fosket 1994).

Several living beings injure and destroy plants quite rigorously, among them specifically herbivores cause enormous biomass destruction. Large herbivore animals which are primarily grazing mammals tend to eat grasses, leaves, and stems. Herbivorous birds, few amphibians, and reptiles are also herbivorous which generally eat seeds, nectar, and fruits of plants. They are ignored because these are large animals that cannot be resisted because of the sessile nature of plants (Zhu et al. 2011). However, small herbivore insects, nematodes, arthropods, and other invertebrates, mostly leaves and nectar eaters, e.g., ant, termite, grasshopper, aphid, caterpillar, cricket, leaf katydid bugs, moth, honeybee, slug, butterfly, snail, and worms are restricted by the plants. Herbivores basically consume plants as a whole or their parts like leaves, stem, flower, fruit, roots, tubers, etc. They can be folivores (leaf eaters) (Prade and Coyle 2023), frugivores (fruit eaters) (Fleming and Kress 2011), granivores (seed eaters) (Hulme and Benkman 2002), xylophages (wood eaters) (Duque-Lazo and Navarro-Cerrillo 2017), and nectivores (nectar feeders) (Ravenscraft and Boggs 2016). These herbivores in the beginning consume few parts of plant which causes primary injury, weakened, and compromised immune response finally provokes vulnerability toward other stresses that transcend complete destruction or even death of the plant. Herbivores are capable enough to vigorously damage flora of plants, therefore the present study was undertaken to summarize the defense strategies undertaken by the plants against herbivores.

Plants have evolved themselves constitutively for their own protection with the course of time with several structural modifications and metabolic changes. Multiple defense tactics have been practiced by plants in retaliation to avoid damage and eradication awarded by extrinsic invaded pathogens including herbivores (Fig. 7.1). Physical defense and chemical defense are two broad categories of defense. Defense is always an energy utilizing process that is why plants develop few anatomical structures as barriers for making difficulty in feeding process of herbivores. These structures which contribute in defense by reducing palatability recognized as

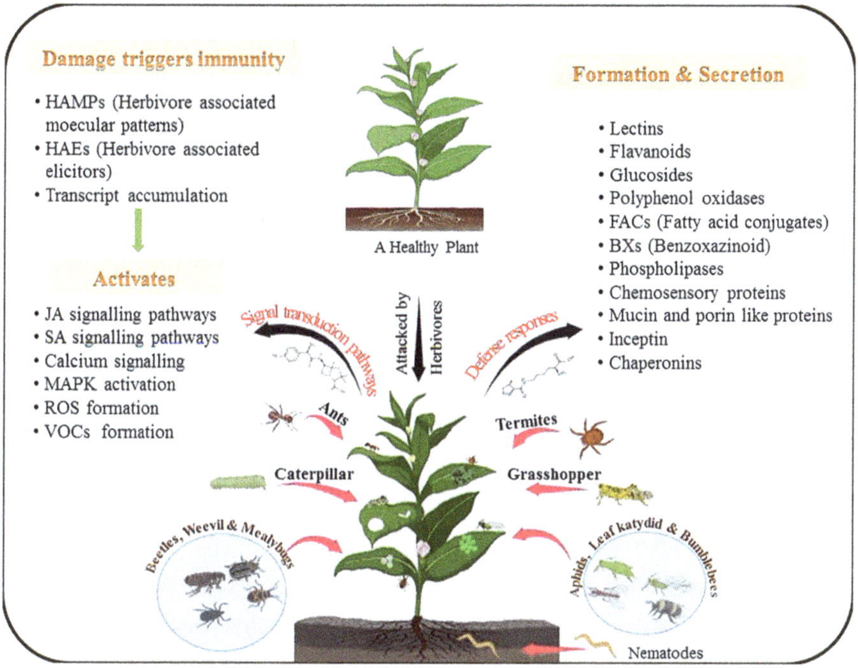

Fig. 7.1 Defense pathways and metabolite synthesis activated by plants against herbivores attack

physical defenders. Whereas, any damage or wound in plant parts compel the plants to restrict their growth and fight at metabolic level by synthesizing multiple complex compounds as secondary metabolite known as chemical defenders (Sestari and Campos 2022).

7.2 Physical Defense Attempted by Plants

Plants physical anatomical structures affirmed the liability of first line of defense. The bark and waxy cuticle on the surface of numerous plant species confront as their intact and impenetrable barrier. Several other vital modifications in their physical or morphological characteristics occur as barrier in response to prominent perturbances created by herbivore pathogens. Structural traits modifications; spinescence (spines and thorns), pubescence (trichomes), and sclerophylly (toughened and hardened leaves) occurs in plant (Mostafa et al. 2022). In response to pathogen, plants even modify their branching style. Shoots at wide axillary angles produce wiry stems (divaricated branched), formation of lignified substances and even plant tissues accumulate various granular minerals within it (War et al. 2012). It has been reported that spinescence that involves plant structures spine, thorns as well as prickles, defend the plants against various arthropod herbivores. Sclerophylly, referred as tough and hard leaves, minimizes the palatability and digestibility of plant tissues within herbivores (Hanley et al. 2007). Trichomes, a layer of hairs on epidermal layer of stems, leaves, and other aerial plant parts, are of two types—non-glandular (straight, spiral, stellate, and hooked) and glandular (War et al. 2012). Glandular trichomes primarily produces sesquiterpenes in *Fragaria x ananassa* (strawberry) (Figueiredo et al. 2013) and *Rubus idaeus* (raspberry) (Graham et al. 2014; Karley et al. 2016) in response to *Tetranychus urticae* (spider mite). The growth of the insect *Helicoverpa zea* (corn earworm) is restrained by glandular trichomes secretions in *Solanum lycopersicum* (potato) plant. Whereas, the attack of *Leptinotarsa decemlineata* (Colorado potato beetle), which extensively affect potato crop plants, is controlled by terpene synthases released from high density non-glandular trichomes, which also restrict oviposition of mites affecting population size on strawberry and raspberry plants (Tian et al. 2012b). Still, trichomes presence on the surface of plants manifest as shield against herbivores, few herbivores make themselves capable enough to breach the resistance of plants. Hence, plants dispose a second surface barrier in front of parasites by producing a waxy coating layer above epidermis the uppermost layer, which is the transformation of various lipids. In *Pisum sativum*, lipids of plant form wax blooms on their epicuticular surfaces, providing resistance toward *Sitona lineatus* (weevil) and *Acyrthosiphum pisum* (pea aphid) insects and prevents them from attaching with the leaves or other surfaces of the plant (White and Eigenbrode 2000). Herbivore diamond back moth *(Plutella xylostella)* of *Brassica* plant species are also trenched by plant lipids and their leaf surface waxes, which reduces the performance of the moth on cabbage (Hariprasad and Van Emden 2010).

7.3 Chemical Defense of Plants

Herbivores while satiating their desire, unknowingly triggers a very complex physiological and biochemical process in plants. After primitive attack of pathogen or herbivores, in response to immunity, plants synthesize and secrete wide range of low-molecular weight chemical compounds as secondary metabolites. These compounds, required in very low concentration to stimulate defense in plant system, are known as elicitors (Patel et al. 2020). For the first time in 1980s, herbivores-induced volatile organic compounds (HI-VOCs) were reported in damaged plants of willow (*Salix sitchensis*), which exhibited resistance against external organisms (Heil 2014). Numerous plants derived as well as pathogen derived proteins, oligosaccharides, glycoprotein and lipids plays significant role as elicitors and commonly known as herbivore-associated molecular patterns (HAMPs) as well as herbivore-associated elicitors (HAEs) (Wang et al. 2023). These intracellular signals activate several phytohormones primarily jasmonic acid (JA), salicylic acid (SA), ethylene pathways, and other biochemical processes such as mitogen activated protein kinase activation, and reactive oxygen species (ROS) formation. The necrotrophic pathogen insects causing wound and tissue damage while feeding induces JA pathway (Arimura et al. 2011) and produces biologically active signaling molecule jasmonates. On the other hand, biotrophic pathogens, phloem-feeding aphids, and spider mites activates SA pathway

(Arimura et al. 2009) that procreate salicylates by repressing JA pathway. However, attack of several feeding insects leads secretion of phytohormone ethylene which act in synergistic as well as in antagonistic way with JA (Von Dahl and Baldwin 2007; Pérez-Llorca et al. 2023). The operating mechanism of these chemical compounds is signal transduction inhibition, nutrient and ion transport inhibition, metabolism inhibition, physiological processes hormonal balance disruption, or membrane disruption (Mithöfer and Boland 2012).

7.4 Elicitors

Plants synthesize various components as secondary metabolites to withstand against pathogens in defense. Even their trace amounts can play significant roles in signaling toward environmental interactions (Bhaskar et al. 2022; Divekar et al. 2022). These chemical compounds when synthesized within plants in response to defense are identified as "Plant derived elicitors." On the other hand, many of the specific compounds derived within pathogen which induce defense activities of plants are considered as "Insect derived elicitors." In other means, insect's oral regurgitates and watery saliva contains active molecules that act as elicitor and later on induces plant defense responses. Thousands of herbivores group possess very peculiar feeding behaviors by which they get recognized on exposure toward plants. They are mainly classified into two groups chewing type and others are piercing-sucking type (Jiménez et al. 2020). The insects which have evolved mouthparts damage plants either by chewing, snipping or tearing belong to category of chewing insects such as caterpillar, beetles and cotton bollworms. Whereas, piercing-sucking type insects possess a specially adapted mouthpart called Stylets (hollow, hard and sharp feeding organ), which help them to prick plant tissues and facilitates phloem sap sieve tubes, it is found in aphids, leafhoppers, nematodes, and nemerteans (Duceppe et al. 2012). These insects terribly injure plants directly via chewing, sucking, or piercing technique and also even in indirect way that cause major troublesome via transmitting other miniature parasitic pathogens such as viruses, bacteria, and fungi (Wang et al. 2023). Elicitors are grouped on the basis of their chemical nature, which is presented in Table 7.1. Table 7.2 summarizes the role of these chemical compounds in plant defense against herbivory.

7.5 Phenolic Compounds

Phenols comprise more than ten thousand compounds which are synthesized either by shikimic acid (phenylpropanoid) pathway (Cheynier et al. 2013) or by the malonic acid (polyketide) pathway (Quideau et al. 2011). Few simple phenols are synthesized from aromatic amino acid phenylalanine into phenylpropanoids like ferulic acid and caffeic acid, while others, such as, coumarin of benzopyrone family synthesize phenylpropanoid lactones, like hydroxycoumarins, umbelliferone and psoroline (furano-coumarin). Similarly, salicylic acid or vanillin derived benzoquinones are also simpler phenols, while flavanoids, lignins, and tannins are complex

Table 7.1 Chemical groups and their corresponding elicitor types

• Chemical groups	• Elicitor types
Phenolic compounds	Coumarin (hydroxycoumarins), Furano-coumarin (psoroline), flavanoids (quercetin), lignin (reservatrol), tannin
Terpenoids	Monoterpenes (p-cymene, camphene, pinene, citronellol), sesquiterpene (*β*-caryophyllene), triterpene (azadirachtin)
N-containing compounds	Alkaloids (indole, nicotine), cyanogenic glycosides (dhurrin), non-protein amino acid (azetidine)
S-containing compounds	Phytoalexin, defensin, glutathione, glucosinolates
Polypeptides	Proteinase and trypsin inhibitors, systemin, inceptins
FACs (fatty acid amino acid conjugates)	Volicitin, *β*-glucosidase, caeliferins

polyphenol derivatives (Balasundram et al. 2006). Phenols are involved primarily in basic metabolisms of plants, for instance, they protect plant from UV radiations (Landry et al. 1995), their lignin forms in vascular plants are vital for secondary cell wall formation (Boerjan et al. 2003). Furthermore, flavanoids are essential for pollen development as well as they attract pollinaters (Van der Meer et al. 1992).

Phenols restrict pathogen not only by repellent activity, but also through toxicity (Kessler et al. 2012b). They are stored generally in vacuoles, sometimes in specialized cells and few are also adjoined with cell wall in inactive form (Pourcel et al. 2007). Their activation occurs through the activity of enzymes, such as glycosidases, polyphenol oxidases, and peroxidases (Constabel and Barbehenn 2008), which elevate defense response by increasing their toxicity. Toxicity encourages lignin production at the damaged site which further prevents nematodes as well as arthropods (Valette et al. 1998). Volatile methyl salicylate compound is found to be involved in indirect defense responses in many plants (Ament et al. 2004, 2010). *Rhopalosiphum padi* (cereal aphid) pathogen avoid phenol rich wheat plant, similarly *Salix dasyclados* (willow plant) repel *Galerucella lineola* (leaf beetle) via increasing phenolic compounds (Fürstenberg-Hägg et al. 2013). Salicylic acid/salicylates, which are benzoquinone derivative present in *Salix* (willow plant), very proficiently arrest the growth and development of larvae of *Operophtera brumata* (oak moth). In cotton plant Gossypol, a phenolic compound defends numerous herbivore such as *Heliothis virescens* (tobacco bollworm) and *Heliothis zea* (bollworm (Ruuhola et al. 2001).

Terpenoids—Terpene is the group that not only protects the plants directly, but also prevents them from pathogens indirectly. Members of this group are derivatives of isoprenoid (five-carbon unit) precursors and uphold extremely diverse metabolically bioactive natural compounds. Wide variety of volatile molecules synthesized due to terpene synthases enzyme actions, thereafter produce various mono-, di-, sesqui-, and triterpenoids (Schilmiller et al. 2009). While volatile terpenoids are extremely lipophilic in nature, they easily get access through plasma membranes and directly affect the insects (Bleeker et al. 2012). Phytoecdysteroid produced in *Polypodium vulgare* fern plant interrupt moulting among insects, since it is analogous to molting hormone (Canals et al. 2005). Similarly, few other amide derived terpenoids

Table 7.2 List of Elicitors, their relative chemical groups, pathogens on which they act, their corresponding host plants and defense mechanism they manifest

• Elicitors	• Compound	• Pathogen	• Host plant	• Function	• References
Benzyl cyanide, indole	Alkaloid	*Pieris* spp. arrest parasitoid *Trichogramma brassicae* eggs	*Brassica oleracea*	Male-derived butterfly antiaphrodisiac mediates induced indirect plant defense.	Fatouros et al. (2008)
Bruchins	Esterified fatty acids	*Bruchus* spp. and *Callosobruchus maculatus*	*Pisum sativum*	Resistant to egg pathogen and callus formation	Doss (2005)
β-caryophyllene	Sesquiterpene	*Sogatella furcifera* planthopper	Rice	Reduce palatability of feeding in the larva	Wang et al. (2015)
β-glu	β-glucosidase	*Pieri brassicae*	*Brassica oleracea*	Increased VOC emission and releases volatiles to attract parasitic *Cotesia glomerata* wasps	Mattiacci et al. (1995)
β-galactofuranose	Polysaccharide	Egyptian cotton leafworm (*S. Littoralis*	Arabidopsis	Membrane depolarization and gene expression	Bricchi et al. (2012, 2013)
Caeliferins	Disulfoxy fatty acids	*Schistocerca Americana*	*Zea mays* (corn)	Induce VOC emissions	Alborn et al. (2007)
CathB3	Cysteine protease	*Myzus persicae*	Tobacco	Decreases aphid production Induction of EDR1-dependent ROS burst	Guo et al. (2020)
CHIT, BGL, THAU, PR10	Chitinase, glucanase, thaumatin-like proteins and *PR10*	*Schizaphis graminum*	*Sorghum bicolor*	Transcriptional regulation against a phloem-feeding aphid in defense	Zhu-Salzman et al. (2004)
Coumarin	Phenol	*Spodoptera litura*	*Glycyrhizza glabra*, Tonka beans	Declined reproductive potential	Mandeep Kaur and Sanehdeep Kaur (2013)

(continued)

Table 7.2 (continued)

• Elicitors	• Compound	• Pathogen	• Host plant	• Function	• References
DIMBOA	Benzoxazinoids	*Rhopalasiphum padi*	*Zea mays*	Decreased aphid growth and survival, results reduced plant palatibility	Ahmad et al. (2011)
E, *S*,—Conophthorin	VOCs	*E. Solidaginis*	*Solidago altissima*	Induces JA pathway, increased VOC emission	Helms et al. (2014, 2017)
E*β*f	(E)-*β*-farnesene	*Chilo partellus*	*Cotesia sesamiae*	HIPV-induced attraction of maize stem borer parasitoids	Tamiru et al. (2015)
E*β*f	(E)-*β*-farnesene	*Myzus persicae*	*Triticum aestivum*	Engineered elevated production of repellent alarm pheromone for aphids	Bruce et al. (2015)
E*β*f	(E)-*β*-farnesene	Sawfly	*Pinus sylvestris*	Oviposition fluid induces VOC emission	Mumm (2003)
FACs	Fatty acid amino acid conjugates	*Spodoptera exigua*	*Zea mays*	Plant elicitor peptides induce plant defenses via nitrogen assimilation in wasps	Alborn et al. (1997) Turlings et al. (1995) Yoshinaga et al. (2014)
Glucosinolates	Aliphatic and indole glucosinolates	*Pieris rapae*, *Mamestra brassicae*	*Brassica oleracea* var. *acephala*	Reduced larval feeding affects growth and sluggish habituation on mature plants	Santolamazza-Carbone et al. (2016)
GOX	Glucose oxidase	European corn borer, *Helicoverpa zea* (corn earworm)	Tomato	Induce JA pathway and initiate delayed activated defenses	Louis et al. (2013) Tian et al. (2012)
Gramine	N,N-Dimethyl-3-aminomethylindole	*Rhopalasiphum padi*	*Hordeum vulgare*	Alkaloid inhibits aphid feeding, finally affect growth and survival	Zúñiga and Corcuera (1986)
GroEL	Chaperonin	*Buchnera aphidicola* in *Macrosiphum euphorbiae*	Arabidopsis	Fecundity of aphids decreases, induction of ROS burst and express PTI marker genes	Chaudhary et al. (2014)

• Elicitors	• Compound	• Pathogen	• Host plant	• Function	• References
HEL/PR4	Hevein-like protein	*Bemisia tabaci*	Arabidopsis	JA pathway regulated defense	Kempema et al. (2007)
Inceptin	Inceptin receptor	*Spodoptera frugiperda*	Maize, cowpea	ATP synthase mediated response	Schmelz et al. (2006)
Kauralexins	Diterpenoid Phytoalexin	*Ostrinia nubilalis*	*Zeamays*	Antimicrobial activity leads to deter feeding of corn borer larvae	Block et al. (2019)
LOXs	Lipoxygenases	*Agelastica alni*	*Alnus glutinosa*	Activation of the octadecanoid pathway causes toxicity to beetle	Tscharntke et al. (2001)
Mp10	Chemosensory protein	*Myzus persicae*	Tobacco	Reduce aphid fecundity	Bos et al. (2010)
MTI-2	Mustard trypsin inhibitor	*Plutella xylostella*	Arabidopsis, and oilseed rape	High mortality rate and delayed larval development significantly	De Leo et al. (2001)
MTI-2	Mustard trypsin inhibitor	*Mamestra brassicae*	Transgenic tobacco and arabidopsis	Larval growth checked and causes mortality	De Leo et al. (2001)
MTI-2	Mustard trypsin inhibitor	*Spodoptera littoralis*	Oilseed rape	Larval development delayed	De Leo et al. (2001)
NI12, NI16, NI28, NI32, NI40, NI43	Disulfide isomerase, Apolipophorin-III, cysteine-rich and chemosensory protein.	*Nilaparvata lugens*	Tobacco (*Nicotiana benthamiana)*	Induces dwarf phenotype and also cell death, callose deposition and, defense-related genes expression, induces chlorosis via *N. Lugens*-specific salivary protein	Rao et al. (2019)
NIMLP	Mucin-like protein	*Nilaparvata lugens*	Rice and tobacco	Formation of salivary sheath; induction of other plant defense responses	Shangguan et al. (2018)
Orysata	Jacalin-related lectin from Rice	*Spodoptera exigua, Myzus persicae, Acyrthosiphon pisum*	*Nicotiana tabacum*	Mortality and retardation of development in chewing and piercing insects.	Zhang et al. (2000) Al Atalah et al. (2014)

(continued)

Table 7.2 (continued)

• Elicitors	• Compound	• Pathogen	• Host plant	• Function	• References
Oxylipins	Lipase	*Schistocerca gregaria*	Arabidopsis	Lipase activity mediates defense against insects	Schäfer et al. (2011)
Oxylipins	Lipases	*Phytophthora infestans*	Potato	Increased 9-lipoxygenase activity and desaturase transcripts accumulation as well as increased phospholipase A2 activity.	Göbel et al. (2001)
Peroxidases	Catalase	*Blissus oxiduus*	Bufallo grassses	ROS mediate the defensive gene activation	Heng-Moss et al. (2004)
PLC	Phospholipase C	*Spodoptera frugiperda* and caterpillars	*Zea mays* and *Cynodon dactylon*	Limit feeding and check weight gain of caterpillar, Induce other defense responses	Acevedo et al. (2018)
PLP bacteria	Porin-like protein	*Spodoptera littoralis*	Arabidopsis	Induce early events related to defense	Guo et al. (2013)
PnTPS1, PtTPS16, PtTPS21	Amino acids	*M. Melolontha*	*P. Trichocarpa and P. Nigra*	Induces root volatiles	Lackus et al. (2018)
PsTPS 1, PsTPS 2, and PsTPS 3	Sesquiterpene synthases	Sawfly	*Pinus sylvestris*	Increased terpene synthase rates	Köpke et al. (2008)
RP309	*Riptoris pedestris* specific salivary protein	*Riptortus pedestris*	Tobacco	Induce ROS burst, cell death and PTI marker genes expression	Dong et al. (2022)
SOD	Superoxide dismutases	*Aphis medicaginis*	*Medicago sativa*	Ca^{2+} signaling initiation, activates NADPH complex i.e. membrane bound	Huang et al. (2007)
Subtilisin	Proteases	*P. Viticola*	*Arabidopsis subtilase, Oryza sativa*	Detection of biotic and external stimulus, initiates programmed cell death	Figueiredo et al. (2014)
Superoxide	ROS	*Diprion pini*	*Pinus sylvestris*	Accumulated ROS	Bittner et al. (2017)
Systemin	Polypeptide	*Spodoptera littoralis*	*Solanum peruvianum*	Induce VOCs, primary systemic signals transport.	Coppola et al. (2017)

• Elicitors	• Compound	• Pathogen	• Host plant	• Function	• References
TePDI	Disulfide isomerase	*Tetranychus evansi*	Tobacco	Induction of ROS burst and callose deposition, defense-related genes expression initiated, and SGT1/HSP90-dependent cell death at last reduces aphid progress	Cui et al. (2023)
Tetranin (Tet1 and Tet2)	Protein	*Tetranychus urticae*	Bean, eggplant	Induce SA, JA as well as ABA biosynthesis Activates cytosolic calcium influx resulting reduced survival of spider mites	Iida et al. (2019)
VgN	N-terminal subunit of vitellogenin	*Nilaparvata lugens*	Rice	Strong defense responses triggered by VgN	Zeng et al. (2023)
VLC	Volatile like compounds	*Diprion pini*	*Pinus sylvestris*	Decreased ethylene	Schroder et al. (2007)
Volicitin (HAEs)	FACs	*Manduca sexta* lepidopterans larvae	Maize, Solanaceae	Trigerred membrane proteins calcium influx and membrane depolarization	Bonaventure et al. (2011)
ZmPep3 (FACs)	Fatty acid amino acid conjugates	*Spodoptera exigua* (beet armyworm)	*Zea mays*	Increased VOC (terpenes and benzoxazinoids) emission impair growth and attract their parasitoids	Huffaker et al. (2013)

Abbreviations: VOC, Volatile organic compounds; FACs, Fatty acid amino acid conjugates; ROS, Reactive oxygen species; DIMBOA, 2,4-dihydroxy-7-methoxy-1,4-benzoxazin-3-one camouflage juvenile hormone of insects (Cruickshank 1971). A terpenoid S(−)-limonene derived from geranyl pyrophosphate deter leafcutter ant (*Atta cephalotes*) in citrus plants (Cherrett 1972). Pine and fir coniferous plants synthesize monoterpenes that cause toxicity against *Scolytus* spp. (bark beetle) (Trapp and Croteau 2001). Isoprene terpenoid repel *Manduca sexta* insects and attract *Diadegma semiclausum* parasitic wasps (Loivamäki et al. 2008). Additionally, the *Phytoseiulus persimilis* predatory mites are attracted towards (3S)-(E)-nerolidol volatile compounds secreted by damaged gerbera plants in response to *Tetranychus urticae* spider mites (Mithfer et al. 2008; Krips et al. 1999), these are few examples of indirect defense.

7.6 N-Containing Compounds

Alkaloids—Synthesis of alkaloids is costlier than the phenolic compounds because plant requires additional energy to make available inorganic nitrogen. Despite of this wide range of different alkaloids (>15,000) are produced by several genera of vascular plants (approx. 20%), e.g., Solanaceae, Liliaceae, Amaryllidaceae, and Leguminous. Morphine, nicotine, caffeine, cocaine, atropine, and strychnine are few alkaloids, which are well known for their metabolic effects. Nicotine is a true alkaloid derived from aspartate and ornithine (Howe and Jander 2008).

Other than this, a specified indole terpene alkaloids group is also reported that possess diversified natural products, majority of them belong to three families: Apocynaceae, Rubiaceae, and Loganiaceae. Approximately 3000 genera produce quite essential biological properties of indoles; for instance, *Rauvolfia, Catharanthus,* and *Strychnos*. Herbiovore induced indoles significantly elevate the secretion of stress hormone abscissic acid and jasmonate-isoleucine conjugates. These alkaloids conduct quick and effective priming agent role, thereby increasing defensive metabolites in healthy tissues and neighboring plant cells (Erb et al. 2015). Apart from these, benzoxazinoid (BXs), an essential plant defense chemical group also belongs to indole. It is found widely in cereals (Poaceae) such as wheat *(Triticum aestivum)*, rye *(Secale cereale)*, and maize (*Zea mays*) (Niemeyer 2009; Wouters et al. 2016) as well as in few dicot families Lamiaceae *(Lamium galeobdolon)*, Ranunculaceae *(Clematis orientalis)*, Acanthaceae *(Aphelandra squarrosa)* (Frey et al. 2009). At primitive stage of the plant growth, the biogenesis of BXs occurs at its peak, and then its level declines and becomes stable within plants (Nomura et al. 2005). In monocots, both roots as well as shoots are primarily involved in synthesis of BXs, while in dicots the synthesis varies significantly among parts of the plants; above grounded parts especially leaves, flower bud as well as flowers are chiefly indulged in the production while root system synthesize it in a lesser amount (Frey et al. 2009). In undamaged plant cells, vacuoles serve as their storage box in the form of glucosides and during emergency β-glucosidases hydrolyze it to enhance toxicity and reactivity within plants (Wouters et al. 2016). BXs inhibit the growth of nematodes, aphids, and whorl-feeding larvae and they also possess allelopathic and antimicrobial properties. Another group of nitrogen containing compounds include hydroxamic acids (HAs), lactams, and benzoxazolinones (Makowska et al. 2015).

DIMBOA is well-known elicitor belong to HAs group derived from indole-3-glycerol phosphate. HAs are defensive against various lepidopteran pests, e.g., European corn borer (*Ostrinia nubilalis*) and several aphid species such as *Sitobion avenae*. These are also effective as allelopathy against weeds by checking their growth in rye and wheat cereals crops (Niemeyer 2009).

Cyanogenic glucosides—These N-containing compounds are found in almost all groups of plants, including dicot as well as monocot, gymnosperms, and pteridophytes. More than 2600 genera of plant species are reported to produce cyanogenic glucosides (Møller 2010). These are derived from amino acids, such as; Dhurrin, derived from tyrosine in *Sorghum bicolor*; linamarin and lotaustralin derived from valine and isoleucine in *Lotus japonicus* (lotus) and cassava from *Manihot esculenta* and amygdalin and prunasin derived from phenylalanine in Rosaceae family including apples, strawberries, plums, peaches, and cherries. These cyanogenic glucosides retard larval growth and development of insect. *Amygdalin and prunasin* provide resistant against the larvae of *Capnodis tenebronis* (flat headed woodborer), while in *Manihot esculenta,* these glucosides increases resistance toward *Cyrtomenus bergi* (cassava burrower bug) (Forslund et al. 2004).

Non-protein amino acid—Several leguminoseae family plants produce these types of non-protein amino acids. An arginine analogue, Canavanine derived from L-homoserine and the arginyl-tRNA synthase of most insects such as bruchid beetle (*Caryedes brasiliensis*) and curculionidae weevil (*Sternechus tuberculatus*) is unable to differentiate between arginine and canavanine, which mislead incorporation of proteins causing deleterious effects (Rosenthal 2001). One of the aromatic amino acid Mimosine produced in the *Leucaena leucocephala* (tropical forage legume) possess insecticidal and nematicidal effects, it suppresses the activity of trehalose, α-amylase, and invertase in dose-dependent manner (Nguyen et al. 2015).

7.7 S-Containing Compounds

Phytoalexins—These are de novo synthesized low-molecular weight compounds, generally possessing antimicrobial effects. Most of the phytoalexins are toxic to nematodes as well as bacteria and fungi (Kaur et al. 2022). These are very common in Brassicaceae (like Brassin) and Poaceae family like Zealexins, Oryzalexins, Kauralexins, Sakuranetin, and Phenyl amides (Arruda et al. 2016). European corn borer (*Ostrinia nubilalis*) growth and proliferation is controlled by these phytoalexins in plants (Huffaker et al. 2011).

Defensin—These are highly stable small cationic, cysteine-rich peptides. *Helicoverpa armigera* a lepidopteran pest responsible for mass crop destruction shows larval growth reduction, inhibited fecundity as well as fertility and delay in metamorphosis in response to defensin (Mulla and Tamhane 2023). Defensins inhibits the insect's digestive enzymes, for example, proteases and α-amylase (Mostafa et al. 2022).

Glutathione—Various detoxification reactions of plants are settled via glutathione. For instance, *Glycine max* (soybean) plants generate glutathione arbitrated hydrogen peroxide in opposition to *Heterodera glycines* nematode. Even lower doses of glutathione can significantly increase H_2O_2 levels, resulting reduced accretion of nematodes (Chen et al. 2020).

Glucosinolates—These glucosides are derived from sulfur as well as nitrogen containing amino acid. They are found extensively in Brassicaceae and Capparidaceae families. Fifty percent of aliphatic glucosinolates are derived from methionine, ten percent of aromatic glucosinolates are synthesized from phenylalanine or tyrosine, such as Sinalbin, and ten percent of indole glucosinolates are derived from tryptophan (Hopkins et al. 2009). These glucosinolates protect cotyledons of *Brassica napus* (oilseed rape) from field slug *(Deroceras reticulatum)* and *Sinapis alba* (white mustard) cotyledons from flee beetle *(Phyllotreta cruciferae)* pathogen (Glen et al. 1990).

7.8 Polypeptides

In polypeptide group, systemin is the first reported polypeptide which acquires hormone-like activity within plants in response to wound (Coppola et al. 2017). In the *Lycopersicon peruvianum* plasma membrane cells, a systemin cell surface receptor SR160 was explored, which belongs to leucine-rich repeat (LRR) receptor of kinase family (Scheer and Ryan 2002). HypSys, a hydroxyl proline rich systemin, was also found in Petunia, which induces defensin genes (Kessler et al. 2013). Systemin awakens the neighboring healthy plants via inducing metabolic changes within the sufferer plant. Molecular changes involve signaling enzymes, transcription factors synthesis at higher level and also increase the rate of pattern-recognition receptors transcription. Thus, neighboring plants show more resistant toward herbivores, rapid response to wounds attenuate the pathogens growth and attract parasitoids at large scale (Coppola et al. 2017). Other peptides such as inceptins (a disulfide-bonded peptide) synthesized in the *Spodoptera frugiperda* gut have also been identified, which play important role in plant defense against herbivore (Acevedo et al. 2018). Amino acids, building blocks of polypeptide, have also shown tolerance against *Amphorophora idaei* (European large raspberry aphid) in *Rubus ideaus* plants (Johnson et al. 2012; Karley et al. 2016).

7.9 Fatty Acid Amino Acid Conjugates (FACs)

FACs are insect elicitors categorized in herbivore-associated molecular patterns (HAMP) (Yoshinaga 2016). The oral regurgitates of lepidopteran larvae possess FACs which induce defense process within plants. The first non-enzymatic FAC elicitor *N*-17-hydroxylinolenoyl-L-glutamine named as "Volicitin" volatile biosynthate was isolated from an insect herbivore oral secretions, *Spodoptera exigua* (beet armyworm). Volicitin application on the host leaves of *Zea mays* induces seedlings to synthesize volatile organic compounds (VOCs) that attract parasitoid wasps, predator of beet armyworm (Turlings et al. 1990; Schmelz et al. 2003). FACs from *Manduca sexta* (hawkmoth), *β*-glucosidase from *Pieris brassicae* (large cabbage white butterfly caterpillar) oral secretions, caeliferins from *Shistocerca spp* (gray bird grasshopper) and bruchins from *Callosobruchus maculates* (cowpea weevil) oviposition fluid have been identified (Paré et al. 2005; Pohnert et al. 1999; Alborn et al. 2000; Halitschke et al. 2001).

7.10 Conclusion

Plants sessile nature prohibits them to move away from threat. Therefore, numerous approaches have been manifested by the plants toward their pathogens. They respond to pathogen attack by sensing chemical stimuli, called elicitors through their special receptors. In environment, there is a wide range of pathogens, among them herbivores have accounted severe noticeable disruption in plants. Under the attack of herbivores, plants not only halt the synthesis of growth metabolites temporarily, but also employ the synthesis and secretion of several secondary metabolites as ultimate powerful defense weapons. These metabolites are synthesized just after the induction of signal transduction pathways governed by several phytohormones such as JA, SA, and ethylene, as well as via Ca^{2+} signaling, ROS formation, MAPK activation, etc. Various elicitors have been reported to participate during plant and herbivore interaction, thereby providing protection against them.

References

Ahmad S, Veyrat N, Gordon-Weeks R et al (2011) Benzoxazinoid metabolites regulate innate immunity against aphids and fungi in maize. Plant Physiol 157:317–327. https://doi.org/10.1104/pp.111.180224

Al Atalah B, Smagghe G, Van Damme EJM (2014) Orysata, a jacalin-related lectin from rice, could protect plants against biting-chewing and piercing-sucking insects. Plant Sci 221–222:21–28. https://doi.org/10.1016/j.plantsci.2014.01.010

Allsopp PG, Cox MC (2002) Sugarcane clones vary in their resistance to sugarcane white grubs. Aust J Agric Res 53:1111. https://doi.org/10.1071/AR02035

Ament K, Kant MR, Sabelis MW et al (2004) Jasmonic acid is a key regulator of spider mite-induced volatile Terpenoid and methyl salicylate emission in tomato. Plant Physiol 135:2025–2037. https://doi.org/10.1104/pp.104.048694

Ament K, Krasikov V, Allmann S et al (2010) Methyl salicylate production in tomato affects biotic interactions: *role of methyl salicylate in tomato defence*. Plant J 62:124–134. https://doi.org/10.1111/j.1365-313X.2010.04132.x

Arimura G-i, Matsui K, Takabayashi J (2009) Chemical and molecular ecology of herbivore-induced plant volatiles: proximate factors and their ultimate functions. Plant Cell Physiol 50:911–923. https://doi.org/10.1093/pcp/pcp030

Arimura G-I, Ozawa R, Maffei ME (2011) Recent advances in plant early signaling in response to herbivory. IJMS 12:3723–3739. https://doi.org/10.3390/ijms12063723

Arruda RL, Paz ATS, Bara MTF et al (2016) An approach on phytoalexins: function, characterization and biosynthesis in plants of the family Poaceae. Cienc Rural 46:1206–1216. https://doi.org/10.1590/0103-8478cr20151164

Balasundram N, Sundram K, Samman S (2006) Phenolic compounds in plants and Agri-industrial by-products: antioxidant activity, occurrence, and potential uses. Food Chem 99:191–203. https://doi.org/10.1016/j.foodchem.2005.07.042

Bhaskar R, Xavier LSE, Udayakumaran G et al (2022) Biotic elicitors: a boon for the in-vitro production of plant secondary metabolites. Plant Cell Tissue Organ Cult 149:7–24. https://doi.org/10.1007/s11240-021-02131-1

Bittner N, Trauer-Kizilelma U, Hilker M (2017) Early plant defence against insect attack: involvement of reactive oxygen species in plant responses to insect egg deposition. Planta 245:993–1007. https://doi.org/10.1007/s00425-017-2654-3

Bleeker PM, Mirabella R, Diergaarde PJ et al (2012) Improved herbivore resistance in cultivated tomato with the sesquiterpene biosynthetic pathway from a wild relative. Proc Natl Acad Sci USA 109:20124–20129. https://doi.org/10.1073/pnas.1208756109

Block AK, Vaughan MM, Schmelz EA, Christensen SA (2019) Biosynthesis and function of terpenoid defense compounds in maize (*Zea mays*). Planta 249:21–30. https://doi.org/10.1007/s00425-018-2999-2

Boerjan W, Ralph J, Baucher M (2003) Lignin biosynthesis. Annu Rev Plant Biol 54:519–546. https://doi.org/10.1146/annurev.arplant.54.031902.134938

Bos JIB, Prince D, Pitino M et al (2010) A functional genomics approach identifies candidate effectors from the aphid species Myzus persicae (green peach aphid). PLoS Genet 6:e1001216. https://doi.org/10.1371/journal.pgen.1001216

Bostock RM, Karban R, Thaler JS et al (2001) Signal interactions in induced resistance to pathogens and insect herbivores. Eur J Plant Pathol 107:103–111. https://doi.org/10.1023/A:1008703904253

Bruce TJA, Aradottir GI, Smart LE et al (2015) The first crop plant genetically engineered to release an insect pheromone for defence. Sci Rep 5:11183. https://doi.org/10.1038/srep11183

Canals D, Irurre-Santilari J, Casas J (2005) The first cytochrome P450 in ferns: evidence for its involvement in phytoecdysteroid biosynthesis in *Polypodium vulgare*. FEBS J 272:4817–4825. https://doi.org/10.1111/j.1742-4658.2005.04897.x

Chaudhary R, Atamian HS, Shen Z et al (2014) GroEL from the endosymbiont *Buchnera aphidicola* betrays the aphid by triggering plant defense. Proc Natl Acad Sci USA 111:8919–8924. https://doi.org/10.1073/pnas.1407687111

Chen D-Y, Chen Q-Y, Wang D-D et al (2020) Differential Transcription and Alternative Splicing in Cotton Underly Specialized Defense Responses Against Pests. Front Plant Sci 11:573131. https://doi.org/10.3389/fpls.2020.573131

Chen X, Li S, Zhao X et al (2020) Modulation of (homo) glutathione metabolism and H2O2 accumulation during soybean cyst nematode infections in susceptible and resistant soybean cultivars. IJMS 21:388. https://doi.org/10.3390/ijms21020388

Cherrett JM (1972) Some factors involved in the selection of vegetable substrate by Atta cephalotes (L.) (hymenoptera: Formicidae) in tropical rain Forest. J Anim Ecol 41:647. https://doi.org/10.2307/3200

Cheynier V, Comte G, Davies KM et al (2013) Plant phenolics: recent advances on their biosynthesis, genetics, and ecophysiology. Plant Physiol Biochem 72:1–20. https://doi.org/10.1016/j.plaphy.2013.05.009

Constabel CP, Barbehenn R (2008) Defensive Roles of Polyphenol Oxidase in Plants. In: Schaller, A. (eds) Induced Plant Resistance to Herbivory. Springer, Dordrecht. https://doi.org/10.1007/978-1-4020-8182-8_12

Coppola M, Cascone P, Madonna V et al (2017) Plant-to-plant communication triggered by systemin primes anti-herbivore resistance in tomato. Sci Rep 7:15522. https://doi.org/10.1038/s41598-017-15481-8

Cruickshank PA (1971) Insect juvenile hormone analogues: effects of some terpenoid amide derivatives. Bull World Health Organ 44:395–396

Cui J-R, Bing X-L, Tang Y-J et al (2023) A conserved protein disulfide isomerase enhances plant resistance against herbivores. Plant Physiol 191:660–678. https://doi.org/10.1093/plphys/kiac489

De Leo F, Bonadé-Bottino M, Ruggiero Ceci L et al (2001) Effects of a mustard trypsin inhibitor expressed in different plants on three lepidopteran pests. Insect Biochem Mol Biol 31:593–602. https://doi.org/10.1016/S0965-1748(00)00164-8

Divekar PA, Narayana S, Divekar BA et al (2022) Plant secondary metabolites as defense tools against herbivores for sustainable crop protection. IJMS 23:2690. https://doi.org/10.3390/ijms23052690

Doares SH, Syrovets T, Weiler EW, Ryan CA (1995) Oligogalacturonides and chitosan activate plant defensive genes through the octadecanoid pathway. Proc Natl Acad Sci USA 92:4095–4098. https://doi.org/10.1073/pnas.92.10.4095

Dong Y, Huang X, Yang Y et al (2022) Characterization of salivary secreted proteins that induce cell death from *Riptortus pedestris* (Fabricius) and their roles in insect-plant interactions. Front Plant Sci 13:912603. https://doi.org/10.3389/fpls.2022.912603

Doss RP (2005) Treatment of pea pods with Bruchin B results in up-regulation of a gene similar to MtN19. Plant Physiol Biochem 43:225–231. https://doi.org/10.1016/j.plaphy.2005.01.016

Duceppe M-O, Cloutier C, Michaud D (2012) Wounding, insect chewing and phloem sap feeding differentially alter the leaf proteome of potato, *Solanum tuberosum*. L Proteome Sci 10:73. https://doi.org/10.1186/1477-5956-10-73

Duque-Lazo J, Navarro-Cerrillo RM (2017) What to save, the host or the pest? The spatial distribution of xylophage insects within the Mediterranean oak woodlands of southwestern Spain. For Ecol Manag 392:90–104. https://doi.org/10.1016/j.foreco.2017.02.047

Erb M, Veyrat N, Robert CAM et al (2015) Indole is an essential herbivore-induced volatile priming signal in maize. Nat Commun 6:6273. https://doi.org/10.1038/ncomms7273

Fathipour Y, Kianpour R, Bagheri A et al (2019) Bottom-up effects of brassica genotypes on performance of diamondback moth, *Plutella xylostella* (Lepidoptera: Plutellidae). Crop Prot 115:135–141. https://doi.org/10.1016/j.cropro.2018.09.020

Fatouros NE, Broekgaarden C, Bukovinszkine'Kiss G et al (2008) Male-derived butterfly anti-aphrodisiac mediates induced indirect plant defense. Proc Natl Acad Sci USA 105:10033–10038. https://doi.org/10.1073/pnas.0707809105

Fernández De Bobadilla M, Vitiello A, Erb M, Poelman EH (2022) Plant defense strategies against attack by multiple herbivores. Trends Plant Sci 27:528–535. https://doi.org/10.1016/j.tplants.2021.12.010

Figueiredo AST, Resende JTV, Morales RGF et al (2013) The role of glandular and non-glandular trichomes in the negative interactions between strawberry cultivars and spider mite. Arthropod Plant Interact 7:53–58. https://doi.org/10.1007/s11829-012-9218-z

Figueiredo A, Monteiro F, Sebastiana M (2014) Subtilisin-like proteases in plantâ€ pathogen recognition and immune priming: a perspective. Front Plant Sci 5. https://doi.org/10.3389/fpls.2014.00739

Fleming TH, Kress WJ (2011) A brief history of fruits and frugivores. Acta Oecol 37:521–530. https://doi.org/10.1016/j.actao.2011.01.016

Forslund K, Morant M, Jørgensen B et al (2004) Biosynthesis of the nitrile glucosides Rhodiocyanoside a and D and the cyanogenic glucosides Lotaustralin and Linamarin in *Lotus japonicus*. Plant Physiol 135:71–84. https://doi.org/10.1104/pp.103.038059

Fosket DE (1994) Biotic factors regulate some aspects of plant development. In: Plant growth and development. Elsevier, pp 517–557

Frey M, Schullehner K, Dick R et al (2009) Benzoxazinoid biosynthesis, a model for evolution of secondary metabolic pathways in plants. Phytochemistry 70:1645–1651. https://doi.org/10.1016/j.phytochem.2009.05.012

Fürstenberg-Hägg J, Zagrobelny M, Bak S (2013) Plant defense against insect herbivores. IJMS 14:10242–10297. https://doi.org/10.3390/ijms140510242

Glen DM, Jones H, Fieldsend JK (1990) Damage to oilseed rape seedlings by the field slug *Deroceras reticulatum* in relation to glucosinolate concentration. Ann Appl Biol 117:197–207. https://doi.org/10.1111/j.1744-7348.1990.tb04207.x

Göbel C, Feussner I, Schmidt A et al (2001) Oxylipin profiling reveals the preferential stimulation of the 9-lipoxygenase pathway in elicitor-treated potato cells. J Biol Chem 276:6267–6273. https://doi.org/10.1074/jbc.M008606200

Graham J, Hackett CA, Smith K et al (2014) Genetic and environmental regulation of plant architectural traits and opportunities for pest control in raspberry. Ann Appl Biol 165:318–328. https://doi.org/10.1111/aab.12134

Guo H, Wielsch N, Hafke JB et al (2013) A porin-like protein from oral secretions of Spodoptera littoralis larvae induces defense-related early events in plant leaves. Insect Biochem Mol Biol 43:849–858. https://doi.org/10.1016/j.ibmb.2013.06.005

Guo H, Zhang Y, Tong J et al (2020) An aphid-secreted salivary protease activates plant defense in phloem. Curr Biol 30:4826–4836.e7. https://doi.org/10.1016/j.cub.2020.09.020

Halitschke R, Schittko U, Pohnert G et al (2001) Molecular interactions between the specialist herbivore *Manduca sexta* (Lepidoptera, Sphingidae) and its natural host *Nicotiana attenuata*. III. Fatty acid-amino acid conjugates in herbivore Oral secretions are necessary and sufficient for herbivore-specific plant responses. Plant Physiol 125:711–717. https://doi.org/10.1104/pp.125.2.711

Hanley ME, Lamont BB, Fairbanks MM, Rafferty CM (2007) Plant structural traits and their role in anti-herbivore defence. Perspect Plant Ecol Evol Syst 8:157–178. https://doi.org/10.1016/j.ppees.2007.01.001

Hariprasad KV, Van Emden HF (2010) Mechanisms of partial plant resistance to diamondback moth (*Plutella xylostella*) in brassicas. Int J Pest Manag 56:15–22. https://doi.org/10.1080/09670870902980834

Heil M (2014) Herbivore-induced plant volatiles: targets, perception and unanswered questions. New Phytol 204:297–306. https://doi.org/10.1111/nph.12977

Helms AM, De Moraes CM, Mescher MC, Tooker JF (2014) The volatile emission of *Eurosta solidaginis* primes herbivore-induced volatile production in *Solidago altissima* and does not directly deter insect feeding. BMC Plant Biol 14:173. https://doi.org/10.1186/1471-2229-14-173

Helms AM, De Moraes CM, Tröger A et al (2017) Identification of an insect-produced olfactory cue that primes plant defenses. Nat Commun 8:337. https://doi.org/10.1038/s41467-017-00335-8

Heng-Moss T, Sarath G, Baxendale F et al (2004) Characterization of oxidative enzyme changes in Buffalograsses challenged by *Blissus occiduus*. J Econ Entomol 97:1086–1095. https://doi.org/10.1093/jee/97.3.1086

Hopkins RJ, Van Dam NM, Van Loon JJA (2009) Role of Glucosinolates in insect-plant relationships and multitrophic interactions. Annu Rev Entomol 54:57–83. https://doi.org/10.1146/annurev.ento.54.110807.090623

Howe GA, Jander G (2008) Plant immunity to insect herbivores. Annu Rev Plant Biol 59:41–66. https://doi.org/10.1146/annurev.arplant.59.032607.092825

Huffaker A, Kaplan F, Vaughan MM et al (2011) Novel acidic Sesquiterpenoids constitute a dominant class of pathogen-induced Phytoalexins in maize. Plant Physiol 156:2082–2097. https://doi.org/10.1104/pp.111.179457

Huffaker A, Pearce G, Veyrat N et al (2013) Plant elicitor peptides are conserved signals regulating direct and indirect antiherbivore defense. Proc Natl Acad Sci USA 110:5707–5712. https://doi.org/10.1073/pnas.1214668110

Hulme PE, Benkman CW (2002) Granivory. Plant–animal interactions: an evolutionary approach. Oxford, pp 185–208

Iida J, Desaki Y, Hata K et al (2019) Tetranins: new putative spider mite elicitors of host plant defense. New Phytol 224:875–885. https://doi.org/10.1111/nph.15813

Jiménez J, Garzo E, Alba-Tercedor J et al (2020) The phloem-pd: a distinctive brief sieve element stylet puncture prior to sieve element phase of aphid feeding behavior. Arthropod Plant Interact 14:67–78. https://doi.org/10.1007/s11829-019-09708-w

Johnson SN, Young MW, Karley AJ (2012) Protected raspberry production alters aphid–plant interactions but not aphid population size. Agri Forest Entomol 14:217–224. https://doi.org/10.1111/j.1461-9563.2011.00561.x

Kahl J, Siemens DH, Aerts RJ et al (2000) Herbivore-induced ethylene suppresses a direct defense but not a putative indirect defense against an adapted herbivore. Planta 210:336–342. https://doi.org/10.1007/PL00008142

Kant MR, Jonckheere W, Knegt B et al (2015) Mechanisms and ecological consequences of plant defence induction and suppression in herbivore communities. Ann Bot 115:1015–1051. https://doi.org/10.1093/aob/mcv054

Karban R (2020) The ecology and evolution of induced responses to herbivory and how plants perceive risk. Ecol Entomol 45:1–9. https://doi.org/10.1111/een.12771

Karley AJ, Mitchell C, Brookes C et al (2016) Exploiting physical defence traits for crop protection: leaf trichomes of *Rubus idaeus* have deterrent effects on spider mites but not aphids: differential effects of leaf trichomes on two herbivores of *Rubus idaeus*. Ann Appl Biol 168:159–172. https://doi.org/10.1111/aab.12252

Kaur M, Kaur S (2013) Tritrophic interactions among coumarin, the herbivore *Spodoptera litura* and a gregarious ectoparasitoid Bracon hebetor. BioControl 58:755–763. https://doi.org/10.1007/s10526-013-9533-z

Kaur S, Samota MK, Choudhary M et al (2022) How do plants defend themselves against pathogens-biochemical mechanisms and genetic interventions. Physiol Mol Biol Plants 28:485–504. https://doi.org/10.1007/s12298-022-01146-y

Kempema LA, Cui X, Holzer FM, Walling LL (2007) Arabidopsis transcriptome changes in response to phloem-feeding Silverleaf whitefly nymphs. Similarities and distinctions in responses to aphids. Plant Physiol 143:849–865. https://doi.org/10.1104/pp.106.090662

Kessler D, Diezel C, Clark DG et al (2013) *Petunia* flowers solve the defence/apparency dilemma of pollinator attraction by deploying complex floral blends. Ecol Lett 16:299–306. https://doi.org/10.1111/ele.12038

Köpke D, Schröder R, Fischer HM et al (2008) Does egg deposition by herbivorous pine sawflies affect transcription of sesquiterpene synthases in pine? Planta 228:427–438. https://doi.org/10.1007/s00425-008-0747-8

Krips OE, Willems PEL, Gols R et al (1999) The response of *Phytoseiulus persimilis* to spider mite-induced volatiles from gerbera: influence of starvation and experience. J Chem Ecol 25:2623–2641. https://doi.org/10.1023/A:1020887104771

Lackus ND, Lackner S, Gershenzon J et al (2018) The occurrence and formation of monoterpenes in herbivore-damaged poplar roots. Sci Rep 8:17936. https://doi.org/10.1038/s41598-018-36302-6

Landry LG, Ccs C, Last RL (1995) Arabidopsis mutants lacking phenolic sunscreens exhibit enhanced ultraviolet-B injury and oxidative damage. Plant Physiol 109:1159–1166. https://doi.org/10.1104/pp.109.4.1159

Loivamäki M, Mumm R, Dicke M, Schnitzler J-P (2008) Isoprene interferes with the attraction of bodyguards by herbaceous plants. Proc Natl Acad Sci USA 105:17430–17435. https://doi.org/10.1073/pnas.0804488105

Louis J, Peiffer M, Ray S et al (2013) Host-specific salivary elicitor(s) of European corn borer induce defenses in tomato and maize. New Phytol 199:66–73. https://doi.org/10.1111/nph.12308

Makowska B, Bakera B, Rakoczy-Trojanowska M (2015) The genetic background of benzoxazinoid biosynthesis in cereals. Acta Physiol Plant 37:176. https://doi.org/10.1007/s11738-015-1927-3

Mattiacci L, Dicke M, Posthumus MA (1995) Beta-glucosidase: an elicitor of herbivore-induced plant odor that attracts host-searching parasitic wasps. Proc Natl Acad Sci USA 92:2036–2040. https://doi.org/10.1073/pnas.92.6.2036

Mescher MC, De Moraes CM (2015) Role of plant sensory perception in plant-animal interactions. J Exp Bot 66:425–433. https://doi.org/10.1093/jxb/eru414

Mitchell C, Brennan RM, Graham J, Karley AJ (2016) Plant defense against herbivorous pests: exploiting resistance and tolerance traits for sustainable crop protection. Front Plant Sci 7. https://doi.org/10.3389/fpls.2016.01132

Mithfer A, Boland W, Maffei ME (2008) Chemical Ecology of Plant–Insect Interactions. In: Parker J (ed) Molecular aspects of plant disease resistance. Wiley-Blackwell, Oxford, UK, pp. 261–291

Mithöfer A, Boland W (2012) Plant defense against herbivores: chemical aspects. Annu Rev Plant Biol 63:431–450. https://doi.org/10.1146/annurev-arplant-042110-103854

Møller BL (2010) Functional diversifications of cyanogenic glucosides. Curr Opin Plant Biol 13:337–346. https://doi.org/10.1016/j.pbi.2010.01.009

Mostafa S, Wang Y, Zeng W, Jin B (2022) Plant responses to herbivory, wounding, and infection. IJMS 23:7031. https://doi.org/10.3390/ijms23137031

Mulla JA, Tamhane VA (2023) Novel insights into plant defensin ingestion induced metabolic responses in the polyphagous insect pest *Helicoverpa armigera*. Sci Rep 13:3151. https://doi.org/10.1038/s41598-023-29250-3

Mumm R (2003) Chemical analysis of volatiles emitted by Pinus sylvestris after induction by insect oviposition. J Chem Ecol 29:1235–1252. https://doi.org/10.1023/A:1023841909199

Nguyen B, Chompoo J, Tawata S (2015) Insecticidal and Nematicidal activities of novel Mimosine derivatives. Molecules 20:16741–16756. https://doi.org/10.3390/molecules200916741

Niemeyer HM (2009) Hydroxamic acids derived from 2-Hydroxy-2 *H*-1,4-Benzoxazin-3(4 *H*)-one: key defense chemicals of cereals. J Agric Food Chem 57:1677–1696. https://doi.org/10.1021/jf8034034

Nomura T, Ishihara A, Yanagita RC et al (2005) Three genomes differentially contribute to the biosynthesis of benzoxazinones in hexaploid wheat. Proc Natl Acad Sci USA 102:16490–16495. https://doi.org/10.1073/pnas.0505156102

Paré PW, Farag MA, Krishnamachari V et al (2005) Elicitors and priming agents initiate plant defense responses. Photosynth Res 85:149–159. https://doi.org/10.1007/s11120-005-1001-x

Patel ZM, Mahapatra R, Jampala SSM (2020) Role of fungal elicitors in plant defense mechanism. In: Molecular aspects of plant beneficial microbes in agriculture. Elsevier, pp 143–158

Pérez-Llorca M, Pollmann S, Müller M (2023) Ethylene and Jasmonates signaling network mediating secondary metabolites under abiotic stress. IJMS 24:5990. https://doi.org/10.3390/ijms24065990

Peumans WJ, Barre A, Houles Astoul C et al (2000) Isolation and characterization of a jacalin-related mannose-binding lectin from salt-stressed rice (*Oryza sativa*) plants. Planta 210:970–978. https://doi.org/10.1007/s004250050705

Pohnert G, Jung V, Haukioja E et al (1999) New fatty acid amides from regurgitant of lepidopteran (Noctuidae, Geometridae) caterpillars. Tetrahedron 55:11275–11280. https://doi.org/10.1016/S0040-4020(99)00639-0

Pourcel L, Routaboul J, Cheynier V et al (2007) Flavonoid oxidation in plants: from biochemical properties to physiological functions. Trends Plant Sci 12:29–36. https://doi.org/10.1016/j.tplants.2006.11.006

Prade P, Coyle DR (2023) Insect pests of forest trees. In: Forest microbiology. Elsevier, pp 195–211

Quideau S, Deffieux D, Douat-Casassus C, Pouységu L (2011) Plant polyphenols: chemical properties, biological activities, and synthesis. Angew Chem Int Ed 50:586–621. https://doi.org/10.1002/anie.201000044

Rao W, Zheng X, Liu B et al (2019) Secretome analysis and in planta expression of salivary proteins identify candidate effectors from the Brown Planthopper *Nilaparvata lugens*. MPMI 32:227–239. https://doi.org/10.1094/MPMI-05-18-0122-R

Ravenscraft A, Boggs CL (2016) Nutrient acquisition across a dietary shift: fruit feeding butterflies crave amino acids, nectivores seek salt. Oecologia 181:1–12. https://doi.org/10.1007/s00442-015-3403-6

Rosenthal GA (2001) L-Canavanine: a higher plant insecticidal allelochemical. Amino Acids 21:319–330. https://doi.org/10.1007/s007260170017

Ruuhola T, Tikkanen O-P, Tahvanainen J (2001) Differences in host use efficiency of larvae of a generalist moth, *Operophtera brumata* on three chemically divergent *Salix species*. J Chem Ecol 27:1595–1615. https://doi.org/10.1023/A:1010458208335

Santolamazza-Carbone S, Sotelo T, Velasco P, Cartea ME (2016) Antibiotic properties of the glucosinolates of *Brassica oleracea var. acephala* similarly affect generalist and specialist larvae of two lepidopteran pests. J Pest Sci 89:195–206. https://doi.org/10.1007/s10340-015-0658-y
Schäfer M, Fischer C, Meldau S et al (2011) Lipase activity in insect Oral secretions mediates defense responses in Arabidopsis. Plant Physiol 156:1520–1534. https://doi.org/10.1104/pp.111.173567
Scheer JM, Ryan CA (2002) The systemin receptor SR160 from *Lycopersicon peruvianum* is a member of the LRR receptor kinase family. Proc Natl Acad Sci USA 99:9585–9590. https://doi.org/10.1073/pnas.132266499
Schilmiller AL, Schauvinhold I, Larson M et al (2009) Monoterpenes in the glandular trichomes of tomato are synthesized from a neryl diphosphate precursor rather than geranyl diphosphate. Proc Natl Acad Sci USA 106:10865–10870. https://doi.org/10.1073/pnas.0904113106
Schmelz EA, Alborn HT, Banchio E, Tumlinson JH (2003) Quantitative relationships between induced jasmonic acid levels and volatile emission in *Zea mays* during Spodoptera exigua herbivory. Planta 216:665–673. https://doi.org/10.1007/s00425-002-0898-y
Schmelz EA, Carroll MJ, LeClere S et al (2006) Fragments of ATP synthase mediate plant perception of insect attack. Proc Natl Acad Sci USA 103:8894–8899. https://doi.org/10.1073/pnas.0602328103
Schroder R, Cristescu SM, Harren FJM, Hilker M (2007) Reduction of ethylene emission from scots pine elicited by insect egg secretion. J Exp Bot 58:1835–1842. https://doi.org/10.1093/jxb/erm044
Sestari I, Campos ML (2022) Into a dilemma of plants: the antagonism between chemical defenses and growth. Plant Mol Biol 109:469–482. https://doi.org/10.1007/s11103-021-01213-0
Shangguan X, Zhang J, Liu B et al (2018) A mucin-like protein of Planthopper is required for feeding and induces immunity response in plants. Plant Physiol 176:552–565. https://doi.org/10.1104/pp.17.00755
Tamiru A, Khan ZR, Bruce TJ (2015) New directions for improving crop resistance to insects by breeding for egg induced defence. Curr Opin Insect Sci 9:51–55. https://doi.org/10.1016/j.cois.2015.02.011
Tian D, Peiffer M, Shoemaker E et al (2012a) Salivary glucose oxidase from caterpillars mediates the induction of rapid and delayed-induced defenses in the tomato plant. PLoS One 7:e36168. https://doi.org/10.1371/journal.pone.0036168
Tian D, Tooker J, Peiffer M et al (2012b) Role of trichomes in defense against herbivores: comparison of herbivore response to woolly and hairless trichome mutants in tomato (*Solanum lycopersicum*). Planta 236:1053–1066. https://doi.org/10.1007/s00425-012-1651-9
Trapp S, Croteau R (2001) D EFENSIVE r ESIN b IOSYNTHESIS in c ONIFERS. Annu Rev Plant Physiol Plant Mol Biol 52:689–724. https://doi.org/10.1146/annurev.arplant.52.1.689
Tscharntke T, Thiessen S, Dolch R, Boland W (2001) Herbivory, induced resistance, and interplant signal transfer in *Alnus glutinosa*. Biochem Syst Ecol 29:1025–1047. https://doi.org/10.1016/S0305-1978(01)00048-5
Turlings TC, Loughrin JH, McCall PJ et al (1995) How caterpillar-damaged plants protect themselves by attracting parasitic wasps. Proc Natl Acad Sci USA 92:4169–4174. https://doi.org/10.1073/pnas.92.10.4169
Valette C, Andary C, Geiger JP, et al (1998) Histochemical and Cytochemical investigations of phenols in roots of Banana infected by the burrowing nematode *Radopholus similis*. Phytopathology 88:1141–1148. https://doi.org/10.1094/PHYTO.1998.88.11.1141
Van Der Meer IM, Stam ME, Van Tunen AJ et al (1992) Antisense inhibition of flavonoid biosynthesis in petunia anthers results in male sterility. Plant Cell 4:253–262. https://doi.org/10.1105/tpc.4.3.253
Verma S, Nizam S, Verma PK (2013) Biotic and abiotic stress signaling in plants. In: Sarwat M, Ahmad A, Abdin M (eds) Stress signaling in plants: genomics and proteomics perspective, vol 1. Springer New York, New York, NY, pp 25–49
Von Dahl CC, Baldwin IT (2007) Deciphering the role of ethylene in plant–herbivore interactions. J Plant Growth Regul 26:201–209. https://doi.org/10.1007/s00344-007-0014-4

Wang Q, Xin Z, Li J et al (2015) (E)-β-caryophyllene functions as a host location signal for the rice white-backed planthopper *Sogatella furcifera*. Physiol Mol Plant Pathol 91:106–112. https://doi.org/10.1016/j.pmpp.2015.07.002

Wang H, Shi S, Hua W (2023) Advances of herbivore-secreted elicitors and effectors in plant-insect interactions. Front Plant Sci 14:1176048. https://doi.org/10.3389/fpls.2023.1176048

War AR, Paulraj MG, Ahmad T et al (2012) Mechanisms of plant defense against insect herbivores. Plant Signal Behav 7:1306–1320. https://doi.org/10.4161/psb.21663

Wei H, Zhikuan J, Qingfang H (2007) Effects of herbivore stress by *Aphis medicaginis* Koch on the malondialdehyde contents and the activities of protective enzymes in different alfalfa varieties. Acta Ecol Sin 27:2177–2183. https://doi.org/10.1016/S1872-2032(07)60048-1

White C, Eigenbrode SD (2000) Leaf surface waxbloom in *Pisum sativum* influences predation and intra-guild interactions involving two predator species. Oecologia 124:252–259. https://doi.org/10.1007/s004420000374

Wouters FC, Blanchette B, Gershenzon J, Vassão DG (2016) Plant defense and herbivore counter-defense: benzoxazinoids and insect herbivores. Phytochem Rev 15:1127–1151. https://doi.org/10.1007/s11101-016-9481-1

Yoshinaga N (2016) Physiological function and ecological aspects of fatty acid-amino acid conjugates in insects†. Biosci Biotechnol Biochem 80:1274–1282. https://doi.org/10.1080/09168451.2016.1153956

Yoshinaga N, Ishikawa C, Seidl-Adams I et al (2014) N-(18-Hydroxylinolenoyl)-l-glutamine: a newly discovered analog of Volicitin in *Manduca sexta* and its elicitor activity in plants. J Chem Ecol 40:484–490. https://doi.org/10.1007/s10886-014-0436-y

Zeng J, Ye W, Hu W et al (2023) The N-terminal subunit of vitellogenin in planthopper eggs and saliva acts as a reliable elicitor that induces defenses in rice. New Phytol 238:1230–1244. https://doi.org/10.1111/nph.18791

Zhu C, Ding Y, Liu H (2011) MiR398 and plant stress responses. Physiol Plant 143:1–9. https://doi.org/10.1111/j.1399-3054.2011.01477.x

Zhu-Salzman K, Salzman RA, Ahn J-E, Koiwa H (2004) Transcriptional regulation of *sorghum defense* determinants against a phloem-feeding aphid. Plant Physiol 134:420–431. https://doi.org/10.1104/pp.103.028324

Zúñiga GE, Corcuera LJ (1986) Effect of gramine in the resistance of barley seedlings to the aphid *Rhopalosiphum padi*. Entomol Exp Appl 40:259–262. https://doi.org/10.1111/j.1570-7458.1986.tb00509.x

Plant Elicitor Peptides: Mechanism of Action and Its Applications in Agriculture

8

Data Ram Saini, Pravin Prakash, Savita Jangde, Krishna Kumar, and Ipsita Maiti

Abstract

In agriculture, to protect crops from insect and pests and for increasing crop yield and quality, the agrochemicals have been used, which implies ecological contamination and potential health risks. A novel line of research in the use of alternative of agrochemicals has been proposed, such as elicitors which are eco-friendly and has less potential health risks. Elicitors, either chemicals or bio-factors, can stimulate plant secondary-metabolite synthesis and amplify the protective mechanisms against invading organisms both locally and systemically. These compounds include both exogenous substances, released by pathogens and endogenous elicitors, i.e., compounds secreted from plants. Some of the classes of endogenous elicitors are cell wall fragments, peptides, and phytohormones. Plant elicitor peptides (Peps) are widely distributed across the plant kingdom and play a role in signaling that contribute to broad-spectrum defense against biotic and abiotic factors. Certain ProPep genes are upregulated in response to insect oral secretions and pathogen invasion, and mature Peps attach with plant elicitor peptide receptors (PEPRs) to trigger calcium signaling, accumulation of reactive species, synthesis of defensive phytohormone, transcriptional reprogramming, and production of defensive proteins. Bioactive Peps are the long sequence of 23–36 amino acids, which are generated from the carboxyl termini of longer pro-peptide precursors (PROPEPs), and typically contain a glycine enriched motif: (S/G)(S)Gxx(G/P)xx(N). Peptides, recognized as important physiological regulator, have found applications in various fields including cosmetics, medicine, plant nutrition, and protection. Nowadays, application of plant elicitor peptides has become a popular research subject area in plant protection

D. R. Saini · P. Prakash (✉) · S. Jangde · K. Kumar · I. Maiti
Institute of Agricultural Sciences, Banaras Hindu University, Varanasi, India
e-mail: pravin.pp@bhu.ac.in

S. Singh, R. Mehrotra (eds.), *Plant Elicitor Peptides*,
https://doi.org/10.1007/978-981-97-6374-0_8

as immune inducer, plant growth regulator, insecticides, and herbicides for their excellent activity and ideal environmental compatibility.

Keywords

Elicitors · Plant elicitor peptides · Induced systemic resistance · Mitogen-activated protein kinase · Peptide receptors

8.1 Introduction

Plants for the commercial exploitation come under the tapestry of agriculture where the optimum quality and quantity of production is the main focus of interest. The crop productivity can be hampered with numerous biotic (insects, bacteria, viruses, fungi, herbivores, and oomycetes) (Vasconsuelo and Boland 2007; Ali et al. 2018) and abiotic (UV-B radiation, ozone, heavy metals, salinity, heat stress, drought, and toxicity or deficiency of nutrients) (Khan et al. 2015) environmental factors known as stresses. Since plants are the primary producers in the food web, they are more prone to invasion or infection by bacteria, fungi, insects, and herbivores (Abdul Malik et al. 2020) during both growing period and storage, which frequently causes economic yield losses (Jamiołkowska 2020) and threat to worldwide global agricultural sustainability and food security (Oerke and Dehne 2004). There are a number of advanced physical (solarization and culture rotation), chemical (fungicides or pesticides), or biological control (antagonist microorganisms) methods as well as development of resistant genotypes (Thakur and Sohal 2013) for combating with pests and diseases. But to get thorough and quicker eradication, people use the chemical management approach (Pal and Gardener 2006) for phytopathogens and insect pests, causing environmental hazards, soil erosion (Zheng et al. 2020), disruption of soil ecosystem, increased cost of cultivation, and pesticide residues in agricultural end-products and gradual pathogen resistance which leads to develop a barrier to the eco-friendly sustainable organic farming and direct health issues. To address the current situation of maintaining food safety and pollution in the environment, there is a search for more eco-safe, economical, and efficient natural alternatives, which allows for upregulation of the natural defense mechanism developed in plants against biotic stresses (Zebelo 2020). It focuses on bio-regulation of plant diseases that can be accomplished by the use of microorganisms or bio-molecules that introduce the enhancement in plant tolerance (Heimpel and Mills 2017). Exposure of biotic stress initiates the complex signaling mechanisms which orchestrates multiple stress responses (Chen et al. 2020) by adjusting their metabolism to achieve a sub-optimal homeostasis for the rapid response against their changing environment (Hopkins and Huner 2009), like invasion of biotrophic and hemibiotrophic pathogens, fungi (Wu and Baldwin 2009; Frías et al. 2013), phloem-feeding insects (Zarate et al. 2007), bacteria (Keswani et al. 2019), and viruses (Zhu et al. 2014) which generates the systemic acquired resistance (SAR) defense responses in plants. At first line of defense, the non-specific defense system like cell wall, cuticles, and phyto-anticipins are preformed to limit the pathogen invasion by acting as both physical and chemical barriers which must be overcome by the pathogens to achieve successful colonization (Jones and Dangl 2006). In addition to non-inducible defenses, plants also have ability to recognize and respond to defense elicitors compounds which are perceived by the innate immune system that allocates induce defense responses (Newman et al. 2013), known as pathogen triggered immunity (PTI). In turn, pathogens have developed various virulence factors (i.e., effectors) to be delivered into host cells to restrict PTI and basal defenses (Jones and Dangl 2006; Wang et al. 2019) for developing an additional layer of defense that permits them to intercept pathogen effectors, leading to effector triggered immunity (ETI), accompanied with hypersensitive reaction (HR). Induced systemic resistance (ISR) is acquired when plant immunity is activated by elicitors or biological factors, commonly known as plant immunity inducers which prime the defense genes or pathogenesis-related (PR) genes in plants (Stangarlin et al. 2011). (Both) Elicitor and the receptor molecules both are key elements necessary for proper functioning of defense signaling and immune system (Abdul Malik et al. 2020). The term "elicitor," originally coined to describe substances that stimulate production of antimicrobial phytoalexins in plants, is now commonly applied to any agents fostering signal transduction for any type of defense response against stresses (Keen and Bruegger 1977; Ebel and Scheel 1997) through inducing physiological changes regulating plant metabolism and stimulating enhancing the biosynthesis of secondary metabolites (Baenas et al. 2014; Jamiołkowska 2020). These molecules known as elicitors recognized through polymorphic family of specific intracellular nucleotide-binding/leucine-rich repeat or pattern recognition receptors (PRRs) in the cell surface and all the immune-signaling pathway events such as ROS evolution, Ca^{2+} channel initiation, the activation of MAPKs cascade, and the release of defensive hormones and genes through transcription factor regulation are taking place. These responses are the hallmark of the damage-triggered immunity (DTI) and pattern-triggered immunity (PTI), related to self and non-self-recognition, respectively (Pastor-Fernández et al. 2023). The general elicitors can evoke general resistance against potential pathogens in both host and non-host plants, on the other hand, race-specific elicitors facilitate resistance gene signaling (Thakur and Sohal 2013; Dalio et al. 2017; Meena and Samal 2019) through the action of complementary pairs of (semi)dominant resistance (R) genes in the host plant and (semi)dominant avirulence (Avr) genes in the pathogen (De Wit 1998). The elicitors can vary on the basis of their source, nature, structure and chemical composition (peptides, proteins, lipids, and oligosaccharides) (D'Ovidio et al. 2004; Pršić and Ongena 2020). In order to activate appropriate protective responses measures, plants need to distinguish between "self" and "non-self" (Nürnberger et al. 1997) by harboring a series of receptor–molecule interactions (He et al. 1993). Non-self-molecules belong to microbe or insect are termed microbe/pathogen/herbivore-associated molecular patterns (MAMPs/PAMPs/HAMPs, respectively) (Yu et al. 2017) and they also trigger the production and release of host-derived compounds known as damage-associated molecular patterns (DAMPs) (Tanaka and Heil 2021). In general, the elicitors can be categorized as endogenous and exogenous groups. Endogenous elicitors (peptides, oligo-galacturonic acid, and xylan fragments derived from the plant cell wall) are released in response to pest and pathogen attack by the host plants, while exogenous elicitors (e.g., elicitins, chitin, β-glucan, virus coat proteins, harpins, lipopolysaccharide (LPS), and RNA replicases) are molecules produced by pathogens and pests (Ramirez-Estrada et al. 2016). The elicitor molecules which are chemically a peptide are coming in the point of interest for their short and efficient nature for inducing general resistance through the pattern specific recognition. Exogenous proteinaceous elicitors are compounds from microorganisms, especially pathogens, which regulate many processes, such as the virulence of pathogens, biotic stresses and abiotic stress expression in plants (Zhou and Zhang 2020; Liu et al. 2021). But the recent studies focus on the plant-synthesized proteinaceous elicitors or plant elicitor peptides providing a broad-spectrum defense against the microbial pathogens and insect herbivores. Generally bioactive part of Peps are long sequences of 23–26 amino acid attached with carboxyl termini of longer PROPEPs typically containing a glycine enriched motif: (S/G)(S)Gxx(G/P)xx(N) (Tavormina et al. 2015). The first study of PEPs is conducted in Arabidopsis with a gene family of eight members and orthologues of AtPEPs have since been recognized through amino acid sequence homology in sequenced angiosperm genomes, including those of staple crops such as rice, soybean, maize, and potato (Huffaker et al. 2006; Lori et al. 2015). Certain ProPep genes are upregulated in response to pathogen invasion and insect oral secretions (Huffaker et al. 2006, 2013), and perception of mature Peps by plant elicitor peptide receptors (PEPRs) to trigger calcium signaling, the accumulation of ROS and reactive nitrogen species (RNS), defensive phytohormone (methyl salicylate, salicylic acid, chitosan, benzothiadiazole, benzoic acid) production, transcriptional (WRKY activation) reprogramming, and the production of defensive proteins and metabolites (volatile compounds) (Qi et al. 2010; Yamaguchi and Huffaker 2011; Ma et al. 2012; Klauser et al. 2013; Huffaker 2015).

8.2 Various Types of Elicitors

In plant system, any compound or any stimulus that helps to trigger any type of inducible defense mechanism in plants is known as elicitor (Nürnberger 1999). Originally in old literature, the term elicitor is only applicable for the molecules which can stimulate the phytoalexin synthesis (Thakur and Sohal 2013), but now a days the term elicitor depicts both of the compounds produced by pathogen (exogenous elicitors) as well as compounds released from plants by pathogenicity (endogenous elicitors) (Ramirez-Estrada et al. 2016). They enhance the synthesis of plant secondary metabolites (Montesano et al. 2003), signal transduction cascades to activate phytohormonal defenses, and resistant genes against biotic as well as abiotic stresses (Ramirez-Estrada et al. 2016). On the basis of origin and molecular structure, elicitors can be classified as physical or chemical, biotic or abiotic, and complex or defined (Thakur and Sohal 2013; Pršić and Ongena 2020). The biotic elicitors are generally of microbial origin as avirulence determinants of pathogenicity or host-derived (microbial enzymes, fungal and bacterial lysates, yeast extracts, and polysaccharides, pectin and its derivatives, intracellular proteins, phytohormones such as methyl jasmonate and salicylic acid, their derivatives, and analogs, etc.) (Ramirez-Estrada et al. 2016). On the other hand, abiotic elicitors are the physical and chemical factors of environment. Elicitors again can be divided into two groups, i.e., "general elicitors" and "race-specific elicitors." While general elicitors can induce defense in both host and non-host plants, race-specific elicitors trigger immune reactions only in specific host cultivars (Vasconsuelo and Boland 2007; Baenas et al. 2014) (Fig. 8.1).

Endogenous or host-derived elicitor molecules which are secreted by plants against a stimulation from pathogens and herbivores through cell damage and disruption are known as primary endogenous danger signal and DAMPs (Gust et al. 2017). Under the biotic stress exposure, cell wall fragments (oligogalacturonides or cellulose fragments) (Hou et al. 2019) and peptides are released which are termed as secondary endogenous danger signal or phytocytokines (Gust et al. 2017). These include ROS, oligosaccharide, and protein fragments as an expression of cell degeneration (Chai and Doke 1987; Pearce et al. 2001). Biotic attack triggers the expression of genes encoded for precursors of plant elicitor peptides (Pearce et al. 1991; Huffaker et al. 2011). There are numerous classes of endogenous elicitors based on their biochemical nature identified from higher plants like oligosaccharides, lipids, saponins, volatile compounds, peptides, and proteins. Among them the plant elicitor peptides are widely studied for their significant role in defense signal transduction.

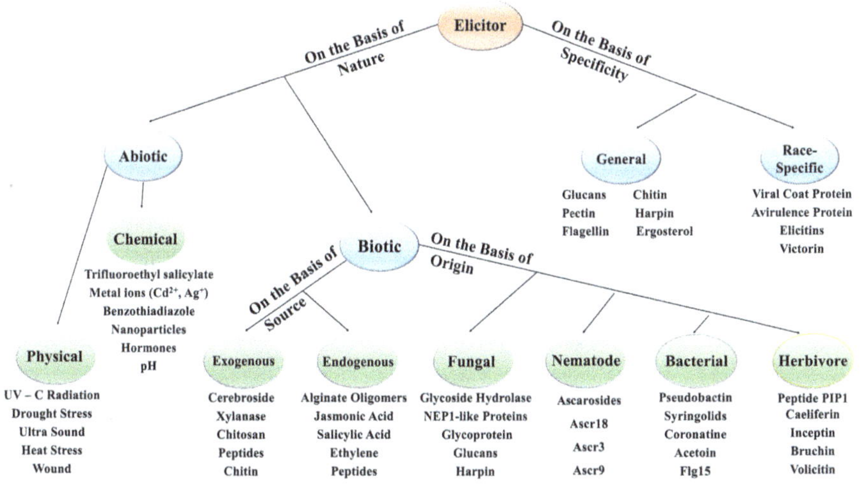

Fig. 8.1 Types of elicitors: on the basis of nature, specificity, source, and origin. *Peptide PIP1* PAMP induced secreted peptide 1, *NEP1* necrosis and ethylene inducing peptide1

8.3 Endogenous Elicitor Peptides

Defense responses are triggered not only by exogenous peptides elicitor molecules (Table 8.1) but also by endogenous host developed peptides molecules (Table 8.2) that are secreted upon injury or infection as danger signals. Plant elicitor peptides (Peps) are naturally occurring endogenous compounds that induce and amplify the first line of inducible plant defense, referred as pattern-triggered immunity (PTI) to protect plants against microbes and herbivores threats. Application of PEP introduces the model host–pathogen immunity in plants relying on molecular recognition and physical interaction by specific receptors transmembrane leucine-rich

Table 8.1 Source, receptor, host plants, and effects of different exogenous elicitor peptides

Exogenous elicitor peptides	Source	Receptor	Host plants	Effects	References
Flagellin (flg22)	Bacteria	Toll-like LRR receptor kinase	Arabidopsis	ROS production, Ca^{2+}/ion fluxes, and medium alkalinization	Bauer et al. 2001
Harpin	Bacteria	115 kDa protein	Tobacco, Arabidopsis	ROS accumulation, and hypersensitive response	Lee et al. 2001
Elicitin	Phytophthora	LRR-receptor-like protein (RLP) elicitin response (ELR)	Potato Tomato Tobacco	HR and immune responses, e.g., electrolyte leakage, the burst of ROS, activation of MAPK cascades, induction of phytohormone and PR genes	Yang et al. 2022
NLPs	Bacteria, Fungi, Oomycetes	BAK1–RLP23–SOBIR1 complex	Arabidopsis	Activation of MAPK signaling cascades, expression of immune-responsive genes, bursts in ROS production, induction of SA, and ethylene	Albert et al. 2015
13-pep glycoprotein	*Phytophthora sojae*	19 kDa plasma membrane-bound protein	Parsley	K^{+}/ H^{+} exchange and Ca^{2+}/influx, ROS, JA, and phytoalexin accumulation	Nennstiel et al. 1998

Table 8.2 Precursor, source, receptor, and effects of different plant elicitor peptides

Plant peptides	Precursor protein	Source	Receptor	Effects	References
AtPeps 1-8	AtproPep1-8	Arabidopsis	AtPEPR1, AtPEPR2	Anti-hemi biotrophic pathogen, anti-chewing insect herbivore, anti-necrotrophic pathogen, root development	Yamaguchi et al. 2010; Klauser et al. 2015
GmPep1-3	GmPROPEP1	Soybean	GmPEPR1a, 2a, 2	Anti-chewing insect herbivore, anti-nematode	Lee et al. 2018
GmPep914	Glyma12g00990	Soybean	–	Expression of defense genes, including Glycine max chalcone synthase1 (Gmachs1), chitinaseb1-1, CYP93A1 and chalcone synthase	Yamaguchi et al. 2011
OsPep3	OsPROPEP3	Rice	OsPEPR1a, OsPEPR1b	Anti-fungal, anti-bacterial, anti-piercing-sucking insect herbivore	Shen et al. 2022
SlPep	SlPROPEP	Tomato	SlPEPR, SlGC17	Anti-necrotrophic pathogen	Yang et al. 2023
SlSystemin	SlproSys	Tomato	SlSYR1, SlSYR2	Anti-necrotrophic pathogen; anti-chewing insect herbivore, target PAL, JA synthesis, and scopoletin synthesis	Wang et al. 2018

(continued)

Table 8.2 (continued)

Plant peptides	Precursor protein	Source	Receptor	Effects	References
ZmPep1, ZmPep3	ZmPROPEP1, ZmPROPEP3	Maize	ZmPEPR1a, ZmPEPR2a	Anti-chewing insect herbivore; anti-necrotrophic pathogen	Huffaker et al. 2011
HypSys	Ntprepro-HypSys	Tobacco	–	Media alkalinization, MAPK Proteinase inhibitors	Pearce et al. 2001
Inceptin	Chloroplastic ATP synthase – γ subunit	Cowpea	–	Fall armyworm larval oral secretion after ingestion of cowpea, Controls kinase, ethylene, JA, ROS, and NO pathways to induce defense response	Schmelz et al. 2007
SubPEP	Putative subtilase (Glyma18g48580)	Soybean	–	Media alkalinization, chalcone synthase, Chitinase1b, CYP93A1, and PDR12 gene expression	Pearce et al. 2010

repeat kinase receptors (LRR-KRs)} displaying a family-specific activity (Ruiz et al. 2018).

The secretion of small defense peptides often involves in the processing of a larger precursor pro-peptide, which differs in structure, indicating various different processing mechanisms. Plant originated peptide type elicitors can be found in different classes based on the structure of their precursor proteins, indicative of different process mechanism to achieve bioactive form for initiating signaling cascade as follows:

1. Peptides synthesized from precursor proteins which lack an N-terminal secretion signal
2. Peptides derived from precursor proteins having an N-terminal secretion signal
3. Cryptic peptides originated from proteins with separate primary functions

The relative positions of bioactive peptide, signal sequence for secretion, and chloroplast localization signal are the determining factors for developing various endogenous plant elicitor peptide families (Yamaguchi and Huffaker 2011).

8.3.1 Peptides Synthesized from Precursor Proteins Which Lack an N-terminal Secretion Signal

8.3.1.1 Systemins

The first isolated plant signaling peptide was from systemin family in tomato. Pearce et al. (1991) found systemin for the host-signal regulating accumulation of protease inhibitors which functions as anti-herbivory defense proteins. The precursor of the 18-aa peptide systemin is a 200-aa protein ProSystemin from whose C-terminus the elicitor is derived (McGurl et al. 1992). Prosystemin accumulates in the cytosol of vascular phloem parenchyma cells, and upon wound-induced processing, the hydrolysis of ProSystemin by subtilisins releases an inactive systemin which is further activated by a leucine-aminopeptidase with the removal of N-terminal amino acid activating functional peptide (Beloshistov et al. 2018). The activation of systemin induces synthesis of jasmonic acid (JA) in the companion-cell-sieve element complex of the vascular bundle, leading to systemic protease inhibitor induction (Li et al. 2001; Narváez-Vásquez and Ryan 2004). Systemin also promotes the secretion of plant volatile compounds that attract natural enemies to kill insect herbivores (Corrado et al. 2007). Analysis of Systemin-treated or transgenic plants that overexpress or suppress the Prosystemin generating gene reveals that systemin in tomato promotes the protease inhibitor and other antinutritive proteins accumulation as well as many other defense responses (Orozco-Cardenas et al. 1993). Constabel et al. (1998) reported that the systemin-like peptides only found in the Solanoideae subfamily of the Solanaceae family such as potato, nightshade, and pepper.

8.3.1.2 Plant Elicitor Peptides (Peps)

Another example for peptides derived from precursor proteins which lack N-terminal secretion signal are the members of the Peps family which regulating the pathogen resistance responses. The first discovered Pep was AtPep1, which was isolated from Arabidopsis leaves utilizing an elicitor-induced alkalinization activity assay with Arabidopsis suspension-cultured cells (Huffaker et al. 2006). Directly derived from the C-terminus of a 92-aa precursor protein AtProPep1 by calcium-dependent metacaspases (Hander et al. 2019), the matured 23-aa AtPep1 peptide specifically binds to two receptors, PEPR1 and PEPR2 (Yamaguchi et al. 2006, 2010). Out of eight homologs of AtPep1, AtPep5 was also biochemically isolated and found to be active; similarly, the others were also synthesized and their ability to bind the PEPR1 receptor and activate alkalinization was confirmed (Huffaker et al. 2006; Yamaguchi et al. 2006). Recently, a homolog in maize, ZmPep1, is also found for regulating maize disease resistance responses (Huffaker et al. 2011). Genes encoding the precursor proteins in both Maize and Arabidopsis are upregulated by

MAMPs. PEPs promote expression of species-specific pathogen defense genes, such as PDF1.2 and PR-1 in Arabidopsis (Huffaker et al. 2006; Huffaker and Ryan 2007) and PR-4 chitinase and SerPIN protease inhibitor in maize (Huffaker et al. 2011).

8.3.2 Peptides Derived from Precursor Proteins Having an N-terminal Secretion Signal

8.3.2.1 Hydroxyproline-rich Systemin (HypSys)

Tobacco plant produces an increased amount of antinutritive protease inhibitor proteins in response to herbivory, which leads to the extraction of two 18-aa glycopeptides from tobacco leaves, hydroxyproline-rich Systemins I and II (NtHypSysI and NtHypSysII) (Pearce et al. 2001). Both peptides are derived from a solitary precursor protein, NtPrePro-HypSys. The aa-sequence of the NtHypSys peptides having their difference in the secretory signals with systemin, the polyprolines are hydroxylated and subsequently glycosylated with pentose sugar chains. Transcripts of PreProHypSys accumulate in phloem parenchyma cells after wounding, and the proprotein localizes to the cell wall matrix (Narváez-Vásquez et al. 2005). Orthologs of NtHypSys have been extracted from solanaceous plants, including tomato, and from a convolvulaceous plant, sweet potato (*Ipomoea batatas*) (Pearce and Ryan 2003; Chen et al. 2008). HypSys treated or transgenic plant shows the induction of defense-related genes which encoding proteins such as enzymes of JA biosynthesis, protease inhibitors or pathogenesis-related genes (Pearce et al. 2001; Narváez-Vásquez et al. 2007; Ren and Lu 2006).

8.3.3 Cryptic Peptide Signals Originated from Proteins with Separate Primary Functions

8.3.3.1 Inceptin

Though there was the evidence of immunoregulatory role of cryptic peptide signals or cryptides in mammals. The first cryptides discovered in immunity regulation in plants are acidic 11–13 aa-di-sulfide bridged peptides known as inceptin family peptides. These inceptin peptides extracted from the oral secretions of fall armyworm (*Spodoptera frugiperda*) larvae which are effective elicitors of ethylene production in cowpea (*Vigna ungiculata*) (Schmelz et al. 2006, 2007). Application of inceptin in cowpea leaves also triggers synthesis of JA, SA, terpene family volatiles, and other secondary metabolites involved in defense that together inhibits fall armyworm growth (Schmelz et al. 2006, 2007, 2009). Inceptins are obtained from the regulatory domain of the γ-subunit of cowpea chloroplastic ATP synthase, which is digested by proteolytic enzymes in the armyworm larval gut (Schmelz et al. 2006). While inceptin-like sequences are found in all plant chloroplastic ATP synthase γ-subunits, its elicitor activity seems to be specific to legumes of Vigna genera and Phaseolus (Schmelz et al. 2006).

8.3.3.2 GmSubPep

A new defense-regulating cryptide was isolated from soybean which induces both the alkalinization response and expression of genes encoding a chitinase and dihydroxy pterocarpan 6a-hydroxylase (CYP93A1), by catalyzing final step of biosynthesis of a soybean phytoalexin, glyceollin (Yamaguchi et al. 2011; Pearce et al. 2010). The 12-aa peptide was discovered to be intrinsicated in the protein-associated domain of a putative extracellular subtilase, Glyma18g48580, and termed as GmSubPep. The gene which encoding GmSubPep is constitutively expressed in all actively growing tissues and it is not only promoted by defense-related phytohormones or wounding (Pearce et al. 2010). It also induces the gene expression encoding chalcone synthase (Yamaguchi et al. 2011; Pearce et al. 2010).

8.4 Mechanism of Plant Elicitor Peptides

Plant immunity is evoked by the perception of both exogenous and endogenous elicitors by cell surface receptors present on hosts. Microbe/pathogen/herbivore-associated molecular patterns (MAMP/PAMP/HAMPs) are well-known exogenous elicitor, whereas the damage-associated molecular patterns (DAMPs) come under endogenous category (Boller and Felix 2009; Ferrari et al. 2013; Albert 2013). The most basic step-in process of plant immunity is the recognition of elicitors from potential invading organisms (including peptides such as, flagellin (flg22), peptidoglycan (PGN), and elongation factor Tu (EF-Tu, elf18) from bacteria, elicitins, ceramides from oomycetes, carbohydrates such as chitin from fungi, coat proteins and RNA replicases from virus), that subsequently induces the physiological responses through signal activation. The recognition of elicitor signal by plasma membrane receptors activates different ion channels, GTP-binding proteins (G-protein) and protein kinase. This activation transfers the elicitor signals to secondary messengers, amplifying the downstream signal transduction (Ebel and Mithöfer 1998; Blume et al. 2000) as following: reversible phosphorylation and dephosphorylation of plasma membrane proteins and cytosolic proteins, cytosolic (Ca^{2+}) spiking, plasma membrane depolarization, Cl^- and K^+ efflux/H^+ influx, extracellular alkalinization and cytoplasmic acidification, mitogen-activated protein kinase (MAPK) activation, NADPH oxidase activation and ROS production, early defense gene expression, ethylene and JA production, late defense response gene expression, and accumulation of secondary metabolites (Zhao et al. 2005). PAMPs/HAMPs/MAMPs are conserved sequenced molecules derived from invading microbes that are specifically attached to the host pattern recognition receptors (PRRs) to trigger defense responses (Zipfel et al. 2004, 2006; Lee et al. 2009). There are various MAMP–PRR interactions characterized in plants, including the bacterially derived peptides such as elf18, Ax21, and flg22 which bind the EFR, Xa21, and FLS2 receptors, respectively (Zipfel et al. 2004, 2006; Lee et al. 2009). Similarly, the PRR CERK1/LysM RLK facilitates the recognition of fungal pathogen by binding with chitin (a fungal-derived oligosaccharide MAMP) which initiates defense signaling (Iizasa et al. 2010). HAMPs like glutamine–fatty acid conjugates, volicitin (Alborn et al. 1997) present in insect oral secretions, and bruchins, a long-chain α, ω-diols found in weevil oviposition-fluid (Doss et al. 2000) are also distinctively recognized and induce plant anti-herbivory defenses. Endogenous elicitors are released by cell disruption such as ROS, oligosaccharide, and protein fragments (Chai and Doke 1987; Albersheim and Anderson 1971; Pearce et al. 1991). In addition to local recognition of these compounds for innate immunity, plants also exploit this receptor-pattern interaction to amplify defense responses by synthesizing similar molecules through regulated specific enzymatic activity.

The plant elicitor peptides (Peps) are most promising endogenous signaling molecules among DAMPs in both monocot and dicot model plants (Huffaker et al. 2011; Bartels et al. 2013; Yamaguchi and Huffaker 2011) that induce defense responses such as (phyto)hormone production, induced expression of resistance genes, the activation of phosphor relays, and the induction of cell secondary messenger and metabolite synthesis. All endogenous peptides work in a loop regulation method with small defensive hormone molecules (jasmonic acid, ethylene and salicylic acid) and loop is maintained by encoding the biosynthetic enzymes for combining with hormones. These phytohormones concentration is increased in response to peptide-elicitation (Heitz et al. 1997; Howe et al. 1996; Krol et al. 2010; Narváez-Vásquez et al. 2007). The genes responsible for the precursors of endogenous peptide elicitors are induced by elicitation treatment (Pearce et al., 1991; Huffaker et al., 2006). Some PROPEP genes are upregulated by pathogen invasion and insect oral secretions (Huffaker et al. 2006, 2013), and mature Peps interact with plant elicitor peptide receptors (PEPRs) to trigger extracellular alkalinization, intracellular Ca^{2+} elevation, the generation of reactive species, MAP kinase activation, phytohormone production, transcriptional reprogramming and the synthesis of defensive proteins and metabolites, callose and lignin deposition, volatile compound production (VOC), accumulation of antimicrobial compounds, and the gene expression of protease inhibitors against herbivore growth (Huffaker 2015; Huffaker et al. 2011, 2013; Klauser et al. 2013; Ma et al. 2012, 2013; Qi et al. 2010; Yamaguchi and Huffaker 2011; Orozco-Cárdenas et al. 2001; Ryan 2000). The downstream signaling pathway of peptide-induced defense responses are established through several mutant studies (Howe et al. 1996; Degenhardt et al. 2010). The process of conferring resistance to the pathogenic microbes is deeply studied in Arabidopsis plant (Zelman and Berkowitz 2023). A family of eight Peps was first found out in *Arabidopsis thaliana*, and orthologues are found recently through amino acid sequence homology in angiosperms and staple crops such as wheat, maize, potato, rice, and soybean (Huffaker et al. 2006; Lori et al. 2015). Peps are synthesized from PROPEPs and are recognized by leucine-rich repeat (LRR) receptor-like kinases referred as Pep receptors (PEPRs). In Arabidopsis, eight PROPEPs (PROPEP1-8) and two PEPRs (PEPR1 and PEPR2) have been identified (Krol et al. 2010; Yamaguchi et al. 2010; Bartels et al. 2013). Most of PROPEPs are transcriptionally activated upon microbial infection or wounding (Bartels and Boller 2015). Individual PROPEPs are thought to be cytoplasmic or to be tonoplast associated, making an idea that Peps are released into the apoplast either actively as a response to danger signals or passively during damage and cellular disintegration (Bartels et al. 2013). There are several proposals about the Pep extraction from Propeps. The cleavage at C-termini last 23 amino acids of PROPEPs sequence produces immature Peps (Yamaguchi and Huffaker 2011). On the other hand, mature Peps are extracted automatically from the C-termini of PROPEPs through meta-caspase-mediated cleavage and transferred to the apoplast (Hander et al. 2019; Shen et al. 2019). Once the processed and mature PEP occurs in the apoplast, they are perceived by PEPRs of other nearby cells and initiate typical PTI responses (Yamaguchi et al. 2006, 2010).

In Arabidopsis, Pep receptors AtPEPR1 and AtPEPR2 recognize the eight AtPep peptides (Yamaguchi et al. 2006, 2010). The AtPEPRs are members of a prolifically duplicated protein family called leucine-rich repeat receptor-like kinases (LRR-RLKs) which had its origin in green algae (Han 2019) and branched out in plant specific evolutionary trend (Liu et al. 2017) as a crucial player in pant defense (Afzal et al. 2008). AtPEPRs have a leucine-rich repeat (LRR) region serving as the ligand-binding domains of these proteins (Afzal et al. 2008), a transmembrane ™ domain, and a kinase domain with a putative guanylyl cyclase (GC) domain that is linked to Ca^{2+} level elevation (Ryan 2000). The FLS2, PEPR, and EFR receptors all make complex with the same co-receptor, BRI1-associated kinase1 (BAK1), which is required for both MAMP and AtPep signaling (Postel et al. 2010; Schulze et al. 2010; HEESE et al. 2007). It is suggested that BAK1 may have function at the intercept of AtPep and MAMP signaling. The mechanism of peptide elicitor binding to the receptor and subsequent signal amplification and induced response is presented in Fig. 8.2.

In Arabidopsis, the studies on the gene expression using MAMPs and AtPeps help to establish the overlapping downstream signaling (Zipfel et al. 2004, 2006; Huffaker et al. 2006). Upon the binding to the receptor and co-receptor, Calcium oscillation, ROS, and phosphorylation of MAPK are major signaling intermediaries that trigger defense responses through Pep signaling. Ca^{2+} elevations are critical as well as the first response for plant defense responses for calcium mediated pathways (Lecourieux et al. 2006; Ryan 2000). Both flg22 as exogenous and the AtPeps as endogenous molecule activate Ca^{2+} signaling by activating of a cyclic nucleotide-gated Ca^{2+} conducting channel (CNGC2) after receptor recognition, that is needed for unhampered signal transduction (Krol et al. 2010; Qi et al. 2010). Though precise mechanisms of CNGC2 activation are an unclear phenomenon, unlike FLS2, the intracellular kinase region of AtPEPR1 was discovered to contain a guanylyl cyclase domain, which produced cGMP, the activator of CNGC-dependent cytosolic Ca^{2+} elevation (Qi et al. 2010). ROS oscillate in cell, interacting with host cell molecules and damaging cell plasma membrane of the pathogen, but also serve as a secondary messenger, such as in SA signaling and in Pep signaling (Ma et al. 2013; Kawano and Bouteau 2013). Transient cytosolic Ca^{2+} spikes both trigger ROS and are triggered by ROS (Kawano and Bouteau 2013) as a loop reaction.

In Arabidopsis, ROS are known to be produced by Pep-induced calcium signaling (Ma et al. 2013) where in tomato, treatment with AtPep1 causes an increase in ROS (Saand et al. 2015). Significant effect of $[Ca^{2+}]$ spiking results in differential activation of transcription factors like WRKY transcription factors and genes

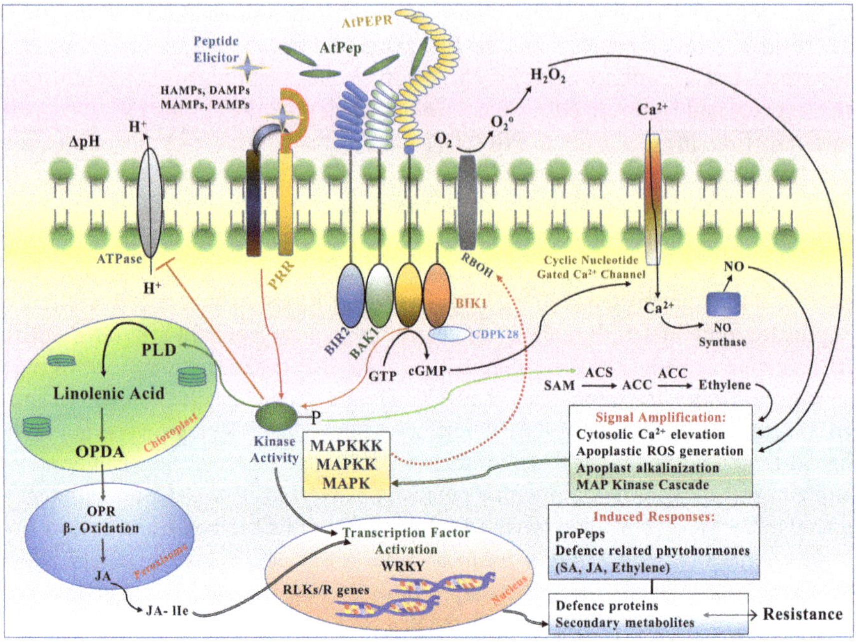

Fig. 8.2 Defense mechanism induced by plant elicitor peptides. *HAMPs* herbivory associated molecular patterns, *DAMPs* damaged associated molecular patterns, *MAMPs* microbes associated molecular patterns, *PAMPs* pathogen associated molecular patterns; *PRR* pattern recognition receptors, *BIR2* BAK1-interacting receptor-like kinase 2, *BAK1* brassinosteroid receptor-associated kinase1, *BIK1* botrytis induced kinase 1, *RBOH* respiratory burst oxidative homologs, *CDPK28* calcium-dependent protein kinase 28, *AtPEP* Arabidopsis plant elicitor peptide, *AtPEPR* Arabidopsis plant elicitor peptide receptor, *RLKs* receptor-like kinase, *R genes* resistance gene, *MAPK* mitogen-activated protein kinases

including MAP Kinase 3 as well as PR-1 and PDF1.2 (Zipfel et al. 2004, 2006; Huffaker et al. 2006; Yamaguchi et al. 2010), which directly regulate all defense gene expression (Dolmetsch et al. 1997; Yang and Poovaiah 2002). Systemin and AtPeps cause calcium influx (Ryan 2000; Moyen et al. 1998) as increment in callose and lignin deposition, a late immune response that mechanically protects plant tissues in a calcium-dependent manner (Zhang et al. 2022). In soybean, GmPep1 triggers the expression of its pro-peptide precursor (GmPROPEP1), a nucleotide-binding site leucine-rich repeat protein (NBS-LRR), a pectin methyl esterase inhibitor (PMEI), Respiratory Burst Oxidase Protein D (RBOHD), and ROS accumulation in leaves (Lee et al. 2018). In brief, pretreatment of Arabidopsis, *Zea mays,* and other plants with Peps stimulate the defense response and develop systemic immunity against diverse microbial pathogens and herbivorous insects (Huffaker et al. 2006, 2011, 2013; Yamaguchi et al. 2010; Liu et al. 2013; Tintor et al. 2013).

8.5 Plant Elicitor Peptides: Defense Responses

Various studies have reported the structure and function of plant elicitor peptides and its receptor (PEPR) system in the model plant Arabidopsis (Tintor et al. 2013; Klauser et al. 2015), *Zea mays* (Huffaker et al. 2013; Lori et al. 2015), soybean (Pearce et al. 2010; Lee et al. 2018), and rice (Shinya et al. 2018; Shen et al. 2022).

Plant peptides have recently revealed themselves as key signaling molecules of stress responses, not only to mechanical wounding by herbivores and pathogen infection but also to nutrient imbalance, drought, and high salinity (Chen et al. 2020). Plant elicitor peptides are small but signature amino acid sequences, which upon recognized by plasma membrane receptor (Huffaker et al. 2006; Yamaguchi et al. 2010) through a signaling cascade activates the transcriptional activity leading to generation of ROS (Jing et al. 2020), synthesis of defense-related proteins, enzymes, phytohormones (ethylene and jasmonic acid), and secondary metabolites (phytoalexins, benzoxazinoids) (Yamaguchi and Huffaker 2011; Bartels and Boller 2015). This signaling also promotes the accumulation of wax and cutin, regulation of plasma membrane dynamics (Jing et al. 2023), and cell wall modification through callose deposition (Clay et al. 2009), root hair growth (Jing et al. 2024), biotic and abiotic stress response, etc. (Fig. 8.3).

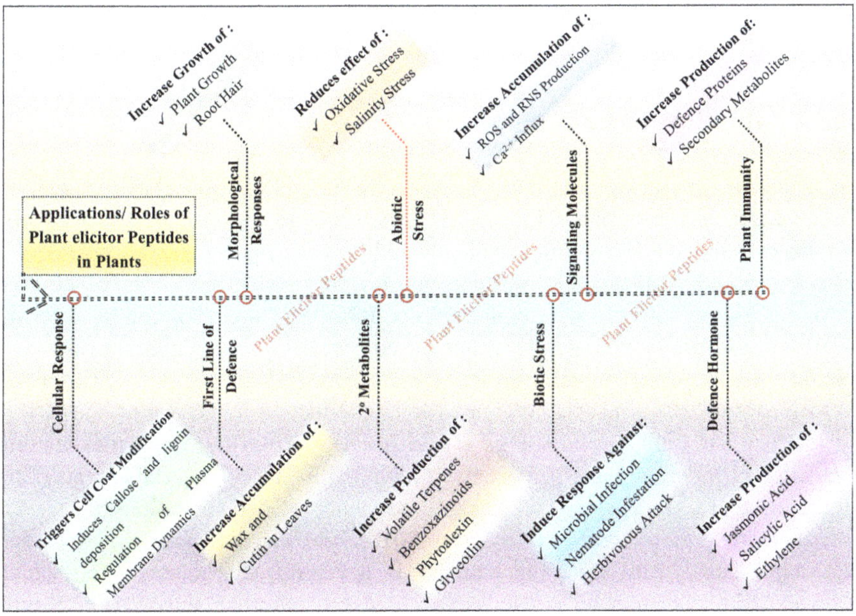

Fig. 8.3 Roles/applications of plant elicitor peptides

8.6 Applications of Plant Elicitor Peptides in Agriculture

Plant elicitor peptides have been found having role in Poaceae, Rosaceae, Solanaceae, Fabaceae, and Brassicaceae. Peps plays role as an endogenous chemical messenger to trigger defense responses in plants. Several studies reported that pretreatment of Peps activates Peps-PEPR signaling system to enhance plant resistance against salinity stress (Wang et al. 2022a), herbivore attacks (Klauser et al. 2015), bacterial or fungal infections (Tintor et al. 2013). There are so many other studies also report that the exogenous application of Peps (as well as flg22) regulated various responses in roots, including increased cytosolic Ca^{2+} levels, ROS accumulation, callose deposition, and increased expression of defense-related genes, which implies that root also have similar immune response as of leaves (Millet et al. 2010; Bartels et al. 2013; Beck et al. 2014; Ma et al. 2014; Poncini et al. 2017). Exogeneous application of PEPs stimulates the pattern-triggered immunity (PTI) pathway (Huffaker et al. 2006) developing the systemic defense (Ross et al. 2014) and microbial pathogenic resistance (Yamaguchi and Huffaker 2011) in plant. There are several studies done on various crops like maize (Huffaker et al. 2011), Arabidopsis (Huffaker et al. 2006), rice (Shen et al. 2022), tobacco, soybean (Lee et al. 2018), tomato (Zelman and Berkowitz 2023), and many others to show their prospective as therapeutic agents due to their broad-spectrum rapid activity, low cytotoxicity to zoological tissues and unique mode of action which hinders emergence of pathogenicity (Holaskova et al. 2015).

8.6.1 Applications of AtPep

Arabidopsis endogenous peptides, members of the AtPep group have frequently been studied and described to induce PTI and to enhance resistance against various groups of pathogens by amplifying the innate immune response (Gully et al. 2015). The exogenous application of AtPep1, a 23 amino acid peptide, increased root development by activating the expression of PROPEP1, transcription of defensin (PDF1.2), and production of H_2O_2 which are component of innate immune response, confers resistance to root pathogen *Pythium irregulare* (Huffaker et al. 2006). Pep1 perceived by PEPR2 receptor increases ROS production by activating the BIK1-respiratory burst oxidase homologs D and F (BIK1-RBOHD/F) enzymes leading induction of roots' immune response and inhibits root growth in Arabidopsis (Jing et al. 2020). This oxidative burst along with membrane depolarization induces the ethylene production in leaves (Krol et al. 2010). Exogeneous application of AtPep activate AtPep1-PEPR signaling pathway in a PEPR2-dependent manner which triggers plant immunity by restricting bacterial entry through activation of S-type anion channel in guard cells leading to stomatal closure (Zheng et al. 2018) and promotes premature cell differentiation, swelling of epidermal and cortex cells and root hair formation. It inhibits root growth through redirect local auxin accumulation by reducing the abundance of PIN2 in the plasma membrane with endocytosis activation and increasing PIN3 expression in the root transition zone of Arabidopsis

(Jing et al. 2019). Inhibition of root elongation is mediated through the reduction of Glutamine Dumpers (AtGDUs) genes expression (regulates amino acid export pathway in root) and extracellular Ca^{2+} dependent pathways in Arabidopsis root (Ma et al. 2014). Shen et al. (2020) also studied that exogenous application of 100 nM AtPep1 induces cell swelling in root apex and inhibition of root growth through increased acidic condition in apo-plastic space by activating the plasma membrane-localized H^+-ATPases (PM H^+-ATPases)—the pump proton in plant cell—to extrude the protons into apoplast in Arabidopsis. Application of 1 μM of pep1 in Arabidopsis seedling stimulating expression of PEPR1 and PEPR2 which reduces the number of stele cells. Furthermore, PEP1 enhances ectopic xylem differentiation, which leads to extra protoxylem or metaxylem in the root maturation zone. PEP1 signaling may interrupt the symplastic connection through a phloem sieve element probably by callose deposition (Dhar et al. 2021). Exogenous application of synthetic AtPep3 @100nM, a C-terminal peptide fragment encoded by AtPROPEP3 and also perceived by PEPR1 receptor, play role in inducing salinity stress tolerance by inhibiting of chlorophyll degradation in plants (Nakaminami et al. 2018).

Previously, some studies identified that plasma membrane-intrinsic Qian Shou kinase 1 (QSK1) and sucrose-induced receptor kinase 1 (SIRK1) as receptor/co-receptor pair involved in the regulation of aquaporins in response to osmotic conditions induced by sucrose. Further studies reveal a member of the elicitor peptide (PEP) family, namely PEP7, as the specific ligand of the receptor kinase SIRK1, secreted to the apoplasm in response to sucrose treatment. Wang et al. (2022b) reported that the PEP7 enhances the phosphorylation of aquaporins in vivo and increased water influx into protoplasts leading to proper growth of lateral roots.

8.6.2 Applications of ZmPep

A plant elicitor peptide, ZmPep3 from maize, is demonstrated to be an effective regulator of anti-herbivore defenses. It is as active as the Lepidopteran elicitor N-linolenoyl-L-glutamine (Gln-18:3) in stimulating volatile release and buildup of defense transcripts and metabolites (Huffaker 2015). ZmPep3 treated plants also induce direct and indirect defenses resistance against *S. exigua* by promoting the accumulation of the 2-hydroxy-4,7-dimethoxy-1,4-benzoxazin-3-one glucoside (HDMBOA-Glc), expression of herbivory defense-related genes and synthesis of ethylene, Jasmonic Acid (JA), and phytoalexins (Huffaker et al. 2013). Another peptide ZmPep1 is also demonstrated that it triggers production of plant defense phytohormones and enhances expression of genes encoding pathogenesis-related proteins (Huffaker et al. 2014). Volatiles synthesized by ZmPep1 works as an attractant to *Cotesia marginiventris* parasitoids and accumulation of proteins and benzoxazinoids that directly suppress larval growth (Huffaker 2015). ZmPep1-treated plants show immunity defense against southern leaf blight and anthracnose stalk rot through stimulating accumulation of 2-hydroxy-4,7-dimethoxy-1,4-benzoxazin-3-one glucoside (HDMBOA-Glc), expression of genes which encoding defense compounds endochitinase A, PR-4, PRms, SerPIN, and benzoxazinoid, and activating de novo synthesis of JA and ethylene in maize leaves (Huffaker et al. 2011). Where ZmPep3 is only specific to the Poaceous family, the other peptides derived from PROPEP orthologs are also identified in Fabaceous and Solanaceous plants (Huffaker et al. 2013).

8.6.3 Applications of OsPep

The BSR1 (receptor-like cytoplasmic kinase broad-spectrum resistance 1) of *Oryza sativa* L. when receives the peptidic elicitor OsPep molecules triggers the ROS burst, ethylene biosynthesis, and diterpenoid phytoalexin accumulation against herbivores (Kanda et al. 2023). Exogenously applied OsPep3 in rice induces transcription of PEP RECEPTORs (PEPRs) and PRECURSORs of PEP (PROPEPs), which develop resistance against brown plant hopper through the biosynthesis of JA, and metabolism of lipid and phenylpropanoid in leaf sheaths (Shen et al. 2022). OsPep3 elicitation raises up immunity in rice against the bacterial pathogen *Xanthomonas oryzae pv. oryzae* and fungal pathogen *Magnaporthe oryzae* as well as provokes immune responses in wheat. Exogenous application of OsPep3 for combating piercing-sucking insect herbivores reveals it as eco-friendly immune stimulator in agriculture for crop protection against a broad spectrum of insect pests and pathogens (Shinya et al. 2018). Synchronized application of OsPep3 and *Mythimna loreyi* oral secretions intensifies an array of defense responses in rice leaf cells, including MAPK cascade activation and generation of defense phytohormones and secondary metabolites (Shinya et al. 2018).

8.6.4 Applications of GmSubPep

The GmSubPep (*Glycine max* Subtilisin Peptide) is a 12 aa peptide, which isolated from soybean leaf triggers defense responses by inducing defense-related genes PDR12 (encoding pleiotropic drug resistance-type transporter), chs (Chalcone synthase), CYP93A1 (cytochrome P450), and Chib-1b (encoding PR-8 chitinase) (Pearce et al. 2010). In addition to GmSubPep, GmPep1, GmPep2, and GmPep3 activate resistance against bacterial, fungal, and broad spectrum of nematodes such as soybean cyst nematode (*Heterodera glycines*) and root-knot nematode (*Meloidogyne incognita*) by inducing systemic resistance through production and accumulation of ROS in apoplast (Lee et al. 2018).

8.6.5 Applications of Other Pep

There are other various Peps which have been studied and reported for their several roles in plant systems and defense signaling in different plant species such as broccoli, oats, soybean, tomato, etc. Exogenous application of 100 nM BoPep4 derived from broccoli leaves, precursor gene BoPROPEP4, increase tolerance to salinity stress in broccoli through reduction of Na^+/K^+ ratio and enhanced accumulation of cutin and wax in leaves (Wang et al. 2022a). Yang et al. (2004) reported that treatment of victorin increased the accumulation of AsCCoAOMT and AsHHT1 transcripts with Avenanthramides phytoalexin accumulation, a specific elicitor in oat lines carrying the Pc-2/Vb gene. SCOOP signature peptides at sub-molar concentration trigger immune responses and alter root development in a MIK2-dependent manner. In recent reports, it is found to strongly activate root immune responses, also cannot induce root callose deposition (Hou et al. 2021). The purified BP178 peptide showed in vitro activity against the bacterial plant pathogen *Dickeya sp.*, the causal agent of the dark brown sheath rot of rice. Transgenic seedlings show enhanced resistance to the fungal pathogen *Fusarium verticillioides*, supporting that the in-plant produced peptide was biologically active (Montesinos et al. 2017). The preventive spray of tomato plants with BP178 controlled bacterial infections, i.e., *Pseudomonas syringae pv. tomato* and *Xanthomonas campestris pv. vesicatoria*, as well as the fungus infections of *Botrytis cinerea* by inducing the expression of several genes encoding pathogenesis-related proteins (PR1-7, PR9, 10, 14) and transcription factors involved in the jasmonic acid, salicylic acid, and ethylene-signaling pathways (Montesinos et al. 2021). Another Pep, systemin which was the first peptide hormone discovered from tomato induces resistance against pathogens (Pearce et al. 1991). Trivilin et al. (2014) studied that the application of SlPep reduces the severity of infection of *Pythium dissoticum* in tomato seedlings. Xu et al. (2018) also studied that SlPEPR, putative tomato Pep receptor (PEPR) termed as PEPR1/2 ortholog receptor-like kinase 1 (PORK1) shows defense response against *B. cinerea* in tomato plants. Similarly, Glutathione also shows defense responses through enhancing light induction of genistein conjugates, glyceollins, apigenin, daidzein, and luteolin in soybean (Graham and Graham 1996). Research is going on the peptidic elicitor group from plants of Rosacea family where the Elicitor activity of PpPep1, PdPep1, and PpPep2 peptides against *Xanthomonas arboricola pv. pruni* (Xap) is studied. Exogenous application of chemically synthesized PpPep1, PdPep1, or PpPep2 induces the expression of the corresponding PROPEP gene in *P. persica* × *P. dulcis* leaves (Ruiz et al. 2018).

8.7 Conclusion

The use of naturally accessible molecules (elicitor peptides) for overcoming biotic stresses (pathogen, pest control) and certain abiotic stresses (salinity) is a newly emerging topic for sustainable and organic future. The introduction of advanced technology and intense study at cellular and molecular level allows for integrated pest and disease management protocols of newer order. Bioactive nature-derived particles with phytosanitary activity have molecular structures that can be altered into new stable compound with similar mode of action. The production of new bio-products through chemical modifications of natural bio-compounds opens a novel door in using the natural elicitor compounds, their isoforms and related forms against biotic stresses. There is a huge example for the use of exogeneous elicitor in plants to ignite the various stress response signals. The new candidate to make its appearance in this panel is plant elicitor peptide or phytocytokines, as the endogenous molecules are easier to synthesize in vitro for their smaller size and effect in lower concentration, hence they are becoming a new alternative to those synthetics. In addition, plant endogenous compounds are easily biodegradable. Commercials of endogenous plant elicitors or green pesticides make plants more tolerant and resistant against upcoming biotic hindrance in a more low-cost manner and have almost no ecological foot-prints as compared with conventional approaches. The complete life cycle of plant is studied to calculate the optimum time of application, dose and probable yield response. But, an array of plant elicitor peptides remains to recognize, which can stimulate the potent and continual immune responses against biotic and abiotic stressors. The interactions between different elicitors are also left to be studied, even though some have been evaluated and reported to show certain impacts. The application or co-application of phytochemicals will let mankind to discover the new emerging opportunities for developing innate and adaptive immunity in plants toward the quality crop production.

References

Abdul Malik NA, Kumar IS, Nadarajah K (2020) Elicitor and receptor molecules: orchestrators of plant defense and immunity. Int J Mol Sci 21(3):963

Afzal AJ, Wood AJ, Lightfoot DA (2008) Plant receptor-like serine threonine kinases: roles in signaling and plant defense. Mol Plant-Microbe Interact 21(5):507–517

Albersheim P, Anderson AJ (1971) Proteins from plant cell walls inhibit polygalacturonases secreted by plant pathogens. Proc Natl Acad Sci 68(8):1815–1819

Albert M (2013) Peptides as triggers of plant defence. J Exp Bot 64(17):5269–5279

Albert I, Böhm H, Albert M, Feiler CE, Imkampe J, Wallmeroth N, Nürnberger T (2015) An RLP23–SOBIR1–BAK1 complex mediates NLP-triggered immunity. Nat Plants 1(10):1–9

Alborn HT, Turlings TC, Jones T, Stenhagen G, Loughrin JH, Tumlinson JH (1997) An elicitor of plant volatiles from beet armyworm oral secretion. Science 276(5314):945–949

Ali S, Ganai BA, Kamili AN, Bhat AA, Mir ZA, Bhat JA, Grover A (2018) Pathogenesis-related proteins and peptides as promising tools for engineering plants with multiple stress tolerance. Microbiol Res 212:29–37

Baenas N, García-Viguera C, Moreno DA (2014) Elicitation: a tool for enriching the bioactive composition of foods. Molecules 19(9):13541–13563

Bartels S, Boller T (2015) Quo vadis, Pep? Plant elicitor peptides at the crossroads of immunity, stress, and development. J Exp Bot 66(17):5183–5193

Bartels S, Lori M, Mbengue M, Van Verk M, Klauser D, Hander T, Boller T (2013) The family of Peps and their precursors in Arabidopsis: differential expression and localization but similar induction of pattern-triggered immune responses. J Exp Bot 64(17):5309–5321

Bauer Z, Gómez-Gómez L, Boller T, Felix G (2001) Sensitivity of different ecotypes and mutants of Arabidopsis thaliana toward the bacterial elicitor flagellin correlates with the presence of receptor-binding sites. J Biol Chem 276(49):45669–45676

Beck M, Wyrsch I, Strutt J, Wimalasekera R, Webb A, Boller T, Robatzek S (2014) Expression patterns of flagellin sensing 2 map to bacterial entry sites in plant shoots and roots. J Exp Bot 65(22):6487–6498

Beloshistov RE, Dreizler K, Galiullina RA, Tuzhikov AI, Serebryakova MV, Reichardt S, Vartapetian AB (2018) Phytaspase-mediated precursor processing and maturation of the wound hormone systemin. New Phytol 218(3):1167–1178

Blume B, Nürnberger T, Nass N, Scheel D (2000) Receptor-mediated increase in cytoplasmic free calcium required for activation of pathogen defense in parsley. Plant Cell 12(8):1425–1440
Boller T, Felix G (2009) A renaissance of elicitors: perception of microbe-associated molecular patterns and danger signals by pattern-recognition receptors. Annu Rev Plant Biol 60:379–406
Chai HB, Doke N (1987) Superoxide anion generation: a response of potato leaves to infection with *Phytophthora infestans*. Phytopathology 77:645–649
Chen YC, Siems WF, Pearce G, Ryan CA (2008) Six peptide wound signals derived from a single precursor protein in Ipomoea batatas leaves activate the expression of the defense gene sporamin. J Biol Chem 283(17):11469–11476
Chen YL, Fan KT, Hung SC, Chen YR (2020) The role of peptides cleaved from protein precursors in eliciting plant stress reactions. New Phytol 225(6):2267–2282
Clay NK, Adio AM, Denoux C, Jander G, Ausubel FM (2009) Glucosinolate metabolites required for an Arabidopsis innate immune response. Science 323(5910):95–101
Constabel CP, Yip L, Ryan CA (1998) Prosystemin from potato, black nightshade, and bell pepper: primary structure and biological activity of predicted systemin polypeptides. Plant Mol Biol 36:55–62
Corrado G, Sasso R, Pasquariello M, Iodice L, Carretta A, Cascone P, Rao R (2007) Systemin regulates both systemic and volatile signaling in tomato plants. Plant Mol Biol 33:669–681
D'Ovidio R, Mattei B, Roberti S, Bellincampi D (2004) Polygalacturonases, polygalacturonase-inhibiting proteins and pectic oligomers in plant–pathogen interactions. Biochim Biophys Acta (BBA)-Proteins Proteom 1696(2):237–244
Dalio RJ, Magalhaes DM, Rodrigues CM, Arena GD, Oliveira TS, Souza-Neto RR, Machado MA (2017) PAMPs, PRRs, effectors and R-genes associated with citrus–pathogen interactions. Ann Bot 119(5):749–774
De Wit PJ (1998) Avirulence and pathogenicity genes of Cladosporium fulvum. In: Genetics and breeding for crop quality and resistance: proceedings of the XV EUCARPIA Congress, Viterbo, Italy, September 20–25. Springer Netherlands, Dordrecht, pp 3–14
Degenhardt DC, Refi-Hind S, Stratmann JW, Lincoln DE (2010) Systemin and jasmonic acid regulate constitutive and herbivore-induced systemic volatile emissions in tomato, Solanum lycopersicum. Phytochem 71(17–18):2024–2037
Dhar S, Kim H, Segonzac C, Lee JY (2021) The danger-associated peptide PEP1 directs cellular reprogramming in the Arabidopsis root vascular system. Mol Cell 44(11):830–842
Dolmetsch RE, Lewis RS, Goodnow CC, Healy JI (1997) Differential activation of transcription factors induced by Ca2+ response amplitude and duration. Nature 386(6627):855–858
Doss RP, Oliver JE, Proebsting WM, Potter SW, Kuy S, Clement SL, DeVilbiss ED (2000) Bruchins: insect-derived plant regulators that stimulate neoplasm formation. Proc Natl Acad Sci 97(11):6218–6223
Ebel J, Mithöfer A (1998) Early events in the elicitation of plant defence. Planta 206:335–348
Ebel J, Scheel D (1997) Signals in host-parasite interactions. In: Plant relationships: Part A. Springer, Berlin, Heidelberg, pp 85–105
Ferrari S, Savatin DV, Sicilia F, Gramegna G, Cervone F, Lorenzo GD (2013) Oligogalacturonides: plant damage-associated molecular patterns and regulators of growth and development. Front Plant Sci 4:49
Frías M, Brito N, González C (2013) The B otrytis cinerea cerato-platanin BcSpl1 is a potent inducer of systemic acquired resistance (SAR) in tobacco and generates a wave of salicylic acid expanding from the site of application. Mol Plant Pathol 14(2):191–196
Graham TL, Graham MY (1996) Signaling in soybean phenylpropanoid responses: dissection of primary, secondary, and conditioning effects of light, wounding, and elicitor treatments. Plant Physiol 110:1123–1133
Gully K, Hander T, Boller T, Bartels S (2015) Perception of Arabidopsis AtPep peptides, but not bacterial elicitors, accelerates starvation-induced senescence. Front Plant Sci 6:126939
Gust AA, Pruitt R, Nürnberger T (2017) Sensing danger: key to activating plant immunity. Trends Plant Sci 22(9):779–791
Han GZ (2019) Origin and evolution of the plant immune system. New Phytol 222(1):70–83

Hander T, Fernández-Fernández ÁD, Kumpf RP, Willems P, Schatowitz H, Rombaut D, Stael S (2019) Damage on plants activates Ca2+-dependent metacaspases for release of immunomodulatory peptides. Science 363(6433):eaar7486
He SY, Huang HC, Collmer A (1993) Pseudomonas syringae pv. syringae harpinPss: a protein that is secreted via the Hrp pathway and elicits the hypersensitive response in plants. Cell 73(7):1255–1266
Heese A, Hann DR, Gimenez-Lbanez S, Jones AM, Kai H, Jia L, Rathjen JP (2007) The receptor-like kinase SERK3. Proc Natl Acad Sci USA 104(29):12217–12222
Heimpel GE, Mills NJ (2017) Biological control. Cambridge University Press, Cambridge
Heitz T, Bergey DR, Ryan CA (1997) A gene encoding a chloroplast-targeted lipoxygenase in tomato leaves is transiently induced by wounding, systemin, and methyl jasmonate. Plant Physiol 114(3):1085–1093
Holaskova E, Galuszka P, Frebort I, Oz MT (2015) Antimicrobial peptide production and plant-based expression systems for medical and agricultural biotechnology. Biotechnol Adv 33(6):1005–1023
Hopkins WG, Huner NPA (2009) Introduction to plant physiology, 4th edn. John Wiley & Son, Inc, NJ
Hou S, Liu Z, Shen H, Wu D (2019) Damage-associated molecular pattern-triggered immunity in plants. Front Plant Sci 10:646
Hou S, Liu D, Huang S, Luo D, Liu Z, Xiang Q, He P (2021) The Arabidopsis MIK2 receptor elicits immunity by sensing a conserved signature from phytocytokines and microbes. Nat Commun 12(1):5494
Howe GA, Lightner J, Browse J, Ryan CA (1996) An octadecanoid pathway mutant (JL5) of tomato is compromised in signaling for defense against insect attack. Plant Cell 8(11):2067–2077
Huffaker A (2015) Plant elicitor peptides in induced defense against insects. Curr Opin Insect Sci 9:44–50
Huffaker A, Ryan CA (2007) Endogenous peptide defense signals in Arabidopsis differentially amplify signaling for the innate immune response. Proc Natl Acad Sci 104(25):10732–10736
Huffaker A, Pearce G, Ryan CA (2006) An endogenous peptide signal in Arabidopsis activates components of the innate immune response. Proc Natl Acad Sci 103(26):10098–10103
Huffaker A, Dafoe NJ, Schmelz EA (2011) ZmPep1, an ortholog of Arabidopsis elicitor peptide 1, regulates maize innate immunity and enhances disease resistance. Plant Physiol 155(3):1325–1338
Huffaker A, Pearce G, Veyrat N, Erb M, Turlings TC, Sartor R, Schmelz EA (2013) Plant elicitor peptides are conserved signals regulating direct and indirect antiherbivore defense. Proc Natl Acad Sci 110(14):5707–5712
Huffaker A, Sims J, Christensen S, Schmelz E (2014) Role of plant elicitor peptides and phytoalexins in enhancing maize resistance to Aspergillus flavus infection. In Meeting Abstract.
Iizasa EI, Mitsutomi M, Nagano Y (2010) Direct binding of a plant LysM receptor-like kinase, LysM RLK1/CERK1, to chitin in vitro. J Biol Chem 285(5):2996–3004
Jamiołkowska A (2020) Natural compounds as elicitors of plant resistance against diseases and new biocontrol strategies. Agron 10(2):173
Jing Y, Zheng X, Zhang D, Shen N, Wang Y, Yang L, Luan S (2019) Danger-associated peptides interact with PIN-dependent local auxin distribution to inhibit root growth in Arabidopsis. Plant Cell 31(8):1767–1787
Jing Y, Shen N, Zheng X, Fu A, Zhao F, Lan W, Luan S (2020) Danger-associated peptide regulates root immune responses and root growth by affecting ROS formation in Arabidopsis. Int J Mol Sci 21(13):4590
Jing Y, Zheng X, Sharifi R, Chen J (2023) Plant elicitor peptide induces endocytosis of plasma membrane proteins in Arabidopsis. Front Plant Sci 14:1328250
Jing Y, Zhao F, Lai K, Sun F, Sun C, Zou X, Luan S (2024) Plant elicitor Peptides regulate root hair development in Arabidopsis. Front Plant Sci 15:1336129
Jones JD, Dangl JL (2006) The plant immune system. Nature 444(7117):323–329

Kanda Y, Shinya T, Maeda S, Mujiono K, Hojo Y, Tomita K, Mori M (2023) BSR1, a rice receptor-like cytoplasmic kinase, positively regulates defense responses to herbivory. Int J Mol Sci 24(12):10395

Kawano T, Bouteau F (2013) Crosstalk between intracellular and extracellular salicylic acid signaling events leading to long-distance spread of signals. Plant Cell Rep 32:1125–1138

Keen NT, Bruegger B (1977) Phytoalexins and chemicals that elicit their production in plants. Am Chem Soc Symp Ser 62:1–26

Keswani C, Prakash O, Bharti N, Vílchez JI, Sansinenea E, Lally RD, Singh HB (2019) Re-addressing the biosafety issues of plant growth promoting rhizobacteria. Sci Total Environ 690:841–852

Khan MIR, Fatma M, Per TS, Anjum NA, Khan NA (2015) Salicylic acid-induced abiotic stress tolerance and underlying mechanisms in plants. Front Plant Sci 6:135066

Klauser D, Flury P, Boller T, Bartels S (2013) Several MAMPs, including chitin fragments, enhance at Pep-triggered oxidative burst independently of wounding. Plant Signal Behav 8(9):e25346

Klauser D, Desurmont GA, Glauser G, Vallat A, Flury P, Boller T, Bartels S (2015) The Arabidopsis Pep-PEPR system is induced by herbivore feeding and contributes to JA-mediated plant defence against herbivory. J Exp Bot 66(17):5327–5336

Krol E, Mentzel T, Chinchilla D, Boller T, Felix G, Kemmerling B, Hedrich R (2010) Perception of the Arabidopsis danger signal peptide 1 involves the pattern recognition receptor AtPEPR1 and its close homologue AtPEPR2. J Biol Chem 285(18):13471–13479

Lecourieux D, Ranjeva R, Pugin A (2006) Calcium in plant defence-signalling pathways. New Phytol 171(2):249–269

Lee J, Klessig DF, Nürnberger T (2001) A harpin binding site in tobacco plasma membranes mediates activation of the pathogenesis-related gene HIN1 independent of extracellular calcium but dependent on mitogen-activated protein kinase activity. Plant Cell 13(5):1079–1093

Lee SW, Han SW, Sririyanum M, Park CJ, Seo YS, Ronald PC (2009) A type I–secreted, sulfated peptide triggers XA21-mediated innate immunity. Science 326(5954):850–853

Lee MW, Huffaker A, Crippen D, Robbins RT, Goggin FL (2018) Plant elicitor peptides promote plant defences against nematodes in soybean. Mol Plant Pathol 19(4):858–869

Li L, Li C, Howe GA (2001) Genetic analysis of wound signaling in tomato. Evidence for a dual role of jasmonic acid in defense and female fertility. Plant Physiol 127(4):1414–1417

Liu Z, Wu Y, Yang F, Zhang Y, Chen S, Xie Q, Zhou JM (2013) BIK1 interacts with PEPRs to mediate ethylene-induced immunity. Proc Natl Acad Sci 110(15):6205–6210

Liu PL, Du L, Huang Y, Gao SM, Yu M (2017) Origin and diversification of leucine-rich repeat receptor-like protein kinase (LRR-RLK) genes in plants. BMC Evol Biol 17:1–16

Liu Y, Gong X, Li M, Si H, Zhou Q, Liu X, Dong J (2021) Effect of osmotic stress on the growth, development and pathogenicity of Setosphaeria turcica. Front Microbiol 12:706349

Lori M, Van Verk MC, Hander T, Schatowitz H, Klauser D, Flury P, Bartels S (2015) Evolutionary divergence of the plant elicitor peptides (Peps) and their receptors: interfamily incompatibility of perception but compatibility of downstream signalling. J Exp Bot 66(17):5315–5325

Ma Y, Walker RK, Zhao Y, Berkowitz GA (2012) Linking ligand perception by PEPR pattern recognition receptors to cytosolic Ca2+ elevation and downstream immune signaling in plants. Proc Natl Acad Sci 109(48):19852–19857

Ma Y, Zhao Y, Walker RK, Berkowitz GA (2013) Molecular steps in the immune signaling pathway evoked by plant elicitor peptides: Ca2+-dependent protein kinases, nitric oxide, and reactive oxygen species are downstream from the early Ca2+ signal. Plant Physiol 163(3):1459–1471

Ma C, Guo J, Kang Y, Doman K, Bryan AC, Tax FE, Qi Z (2014) AtPEPTIDE RECEPTOR2 mediates the AtPEPTIDE1-induced cytosolic Ca2+ rise, which is required for the suppression of Glutamine Dumper gene expression in Arabidopsis roots. J Integr Plant Biol 56(7):684–694

McGurl B, Pearce G, Orozco-Cardenas M, Ryan CA (1992) Structure, expression, and antisense inhibition of the systemin precursor gene. Science 255(5051):1570–1573

Meena M, Samal S (2019) Alternaria host-specific (HSTs) toxins: an overview of chemical characterization, target sites, regulation and their toxic effects. Toxicol Rep 6:745–758

Millet YA, Danna CH, Clay NK, Songnuan W, Simon MD, Werck-Reichhart D, Ausubel FM (2010) Innate immune responses activated in Arabidopsis roots by microbe-associated molecular patterns. Plant Cell 22(3):973–990
Montesano M, Brader G, Palva ET (2003) Pathogen derived elicitors: searching for receptors in plants. Mol Plant Pathol 4(1):73–79
Montesinos L, Bundó M, Badosa E, San Segundo B, Coca M, Montesinos E (2017) Production of BP178, a derivative of the synthetic antibacterial peptide BP100, in the rice seed endosperm. BMC Plant Biol 17:1–14
Montesinos L, Gascón B, Ruz L, Badosa E, Planas M, Feliu L, Montesinos E (2021) A bifunctional synthetic peptide with antimicrobial and plant elicitation properties that protect tomato plants from bacterial and fungal infections. Front Plant Sci 12:756357
Moyen C, Hammond-Kosack K, Jones J, Knight MR, Johannes E (1998) Systemin triggers an increase of cytoplasmic calcium in tomato mesophyll cells: Ca2+ mobilization from intra-and extracellular compartments. Plant Cell Environ 21(11):1101–1111
Nakaminami K, Okamoto M, Higuchi-Takeuchi M, Yoshizumi T, Yamaguchi Y, Fukao Y, Hanada K (2018) AtPep3 is a hormone-like peptide that plays a role in the salinity stress tolerance of plants. Proc Natl Acad Sci 115(22):5810–5815
Narváez-Vásquez J, Ryan CA (2004) The cellular localization of prosystemin: a functional role for phloem parenchyma in systemic wound signaling. Planta 218:360–369
Narváez-Vásquez J, Pearce G, Ryan CA (2005) The plant cell wall matrix harbors a precursor of defense signaling peptides. Proc Natl Acad Sci 102(36):12974–12977
Narváez-Vásquez J, Orozco-Cárdenas ML, Ryan CA (2007) Systemic wound signaling in tomato leaves is cooperatively regulated by systemin and hydroxyproline-rich glycopeptide signals. Plant Mol Biol 65:711–718
Nennstiel D, Scheel D, Nürnberger T (1998) Characterization and partial purification of an oligopeptide elicitor receptor from parsley (Petroselinum crispum). FEBS Lett 431(3):405–410
Newman MA, Sundelin T, Nielsen JT, Erbs G (2013) MAMP (microbe-associated molecular pattern) triggered immunity in plants. Front Plant Sci 4:50369
Nürnberger T (1999) Signal perception in plant pathogen defense. Cell Mol Life Sci CMLS 55:167–182
Nürnberger T, Wirtz W, Nennstiel D, Hahlbrock K, Jabs T, Zimmermann S, Scheel D (1997) Signal perception and intracellular signal transduction in plant pathogen defense. J Recept Signal Transduct 17(1–3):127–136
Oerke EC, Dehne HW (2004) Safeguarding production—losses in major crops and the role of crop protection. Crop Prot 23(4):275–285
Orozco-Cardenas M, McGurl B, Ryan CA (1993) Expression of an antisense prosystemin gene in tomato plants reduces resistance toward Manduca sexta larvae. Proc Natl Acad Sci 90(17):8273–8276
Orozco-Cárdenas ML, Narváez-Vásquez J, Ryan CA (2001) Hydrogen peroxide acts as a second messenger for the induction of defense genes in tomato plants in response to wounding, systemin, and methyl jasmonate. Plant Cell 13(1):179–191
Pal KK, Gardener BM (2006) Biological control of plant pathogens. Plant Health Instr 2:1117–1142
Pastor-Fernández J, Sánchez-Bel P, Flors V, Cerezo M, Pastor V (2023) Small signals lead to big changes: the potential of peptide-induced resistance in plants. J Fungus 9(2):265
Pearce G, Ryan CA (2003) Systemic signaling in tomato plants for defense against herbivores: isolation and characterization of three novel defense-signaling glycopeptide hormones coded in a single precursor gene. J Biol Chem 278(32):30044–30050
Pearce G, Strydom D, Johnson S, Ryan CA (1991) A polypeptide from tomato leaves induces wound-inducible proteinase inhibitor proteins. Science 253(5022):895–897
Pearce G, Moura DS, Stratmann J, Ryan CA (2001) Production of multiple plant hormones from a single polyprotein precursor. Nature 411(6839):817–820
Pearce G, Yamaguchi Y, Barona G, Ryan CA (2010) A subtilisin-like protein from soybean contains an embedded, cryptic signal that activates defense-related genes. Proc Natl Acad Sci 107(33):14921–14925

Poncini L, Wyrsch I, Dénervaud Tendon V, Vorley T, Boller T, Geldner N, Lehmann S (2017) In roots of Arabidopsis thaliana, the damage-associated molecular pattern AtPep1 is a stronger elicitor of immune signalling than flg22 or the chitin heptamer. PLoS One 12(10):e0185808

Postel S, Küfner I, Beuter C, Mazzotta S, Schwedt A, Borlotti A, Nürnberger T (2010) The multifunctional leucine-rich repeat receptor kinase BAK1 is implicated in Arabidopsis development and immunity. Eur J Cell Biol 89(2–3):169–174

Pršić J, Ongena M (2020) Elicitors of plant immunity triggered by beneficial bacteria. Front Plant Sci 11:594530

Qi Z, Verma R, Gehring C, Yamaguchi Y, Zhao Y, Ryan CA, Berkowitz GA (2010) Ca2+ signaling by plant Arabidopsis thaliana Pep peptides depends on AtPepR1, a receptor with guanylyl cyclase activity, and cGMP-activated Ca2+ channels. Proc Natl Acad Sci 107(49):21193–21198

Ramirez-Estrada K, Vidal-Limon H, Hidalgo D, Moyano E, Golenioswki M, Cusidó RM, Palazon J (2016) Elicitation, an effective strategy for the biotechnological production of bioactive high-added value compounds in plant cell factories. Molecules 21(2):182

Ren F, Lu YT (2006) Overexpression of tobacco hydroxyproline-rich glycopeptide systemin precursor A gene in transgenic tobacco enhances resistance against Helicoverpa armigera larvae. Plant Sci 171(2):286–292

Ross A, Yamada K, Hiruma K, Yamashita-Yamada M, Lu X, Takano Y, Saijo Y (2014) The Arabidopsis PEPR pathway couples local and systemic plant immunity. EMBO J 33(1):62–75

Ruiz C, Nadal A, Montesinos E, Pla M (2018) Novel Rosaceae plant elicitor peptides as sustainable tools to control Xanthomonas arboricola pv. pruni in Prunus spp. Mol Plant Pathol 19(2):418–431

Ryan CA (2000) The systemin signaling pathway: differential activation of plant defensive genes. Biochim Biophys Acta (BBA)-Protein Struct Mol Enzymol 1477(1–2):112–121

Saand MA, Xu YP, Li W, Wang JP, Cai XZ (2015) Cyclic nucleotide gated channel gene family in tomato: genome-wide identification and functional analyses in disease resistance. Front Plant Sci 6:135505

Schmelz EA, Carroll MJ, LeClere S, Phipps SM, Meredith J, Chourey PS, Teal PE (2006) Fragments of ATP synthase mediate plant perception of insect attack. Proc Natl Acad Sci 103(23):8894–8899

Schmelz EA, LeClere S, Carroll MJ, Alborn HT, Teal PE (2007) Cowpea chloroplastic ATP synthase is the source of multiple plant defense elicitors during insect herbivory. Plant Physiol 144(2):793–805

Schmelz EA, Engelberth J, Alborn HT, Tumlinson JH, Teal PE (2009) Phytohormone-based activity mapping of insect herbivore-produced elicitors. Proc Natl Acad Sci 106(2):653–657

Schulze B, Mentzel T, Jehle AK, Mueller K, Beeler S, Boller T, Chinchilla D (2010) Rapid heteromerization and phosphorylation of ligand-activated plant transmembrane receptors and their associated kinase BAK1. J Biol Chem 285(13):9444–9451

Shen W, Liu J, Li JF (2019) Type-II metacaspases mediate the processing of plant elicitor peptides in Arabidopsis. Mol Plant 12(11):1524–1533

Shen N, Jing Y, Tu G, Fu A, Lan W (2020) Danger-associated peptide regulates root growth by promoting protons extrusion in an AHA2-dependent manner in arabidopsis. Int J Mol Sci 21(21):7963

Shen W, Zhang X, Liu J, Tao K, Li C, Xiao S, Li JF (2022) Plant elicitor peptide signalling confers rice resistance to piercing-sucking insect herbivores and pathogens. Plant Biotechnol J 20(5):991–1005

Shinya T, Yasuda S, Hyodo K, Tani R, Hojo Y, Fujiwara Y, Galis I (2018) Integration of danger peptide signals with herbivore-associated molecular pattern signaling amplifies anti-herbivore defense responses in rice. Plant J 94(4):626–637

Stangarlin JR, Kuhn OJ, Assi L, Schwan-Estrada KR (2011) Control of plant diseases using extracts from medicinal plants and fungi. Science against microbial pathogens: communicating current research and technological advances, vol 2. Formatex, Badajoz, pp 1033–1042

Tanaka K, Heil M (2021) Damage-associated molecular patterns (DAMPs) in plant innate immunity: applying the danger model and evolutionary perspectives. Annu Rev Phytopathol 59:53–75

Tavormina P, De Coninck B, Nikonorova N, De Smet I, Cammue BP (2015) The plant peptidome: an expanding repertoire of structural features and biological functions. Plant Cell 27(8):2095–2118

Thakur M, Sohal BS (2013) Role of elicitors in inducing resistance in plants against pathogen infection: a review. Int Sch Res Notices 2013

Tintor N, Ross A, Kanehara K, Yamada K, Fan L, Kemmerling B, Saijo Y (2013) Layered pattern receptor signaling via ethylene and endogenous elicitor peptides during Arabidopsis immunity to bacterial infection. Proc Natl Acad Sci 110(15):6211–6216

Trivilin AP, Hartke S, Moraes MG (2014) Components of different signalling pathways regulated by a new orthologue of A t PROPEP 1 in tomato following infection by pathogens. Plant Pathol 63(5):1110–1118

Vasconsuelo A, Boland R (2007) Molecular aspects of the early stages of elicitation of secondary metabolites in plants. Plant Sci 172(5):861–875

Wang L, Einig E, Almeida-Trapp M, Albert M, Fliegmann J, Mithöfer A, Felix G (2018) The systemin receptor SYR1 enhances resistance of tomato against herbivorous insects. Nat Plants 4(3):152–156

Wang J, Wu D, Wang Y, Xie D (2019) Jasmonate action in plant defense against insects. J Exp Bot 70(13):3391–3400

Wang A, Guo J, Wang S, Zhang Y, Lu F, Duan J, Ji W (2022a) BoPEP4, a c-terminally encoded plant elicitor peptide from broccoli, plays a role in salinity stress tolerance. Int J Mol Sci 23(6):3090

Wang J, Xi L, Wu XN, König S, Rohr L, Neumann T, Schulze WX (2022b) PEP7 acts as a peptide ligand for the receptor kinase SIRK1 to regulate aquaporin-mediated water influx and lateral root growth. Mol Plant 15(10):1615–1631

Wu J, Baldwin IT (2009) Herbivory-induced signalling in plants: perception and action. Plant Cell Environ 32(9):1161–1174

Xu S, Liao CJ, Jaiswal N, Lee S, Yun DJ, Lee SY, Mengiste T (2018) Tomato PEPR1 ORTHOLOG RECEPTOR-LIKE KINASE1 regulates responses to systemin, necrotrophic fungi, and insect herbivory. Plant Cell 30(9):2214–2229

Yamaguchi Y, Huffaker A (2011) Endogenous peptide elicitors in higher plants. Curr Opin Plant Biol 14(4):351–357

Yamaguchi Y, Pearce G, Ryan CA (2006) The cell surface leucine-rich repeat receptor for At Pep1, an endogenous peptide elicitor in Arabidopsis, is functional in transgenic tobacco cells. Proc Natl Acad Sci 103(26):10104–10109

Yamaguchi Y, Huffaker A, Bryan AC, Tax FE, Ryan CA (2010) PEPR2 is a second receptor for the Pep1 and Pep2 peptides and contributes to defense responses in Arabidopsis. Plant Cell 22(2):508–522

Yamaguchi Y, Barona G, Ryan CA, Pearce G (2011) GmPep914, an eight-amino acid peptide isolated from soybean leaves, activates defense-related genes. Plant Physiol 156(2):932–942

Yang TPBW, Poovaiah BW (2002) Hydrogen peroxide homeostasis: activation of plant catalase by calcium/calmodulin. Proc Natl Acad Sci 99(6):4097–4102

Yang Q, Xuan Trinh H, Imai S, Ishihara A, Zhang L, Nakayashiki H, Mayama S (2004) Analysis of the involvement of hydroxyanthranilate hydroxycinnamoyltransferase and caffeoyl-CoA 3-O-methyltransferase in phytoalexin biosynthesis in oat. Mol Plant-Microbe Interact 17(1):81–89

Yang K, Wang Y, Zhao H, Shen D, Dou D, Jing M (2022) Novel Elicitin from Pythium oligandrum confers disease resistance against Phytophthora capsici in Solanaceae plants. J Agric Food Chem 70(51):16135–16145

Yang R, Liu J, Wang Z, Zhao L, Xue T, Meng J, Luan Y (2023) The SlWRKY6-SlPROPEP-SlPep module confers tomato resistance to Phytophthora infestans. Sci Hortic 318:112117

Yu X, Feng B, He P, Shan L (2017) From chaos to harmony: responses and signaling upon microbial pattern recognition. Annu Rev Phytopathol 55:109–137
Zarate SI, Kempema LA, Walling LL (2007) Silverleaf whitefly induces salicylic acid defenses and suppresses effectual jasmonic acid defenses. Plant Physiol 143(2):866–875
Zebelo SA (2020) Decrypting early perception of biotic stress on plants. In: Mérillon JM, Ramawat K (eds) Co-Evolution of secondary metabolites. Springer Nature, Switzerland, pp 577–592
Zelman AK, Berkowitz GA (2023) Plant elicitor peptide (Pep) signaling and pathogen defense in tomato. Plants 12(15):2856
Zhang J, Li Y, Bao Q, Wang H, Hou S (2022) Plant elicitor peptide 1 fortifies root cell walls and triggers a systemic root-to-shoot immune signaling in Arabidopsis. Plant Signal Behav 17(1):2034270
Zhao J, Davis LC, Verpoorte R (2005) Elicitor signal transduction leading to production of plant secondary metabolites. Biotechnol Adv 23(4):283–333
Zheng X, Kang S, Jing Y, Ren Z, Li L, Zhou JM, Luan S (2018) Danger-associated peptides close stomata by OST1-independent activation of anion channels in guard cells. Plant Cell 30(5):1132–1146
Zheng F, Chen L, Zhang P, Zhou J, Lu X, Tian W (2020) Carbohydrate polymers exhibit great potential as effective elicitors in organic agriculture: a review. Carbohydr Polym 230:115637
Zhou JM, Zhang Y (2020) Plant immunity: danger perception and signaling. Cell 181(5):978–989
Zhu F, Xi DH, Yuan S, Xu F, Zhang DW, Lin HH (2014) Salicylic acid and jasmonic acid are essential for systemic resistance against tobacco mosaic virus in Nicotiana benthamiana. Mol Plant-Microbe Interact 27(6):567–577
Zipfel C, Robatzek S, Navarro L, Oakeley EJ, Jones JD, Felix G, Boller T (2004) Bacterial disease resistance in Arabidopsis through flagellin perception. Nature 428(6984):764–767
Zipfel C, Kunze G, Chinchilla D, Caniard A, Jones JD, Boller T, Felix G (2006) Perception of the bacterial PAMP EF-Tu by the receptor EFR restricts Agrobacterium-mediated transformation. Cell 125(4):749–760

9 Legume Health: Unveiling the Potential of Plant Elicitor Peptides

Krutika S. Abhyankar and Monisha Kottayi

Abstract

Legumes are one of the most important economic crops cultivated for their nutritional value which includes proteins, dietary fibers, phytosterols, polyphenols, and micronutrients. However, they are challenged by many factors like water scarcity, high salinity, metal toxicity, and biotic factors like microorganisms and parasitic nematodes. To protect themselves against these potential threats, plants produce many signaling molecules like peptide elicitors called plant elicitor peptides. Plant elicitor peptides (PEPs) are a class of elicitors which elicits defense responses by activating defense pathways in terms of elevated expression of defense-related genes, hormone production, and induction of secondary messenger synthesis. They belong to a class of endogenous elicitors which lead to enhanced immunity against various abiotic and biotic stresses. These peptides are perceived by membrane receptors in the plant cells, which bind to the peptide ligands to initiate the signaling cascades. The exploration of PEP's represents a good alternative with potential in crop protection. This chapter deals with the peptide elicitors used to combat abiotic and biotic stress in leguminous plants.

Keywords

Peptide elicitors · Plant defense · Legumes · Abiotic stress · Biotic stress

K. S. Abhyankar (✉) · M. Kottayi
Division of Biomedical and Life Science, School of Science, Navrachana University, Vadodara, India
e-mail: bholekrutika@gmail.co.in

S. Singh, R. Mehrotra (eds.), *Plant Elicitor Peptides*,
https://doi.org/10.1007/978-981-97-6374-0_9

Table 9.1 Nutritional value of pulse grains (per 100 g)

Pulses	Energy (KCal)	Protein (g)	Fat (g)	Carbohydrate (g)	Total dietary Fiber (%)
Chickpea	368	21.0	5.7	61	22.7
Pigeon pea	342	21.7	1.49	62	15.5
Urd bean	347	24.0	1.6	63.4	–
Mung bean	345	25.0	1.1	62.6	16.3
Lentil	346	27.2	1.0	60	11.5
Field pea	345	25.1	0.8	61.8	13.4
Rajma	345	23.0	1.3	62.7	17.7
Cowpea	346	28.0	0.3	63.4	18.2
Horse gram	321	23.6	2.3	59.1	15.0
Moth bean	330	24.0	1.5	61.9	–

Source: "Pulses for human health and nutrition," 2024 Technical Bulletin, IIPR, Kanpur (IIPR, 2024)

9.1 Introduction

9.1.1 Legumes Cultivated in India

Legumes form an important component of our daily diet as they are rich in proteins, carbohydrates, lysine, their low glycemic index, and mineral composition, which includes calcium, potassium, magnesium, and phosphorous. They are considered one of the healthy foods for humanity due to their valuable component, such as protein, carbohydrates, folic acid, fibers, in addition, legumes seeds are low fats foods (Table 9.1), and therefore, legumes are very important crops worldwide (Gebrelibanos et al. 2013). Legumes also form an important source of fodder in the form of green manure or as components of animal feed.

Legumes are the second most consumed food crop after cereals around the globe. Pulses are a cheap source of protein; hence, they are gaining popularity as the staple food in many parts of India. They are consumed mainly for their potential health benefits which are due to phytochemicals like lectins, antimicrobial peptides, flavonoids, phenolics phytates, saponins, tannins, phytosterols, and oxalates (Venkidasamy et al. 2019). Legumes play a very important role in human as well as animal health as a rich source of plant protein (Tharanathan and Mahadevamma 2003). They have an inherent ability to fix atmospheric nitrogen and mitigate greenhouse gas emissions (Lemke et al. 2007).

9.2 Factors Affecting the Productivity of Legumes

Many factors like environmental, biological, genetic, socio-economic, soil and climatic constraints, and failure of breeders to produce improved varieties are reasons for low productivity of legumes. Unpredictable rainfall patterns result in decreased crop yields (Herrero et al. 2010). However, many abiotic and biotic stresses are the major concern in the cultivation of legumes.

9.2.1 Abiotic Stress: The Major Abiotic Stress Faced by Legumes Are Water, Temperature, and Salinity

9.2.1.1 Water Stress

Plants experience drought when they face extreme scarcity of water. This is caused mainly due to inadequate rainfall. Drought is associated with low germination rates and decreased photosynthetic activity (Chowdhury et al. 2016). With the increase in global climate change, legume productivity is severely affected. Drought drastically disturbs the germination percentage, morphological characters, and reproductive physiology of the plants. Soybean and chickpea plants showed poor germination percentage when faced with water deficit (Heatherly 1993; Li et al. 2018; Nadeem et al. 2019). In chickpea, the chlorophyll content, chlorophyll fluorescence, and photosynthesis were drastically affected by drought (Mafakheri et al. 2010). Faba beans showed changes in chlorophyll fluorescence and antioxidant enzyme activities when subject to drought stress (Abid et al. 2017). Arrese-Igor et al. reported that drought stress induces oxidative damage in legumes which affects nodule formation and thereby, biological nitrogen fixation (Latef and Ahmad 2015; Arrese-Igor Sánchez et al. 2011).

9.2.1.2 Temperature Stress: Heat

The increase in the temperature of soil or of the atmosphere beyond a threshold level leads to heat stress. When subject to higher temperatures, legumes show symptoms like discoloration of the veins, leaf burns, and senescence (Ismail and Hall 1999; Vollenweider and Günthardt-Goerg 2005). A temperature $\geq$30 °C is associated with decreased germination and poor pollen viability, shedding of pollen along with reduction in pollen elongation (Fahad et al. 2016; Kumari et al. 2021) Chickpea, common bean, mung bean, and cowpea showed an altered reproductive development when subjected to heat stress (Kaushal et al. 2013; Guilioni et al. 1997; Satyanarayana et al. 2002; Sharma et al. 2016; Musavi et al. 1992). Such plants show a decreased vegetative and pod-filling period. High temperature negatively regulates flowering, pollen and ovule viability, overall fertilization, and the quality of grains (Kumari Sita et al. 2017).

9.2.1.3 Temperature Stress: Cold

Cold stress is a potential deterrent to legume growth and development. Cold stress is experienced by plants when they experience temperatures in the range of 0–15 °C (chilling stress) or below 0 °C (freezing stress). It primarily affects the physiological and cellular responses in legumes. The potential target is the plasma membrane, causing leakage of various ions, lipids, and proteins across the membrane (Cheng and Hu 2010). Cold stress in chickpea led to increase in susceptibility and seedling death. Experiments on 5 days old mung bean seedlings exposed to 4 °C caused severe chilling injury. It concomitantly reduces the rate of photosynthesis, synthesis of other metabolites, and the overall physiology of the plants. This includes low germination, morphological changes like stunted growth, yellowing of leaves, and wilting (Iqbal et al. 2010).

9.2.1.4 Salinity Stress

Salinity is one of the important abiotic stresses studied in legumes and is caused due to an abundance of salts in soil and water via irrigation, environment, or other agricultural practices. Salt stress induces a wide range of changes in plants which affect the overall growth and reproduction. Its effects are seen in seed germination, photosynthesis, transport across membranes, and the yield of legumes (Zhou et al. 2023). NaCl treatment given to soybeans for 7 days proved that overall plant growth was inhibited upon a continuous exposure to salt (Ning et al. 2018). Broad common beans displayed a reduction in plant height and leaf area and number when subject to increased concentration of NaCl (Torche et al. 2018). Similarly, seedling growth and development was inhibited in pea plants subject to salt stress (Tokarz et al. 2020). High salt concentrations disturb the Na^+/K^+ ratio in legumes as observed in mungbean and chickpea where a significant reduction in the Na^+/K^+ ratio was seen upon salt stress induction (Nandwal et al. 2000; Garg and Bhandari 2016). However, in soybean, salinity stress led to accumulation of Mg^{2+}, K^+, and Ca^{2+} in the leaves (Essa 2002).

9.2.2 Biotic Stress

Biotic stress has been found to be a major cause of concern for yield losses of legumes. Biotic factors like bacteria, fungi, and insects cause major crop losses. It has been reported that soybean productivity is significantly affected by the root-knot nematodes causing 44.7% reduction in yield (Rhouma et al. 2024). MYMD is a virus which causes economic loss in its productivity by 85%. Mung beans are also affected by powdery mildew and *Cercospora* leaf spot disease (Sehrawat et al. 2019). Pigeonpea faced significant crop losses of up to 95% when challenged by sterility mosaic virus and *Fusarium* (Kudapa et al. 2012). Groundnut production witnessed a drastic reduction when challenged by aphids, mites, and pod borers. It is also affected by fungi like *Aspergillus* and Sclerotium rolfsii causing 50% reduction in its yield. *Maruca vitrata* commonly known as the pod borer is a major pest of cowpea and pigeonpea (Bosamia et al. 2020). *Melanagromyza obtusa Malloch* known as the pod fly causes extensive damage to pigeonpea in India. Pigeonpea, field pea, and lentil are mostly affected by the spiny pod borer, while chickpea is affected by the leaf miner (Sharma et al. 1999). *Helicoverpa armigera* causes an estimated loss of over $ 2 billion annually in food legumes, although over $ 1 billion worth of insecticides are being used for its control (Arora et al. 2005). Bacterial diseases cause equally damaging diseases in legumes. *Pseudomonas syringae* which typically overwinters in the weeds and legume debris leads to cell death and necrosis (Melotto et al. 2008; Dell'Olmo et al. 2023). *P. savastanoi* the causal agent of bacterial blight of soybean affects the seed quality, oil content, and germination can be seriously compromised.

In the current scenario, the methods employed for crop protection include practices like exclusion, cultural methods, eradication and the use of insecticides, fungicides, bactericides, and pesticides. However, recent studies have proven that these strategies are not safe for the soil health as well as human health and pose serious threats to the existence of non-target organisms which may be beneficial to the ecosystem. Hence, a method of self-defense exploiting the plant's own defense response machinery is required which will have a targeted effect on the pathogens and pests and help to combat abiotic stress (Zelman and Berkowitz 2023). This refers to the use of plant-derived compounds which could activate defense responses and provide a long-term protection to the plants. Plant elicitor peptides (Peps) are one such class of compounds which, being biological in origin, pose no threats to the plants or the environment. These are endogenous polypeptides made up of amino acids which help in triggering the innate immune defense responses in plants.

9.3 Peptide Elicitors in Legumes

Elicitors are compounds which activate chemical defense in plants. They are categories into exogenous and endogenous elicitors. Exogenous elicitors are compounds produced by pathogens, whereas endogenous elicitors are molecules released from plants in response to pathogenic attack (Ramirez-Estrada et al. 2016). The assessment of endogenous elicitors' biological significance is considerably more difficult since they appear to be produced and identified in a wider range of situations. Both locally and systemically, endogenous elicitors stimulate and intensify the defense mechanisms against invasive invaders. Present knowledge has revealed that endogenous peptide elicitors are present in many plant species and play a crucial role in controlling immunity against both infections and herbivores (Yamaguchi and Huffaker 2011). There are different types of endogenous elicitors belonging to polysaccharide, intracellular proteins, and other small bioactive molecules. The present chapter elucidates the role of plant elicitor peptides (Peps). Peps are endogenous elicitors which are small peptides produced by plants and contribute to immunity against attack by bacteria, fungi, as well as herbivores (Bartels and Boller 2015). It was demonstrated that a small member of Peps from *Glycine max* (Soybean) GmPep3, *Medicago truncatula* MtPep1, and *Arachis hypogaea* (Peanut) AhPep1 from Fabaceae family and Solanaceae members such as *Solanum melongena* (Aubergine) SmPep1, *Capsicum annuum* (Pepper) CaPep1, and *Solanum tuberosum* (Potato) StPep1 could cause the release of volatiles, that is considered to be a part of common defense mechanism against herbivore attack (Huffaker et al. 2013). Peps mature from bigger precursor proteins known as PROPEPs and are detected by Pep receptors (PEPRs), which are leucine-rich repeat (LRR) receptor-like kinases (Lori et al. 2015). On perceiving threats, Peps activates defense mechanism by triggering synthesis of defense hormones, by activating defense gene expression, enhancing signaling pathway involved in phosphorylation, and by production of secondary messengers. Peps are found and active in angiosperms throughout the evolutionary tree, including numerous significant crop plants. They may contribute to the regulation of plant development in addition to their potential role in the response to biotic stress (Bartels et al. 2013; Bartels and Boller 2015). In current work, more emphasis is envisaged to understand the Peps derived from legumes and their mechanism of action. Legumes are cultivated widely as it is an essential part of its agricultural landscape and dietary habits. However, they are routinely exposed to biotic and abiotic stresses. The huge list of biotic stress is related to the pests and disease found in the legumes. Some of the pests include aphids and pod borers. While fungal diseases like fusarium wilt and powdery mildew, bacterial blight and viral disease like mosaic viruses contribute to a great damage in cultivated legume crop. The following abiotic stress includes drought, salinity, temperature extremes, and soil nutrient depletion. Therefore, exploiting peptide elicitors from the legume could help in developing strategies to enhance disease resistance, abiotic stress tolerance, and overall crop performance in legumes species and could have significant agricultural implications. Therefore, endogenous peptides found in varied legume species are discussed in detail. The list of few endogenous Peps found in the legumes is represented in Table 9.2.

9.3.1 Plant Peptides in Coping with Abiotic Stress in Legumes

Abiotic stresses such as drought, extreme temperatures, salinity, and waterlogging have detrimental effects on the plant physiology and biochemical landscape of crops. The physiological responses of plants under abiotic stresses include, among other things, photoinhibition, loss of osmoregulation, and inhibition of biological nitrogen fixation. These responses result from the accumulation of stress-induced osmolytes (such as proline, polyamines, glycine betaine) and reactive oxygen species (ROS) in plant cells (Singh et al. 2023).

9.3.1.1 CLAVATA3/ESR-Related Peptide (CLE Peptide)

Several CLE peptides play roles in symbiosis, parasitism, and responses to abiotic cues. These peptides are involved in various processes throughout the plant life cycle, such as maintaining stem cell homeostasis and establishing symbiosis with rhizobia in legumes. Interestingly, CLE-like genes with slightly different sequences have also been discovered in nematodes that parasitize plants. Nematode CLE-like peptides seem to mimic plant host CLE ligands, activating the host plant's CLE signaling pathways to facilitate successful infection (Yamaguchi et al. 2016).

Soybean (*Glycine max*) also produces CLE peptides that play crucial roles in plant development and responses. These peptides help maintain stem cell populations in the shoot apical meristem (SAM) and root apical meristem (RAM), supporting shoot and root development, lateral organ formation, and vascular differentiation. In soybeans, CLE peptides are translated from prepropeptides containing the conserved CLE domain at the C-terminus. Soybean CLE signaling pathways likely integrate environmental cues and may influence responses to stress, nutrient

Table 9.2 Peps related to Biotic and abiotic stress in legumes

Factors	Name of the organism/stress causing agent	Peps elicitor	Name of the legume	Size/ characteristic of the peptide	Mechanism	References
Symbiosis, parasitism, and responses to abiotic stress	Information not available	CLE peptide	*Glycine max L.*	CLE peptides are encoded by CLE genes and have a conserved 14-amino-acid CLE domain near the C-terminus. The functional CLE peptide is derived from proteolytic cleavage of the C-terminal region of the pre-pro-protein	Assisting in the growth of shoots and roots, the formation of lateral organs, and the differentiation of vascular tissues	Jones et al. (2022), Yamaguchi et al. (2016)
Abiotic stress	Nitrogen deficiency	CEP	*Medicago truncatula*	The mature CEP peptide is typically 12–13 amino acids long and possesses a tri-arabinose moiety attached to a highly conserved hydroxylated central proline residue	The MtCEP1/MtCRA2 pathway maintains a balance between root and nodule development in low-N conditions by decreasing auxin production and blocking ethylene signaling	Roy et al. (2022), Hastwell et al. (2017), Zhu et al. (2020)
Plant growth, development, and abiotic stress responses	Information not available	Phytosulfokines	*Medicago truncatula*	Information not available	It impacts root elongation and the formation of lateral roots by stimulating cell division and enlargement within the root structure	Di et al. (2022)
Biotic stress	Information not available	GmSubPep	*Glycine max L.*	12 amino acids NTPPRRAKSRPH	Media alkalinizationChitinase1b, CYP93A1, chalcone synthase and PDR12 gene expression	Pearce et al. (2010)

(continued)

Table 9.2 (continued)

Factors	Name of the organism/stress causing agent	Peps elicitor	Name of the legume	Size/ characteristic of the peptide	Mechanism	References
Biotic stress	Information not available	Pep914 Pep890	*Glycine max L.*	8 amino acids DHPRGGNY and DLPRGGNY	Media alkalinization CYP93A1, Chib1-1, and chalcone synthase gene expression	Yamaguchi et al. (2011)
Biotic stress	*Meloidogyne incognita* *Heterodera glycines*	GmPep1, GmPep2 and GmPep3	*Glycine max L.*	GmPep1: ASLMATRGSRGSKISDGSGPQHN GmPep2: ASSMARRGNRGSRISHGSGPQHN GmPep3: PSHGSVGGKRGSPISQGKGGQHN	They activate plant defenses through systemic transcriptional reprogramming and reactive oxygen species (ROS) signaling, making them potential strategies for nematode management	Lee et al. (2018)
Biotic stress	*Xanthomonas axonopodis pv. glycines* *Phakopsora pachyrhizi's*	IAP (Gm0025x 00667(75–100) and Gm0026x00785 (77–103))	*Glycine max L.*	RWRFLRKISSVHMFSVKALDDFRQL and HKMDLHWYLRTLEEVVIRALQRFQFR	Information not available	Brand et al. (2012)
Biotic stress	*Pseudomonas syringae pv glycinae* *Phytophtora sojae.*	Matrix metalloproteinases	*Glycine max L.*	Information not available	Information not available	Liu et al. (2001)

Biotic stress	*Ensifer meliloti*	Nodule-Specific Cysteine-Rich (NCR) Peptides NCR035, NCR055, NCR211, and NCR247	*Medicago truncatula*	Information not available	Information not available	Van de Velde et al. (2010) , Kim et al. (2015), Mikuláss et al. (2016)
Biotic stress	*Escherichia coli Sinorhizobium meliloti.*	NCR169	*Medicago truncatula*	Information not available	Lysine-rich region of NCR169 carrying positive charge possibly plays a role in its antimicrobial activity	Isozumi et al. (2021)
Biotic stress	Information not available	BLAD polypeptide	*Lupinus albus*	20 kDa polypeptide of β-conglutin, characterized as a fragment of the amino acid sequence of β-conglutin	Antifungal activity: The BLAD polypeptide disrupts the inner cell membrane of the fungus, hindering crucial cellular functions and inhibiting fungal growth	Monteiro et al. (2015)
Biotic stress	*Candida albicans and Candida guilliermondii Fusarium oxysporum Fusarium solani*	PvD1	*Phaseolus vulgaris L.*	6 kDa	Antifungal activity: The diverse mechanisms of PvD1, such as membrane disturbance, metabolic interference, and ROS/NO induction, position it as a promising option for fighting fungal infections	Mello et al. (2011), Games et al. (2008)

(continued)

Table 9.2 (continued)

Factors	Name of the organism/stress causing agent	Peps elicitor	Name of the legume	Size/ characteristic of the peptide	Mechanism	References
Biotic stress	*Sitophilus oryzae* *Sitophilus zeamais* *Sitophilus granaries* *Acyrthosiphon pisum*	PA1b	*Pisum sativum*	ASCN GVCSPFEMPP CGSSACRCIP VGLVVGYCRH PSG	The entomotoxin PA1b induces apoptosis in insects by interacting with PA1b/V-ATPase, leading to insect mortality	Eyraud et al. (2017), Higgins et al. (1986)
Biotic stress	*Helicoverpa armigera*	Cyclotides	*Clitoria ternatea*	Gly residues at the proto-N-terminus and Asn residues at the proto-C-terminus of the cyclotide domain within the precursor proteins	Insecticidal activity: Cyclotides kill insects by disrupting their cell membranes, which ultimately leads to insect mortality	Barbeta et al. (2008), Jennings et al. (2001)
Biotic stress	*Callosobruchus. chinensis* *Callosobruchus. maculatus* *Tenebrio molitor* *Zaborotes subfasciatus*	VrD1	*Vigna radiata*	~ 46 amino acids	Insecticidal activity: The mechanism of action of VrD1 against insect pests involves inhibiting alpha-amylase activity, which disrupts their ability to digest carbohydrates and subsequently affects their survival and development	Chen et al. (2002), Lin et al. (2005), Lin et al. (2007), Liu et al. (2006)

Biotic stress	*Acanthoscelides obtectus Zabrotes subfasciatus.*	VuD1	*Vigna unguiculata*	Information not available	The insecticidal efficacy of VuD1 operates through the inhibition of alpha-amylase activity in insect pests. This interference hampers their carbohydrate digestion, thereby impacting their survival and developmental processes	Pelegrini et al. (2008)

availability, and symbiotic interactions, such as nodulation with rhizobia (Jones et al. 2022).

9.3.1.2 Cysteine Protease 1 (CEP1)

For optimal growth and development, plants need to acquire macronutrients (such as nitrogen, phosphorus, potassium, and sulfur) and micronutrients from the soil. A small signaling peptide hormone, CEP1, plays a crucial role in plant growth and development. It enhances the nitrate uptake rate per unit root length in *Medicago truncatula* plants when deprived of nitrogen (N) in the high-affinity transport range. Additionally, CEP1 increases the uptake of phosphate and sulfate in both *Medicago truncatula* and *Arabidopsis thaliana* plants. CEP1 is part of a family of signaling peptides that regulate nutrient uptake and utilization. Transcriptome analysis shows that CEP1 treatment induces the expression of genes related to nutrient transporters and signaling pathways downstream of CEP1 signaling. Thus, CEP1 is a fascinating peptide hormone that orchestrates nutrient uptake and utilization in plants, especially under nitrogen deficiency stress (Roy et al. 2022). Synthetic Medicago CEP1 inhibits lateral root initiation but promotes nodulation under nitrogen deficiency in Medicago (Mohd-Radzman et al. 2015). However, the number of root nodules is negatively regulated by tri-arabinosylation of Medicago CEP1 (Patel et al. 2018).

9.3.1.3 Phytosulfokines (PSKs)

It belongs to the class of tyrosine-sulfated pentapeptides that play essential roles in plant growth, development, and stress responses. A novel PSK-ε has been identified in *Medicago truncatula*, and its precursor protein, MtPSKε, has been reported to show high expression in root tips and emerging lateral roots. PSK-ε likely influences root elongation and lateral root formation by inducing cell division and expansion in roots. Treatment with the PSK-ε peptide and overexpression of MtPSKε together increase the number of nodules in *M. truncatula* (Di et al. 2022).

9.3.1.4 Legume Peptide Elicitors: Targeting Bacterial and Fungal Pathogens

Peptides derived from legume plants can trigger immune responses against bacterial and fungal pathogens. These peptides serve as signaling molecules, activating defense mechanisms in plants to protect against pathogen invasion. Studies in this field aim to understand the mechanisms of action of these peptides and their potential applications in agriculture for developing sustainable strategies for crop protection against diseases caused by bacteria and fungi. Small peptides known as antibacterial peptides (AMPs) exhibit antibacterial properties (Benko-Iseppon et al. 2010; Tam et al. 2015; Tavormina et al. 2015). These peptides are present in various plant tissues, including seeds, pods, fruits, leaves, flowers, tubers, and roots (Benko-Iseppon et al. 2010). Detailed discussions on specific peptide elicitors for microbial pathogens, their characteristics, and mechanisms are provided.

9.3.1.4.1 *Glycine max* Subtilase Peptide (GmSubPep)

Glycine max subtilase peptide (GmSubPep) plays a crucial role in soybean defense mechanisms. Soybean yield is significantly affected by various biotic stresses, including insect attacks and pathogen infections (by fungi, oomycetes, bacteria, viruses, and nematodes), leading to the production of small signaling peptides like systemin in tomato (Pearce et al. 2010). GmSubPep, a plant defense peptide signal specific to soybeans, is embedded within the subtilisin-like protein (subtilase) encoded by the Glyma18g48580 gene. Subtilase is secreted into the apoplast and undergoes cleavage to produce GmSubPep upon contact with fungal and bacterial pathogens (Pearce et al. 2010). The endogenous GmSubPep consists of 12 amino acids (NTPPRRAKSRPH) and is part of the 789-amino acid subtilase protein in soybean. Treatment with synthetic GmSubPep results in rapid alkalinization of the culture medium and induces the expression of certain defense genes. Soybean suspensions exhibit pH alterations and an alkalizing response within 15 minutes of exposure to the peptide, indicating its bioactivity at extremely low concentrations (Pearce et al. 2010). This activation of defense genes includes CYP93A1, involved in phytoalexin synthesis, a pathogenesis-related gene (Suzuki et al. 1996), chitinaseb1-1 [Chib1-1], and PDR12, a salicylic-acid-inducible ATP-binding cassette transporter (Watanabe et al. 1999; Eichhorn et al. 2006), and *Glycine max* chalcone synthase1 (Gmachs1), which is crucial for phytoalexin production (Akada et al. 1990). Therefore, GmSubPep is recognized for its ability to trigger extracellular alkalinization and induce the expression of defense and stress-related genes.

9.3.1.4.2 Pep914 and Pep890

Pep914 and Pep890 are two particularly intriguing peptides found in soybean leaves. These small peptides, consisting of eight amino acids each (DHPRGGNY and DLPRGGNY), are encoded by the genes GmPROPEP914 and GmPROPEP89, respectively, which serve as precursors for GmPep914 and GmPep89 (Yamaguchi et al. 2011). Both peptides are capable of eliciting defense responses in soybeans. Endogenous soybean Pep914 is derived from the C-terminal end of the 52-amino acid PROPEP914 protein, while the conserved soybean peptide Pep890 originates from the C-terminal end of the precursor protein PROPEP890, which shares homology with PROPEP914. Both Pep914 and Pep890 induce the expression of numerous genes associated with pathogen defense and enzymes involved in the production of the antimicrobial metabolite phytoalexin.

9.3.1.5 Intragenic Antimicrobial Peptides (IAPs)

9.3.1.5.1 Gm0025x00667(75–100) and Gm0026x00785(77–103)

Intragenic antimicrobial peptides (IAPs), such as Gm0025x00667(75–100) and Gm0026x00785(77–103), have been the focus of significant advancement in silico studies, offering insights into the biologically active peptides embedded within expressed proteins, irrespective of their structural and functional roles in the host organism. The discussion on these bioactive peptides has been made feasible by the wealth of genomic data currently available for exploration (Brand et al. 2012). Moreover, proteins from various organisms have been found to harbor structural features known as intragenic antimicrobial peptides, bearing striking resemblance to AMPs (Brand et al. 2012; Ramada et al. 2017). Two putative IAPs, Gm0025x00667(75–100) and Gm0026x00785(77–103), were identified from soybean, and subsequent functional studies were conducted. The primary structures of Gm0025x00667(75–100) and Gm0026x00785(77–103) were found to be RWRFLRKISSVHMFSVKALDDFRQL and HKMDLHWYLRTLEEVVIRALQRFQFR, respectively. Notably, Gm0025x00667(75–100) was derived from the flavonoid 3-hydroxylase enzyme fragment, while Gm0026x00785(77–103) originated from lipoate-protein ligase B. Both IAPs were synthesized in vitro and assessed for their resistance against plant pathogens. Inhibition of *Xanthomonas axonopodis pv. glycines*, the causal agent of bacterial pustule disease in soybeans, was observed upon treatment with both IAPs (Brand et al. 2012). *Phakopsora pachyrhizi* infection, responsible for Asian soybean rust disease, was also mitigated by the application of both IAPs in ex vivo tests. Furthermore, transgenic soybean plants expressing either IAP exhibited enhanced resistance to *P. pachyrhizi*-induced Asian rust (Brand et al. 2012).

9.3.1.6 Matrix Metalloproteinases (MMP)

MMPs constitute a group of zinc-dependent endopeptidases speculated to play roles in plant growth, development, and stress responses (Liu et al. 2017). The first plant MMP identified was the soybean SMEP1 protein, found to be exclusively transcribed in mature leaves (Graham et al. 1991; McGeehan et al. 1992; Pak et al. 1997). Additionally, a soybean gene known as GmMMP2 was discovered to be activated during both compatible and incompatible interactions with the bacterial pathogen *Pseudomonas syringae pv. glycinae* and the oomycete pathogen Phytophthora sojae. The activity of GmMMP2 was observed to increase in soybean cell suspension cultures, suggesting its potential involvement in a novel defense mechanism (Liu et al. 2001).

9.3.1.7 Nodule-Specific Cysteine-Rich Peptides (NCR)

NCRs were initially discovered through transcriptomic analysis of *Medicago truncatula* nodules, introducing a novel protein family (Mergaert et al. 2003). These peptides typically consist of 60–90 amino acids, featuring conserved N-terminal signal domains and 4–6 cysteine residues in the mature peptide, forming disulfide bonds (Mergaert et al. 2003). Despite structural similarities, the sequences of NCR peptides exhibit considerable variation and heterogeneity, characterized by distinct sequence motifs and repeats (Mergaert et al. 2003). They can be categorized based on charge distribution as cationic, anionic, or neutral (Montiel et al. 2017). The antibacterial activity of NCR peptides varies depending on their sequences and pKa values, making them attractive candidates as potential antimicrobial agents (Ördögh et al. 2014). Presently, approximately 300 different sequences of genes encoding NCR peptides have been reported in *Medicago truncatula*, *Pisum sativum*, *Vicia faba L.*, and *Trifolium repens L.*, while *Glycyrrhiza uralensis*, *Oxytropis lambertii*

Pursh, *Astragalus canadensis L.*, and others account for about 600 gene sequences (Mergaert et al. 2003; Montiel et al. 2017). In the model legume *M. truncatula* alone, over 700 NCR genes have been identified, with 639 NCR peptides detected (Montiel et al. 2017; Roux et al. 2014). Recent studies have also revealed 360 genes encoding NCR peptides expressed in nodules of *Pisum sativum* (Zorin et al. 2022).

9.3.1.7.1 NCR035, NCR055, NCR211, and NCR247

Studies on NCR peptides identified in *M. truncatula*, such as NCR035, NCR055, NCR211, and NCR247, have shown inhibitory activity against the growth of *Sinorhizobium meliloti* (Van de Velde et al. 2010; Kim et al. 2015; Mikuláss et al. 2016). Although *S. meliloti* typically forms symbiotic associations in natural settings, NCR peptides are present in low concentrations and do not result in rhizobia death. However, under in vitro conditions, NCR peptides exhibit antimicrobial activity (Maróti et al. 2015; Kereszt et al. 2018). Nevertheless, the precise role of NCR peptides' antimicrobial activity in bacteroid development remains unknown (Lima et al. 2020).

9.3.1.7.2 NCR169

NCR169, a cationic peptide with a pI of 8.45 and composed of 38 amino acids, plays a vital role in bacteroid differentiation (Isozumi et al. 2021). Residues of NCR169 are in the peribacteroid space, situated between the bacteroid membrane and symbiosome membrane. These peptides disrupt the bacterial cell cycle, shifting it from a regular cycle of single DNA replication followed by cell division to an endoreduplication cycle, resulting in giant, polyploid symbiotic bacterial cells that effectively fix atmospheric nitrogen for the plant's benefit (Dendene et al. 2022). Loss of NCR169 leads to nitrogen fixation impairment in *M. truncatula* dnf7 mutants, causing impaired bacterial differentiation and early senescence in symbiotic cells (Horváth et al. 2015). However, the delicate balance between senescence and immune repression within root nodules can be influenced by environmental factors; nodule senescence decreases, and defense mechanisms strengthen when grown in non-sterile conditions (Berrabah et al. 2023).

Structural and functional studies of NCR peptides have revealed two NMR structures of NCR169 with different disulfide linkage patterns. Both structures feature a consensus C-terminal β-sheet attached to an extended N-terminal region, with disulfide bonds contributing to NCR169's structural stability and solubility. Additionally, NCR169 exhibits antimicrobial properties against bacteria such as *Escherichia coli* and *Sinorhizobium meliloti*. One oxidized form of NCR169 was observed to bind to negatively charged bacterial phospholipids, with the lysine-rich region possibly playing a role in its antimicrobial activity. Notably, the active region remains disordered even in the presence of bound phospholipids, indicating the importance of its disordered conformation for its function. Observations of bacterial morphology have provided insights into the mechanism of action of NCR169 (Isozumi et al. 2021).

9.3.1.8 BLAD Polypeptide

It originates from sweet lupine seeds, particularly during the germination of *Lupinus albus*, a significant physiological process that may be susceptible to pathogen attacks. To counteract pathogens during this crucial period, seeds synthesize various compounds. In *Lupinus* species, the structure and concentration of β-conglutin transiently change during the initial stages of germination. This transformation involves the emergence of a new set of polypeptides, while other large molecular weight compounds gradually decrease as germination progresses. Consequently, there is an increase in the number of small molecular mass compounds due to limited proteolysis activity observed during seed germination (Monteiro et al. 2015).

During this critical germination phase, a 20-kDa polypeptide known as BLAD accumulates exclusively in the cotyledons of *Lupinus* species and exhibits antifungal activity. BLAD is a stable intermediate product of β-conglutin catabolism (Monteiro et al. 2015). This polypeptide targets the fungal cell wall, causing structural damage that weakens the integrity of the cell, making it more susceptible to damage. Additionally, BLAD interferes with the inner cell membrane of the fungus, compromising this critical barrier and disrupting essential cellular processes, thereby impeding fungal growth (Monteiro et al. 2015).

Furthermore, based on the stability of the peptide, two products were developed: ProBlad® Verde by the American company Sym-Agro and Problad Plus™ by the Portuguese company Consume em Verde (Pinheiro et al. 2018).

9.3.1.9 Narrow Leafed Lupin (NLL) β-Conglutin Proteins

NLL β-conglutin (vicilin or 7S acid globulin), a protein family with multiple genes, serves as a prominent protein in NLL seeds, displaying diverse properties and recently discovered benefits. It is recognized for its antifungal properties, which enhance plant resistance to necrotrophic oomycete pathogens. Moreover, it signifies a functional interplay between storage protein mobilization and seed oxidative metabolism, serving as a regulatory and signaling hub that controls the underlying physiological processes of seed dormancy breakage, release during germination, and seedling growth (Jimenez-Lopez 2020).

9.3.1.10 PvD1 Phaseolus Vulgaris Antifungal

Defensin derived from *Phaseolus vulgaris (L.)* seeds, designated as PvD1, has been isolated and characterized. Both natural and synthetic PvD1 peptides adopt a canonical CSαβ motif, which is stabilized by conserved disulfide bonds. Synthetic PvD1 maintains its biological activity against four different *Candida* species and exhibits no in vivo toxicity, suggesting its safety profile (Skalska et al. 2020).

Ultrastructural analysis of *Candida albicans* and *Candida guilliermondii* cells treated with PvD1 reveals disorganization of both cytoplasmic content and the plasma membrane, which likely contributes to the antifungal effect of PvD1. PvD1 inhibits glucose-stimulated acidification of the medium by yeast cells and filamentous fungi, thereby interfering with metabolic processes and enhancing its antifungal properties. Additionally, PvD1 induces the production of reactive oxygen species (ROS) and nitric oxide (NO) in *C. albicans* and *F. oxysporum* cells, which are involved in defense responses against fungal pathogens (Mello et al. 2011). In summary, PvD1's multifaceted mechanisms, including membrane disruption, metabolic interference, and induction of ROS/NO, position it as a promising candidate for combating fungal infections (Mello et al. 2011).

9.3.1.11 Legume Peptide Elicitors: Targeting Nematode

Peps activate several defensive pathways, including those previously linked to nematode resistance, making them a potential source of broad-spectrum resistance against nematodes.

9.3.1.11.1 GmPep1, GmPep2, and GmPep3

Three Peps derived from soybean (*Glycine max*), namely GmPep1, GmPep2, and GmPep3, were found to significantly reduce the reproduction of two destructive agricultural pests: the root-knot nematode (*Meloidogyne incognita*) and the soybean cyst nematode (*Heterodera glycines*), exerting a remarkable impact on pathogen control. Treatment with these peptides also conferred protection to plants against the detrimental effects of root-knot nematodes on above-ground growth and enhanced the basal expression levels of genes involved in nematode defense.

Specifically, GmPep1 induced the expression of its precursor molecule, GmPROPEP1, along with genes encoding a nucleotide-binding site leucine-rich repeat protein (NBS-LRR), a pectin methylesterase inhibitor (PMEI), Respiratory Burst Oxidase Protein D (RBOHD), and the accumulation of reactive oxygen species (ROS) in leaves. Moreover, treatments with GmPep2 and GmPep3 seeds resulted in increased expression of RBOHD and ROS accumulation in both roots and leaves. These findings suggest that GmPeps activate plant defenses by eliciting systemic changes in gene expression and signaling through ROS, indicating that seed treatments with Pep could be a promising strategy for nematode management (Lee et al. 2018).

9.3.1.12 Insecticidal Peptide Elicitors

Most plant-derived insecticidal peptides, including cyclic peptides, pea albumin, and defense and recombinant peptides (Grover et al. 2021), originate from botanical families such as Rubiaceae, Leguminosae, Violaceae, Solanaceae, and Cucurbitaceae (Craik 2010). Over 47 cyclic peptides from *Clitoria ternatea* have been identified to possess insecticidal effects (Gilding et al. 2016).

9.3.1.12.1 Plant Knottins/ Cystine Knot Peptides

Knottin peptides represent a distinctive and functionally varied class of low molecular weight peptides stabilized by disulfide bonds, featuring a unique architectural design naturally stabilized by a remarkable cystine knot (Postic et al. 2018). These peptides, typically less than 6 kDa in size, are characterized by three cysteine bonds, forming a rigid knot structure that imparts high thermal, molecular, and chemical stability. The cysteine knot configuration in the knottin family commonly involves bonds between C1 – C-4, C-2 – C-5, and C-3 – C-6 (Craik et al. 2001).

Notable molecular families with this knotted structure include cyclotides, disulfide-rich protease inhibitors such as trypsin, carboxypeptidase, and α-amylase inhibitors, Pea albumin peptide (PA1b), some plant defensins, and hevein-like peptides (Moore et al. 2012; Kumari Sita et al. 2017). Knottins exhibit various pharmacological activities, including interaction with plasma membranes and ion channels, leading to antimicrobial activity observed in their botanical sources known for pathogen resistance (Molesini et al. 2017). Additionally, knottins have demonstrated antihelminthic, antimalarial, anti-HIV, uterotonic, insecticidal, anti-tumor, and antifungal activities (Postic et al. 2018).

9.3.1.12.2 Pea Albumin 1 Subunit b Peptides (PA1b)

PA1b, derived from pea seeds and other legume seeds, particularly *Pisum sativum*, is recognized as a rich source of cysteine-rich knotted peptides with potent bioinsecticidal properties (Delobel et al. 1998; Gressent et al. 2011). Structurally, PA1b is a 37-amino acid peptide containing six cysteines forming three intramolecular disulfide bridges (Higgins et al. 1986). Its robust insecticidal activity against cereal weevils (*Sitophilus oryzae, S. zeamais, S. granarius*) and the pea aphid (*Acyrthosiphon pisum*) has been attributed to its ability to induce apoptosis through PA1b/V-ATPase interaction (Eyraud et al. 2017). However, despite its effectiveness as an insecticide, research on the potential therapeutic applications of PA1b and its analogues derived from legume plants, including *Cajanus* species, is limited (Barashkova and Rogozhin 2020). Furthermore, PA1b has demonstrated oral activity and remarkable insecticidal efficacy against various insects, including cereal weevils, aphids, and mosquitoes specifically *Culex pipiens* and *Aedes aegyptii* (Gressent et al. 2011; Rahioui et al. 2014).

In insects, PA1b acts by binding to the subunits c and e of the plasma membrane H-ATPase (V-ATPase) in the midgut, leading to the formation of apoptotic bodies and morphological changes characteristic of apoptosis (Eyraud et al. 2017; Beyenbach and Wieczorek 2006; Vaidyanathan and Scott 2006; Wyllie et al. 1973). The induction of apoptosis in insect gut cells upon PA1b ingestion is crucial for its insecticidal effect, and this process is mediated by the binding of PA1b to the V-ATPase receptor (Lu and Zhou 2012). Due to its specificity for insects, PA1b is considered safe for mammals and non-target organisms, making it a promising pesticide for organic farming (Gressent et al. 2011; Rahioui et al. 2014).

9.3.1.12.3 Cyclotides

Cyclotides, a family of cyclized knottins, are categorized into three subfamilies: bracelet, mobius, and trypsin inhibitors (Molesini et al. 2017). These cyclotides seem to play a role in plant defense due to their pesticidal properties, including insecticidal (Barbeta et al. 2008; Jennings et al. 2001), nematocidal (Colgrave et al. 2008), and molluscicidal (Plan et al. 2008) activities. The significant discovery of numerous ultrastable circular knottins (cyclotides) in *C. ternatea* has opened potential innovations to explore the dietary roles of these cystine knot peptides. Cyclotides were isolated from the butterfly pea, *Clitoria ternatea*, with their domain corresponding to the mature cyclotide from this Fabaceae plant embedded within an albumin precursor protein. The peptide exhibited the classic knotted cyclotide fold. Cter M showed insecticidal activity against the cotton budworm *Helicoverpa armigera* and interacted with phospholipid membranes, indicating its activity might involve membrane disruption.

The cyclotide-rich extract was utilized to produce the eco-friendly biopesticide Sero-X, approved for use in Australia in 2017. Sero-X demonstrates a unique mode of action in its pesticidal activity. It shows direct toxicity to small, soft-bodied larvae and nymphs. Another aspect of its pesticidal activity results from pests being starved and refraining from feeding on Sero-X treated crops. Additionally, the residue of Sero-X acts as a deterrent for oviposition, contributing to overall agricultural improvements. Its nontoxic and bee-friendly characteristics have led to its approval for use on cotton and macadamia nut plants in Australia, effectively controlling *Helicoverpa armigera*, *Bemesia tabaci*, and *Nezara viridula* (Oguis et al. 2019).

9.3.1.12.4 *Vigna Radiata* Cysteine-Rich Protein (VrCRP) / *V. Radiata* Defensin 1 (VrD1)

VrCRP plays a crucial role in bruchid resistance in mung bean (*Vigna radiata*). Comprising 73 amino acids and 8 cysteine residues, the mature VrCRP is a basic cysteine-rich protein with a predicted molecular mass of 5.9 kDa and an isoelectric point of 8.23. Isolated from the seed coat of the bruchid-resistant wild mung bean genotype TC 1966, VrCRP has been reported to have insecticidal (lethal to *C. chinensis*), fungicidal, and bactericidal activities. Studies indicate that mung bean seeds containing 0.2% VrCRP completely inhibit the development of bruchid larvae. The isolation and transgenic transfer of this multifunctional VrCRP gene from plants to popular high-yielding food legume genotypes can confer bruchid resistance (Chen et al. 2002).

Additionally, the VrD1 protein (*V. radiata* defensin 1), previously known as VrCRP and isolated from seeds of VC6089, is reported to confer resistance to mung bean against *C. maculatus* (Chen et al. 2002). A concentration of 0.1% VrD1 effectively acts as an insecticide by inhibiting the growth of *C. maculatus* (Lin et al. 2005). Moreover, both VrD1 (*Vigna radiata* defensin) and VuD1 (*V. unguiculata* defensin) exhibit insect α-amylase inhibitory activity, which is detrimental to insect pests such as *Tenebrio molitor*, *Z. subfasciatus*, *Acanthoscelides obtectus*, and *Zabrotes subfasciatus* (Lin et al. 2007; Liu et al. 2006; Pelegrini et al. 2008). As defensins with α-amylase inhibitory activity interfere with insect digestion, leading to energy deprivation from starch (De Oliveira Carvalho and Gomes 2009), VrD1 at 0.2% has been noted to exhibit higher amylase inhibitory activity compared to vignatic acid A (1.0%) and some specific lectins (2.0%) (Murdock et al. 1990; Sugawara et al. 1996). Therefore, VrD1-based transgenic crops or VrD1-based bio-insecticides could be significant components of bruchid management programs and aid in agricultural practices (Lin et al. 2005).

9.4 Peptide Elicitors Significance in Legumes to Combat Varied Stress

Through the exploration of diverse PEPs such as Gmsub pep, pep914, pep890, MMP, NCR peptides, BLAD polypeptide, PVD1, GmPep1, Gmpep2, Gmpep3, plant knottins, cyclotides, CLE peptides, and CEP, it becomes evident that these molecules play pivotal roles in enhancing plant defense mechanisms against both biotic and abiotic stresses. These PEPs act as molecular messengers, eliciting robust defense responses in leguminous plants. They trigger the activation of defense pathways, leading to elevated expression of defense-related genes, synthesis of hormones, and induction of secondary messengers. By fortifying plant immunity, PEPs help legumes withstand challenges such as water scarcity, high salinity, microbial infections, and parasitic nematodes. Furthermore, the diverse array of PEPs offers promising opportunities for crop protection and improvement. Their potential in enhancing stress tolerance and resilience in leguminous plants opens avenues for sustainable agriculture practices. Antimicrobial peptides (AMPs) from soybeans were demonstrated to possess in vitro antimicrobial activities and are potential candidates for biopesticides against soybean disease. Continued research into the mechanisms of action and optimization of PEP application methods will further unlock the full potential of these peptide elicitors in mitigating stresses and ensuring global food security. Additionally, studies on PA1b suggest that it possesses attractive option for sustainable pest management. PA1b is assumed to be a promising bioinsecticide with no allergenicity or toxicity to hosts. In essence, PEPs represent a valuable tool in the agricultural arsenal, offering innovative solutions to the complex challenges faced by legume cultivation in diverse environments. Their significance underscores the importance of harnessing nature's own defense mechanisms to safeguard crop productivity and sustainability in the face of evolving environmental pressures.

9.5 Versatility and Potential of Peptide Elicitors (PEPs)

PEPs exhibit basic biopesticide activity, adding another dimension to their utility beyond antimicrobial and therapeutic applications. Their ability to elicit defense responses in plants against biotic stressors makes them valuable tools in sustainable agriculture practices, offering eco-friendly alternatives to synthetic pesticides. The utilization of PEPs extends beyond plant defense mechanisms, showcasing their potential in various life forms. NCR247 and its derivatives exhibit potent antimicrobial activity against a wide range of bacteria (*Enterococcus faecalis, Staphylococcus aureus, Klebsiella pneumoniae, Acinetobacter baumannii, Pseudomonas aeruginosa, Escherichia coli, Listeria monocytogenes,* and *Salmonella enterica*), suggesting their applicability in human health as potential antimicrobial agents without exhibiting cytotoxicity (Jenei et al. 2020). Similarly, cationic NCR peptides demonstrate therapeutic potential for treating candidiasis, highlighting their relevance in combating fungal infections (Ördögh et al. 2014).

Cyclotides, known for their diverse biological activities, including antimicrobial and anti-cancer properties, present opportunities for agricultural and pharmaceutical applications (Gran 1970; Gustafson et al. 1994; Tam et al. 1999; Svangård et al. 2007). By leveraging cyclotide gene transcripts, crop plants could be engineered to produce improved cyclotides for various purposes, including the production of designer peptide drugs (Poth et al. 2011). Moreover, NLL β-conglutins not only offer nutraceutical benefits but also show promise in cancer therapy. These proteins have been found to inhibit metastasis and recurrence in breast cancer cell lines, underscoring their potential as innovative ingredients for functional food products (Escudero-Feliu et al. 2023).

In essence, the multifaceted attributes of PEPs, ranging from antimicrobial and therapeutic properties to biopesticide activity, underscore their versatility and potential in addressing various challenges across different life forms, thereby opening avenues for diverse applications in agriculture, healthcare, and beyond.

References

Abid G, M'hamdi M, Mingeot D, Aouida M, Aroua I, Muhovski Y, Sassi K, Souissi F, Mannai K, Jebara M (2017) Effect of drought stress on chlorophyll fluorescence, antioxidant enzyme activities and gene expression patterns in faba bean (Vicia faba L.). Arch Agron Soil Sci 63(4):536–552

Akada S, Kung SD, Dube SK (1990) The nucleotide sequence of gene 3 of the soybean chalcone synthase multigene family. Nucleic Acids Res 18(19):5899

Arora R, Sharma HC, Dreissche EV, Sharma KK (2005) Biological activity of lectins from grain legumes and garlic against the legume pod borer, Helicoverpa armigera. J SAT Agric Res 1(1):1–3

Arrese-Igor Sánchez C, García EG, Bilbao DM, Fernández RL, Rodríguez EL, Quintana EG (2011) Physiological responses of legume nodules to drought. Plant Stress 5(Special issue 1):24–31

Barashkova AS, Rogozhin EA (2020) Isolation of antimicrobial peptides from different plant sources: does a general extraction method exist? Plant Methods 16(1):143

Barbeta BL, Marshall AT, Gillon AD, Craik DJ, Anderson MA (2008) Plant cyclotides disrupt epithelial cells in the midgut of lepidopteran larvae. Proc Natl Acad Sci 105(4):1221–1225

Bartels S, Boller T (2015) Quo vadis, pep? Plant elicitor peptides at the crossroads of immunity, stress, and development. J Exp Bot 66(17):5183–5193

Bartels S, Lori M, Mbengue M, Van Verk M, Klauser D, Hander T, Böni R, Robatzek S, Boller T (2013) The family of Peps and their precursors in Arabidopsis: differential expression and localization but similar induction of pattern-triggered immune responses. J Exp Bot 64(17):5309–5321

Benko-Iseppon AM, Lins Galdino S, Calsa T Jr, Akio Kido E, Tossi A, Carlos Belarmino L, Crovella S (2010) Overview on plant antimicrobial peptides. Curr Protein Pept Sci 11(3):181

Berrabah F, Bernal G, Elhosseyn AS, El Kassis C, L'Horset R, Benaceur F, Wen J, Mysore KS, Garmier M, Gourion B, Ratet P (2023) Insight into the control of nodule immunity and senescence during Medicago truncatula symbiosis. Plant Physiol 191(1):729–746

Beyenbach KW, Wieczorek H (2006) The V-type H+ atpase: molecular structure and function, physiological roles and regulation. J Exp Biol 209(4):577–589

Bosamia TC, Dodia SM, Mishra GP, Ahmad S, Joshi B, Thirumalaisamy PP, Kumar N, Rathnakumar AL, Sangh C, Kumar A, Thankappan R (2020) Unraveling the mechanisms of

resistance to Sclerotium rolfsii in peanut (Arachis hypogaea L.) using comparative RNA-Seq analysis of resistant and susceptible genotypes. PLoS One 15(8):e0236823
Brand GD, Magalhaes MT, Tinoco ML, Aragao FJ, Nicoli J, Kelly SM, Cooper A, Bloch Jr C (2012) Probing protein sequences as sources for encrypted antimicrobial peptides
Chen KC, Lin CY, Kuan CC, Sung HY, Chen CS (2002) A novel defensin encoded by a mungbean cdna exhibits insecticidal activity against bruchid. J Agric Food Chem 50(25):7258–7263
Cheng H, Hu Y (2010) Municipal solid waste (MSW) as a renewable source of energy: current and future practices in China. Bioresour Technol 101(11):3816–3824
Chowdhury JA, Karim MA, Khaliq QA, Ahmed AU, Khan MSA (2016) Effect of drought stress on gas exchange characteristics of four soybean genotypes
Colgrave ML, Kotze AC, Huang YH, O'Grady J, Simonsen SM, Craik DJ (2008) Cyclotides: natural, circular plant peptides that possess significant activity against gastrointestinal nematode parasites of sheep. Biochemistry 47(20):5581–5589
Craik DJ (2010) Discovery and applications of the plant cyclotides. Toxicon 56(7):1092–1102
Craik DJ, Daly NL, Waine C (2001) The cystine knot motif in toxins and implications for drug design. Toxicon 39(1):43–60
De Oliveira Carvalho A, Gomes VM (2009) Plant defensins—prospects for the biological functions and biotechnological properties. Peptides 30(5):1007–1020
Dell'Olmo E, Tiberini A, Sigillo L (2023) Leguminous seedborne pathogens: seed health and sustainable crop management. Plan Theory 12(10):2040
Delobel B, Grenier AM, Gueguen J, Ferrasson E, Mbaiguinam M (1998) Utilisation d'un polypeptide dérivé d'une albumine PA1b de légumineuse comme insecticide. French Patent 98(05877):11
Dendene S, Frascella A, Nicoud Q, Timchenko T, Mergaert P, Alunni B, Biondi EG (2022) Cell cycle and terminal differentiation in Sinorhizobium meliloti. In: Cell cycle regulation and development in Alphaproteobacteria. Springer, Cham, pp 221–244
Di Q, Li Y, Zhang D, Wu W, Zhang L, Zhao X, Luo L, Yu L (2022) A novel type of phytosulfokine, PSK-ε, positively regulates root elongation and formation of lateral roots and root nodules in Medicago truncatula. Plant Signal Behav 17(1):2134672
Eichhorn H, Klinghammer M, Becht P, Tenhaken R (2006) Isolation of a novel ABC-transporter gene from soybean induced by salicylic acid. J Exp Bot 57(10):2193–2201
Escudero-Feliu J, García-Costela M, Moreno-sanjuan S, Puentes-Pardo JD, Arrabal SR, González-Novoa P, Núñez MI, Carazo Á, Jimenez-Lopez JC, León J (2023) Narrow leafed lupin (Lupinus angustifolius L.) B-conglutin seed proteins as a new natural cytotoxic agents against breast cancer cells. Nutrients 15(3):523
Essa TA (2002) Effect of salinity stress on growth and nutrient composition of three soybean (Glycine max L. Merrill) cultivars. J Agron Crop Sci 188(2):86–93
Eyraud V, Balmand S, Karaki L, Rahioui I, Sivignon C, Delmas AF, Royer C, Rahbé Y, Da Silva P, Gressent F (2017) The interaction of the bioinsecticide PA1b (pea albumin 1 subunit b) with the insect V-atpase triggers apoptosis. Sci Rep 7(1):4902
Fahad S, Hussain S, Saud S, Khan F, Hassan S, Amanullah, Nasim W, Arif M, Wang F, Huang J (2016) Exogenously applied plant growth regulators affect heat-stressed rice pollens. J Agron Crop Sci 202(2):139–150
Games PD, Dos Santos IS, Mello ÉO, Diz MS, Carvalho AO, de Souza-Filho GA, Da Cunha M, Vasconcelos IM, Ferreira BDS, Gomes VM (2008) Isolation, characterization and cloning of a cdna encoding a new antifungal defensin from Phaseolus vulgaris L. seeds. Peptides 29(12):2090–2100
Garg N, Bhandari P (2016) Silicon nutrition and mycorrhizal inoculations improve growth, nutrient status, K+/Na+ ratio and yield of Cicer arietinum L. genotypes under salinity stress. Plant Growth Regul 78:371–387
Gebrelibanos M, Tesfaye D, Raghavendra Y, Sintayeyu B (2013) Nutritional and health implications of legumes. Int J Pharm Sci Res 4(4):1269

Gilding EK, Jackson MA, Poth AG, Henriques ST, Prentis PJ, Mahatmanto T, Craik DJ (2016) Gene coevolution and regulation lock cyclic plant defence peptides to their targets. New Phytol 210(2):717–730
Graham JS, Xiong J, Gillikin JW (1991) Purification and developmental analysis of a metalloendoproteinase from the leaves of Glycine max. Plant Physiol 97(2):786–792
Gran L (1970) An oxytocic principle found in Oldenlandia affinis DC. Medd Nor Farm Selsk 12(173):80
Gressent F, Da Silva P, Eyraud V, Karaki L, Royer C (2011) Pea albumin 1 subunit b (PA1b), a promising bioinsecticide of plant origin. Toxins 3(12):1502–1517
Grover T, Mishra R, Gulati P, Mohanty A (2021) An insight into biological activities of native cyclotides for potential applications in agriculture and pharmaceutics. Peptides 135:170430
Guilioni L, Wery J, Tardieu F (1997) Heat stress-induced abortion of buds and flowers in pea: is sensitivity linked to organ age or to relations between reproductive organs? Ann Bot 80(2):159–168
Gustafson KR, Sowder RC, Henderson LE, Parsons IC, Kashman Y, Cardellina JH, Mcmahon JB, Buckheit RW Jr, Pannell LK, Boyd MR (1994) Circulins a and B. Novel human immunodeficiency virus (HIV)-inhibitory macrocyclic peptides from the tropical tree Chassalia parvifolia. J Am Chem Soc 116(20):9337–9338
Hastwell AH, de Bang TC, Gresshoff PM, Ferguson BJ (2017) CLE peptide-encoding gene families in Medicago truncatula and Lotus japonicus, compared with those of soybean, common bean and Arabidopsis. Sci Rep 7(1):9384
Heatherly LG (1993) Drought stress and irrigation effects on germination of harvested soybean seed. Crop Sci 33(4):777–781
Herrero M, Thornton PK, Notenbaert AM, Wood S, Msangi S, Freeman HA, Bossio D, Dixon J, Peters M, van de Steeg J, Lynam J (2010) Smart investments in sustainable food production: revisiting mixed crop-livestock systems. Science 327(5967):822–825
Higgins TJ, Chandler PM, Randall PJ, Spencer D, Beach LR, Blagrove RJ, Kortt AA, Inglis AS (1986) Gene structure, protein structure, and regulation of the synthesis of a sulfur-rich protein in pea seeds. J Biol Chem 261(24):11124–11130
Horváth B, Domonkos Á, Kereszt A, Szűcs A, Ábrahám E, Ayaydin F, Bóka K, Chen Y, Chen R, Murray JD, Udvardi MK (2015) Loss of the nodule-specific cysteine rich peptide, NCR169, abolishes symbiotic nitrogen fixation in the Medicago truncatula dnf7 mutant. Proc Natl Acad Sci 112(49):15232–15237
Huffaker A, Pearce G, Veyrat N, Erb M, Turlings TC, Sartor R, Shen Z, Briggs SP, Vaughan MM, Alborn HT, Teal PE (2013) Plant elicitor peptides are conserved signals regulating direct and indirect antiherbivore defense. Proc Natl Acad Sci 110(14):5707–5712
Iqbal MK, Shafiq T, Ahmed K (2010) Characterization of bulking agents and its effects on physical properties of compost. Bioresour Technol 101(6):1913–1919
Ismail AM, Hall AE (1999) Reproductive-stage heat tolerance, leaf membrane thermostability and plant morphology in cowpea. Crop Sci 39(6):1762–1768
Isozumi N, Masubuchi Y, Imamura T, Mori M, Koga H, Ohki S (2021) Structure and antimicrobial activity of NCR169, a nodule-specific cysteine-rich peptide of Medicago truncatula. Sci Rep 11(1):9923
Jenei S, Tiricz H, Szolomájer J, Tímár E, Klement É, Al Bouni MA, Lima RM, Kata D, Harmati M, Buzás K, Földesi I (2020) Potent chimeric antimicrobial derivatives of the Medicago truncatula NCR247 symbiotic peptide. Front Microbiol 11:512394
Jennings C, West J, Waine C, Craik D, Anderson M (2001) Biosynthesis and insecticidal properties of plant cyclotides: the cyclic knotted proteins from Oldenlandia affinis. Proc Natl Acad Sci 98(19):10614–10619
Jimenez-Lopez JC (2020) Narrow-leafed lupin (Lupinus angustifolius L.) B-conglutin: a multifunctional family of proteins with roles in plant defence, human health benefits, and potential uses as functional food. Legume Science 2(2):e33
Jones CH, Hastwell AH, Gresshoff PM, Ferguson BJ (2022) Soybean CLE peptides and their CLAVATA-like signaling pathways. Adv Bot Res 102:153–175

Kaushal N, Awasthi R, Gupta K, Gaur P, Siddique KH, Nayyar H (2013) Heat-stress-induced reproductive failures in chickpea (Cicer arietinum) are associated with impaired sucrose metabolism in leaves and anthers. Funct Plant Biol 40(12):1334–1349

Kereszt A, Mergaert P, Montiel J, Endre G, Kondorosi É (2018) Impact of plant peptides on symbiotic nodule development and functioning. Front Plant Sci 9:396805

Kim M, Chen Y, Xi J, Waters C, Chen R, Wang D (2015) An antimicrobial peptide essential for bacterial survival in the nitrogen-fixing symbiosis. Proc Natl Acad Sci 112(49):15238–15243

Kudapa H, Bharti AK, Cannon SB, Farmer AD, Mulaosmanovic B, Kramer R, Bohra A, Weeks NT, Crow JA, Tuteja R, Shah T (2012) A comprehensive transcriptome assembly of pigeonpea (Cajanus cajan L.) using sanger and second-generation sequencing platforms. Mol Plant 5(5):1020–1028

Kumari Sita KS, Akanksha Sehgal AS, Rao BH, Nair RM, Prasad PVV, Shiv Kumar SK, Gaur PM, Muhammad Farooq MF, Siddique KHM, Varshney RK, Harsh Nayyar HN (2017) Food legumes and rising temperatures: effects, adaptive functional mechanisms specific to reproductive growth stage and strategies to improve heat tolerance

Kumari VV, Roy A, Vijayan R, Banerjee P, Verma VC, Nalia A, Pramanik M, Mukherjee B, Ghosh A, Reja MH, Chandran MAS (2021) Drought and heat stress in cool-season food legumes in sub-tropical regions: consequences, adaptation, and mitigation strategies. Plan Theory 10(6):1038

Latef AAHA, Ahmad P (2015) Legumes and breeding under abiotic stress: an overview. In: Legumes under environmental stress: yield, improvement and adaptations, pp 1–20

Lee MW, Huffaker A, Crippen D, Robbins RT, Goggin FL (2018) Plant elicitor peptides promote plant defences against nematodes in soybean. Mol Plant Pathol 19(4):858–869

Lemke RL, Zhong Z, Campbell CA, Zentner R (2007) Can pulse crops play a role in mitigating greenhouse gases from north American agriculture? Agron J 99(6):1719–1725

Li P, Zhang Y, Wu X, Liu Y (2018) Drought stress impact on leaf proteome variations of faba bean (Vicia faba L.) in the Qinghai–Tibet plateau of China. 3 Biotech 8:1–12

Lima RM, Kylarová S, Mergaert P, Kondorosi É (2020) Unexplored arsenals of legume peptides with potential for their applications in medicine and agriculture. Front Microbiol 11:546514

Lin C, Chen CS, Horng SB (2005) Characterization of resistance to Callosobruchus maculatus (Coleoptera: Bruchidae) in mungbean variety VC6089A and its resistance-associated protein vrd1. J Econ Entomol 98(4):1369–1373

Lin KF, Lee TR, Tsai PH, Hsu MP, Chen CS, Lyu PC (2007) Structure-based protein engineering for α-amylase inhibitory activity of plant defensin. Prot Struc Func Bioinform 68(2):530–540

Liu Y, Dammann C, Bhattacharyya MK (2001) The matrix metalloproteinase gene gmmmp2 is activated in response to pathogenic infections in soybean. Plant Physiol 127(4):1788–1797

Liu YJ, Cheng CS, Lai SM, Hsu MP, Chen CS, Lyu PC (2006) Solution structure of the plant defensin vrd1 from mung bean and its possible role in insecticidal activity against bruchids. Prot Struc Func Bioinform 63(4):777–786

Liu S, Liu Y, Jia Y, Wei J, Wang S, Liu X, Zhou Y, Zhu Y, Gu W, Ma H (2017) Gm1-MMP is involved in growth and development of leaf and seed, and enhances tolerance to high temperature and humidity stress in transgenic Arabidopsis. Plant Sci 259:48–61

Lori M, Van Verk MC, Hander T, Schatowitz H, Klauser D, Flury P, Gehring CA, Boller T, Bartels S (2015) Evolutionary divergence of the plant elicitor peptides (Peps) and their receptors: interfamily incompatibility of perception but compatibility of downstream signalling. J Exp Bot 66(17):5315–5325

Lu N, Zhou Z (2012) Membrane trafficking and phagosome maturation during the clearance of apoptotic cells. Int Rev Cell Mol Biol 293:269–309

Mafakheri A, Siosemardeh AF, Bahramnejad B, Struik PC, Sohrabi Y (2010) Effect of drought stress on yield, proline and chlorophyll contents in three chickpea cultivars. Aust J Crop Sci 4(8):580–585

Maróti G, Downie JA, Kondorosi É (2015) Plant cysteine-rich peptides that inhibit pathogen growth and control rhizobial differentiation in legume nodules. Curr Opin Plant Biol 26:57–63

Mcgeehan G, Burkhart W, Anderegg R, Becherer JD, Gillikin JW, Graham JS (1992) Sequencing and characterization of the soybean leaf metalloproteinase: structural and functional similarity to the matrix metalloproteinase family. Plant Physiol 99(3):1179–1183

Mello EO, Ribeiro SF, Carvalho AO, Santos IS, Da Cunha M, Santa-Catarina C, Gomes VM (2011) Antifungal activity of Pv D1 defensin involves plasma membrane permeabilization, inhibition of medium acidification, and induction of ROS in fungi cells. Curr Microbiol 62:1209–1217

Melotto M, Underwood W, He SY (2008) Role of stomata in plant innate immunity and foliar bacterial diseases. Annu Rev Phytopathol 46:101–122

Mergaert P, Nikovics K, Kelemen Z, Maunoury N, Vaubert D, Kondorosi A, Kondorosi E (2003) A novel family in Medicago truncatula consisting of more than 300 nodule-specific genes coding for small, secreted polypeptides with conserved cysteine motifs. Plant Physiol 132(1):161–173

Mikuláss KR, Nagy K, Bogos B, Szegletes Z, Kovács E, Farkas A, Váró G, Kondorosi É, Kereszt A (2016) Antimicrobial nodule-specific cysteine-rich peptides disturb the integrity of bacterial outer and inner membranes and cause loss of membrane potential. Ann Clin Microbiol Antimicrob 15:1–5

Mohd-Radzman NA, Binos S, Truong TT, Imin N, Mariani M, Djordjevic MA (2015) Novel mtcep1 peptides produced in vivo differentially regulate root development in Medicago truncatula. J Exp Bot 66(17):5289–5300

Molesini B, Treggiari D, Dalbeni A, Minuz P, Pandolfini T (2017) Plant cystine-knot peptides: pharmacological perspectives. Br J Clin Pharmacol 83(1):63–70

Monteiro S, Carreira A, Freitas R, Pinheiro AM, Ferreira RB (2015) A nontoxic polypeptide oligomer with a fungicide potency under agricultural conditions which is equal or greater than that of their chemical counterparts. PLoS One 10(4):e0122095

Montiel J, Downie JA, Farkas A, Bihari P, Herczeg R, Bálint B, Mergaert P, Kereszt A, Kondorosi É (2017) Morphotype of bacteroids in different legumes correlates with the number and type of symbiotic NCR peptides. Proc Natl Acad Sci 114(19):5041–5046

Moore SJ, Leung CL, Cochran JR (2012) Knottins: disulfide-bonded therapeutic and diagnostic peptides. Drug Discov Today Technol 9(1):e3–e11

Murdock LL, Huesing JE, Nielsen SS, Pratt RC, Shade RE (1990) Biological effects of plant lectins on the cowpea weevil. Phytochemistry 29(1):85–89

Musavi MT, Ahmed W, Chan KH, Faris KB, Hummels DM (1992) On the training of radial basis function classifiers. Neural Netw 5(4):595–603

Nadeem M, Li J, Yahya M, Sher A, Ma C, Wang X, Qiu L (2019) Research progress and perspective on drought stress in legumes: a review. Int J Mol Sci 20(10):2541

Nandwal AS, Godara M, Kamboj DV, Kundu BS, Mann A, Kumar B, Sharma SK (2000) Nodule functioning in trifoliate and pentafoliate mungbean genotypes as influenced by salinity. Biol Plant 43:459–462

Ning L, Kan G, Shao H, Yu D (2018) Physiological and transcriptional responses to salt stress in salt-tolerant and salt-sensitive soybean (Glycine max [L.] Merr.) seedlings. Land Degrad Dev 29(8):2707–2719

Oguis GK, Gilding EK, Jackson MA, Craik DJ (2019) Butterfly pea (Clitoria ternatea), a cyclotide-bearing plant with applications in agriculture and medicine. Front Plant Sci 10:448370

Ördögh L, Vörös A, Nagy I, Kondorosi É, Kereszt A (2014) Symbiotic plant peptides eliminate Candida albicans both in vitro and in an epithelial infection model and inhibit the proliferation of immortalized human cells. Biomed Res Int

Pak JH, Liu CY, Huangpu J, Graham JS (1997) Construction and characterization of the soybean leaf metalloproteinase cdna. FEBS Lett 404(2–3, 283):–288

Patel N, Mohd-Radzman NA, Corcilius L, Crossett B, Connolly A, Cordwell SJ, Ivanovici A, Taylor K, Williams J, Binos S, Mariani M (2018) Diverse peptide hormones affecting root growth identified in the Medicago truncatula secreted peptidome. Mol Cell Proteomics 17(1):160–174

Pearce G, Yamaguchi Y, Barona G, Ryan CA (2010) A subtilisin-like protein from soybean contains an embedded, cryptic signal that activates defense-related genes. Proc Natl Acad Sci 107(33):14921–14925

Pelegrini PB, Lay FT, Murad AM, Anderson MA, Franco OL (2008) Novel insights on the mechanism of action of α-amylase inhibitors from the plant defensin family. Prot Struc Func Bioinform 73(3):719–729
Pinheiro AM, Carreira A, Ferreira RB, Monteiro S (2018) Fusion proteins towards fungi and bacteria in plant protection. Microbiology 164(1):11–19
Plan MRR, Saska I, Cagauan AG, Craik DJ (2008) Backbone cyclised peptides from plants show molluscicidal activity against the rice pest Pomacea canaliculata (golden apple snail). J Agric Food Chem 56(13):5237–5241
Postic G, Gracy J, Périn C, Chiche L, Gelly JC (2018) KNOTTIN: the database of inhibitor cystine knot scaffold after 10 years, toward a systematic structure modeling. Nucleic Acids Res 46(D1):D454–D458
Poth AG, Colgrave ML, Lyons RE, Daly NL, Craik DJ (2011) Discovery of an unusual biosynthetic origin for circular proteins in legumes. Proc Natl Acad Sci 108(25):10127–10132
Pulses for human health and nutrition (2024) Technical bulletin. IIPR, Kanpur
Rahioui I, Eyraud V, Karaki L, Sasse F, Carre-Pierrat M, Qin A, Zheng MH, Toepfer S, Sivignon C, Royer C, Da Silva P (2014) Host range of the potential biopesticide pea albumin 1b (PA1b) is limited to insects. Toxicon 89:67–76
Ramada MHS, Brand GD, Abrão FY, Oliveira M, Filho JC, Galbieri R, Gramacho KP, Prates MV, Bloch C Jr (2017) Encrypted antimicrobial peptides from plant proteins. Sci Rep 7(1):13263
Ramirez-Estrada K, Vidal-Limon H, Hidalgo D, Moyano E, Golenioswki M, Cusidó RM, Palazon J (2016) Elicitation, an effective strategy for the biotechnological production of bioactive high-added value compounds in plant cell factories. Molecules 21(2):182
Rhouma A, Hajji-Hedfi L, Okon OG, Bassey HO (2024) Investigating the effectiveness of endophytic fungi under biotic and abiotic agricultural stress conditions. J Oasis Agric Sustain Dev 6(01):123–138
Roux B, Rodde N, Jardinaud MF, Timmers T, Sauviac L, Cottret L, Carrère S, Sallet E, Courcelle E, Moreau S, Debellé F (2014) An integrated analysis of plant and bacterial gene expression in symbiotic root nodules using laser-capture microdissection coupled to RNA sequencing. Plant J 77(6):817–837
Roy S, Griffiths M, Torres-Jerez I, Sanchez B, Antonelli E, Jain D, Krom N, Zhang S, York LM, Scheible WR, Udvardi M (2022) Application of synthetic peptide CEP1 increases nutrient uptake rates along plant roots. Front Plant Sci 12:793145
Satyanarayana V, Vara Prasad PV, Murthy VRK, Boote KJ (2002) Influence of integrated use of farmyard manure and inorganic fertilizers on yield and yield components of irrigated lowland rice. J Plant Nutr 25(10):2081–2090
Sehrawat N, Yadav M, Sharma AK, Kumar V, Bhat KV (2019) Salt stress and mungbean [Vigna radiata (L.) Wilczek]: effects, physiological perspective and management practices for alleviating salinity. Arch Agron Soil Sci 65(9):1287–1301
Sharma HC, Saxena KB, Bhagwat VR (1999) The legume pod borer, Maruca vitrata: bionomics and management
Sharma L, Priya M, Bindumadhava H, Nair RM, Nayyar H (2016) Influence of high temperature stress on growth, phenology and yield performance of mungbean [Vigna radiata (L.) Wilczek] under managed growth conditions. Sci Hortic 213:379–391
Singh P, Thakur S, Kumar S, Mondal B, Rathore M, Das A (2023) Genetic engineering for enhancing abiotic stress tolerance in pulses. In: Legumes: physiology and molecular biology of abiotic stress tolerance. Singapore, Springer, pp 345–367
Skalska J, Andrade VM, Cena GL, Harvey PJ, Gaspar D, Mello ÉO, Henriques ST, Valle J, Gomes VM, Conceição K, Castanho MA (2020) Synthesis, structure, and activity of the antifungal plant Defensin Pv D1. J Med Chem 63(17):9391–9402
Sugawara F, Ishimoto M, Le-Van N, Koshino H, Uzawa J, Yoshida S, Kitamura K (1996) Insecticidal peptide from mungbean: a resistant factor against infestation with azuki bean weevil. J Agric Food Chem 44(10):3360–3364
Suzuki G, Ohta H, Kato T, Igarashi T, Sakai F, Shibata D, Takano A, Masuda T, Shioi Y, Takamiya KI (1996) Induction of a novel cytochrome P450 (CYP93 family) by methyl jasmonate in soybean suspension-cultured cells. FEBS Lett 383(1–2):83–86

Svangård E, Burman R, Gunasekera S, Lövborg H, Gullbo J, Göransson U (2007) Mechanism of action of cytotoxic cyclotides: cycloviolacin O2 disrupts lipid membranes. J Nat Prod 70(4):643–647
Tam JP, Lu YA, Yang JL, Chiu KW (1999) An unusual structural motif of antimicrobial peptides containing end-to-end macrocycle and cystine-knot disulfides. Proc Natl Acad Sci 96(16):8913–8918
Tam JP, Wang S, Wong KH, Tan WL (2015) Antimicrobial peptides from plants. Pharmaceuticals 8(4):711–757
Tavormina P, De Coninck B, Nikonorova N, De Smet I, Cammue BP (2015) The plant peptidome: an expanding repertoire of structural features and biological functions. Plant Cell 27(8):2095–2118
Tharanathan RN, Mahadevamma S (2003) Grain legumes—a boon to human nutrition. Trends Food Sci Technol 14(12):507–518
Tokarz B, Wójtowicz T, Makowski W, Jędrzejczyk RJ, Tokarz KM (2020) What is the difference between the response of grass pea (Lathyrus sativus L.) to salinity and drought stress?—a physiological study. Agronomy 10(6):833
Torche Y, Blair M, Saida C (2018) Biochemical, physiological and phenological genetic analysis in common bean (Phaseolus vulgaris L.) under salt stress. Ann Agric Sci 63(2):153–161
Vaidyanathan R, Scott TW (2006) Apoptosis in mosquito midgut epithelia associated with West Nile virus infection. Apoptosis 11:1643–1651
Van de Velde W, Zehirov G, Szatmari A, Debreczeny M, Ishihara H, Kevei Z, Farkas A, Mikulass K, Nagy A, Tiricz H, Satiat-Jeunemaitre B (2010) Plant peptides govern terminal differentiation of bacteria in symbiosis. Science 327(5969):1122–1126
Venkidasamy B, Selvaraj D, Nile AS, Ramalingam S, Kai G, Nile SH (2019) Indian pulses: a review on nutritional, functional and biochemical properties with future perspectives. Trends Food Sci Technol 88:228–242
Vollenweider P, Günthardt-Goerg MS (2005) Diagnosis of abiotic and biotic stress factors using the visible symptoms in foliage. Environ Pollut 137(3):455–465
Watanabe A, Nong VH, Zhang D, Arahira M, Yeboah NA, Udaka K, Fukazawa C (1999) Molecular cloning and ethylene-inducible expression of Chib1 chitinase from soybean (Glycine max (L.) Merr.). Biosci Biotechnol Biochem 63(2):251–256
Wyllie AH, Kerr JF, Currie AR (1973) Cell death in the normal neonatal rat adrenal cortex. J Pathol 111(4):255–261
Yamaguchi Y, Huffaker A (2011) Endogenous peptide elicitors in higher plants. Curr Opin Plant Biol 14(4):351–357
Yamaguchi Y, Barona G, Ryan CA, Pearce G (2011) Gmpep914, an eight-amino acid peptide isolated from soybean leaves, activates defense-related genes. Plant Physiol 156(2):932–942
Yamaguchi YL, Ishida T, Sawa S (2016) CLE peptides and their signaling pathways in plant development. J Exp Bot 67(16):4813–4826
Zelman AK, Berkowitz GA (2023) Plant elicitor peptide (pep) signaling and pathogen defense in tomato. Plan Theory 12(15):2856
Zhou C, McCarthy SA, Durbin R (2023) YaHS: yet another hi-C scaffolding tool. Bioinformatics 39(1):btac808
Zhu F, Deng J, Chen H, Liu P, Zheng L, Ye Q, Li R, Brault M, Wen J, Frugier F, Dong J (2020) A CEP peptide receptor-like kinase regulates auxin biosynthesis and ethylene signaling to coordinate root growth and symbiotic nodulation in Medicago truncatula. Plant Cell 32(9):2855–2877
Zorin EA, Kliukova MS, Afonin AM, Gribchenko ES, Gordon ML, Sulima AS, Zhernakov AI, Kulaeva OA, Romanyuk DA, Kusakin PG, Tsyganova AV (2022) A variable gene family encoding nodule-specific cysteine-rich peptides in pea (Pisum sativum L.). Front Plant Sci 13:884726

Harnessing Plant Innate Immunity for Improved Biomass Production in Bioenergy Crops

10

Senthil Nagappan and Dig Vijay Singh

Abstract

Harnessing plant innate immunity offers a sustainable and efficient approach to increase biomass production in bioenergy crops. By enhancing natural defense mechanisms, such as pattern-triggered immunity (PTI) and effector-triggered immunity (ETI), through genetic engineering or selective breeding, crops can become more resistant to diseases, reducing the need for chemical inputs and allowing more resources to be allocated to growth. Additionally, manipulating hormonal pathways and promoting beneficial microbial interactions can further boost biomass yields and stress tolerance. Using chemical priming agents and natural systemic acquired resistance (SAR) inducers can also activate systemic defenses, enhancing overall plant resilience. Bioenergy crops like switchgrass, napier grass, and miscanthus offer high yields, adaptability to various climates, and environmental benefits such as carbon sequestration and soil health improvement. However, challenges such as land use competition, sustainability concerns, and economic viability remain. By optimizing cultivation practices and conversion technologies, bioenergy crops can play a significant role in reducing reliance on fossil fuels and mitigating greenhouse gas emissions, thereby contributing to a more sustainable energy future.

Keywords

Bioenergy crops · Immunity · Genetic engineering · Systemic acquired resistance · Microbial interactions

S. Nagappan (✉) · D. V. Singh
Sardar Swaran Singh National Institute of Bioenergy, Kapurthala, India

S. Singh, R. Mehrotra (eds.), *Plant Elicitor Peptides*,
https://doi.org/10.1007/978-981-97-6374-0_10

10.1 Introduction

Increasing biomass production in bioenergy crops by harnessing plant innate immunity is a promising strategy. Plants have natural defense mechanisms, such as pattern-triggered immunity (PTI) and effector-triggered immunity (ETI), which protect them from pathogens and stress (Bigeard et al. 2015). By enhancing these immune responses through genetic engineering or selective breeding, crops can become more resistant to diseases, reducing the need for chemical inputs and allowing more resources to be allocated to growth (Abdul Aziz et al. 2022). For example, overexpressing pattern recognition receptors (PRRs) and resistance (R) proteins can strengthen the plant's immune system, leading to healthier and more robust biomass production. Additionally, manipulating hormonal pathways like those involving salicylic acid and jasmonic acid can balance growth and defense, further optimizing biomass yield (van Butselaar and Van den Ackerveken 2020; Shyu and Brutnell 2015). Furthermore, integrating beneficial microbial interactions can significantly boost biomass production. Plants form symbiotic relationships with microorganisms such as mycorrhizal fungi and plant growth-promoting rhizobacteria, which enhance nutrient uptake and stress tolerance (Boostani et al. 2014). By promoting these beneficial interactions, plants can better withstand environmental challenges and allocate more energy toward growth (Arif et al. 2020). Chemical priming agents and natural SAR inducers can also be used to activate systemic acquired resistance, preparing the entire plant to better defend itself against pathogens (Hönig et al. 2023).

Moreover, scientists are also more focused on using transgenics or specialized hybrids as energy crops in order to achieve the goals of sustainable energy through the production of biofuels (Ndimba et al. 2013). Using biotechnology to improve crop properties and the effectiveness of biochemical conversion processes can significantly lower the cost of producing biofuels (Li et al. 2014). Additionally, the acceptability of such energy sources by society and the environment must be taken into account in the development of sustainable biomass energy. Traditional breeding methods may be effective, but combining them with the state-of-the-art biotechnological tools will improve yield significantly and benefits in terms of increased crop productivity, yield stability, resistance to pests, diseases, and environmental conditions (Dalla Costa et al. 2019). Thus, utilizing cutting-edge biotechnological techniques could result in the transfer of important genes in different species not only to enhance biomass and biofuel yield but also improved tolerance to diverse biotic and abiotic stress conditions. Moreover, innate immunity, phytohormone signaling, microbial interaction with advanced breeding techniques and biotechnology can lead to the development of bioenergy crops with enhanced innate immunity, resulting in increased biomass production and improved sustainability. Therefore, exploiting different genetic engineering techniques along with crop-based immunity, signaling molecules, and microbial community-based beneficial interaction can serve as the important strategy to improve biomass yield but also enhances the economic viability of crops toward bioenergy production.

10.2 Bioenergy Crops

Bioenergy crops, such as switchgrass, napier grass, miscanthus, corn, sugarcane, sorghum, willow, poplar, and jatropha, are cultivated specifically for their potential to produce biomass for biofuels and renewable energy (Yadav et al. 2019) (Table 10.1). These crops offer numerous advantages, including high biomass yields, low input requirements, adaptability to various climates and soil types, and potential for environmental benefits like carbon sequestration and soil health improvement (Stoof et al. 2015). They are used in various applications such as ethanol and biodiesel production, biogas generation, biochar creation, and direct combustion for heat and power (Barot 2022). Ideal bioenergy crops are those that provide high efficiency in converting biomass to biofuel while requiring minimal fertilizers, pesticides, and water, and can grow on marginal lands (Mehmood et al. 2017). However, there are challenges associated with bioenergy crop production, including land use competition with food crops, sustainability concerns like deforestation and biodiversity loss, economic viability, and the need for technological advancements to improve biomass conversion efficiency (Rial 2024). Balancing these factors is crucial for the successful integration of bioenergy crops into

Table 10.1 Different generation of energy crops

First-generation bioenergy crops	Second-generation bioenergy crops	Third-generation bioenergy crops	Fourth-generation bioenergy crops
Corn, sugarcane, wheat, soybeans, rapeseed, sugar-beet, sunflower, sorghum	Switchgrass, miscanthus, jatropha willow, poplar, eucalyptus, agricultural waste, and forest residue	Microalgae, cyanobacteria, macroalgae	Genetically modified algae, genetically engineered crops sugarcane, switchgrass, miscanthus
Different products			
Bioethanol, biodiesel	Ethanol, bio-oils, biogas	Bioethanol, biodiesel, biogas, biohydrogen	Enhance biofuels yield with specific characteristics,
Advantages			
Well-established technology and infrastructure	Reduced food: fuel competition, utilize marginal lands and lower emission of greenhouse gases	Sustainable, high productivity, uses wastewater and atmospheric CO_2 as well as industrial flue gas.	Increase biomass production, yield, crop resilience, carbon capture and reduce pretreatment cost
Disadvantages			
Compete with food production, negative environmental impacts	Higher operational cost, complex and expensive processes of conversion, scalability problem, and technological issue	Exorbitant microalgae production cost, expensive processing, variable productivity, challenges related to technology and scalability	Ethical, regulatory and societal challenges, potential ecological risks

sustainable energy systems. Therefore, optimizing cultivation practices and enhancing conversion technologies, bioenergy crops can play a significant role in reducing reliance on fossil fuels and mitigating greenhouse gas emissions, thereby contributing to a more sustainable energy future.

10.3 Plant Innate Immunity in Bioenergy Crops

Plant innate immunity consists of two main branches: pattern-triggered immunity (PTI): Activated by the recognition of pathogen-associated molecular patterns (PAMPs) by pattern recognition receptors (PRRs). Effector-triggered immunity (ETI): activated by the detection of specific pathogen effectors by plant resistance (R) proteins. For example, pattern-triggered immunity (PTI) in switchgrass involves the plant's recognition of microbial-associated molecular patterns (MAMPs), such as the elicitor flg22, which triggers the activation of defense responses (Edwards et al. 2023). Edwards et al. (2023) study measured reactive oxygen species (ROS) burst profiles in response to flg22 across a diverse panel of switchgrass genotypes, revealing significant variation in PTI responses (Edwards et al. 2023). The genetic analysis identified loci associated with these responses and showed that PTI responses, characterized by different modes of ROS burst (magnitude, timing, and acute vs. gradual responses), have significant narrow-sense heritability (Edwards et al. 2023). Furthermore, the genetic covariances between PTI responses and the composition of root-associated microbiota suggest that the plant's immune response and microbial community structure are interconnected, with significant positive and negative correlations observed between specific PTI axes and microbial abundances. This indicates a complex genetic interplay between switchgrass immunity and its root microbiome (Edwards et al. 2023). This symbiotic relationship between a robust immune system and a healthy microbiome can contribute to increased biomass yield, making switchgrass a more viable and productive bioenergy crop. Effector-triggered immunity (ETI) is a critical second line of defense in plants, activated when pathogen effectors are recognized by resistance (R) genes in the plant. This recognition leads to a robust immune response, often involving localized programmed cell death known as the hypersensitive response, which limits pathogen spread and confers resistance. In switchgrass, ETI plays a vital role in defending against pathogens, including those that secrete necrotrophic effectors which trigger cell death and facilitate pathogen colonization (Songsomboon et al. 2022).

10.3.1 Enhancing Disease Resistance

By enhancing disease resistance, plants can allocate more resources to growth rather than defense. Overexpressing PRR or R proteins in bioenergy crops can significantly enhance their resistance to pathogens, thereby improving their productivity and sustainability. The VpTNL1 gene, encoding a TIR-NB-ARC-LRR protein, was found to confer increased resistance to powdery mildew and bacterial pathogens in transgenic Arabidopsis and tobacco plants (Wen et al. 2017). Drawing inspiration from the successful overexpression of the VpTNL1 gene from wild Chinese grapevine in Arabidopsis and tobacco, similar strategies can be applied to crops like switchgrass, poplar, and Miscanthus (Wen et al. 2017). This genetic modification can bolster the plants' immunity against fungi and bacteria, ensuring robust growth and higher yields, which are essential for sustainable biofuel production. Molecular approaches, including CRISPR/Cas9, offer advanced tools for enhancing plant immunity against pathogens. Compared to earlier techniques like zinc finger nuclease (ZFN) and transcription activator-like effector nucleases (TALENs), CRISPR/Cas9 provides efficient and precise genome editing capabilities (Zaynab et al. 2020). CRISPR/Cas9 technology has revolutionized genetic manipulation in crops, offering a versatile platform for targeted genome editing. By guiding a DNA endonuclease with RNA, CRISPR/Cas9 can create specific modifications in the plant genome to confer resistance against bacteria, viruses, insects, and fungi. Few studies have explored the use of the CRISPR/Cas9 method to combat bacterial diseases. In one study, CRISPR-Cas9 was employed to mutate the OsSWEET13 gene in rice, providing resistance to bacterial blight caused by the γ-proteobacteria pv. Oryzae (Zaynab et al. 2020). This technology opens up new possibilities for studying plant–pathogen interactions and developing bioenergy crop varieties with enhanced resistance.

10.3.2 Manipulating Hormonal Pathways

Plant hormones are a complex series of chemicals that regulate meristematic cell division and cell elongation necessary processes involved in plant growth and development (Depuydt and Hardtke 2011). These chemical signals control the synthesis of microtubules and cell plates, the deposition of cell wall constituents, remodeling all essential processes for growth and the subsequent buildup of biomass. Scientists during the green revolution selected cultivars with shorter height and higher grain yields, but with lower levels of endogenous hormones such as auxin (Vanneste and Friml 2009). These hormones are important growth regulators that act at the cellular and developmental levels. In order to promote proper growth and development, it is crucial for plants to create and maintain adequate amounts of phytohormones. Therefore, modifying their biosynthesis or signaling intermediates will provide useful points of control that can be leveraged to modify crop biomass and architecture.

Plant hormones like salicylic acid (SA), jasmonic acid (JA), and ethylene play crucial roles in immunity and growth. Balancing these pathways can optimize growth (Table 10.2) and defense.

10.3.2.1 SA Pathway

Generally, salicyclic acid is associated with systemic acquired resistance (SAR) and responses to biotrophic pathogens. Applying salicylic acid (SA) to switchgrass (Panicum virgatum L.) has shown significant benefits in enhancing germination and seedling growth parameters (Özyazıcı et al. 2023). A study by Özyazıcı et al. (2023)

Table 10.2 Different phytohormones and their role under favorable and unfavorable conditions

Phytohormone	Role under favorable conditions	Role under stress conditions	References
Gibberellic acid	Increase photosynthetic efficiency cell division and regulate growth and increase stem elongation	Reduce growth in order to allow plant adapt and survive, increase grain yield, maintain hormonal homeostasis, regulate the metabolic pathways and antioxidative enzymes	(Albacete et al. 2014; Ashfaq and Khan 2017; Fahad et al. 2014; Sharma et al. 2015; Babenko et al. 2015; Liu et al. 2020; Wang et al. 2021a)
Cytokinins	Improve plant growth, nutrient mobilization, biomass survival and increase photosynthetic capacity	Increase growth, heat shock proteins and stimulate stomatal closure, Slow leaf senescence, reduces membrane damage, and prevent productivity loss	
Auxins	Cell elongation, vascular differentiation, Carbohydrate mobilization and utilization, improve plant health	Promote stress-induced morphological response, induce ROS detoxification enzymes, modulate expression of stress-responsive genes	
Abscisic acid	Improve cellular growth under normal circumstances,	stomatal closure to reduce transpiration, water loss and eventually restrict cellular growth; enhance photosynthesis, and translocation of assimilates	
Salicylic acid	Protect against pathogenic organisms, improve plant growth	Activate innate defense response, enhance physiological processes and improve tolerance. Provide tolerance toward drought, chilling, heat and metal stress; increased concentration of indole acetic acid and decreased abscisic acid concentration	
Ethylene	Seed germination, root, stem and plant growth, fruit ripening	Promote leaf senescence, closure of stomata during drought conditions, enhance water uptake, maintain ion Homeostasis, activate pathogenesis-related genes	
Jasmonic acid	Regulate growth, enhance production of secondary metabolites, stimulate activity against pathogens, protect plant by stumulating systemic acquired resistance	Activate protective mechanisms, modulate expression of different genes, increase enzymatic activity and interaction with other pathways to enhance resistance and resilience	

demonstrated that SA seed priming at various concentrations significantly improved germination percentage, mean germination time, germination index, and seedling vigor index across three switchgrass cultivars (Özyazıcı et al. 2023). The study by Luo et al. (2019) on the integrated transcriptome analysis in poplar reveals that plant hormones jasmonic acid (JA) and salicylic acid (SA) play crucial roles in coordinating growth and defense responses upon fungal infection (Luo et al. 2019). This research identified 943 common responsive genes involved in stress responses, metabolism, and growth. The study highlights the distinct gene expression profiles mediated by JA and SA, indicating their differential effects on fungal defense mechanisms.

10.3.2.2 JA Pathway: Linked to Resistance Against Necrotrophic Pathogens and Herbivores

Fungal necrotrophic pathogens cause significant crop losses globally and can infect a wide variety of plants, including bioenergy crops such as switchgrass, poplar, and miscanthus. The perception of these pathogens by plant-specific receptors triggers a complex signaling network involving mitogen-activated protein kinase (MAPK) cascades, which play a pivotal role in activating both hormone-dependent and hormone-independent defense responses. Hormones like jasmonates (JA) and ethylene (ET) are key players in the plant's defense against necrotrophic pathogens. Upon infection, JA and ET signaling pathways are activated, leading to the expression of defense-related genes that encode for antifungal proteins, such as plant defensins, and enzymes involved in the synthesis of defensive secondary metabolites like phytoalexins, which inhibit pathogen growth. MAPK cascades integrate signals from pathogen-associated molecular patterns (PAMPs) and various stress signals, leading to rapid defense responses, including cell wall strengthening and the production of reactive oxygen species (ROS) that inhibit pathogen growth. For bioenergy crops, enhancing these signaling pathways can significantly improve resistance to necrotrophic pathogens, leading to healthier plants and higher biomass yields. Healthier bioenergy crops can allocate more resources to growth rather than defense, resulting in increased biomass production. This optimization directly translates to more raw material for bioenergy production and reduces the need for chemical fungicides, making bioenergy crop production more sustainable and environmentally friendly. Moreover, reducing crop losses due to fungal pathogens ensures a more stable and reliable supply of biomass for bioenergy, which is crucial for the economic viability of bioenergy as a renewable energy source.

10.3.3 Exploiting Systemic Acquired Resistance (SAR)

Systemic acquired resistance (SAR) is a pivotal defense mechanism in bioenergy crops, offering long-lasting protection against a broad spectrum of pathogens (Fig. 10.1). SAR can be described as a sophisticated "priming" mechanism where a localized infection or application of specific agents prepares the entire plant for future attacks. This response not only enhances the plant's immunity but also helps

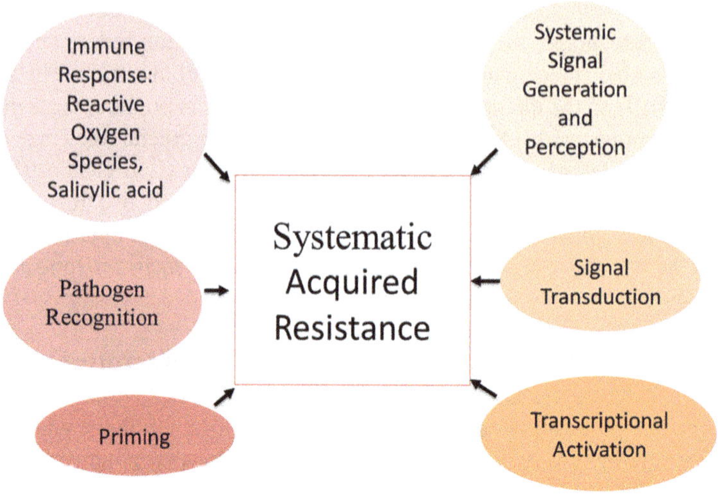

Fig. 10.1 Factors responsible for activation of systemized acquired resistance in plants

maintain the overall health and productivity of bioenergy crops, which are often grown in extensive monocultures and thus more vulnerable to widespread pathogen attacks. The implementation of SAR in bioenergy crops can significantly reduce reliance on chemical pesticides, promoting more sustainable agricultural practices and ensuring a stable supply of biomass for bioenergy production.

One effective approach to induce SAR in bioenergy crops is the application of chemical priming agents such as benzothiadiazole (BTH). BTH mimics the natural defense signaling molecules in plants, activating SAR without adversely affecting plant growth. Studies have demonstrated that BTH application in bioenergy crops can enhance resistance to various pathogens, leading to healthier plants and higher biomass yields. By employing chemical priming agents, farmers can proactively boost the immune systems of bioenergy crops, ensuring a robust defense against potential outbreaks of diseases, thereby safeguarding the biomass yield critical for bioenergy production.

Salicylic acid (SA) plays a central role in SAR, being synthesized and accumulated in response to pathogen attacks. This accumulation leads to the activation of pathogenesis-related (PR) genes, which bolster the plant's defenses. SA and its functional analogs can induce SAR, providing broad-spectrum resistance and generating phloem-mobile signals that spread the resistance response throughout the plant (ScienceDirect.com 2018, 2019). Recent studies have highlighted N-hydroxypipecolic acid (NHP) as another critical molecule in SAR, contributing to the accumulation of extracellular NAD(P) (eNAD(P)) in local leaf tissues. These mobile signals are quickly transported to systemic leaves, priming the plant for enhanced defense (Nature 2023).

Defense priming and transcriptional reprogramming are key components of SAR, preparing the plant for a faster and stronger response to subsequent pathogen attacks. Specific marker and readout genes are activated during interactions with pathogens, such as Pseudomonas cannabina pv. alisalensis (Nature 2024a), and extensive changes in gene expression occur to bolster both local and systemic defenses (Nature 2017). Other molecular players, such as free radicals, also mediate SAR, although their exact roles need further investigation (ScienceDirect.com 2014). Moreover, protein glycosylation changes during SAR indicate the involvement of glycoproteins in the plant's immune response (ScienceDirect.com 2022a). Understanding these intricate signaling networks is crucial for developing strategies to enhance plant immunity, particularly for crops, by leveraging molecular insights and biotechnological innovations.

Additionally, natural SAR inducers pose a sustainable and eco-friendly alternative to chemical priming. Natural compounds, such as certain plant extracts and beneficial microbes, can trigger SAR in bioenergy crops. These natural inducers not only enhance disease resistance but also contribute to the overall resilience and adaptability of bioenergy cropping systems. Integrating natural SAR inducers into bioenergy crop management practices can lead to reduced input costs, minimized environmental impact, and improved crop health, ultimately supporting the economic viability and sustainability of bioenergy production.

10.4 Engineering Beneficial Microbial Interactions

Inoculating plants with beneficial microbes, such as mycorrhizal fungi and plant growth-promoting rhizobacteria (PGPR), has emerged as a promising strategy to enhance plant immunity and biomass production (Table 10.3). These microbes form symbiotic relationships with plants, providing various benefits that positively impact plant health and growth. For example, mycorrhizal fungi improve nutrient uptake, particularly phosphorus, and water absorption, while also triggering systemic resistance in plants, thereby enhancing their ability to fend off pathogens (Smith and Read 2010). Similarly, PGPR can induce systemic resistance in plants, promoting the production of defense-related enzymes and metabolites (Pieterse et al. 2014). This dual mechanism of nutrient enhancement and disease resistance contributes to improved plant immunity and overall health.

Furthermore, the interaction between plants and beneficial microbes can significantly increase biomass production. PGPR, for instance, facilitate plant growth through various mechanisms, including nitrogen fixation, phosphate solubilization, and the production of phytohormones such as auxins, cytokinins, and gibberellins. These activities promote root growth and enhance nutrient uptake, leading to improved plant growth and biomass (Bhattacharyya and Jha 2012). Co-inoculation strategies, such as combining arbuscular mycorrhizal fungi with phosphate-solubilizing bacteria, have demonstrated enhanced growth and production in plants. This synergy between different microbial species leverages their complementary traits to improve phosphorus availability and overall plant health (Taktek et al. 2021). Inoculating bioenergy plants with beneficial microbes offers a sustainable and effective approach to enhance plant immunity and biomass production. The

Table 10.3 Microbial community interaction with different plants/crops

Crop/Plant	Microbes	Role under different conditions	References
Wheat	*Azospirillum brasilense*	Phytohormone production	(Spaepen 2014)
	Azospirillum brasilense	Biofilm formed against infection	(Assmus et al. 1995)
	B. subtilis	Growth promotion	(Egorshina et al. 2012)
	Pseudomonas strains	enhanced growth; Spike length	(Iqbal and Hasnain 2013)
	Burkholderia phytofirmans	Increase yield by enhancing photosynthetic rate, water-use efficiency and chlorophyll content	(Naveed et al. 2014)
Rice	*Herbaspirillum seropedicae*	Increased yield	(Baldani et al. 2000)
	Herbaspirillum strains	Fix nitrogen	(Elbeltagy et al. 2001)
Sorghum	Phosphate-solubilizing microbes	Disease resistance	(Rizvi et al. 2021)
	Bacillus sp.	Phytoremediation	(Luo et al. 2012)
	Enterobacter sp. *UYSB89 and Kosakonia* sp. *UYSB139*	promotes growth by producing IAA	(Heijo et al. 2021)
Sugarcane	*G. diazotrophicus*	Plant protection by stimulating defense system	(Arencibia et al. 2006)
	Burkholderia australis	Increased yield	(Paungfoo-Lonhienne et al. 2014)
Poplar	*Enterobacter* sp.	Increased biomass	(Rogers et al. 2012)
Miscanthus seedlings	*Herbaspirillum frisingense*	Growth promotion	(Straub et al. 2013)
	Clostridium, Enterobacter	Induces salinity tolerance	(Ye et al. 2005)
Switch grass	*Burkholderia phytofirmans*	Enhances growth	(Kim et al. 2012)

synergistic interactions between mycorrhizal fungi, PGPR, and other beneficial microbes play a crucial role in improving plant health and productivity. Further research and field trials are needed to optimize these microbial consortia for various crops and environmental conditions, ultimately contributing to more sustainable and productive agricultural practices.

Designing synthetic microbial communities tailored for bioenergy crops presents a promising avenue for addressing climate change and ensuring global food security. By harnessing the principles of synthetic biology, researchers can engineer consortia with specific traits that enhance plant growth, nutrient uptake, and biomass production. For instance, the use of cross-feeding strains in microbial consortia can lead to improved system performance and robustness, surpassing the capabilities of single-strain biomanufacturing (Nature 2024b). These synthetic communities can efficiently convert organic matter into biofuels or other valuable products, offering sustainable solutions for bioenergy crop cultivation. Synthetic biology also enables the optimization of microbial consortia to thrive in specific environmental conditions and support various crop types. For example, synthetic light-driven microbial consortia provide a versatile platform for microbial growth and bio-chemicals production, which can benefit bioenergy crop cultivation (ScienceDirect.com 2023). Moreover, computational modeling of metabolism in microbial communities enhances the design and optimization of synthetic communities for optimal benefits in bioenergy crops (ScienceDirect.com 2021). This approach allows researchers to predict and optimize the performance of synthetic communities under different conditions, leading to more efficient and sustainable bioenergy production. Additionally, the division of labor for substrate utilization in natural and synthetic microbial communities can enhance nutrient cycling and biomass production in bioenergy crops. By engineering microbial communities with specific metabolic pathways and interactions, researchers can create synthetic communities that efficiently convert organic matter into biofuels or other valuable products (ScienceDirect.com 2022b). Overall, designing synthetic microbial communities for bioenergy crops holds significant promise for enhancing agricultural sustainability, mitigating the effects of climate change, and ensuring global food and energy security.

10.5 Breeding and Biotechnology

Combining traditional breeding techniques with modern biotechnology (Fig. 10.2) can develop crops with enhanced biomass production. Marker-assisted selection (MAS) involves identifying specific DNA markers that are associated with beneficial traits in crops, such as high biomass yield, disease resistance, and enhanced energy content. By using these markers, breeders can screen and select plants with the desired traits at an early stage, significantly accelerating the breeding process compared to traditional methods. This approach not only saves time but also increases the accuracy of selecting superior genotypes for further cultivation and breeding programs. Research has shown that the development and refinement of genomic-enabled breeding methods for energycane can substantially improve the efficiency of breeding these bioenergy crops (Olatoye et al. 2019). Through MAS, breeders can enhance traits such as biomass yield and stress tolerance, making these crops more viable for large-scale bioenergy production. In forestry, MAS has been utilized to accelerate the breeding of species for bioenergy purposes by identifying DNA markers linked to traits like fast growth and high biomass accumulation (Ahmar et al. 2021). Genomic prediction and MAS have been applied to high biomass sorghum, demonstrating the superiority of these methods over traditional selection techniques in terms of accuracy and efficiency (de Oliveira et al. 2018). This has led to the development of sorghum varieties with improved biomass yield

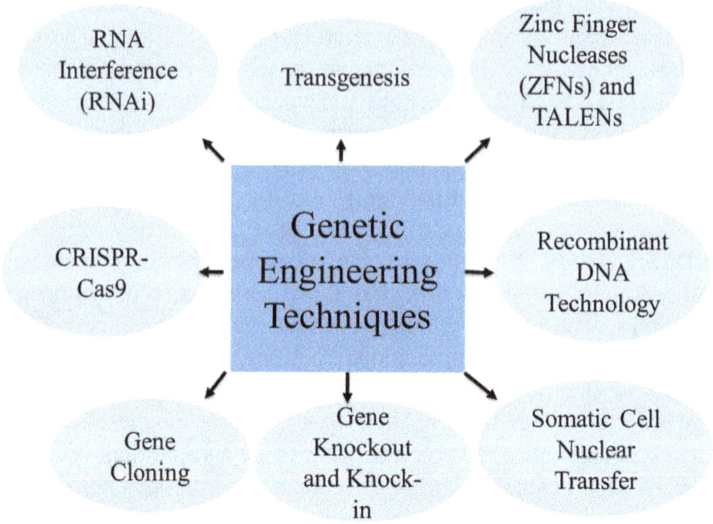

Fig. 10.2 Different genetic engineering techniques for crop modification

and bioenergy potential. Despite its advantages, MAS faces several challenges, including the need for extensive genetic mapping and the identification of reliable markers for all desired traits. Additionally, the integration of MAS with other breeding techniques, such as genomic selection and genome editing, is essential to fully realize its potential. As genomic technologies advance, the efficiency and precision of MAS in bioenergy crop breeding are expected to improve. This will contribute to the development of more robust and productive bioenergy crops, ultimately supporting the sustainable production of biofuels.

Genetic engineering has emerged as a transformative tool in enhancing biomass production and promoting sustainability in forestry. By introducing or modifying specific genes, scientists can significantly improve traits such as biotic and abiotic stress tolerance, wood quality, root formation, and phytoremediation capabilities in trees. This approach not only supports the sustainable production of bioenergy but also helps maintain biodiversity and mitigate the effects of climate change. For instance, genetic engineering techniques have enabled the introduction of genes that confer resistance to pests, diseases, and environmental stresses such as drought and salinity, which are crucial for developing resilient tree species that can thrive in various climatic conditions and continue to provide raw materials for bioenergy production (Parmar et al. 2017). Additionally, advancements in tree genetic engineering include enhancing wood quality and promoting better root formation, which are essential for sustainable forestry and bioenergy purposes. Genetic modifications can lead to improved growth rates and higher biomass yields, which are vital for both timber and bioenergy purposes. Moreover, genetic engineering has also been applied to horticultural crops, offering numerous applications in improving biotic and abiotic stress tolerance and enhancing produce quality. Over the past two decades, a significant number of transgenic horticultural crops have been developed, showcasing the potential of genetic modifications in this field. Genes such as natural and synthetic Cry genes, protease inhibitors, trypsin inhibitors, and cystatin genes have been incorporated to develop insect and nematode-resistant crops. Furthermore, genes like chitinase, glucanase, osmotin, and defensin have been used to provide resistance against fungal and bacterial diseases (Harfouche et al. 2010). Genetic modifications have also been utilized to improve various traits related to quality, texture, and shelf-life of horticultural crops. Techniques like RNA interference (RNAi) and CRISPR/Cas9 have been applied to modify traits such as color, increase shelf-life, and reduce post-harvest losses. The integration of genetic engineering with traditional breeding techniques and advanced genome editing tools like CRISPR/Cas9 holds great promise for the future, enabling the development of superior crop cultivars with enhanced stress tolerance and improved quality traits (Zhou and Luo 2013).

10.6 Integrated Pest Management (IPM)

Implementing IPM strategies can reduce pest pressures and enhance bioenergy crop biomass. Biological control using natural predators and parasitoid can be done. Cultural Practices: crop rotation, intercropping, and proper spacing can minimize pest outbreaks in bioenergy crops. Biological control through habitat management is a sustainable method for managing insect pests in agricultural crops. This approach involves enhancing the presence of natural enemies, such as predators and parasitoids, by creating favorable habitats through practices like crop rotation, intercropping, and maintaining diverse landscapes. For example, plant characteristics such as flower shape, color, and blooming period can provide essential resources like nectar and pollen to natural enemies, increasing their longevity and fecundity. Agricultural practices, including tillage, crop rotation, and intercropping, can significantly influence these natural enemies, improving their effectiveness in controlling insect pests. The push–pull strategy, which utilizes semiochemicals like herbivore-induced plant volatiles (HIPVs), is one innovative method that prevents pest infestations while attracting natural enemies (Akter et al. 2019). This method exemplifies how manipulating the agricultural landscape and plant characteristics can enhance biological pest control.

Conservation agriculture (CA) practices also play a crucial role in biological control by maintaining ecosystem functions and biodiversity. CA emphasizes minimal soil disturbance, permanent soil cover, and crop diversification, which collectively improve soil health and reduce the negative impacts of conventional, resource-intensive farming practices. These practices alter moisture and temperature regimes, potentially affecting insect pest populations and their natural enemies. Studies have shown that the rotation-intercropping ecosystem, such as wheat-maize-cotton systems, increases the abundance of dominant natural enemies, like Propylea japonica, thereby promoting pest control (Ouyang et al. 2014). Additionally, temporal crop diversification in cereal-based systems has been found to influence the presence and activity of generalist predators, enhancing their biocontrol potential (Puliga et al. 2019). Intercropping flowering plants in orchards, for instance, supports predators by providing shelter, nectar, and alternative prey, thus facilitating effective pest control (Zhang et al. 2021). These findings underscore the importance of integrating habitat management and conservation agriculture practices to enhance biological control and reduce pesticide use in agricultural systems.

10.7 Future Directions

The future direction for increasing biomass production in bioenergy crops involves leveraging advancements in omics technologies to enhance plant innate immunity and stress tolerance. Proteomics, combined with genomics and other omics branches, accelerates gene discovery and crop improvement (Eldakak et al. n.d.). Future efforts will focus on building comprehensive omics databases and employing these data to breed next-generation bioenergy crops with improved biomass production. Omics technologies facilitate marker-assisted selection, gene discovery, and expression profiling, which are crucial for enhancing crop traits (Pathak et al. 2018). These technologies enable the identification and introgression of essential traits, improving crop resilience to abiotic and biotic stresses. Applying omics approaches to identify stress-responsive genes and pathways in crops like soybean, and developing stress-tolerant varieties through genetic engineering and breeding strategies, will ensure sustainable bioenergy production. Future research will aim to enhance the yield and stress tolerance of bioenergy crops, reducing the environmental impact of agriculture and supporting bioenergy sustainability (Altaf et al. 2023). Integrating data from various biological levels to create comprehensive models for predicting and enhancing biomass production is crucial for advancing bioenergy crops. Current bioenergy crop models, such as those reviewed by Nair et al., simulate field-scale production by incorporating factors like leaf area dynamics, phenology, radiation interception, and biomass partitioning. However, these models often require further refinement in parameterization and validation due to the scarcity of detailed data on climate, soil, and crop management practices. The development of high-quality field data and more sophisticated models can significantly improve predictions of biomass yields and help in planning sustainable bioenergy systems (Nair et al. 2012). Future directions should focus on creating integrated modeling frameworks that combine multiple data sources and methodologies. Welfle et al. (2020) emphasize the importance of using multiple models in parallel to capture the complexity of bioenergy systems. Integrated assessment models (IAMs) and energy system models, though widely used in policy development, often miss the nuances of specific bioenergy processes and environmental impacts. Combining these with specialist bioenergy models can provide a more holistic understanding of biomass production and its sustainability, thus offering more robust conclusions for policy and practical applications (Welfle et al. 2020).

Advancements in technology, particularly machine learning (ML), offer new opportunities to enhance the accuracy of bioenergy models. According to Wang et al. (2021b) ML can handle the complexity of bioenergy systems by classifying, regressing, and optimizing various biological and environmental variables. The application of ML in predicting bioenergy outcomes, such as lignocellulosic biofuels and microalgae cultivation, demonstrates its potential to boost research and development in this field. Integrating ML algorithms with large, high-resolution geo-referenced datasets can improve predictions and optimize biomass production, leading to more efficient and sustainable bioenergy systems (Wang et al. 2021b).

10.8 Conclusion

Harnessing plant innate immunity presents a promising, sustainable, and efficient strategy for enhancing biomass production in bioenergy crops. By leveraging the natural defense mechanisms of plants, we can develop crops that are more resistant to biotic and abiotic stresses, ultimately leading to higher yields and more reliable production systems. This approach aligns with the principles of sustainable agriculture, reducing the reliance on chemical inputs and promoting environmental health. Research has demonstrated that manipulating immune signaling pathways and resistance genes can significantly improve crop resilience. These genetic interventions can enhance the plant's ability to withstand pathogens, pests, and environmental challenges, thereby safeguarding biomass yields. Additionally, integrating advanced technologies such as genome editing and synthetic biology can further optimize these immune responses, tailoring them to specific crop needs and local growing conditions. The future of bioenergy crop production lies in multidisciplinary efforts that combine plant immunology, genomics, and bioengineering. By fostering collaborations between scientists, policymakers, and industry stakeholders, we can accelerate the development and deployment of immune-enhanced bioenergy crops. This will not only contribute to meeting the growing demand for renewable energy but also support sustainable agricultural practices that are crucial for global food security and environmental conservation. Embracing plant innate immunity as a core component of bioenergy crop improvement strategies promises a resilient and productive future for bioenergy systems.

References

Abdul Aziz M, Brini F, Rouached H, Masmoudi K (2022) Genetically engineered crops for sustainably enhanced food production systems. Front Plant Sci 13:1027828. https://doi.org/10.3389/fpls.2022.1027828

Ahmar S, Ballesta P, Ali M, Mora-Poblete F (2021) Achievements and challenges of genomics-assisted breeding in forest trees: From marker-assisted selection to genome editing. Int J Mol Sci, MDPI. https://doi.org/10.3390/ijms221910583

Akter MS, Siddique SS, Momotaz R, Arifunnahar M, Alam KM, Mohiuddin SJ (2019) Biological control of insect pests of agricultural crops through habitat management. Plant Breeding Division, Bangladesh Agricultural Research Institute, Dhaka, Bangladesh. https://doi.org/10.4236/jacen.2019.81001

Albacete AA, Martínez-Andújar C, Pérez-Alfocea F (2014) Hormonal and metabolic regulation of source–sink relations under salinity and drought: from plant survival to crop yield stability. Biotechnol Adv 32(1):12–30. https://doi.org/10.1016/j.biotechadv.2013.10.005

Altaf MT, Liaqat W, Iqbal J, Baig MMA, Ali A, Nadeem MA, Baloch FS (2023) Exploring omics approaches to enhance stress tolerance in soybean for sustainable bioenergy production. In: Aasim M, Baloch FS, Nadeem MA, Habyarimana E, Ahmed S, Chung G (eds) Biotechnology and omics approaches for bioenergy crops. Springer, Singapore. https://doi.org/10.1007/978-981-99-4954-0_7

Arencibia AD, Vinagre F, Estevez Y, Bernal A, Perez J, Cavalcanti J, Santana I, Hemerly AS (2006) Gluconoacetobacterdiazotrophicuselicitate a sugarcane defense response against a pathogenic bacteria Xanthomonasalbilineans. Plant Signal Behav 1(5):265–273. https://doi.org/10.4161/psb.1.5.3390

Arif I, Batool M, Schenk PM (2020) Plant microbiome engineering: expected benefits for improved crop growth and resilience. Trends Biotechnol 38(12):1385–1396. https://doi.org/10.1016/j.tibtech.2020.04.015

Ashfaq M, Khan S (2017) Role of phytohormones in improving the yield of oilseed crops. Oilseed Crops:165–183. https://doi.org/10.1002/9781119048800.ch9

Assmus B, Hutzler P, Kirchhof G, Amann R, Lawrence JR, Hartmann A (1995) In situ localization of Azospirillumbrasilense in the rhizosphere of wheat with fluorescently labeled, rRNA-targeted oligonucleotide probes and scanning confocal laser microscopy. Appl Environ Microbiol 61(3):1013–1019. https://doi.org/10.1128/aem.61.3.1013-1019.1995

Babenko LM, Kosakivska IV, Skaterna TD (2015) Jasmonic acid: role in biotechnology and the regulation of plants biochemical processes. Biotechnol Acta 8(2):36–51

Baldani VLD, Reis FB, Silva LG, Reis VM, Teixeira KDS, Döbereiner J (2000) Biological nitrogen fixation (BNF) in non-leguminous plants: the role of endophytic diazotrophs. In: Pedrosa FO, Hungria M, Yates G, Newton WE (eds) Nitrogen fixation: from molecules to crop productivity. Springer, Dordrecht, pp 397–400. https://doi.org/10.1007/0-306-47615-0_216

Barot S (2022) Biomass and bioenergy: resources, conversion and application. In: Nayan Kumar P (ed) Renewable Energy for Sustainable Growth Assessment. Wiley, pp 243–262. https://doi.org/10.1002/9781119785460.ch9

Bhattacharyya PN, Jha DK (2012) Plant growth-promoting rhizobacteria (PGPR): emergence in agriculture. World J Microbiol Biotechnol 28(4):1327–1350. https://doi.org/10.1007/s11274-011-0979-9

Bigeard J, Colcombet J, Hirt H (2015) Signaling mechanisms in pattern-triggered immunity (PTI). Mol Plant 8(4):521–539. https://doi.org/10.1016/j.molp.2014.12.022

Boostani HR, Chorom M, Moezzi AA, Enayatizamir N (2014) Mechanisms of plant growth promoting rhizobacteria (PGPR) and mycorrhizae fungi to enhancement of plant growth under salinity stress: a review. Sci J Biol Sci 3(11):98–107

Dalla Costa L, Malnoy M, Lecourieux D, Deluc L, Ouaked-Lecourieux F, Thomas M, Torregrosa LJM (2019) The state-of-the-art of grapevine biotechnology and new breeding technologies (NBTS). Oeno One 53(2):189–212. https://doi.org/10.20870/oeno-one.2019.53.2.2405

de Oliveira AA, Pastina MM, de Souza VF (2018) Genomic prediction applied to high-biomass sorghum for bioenergy production. Molecular Breeding, Springer. https://doi.org/10.1534/genetics.108.098277

Depuydt S, Hardtke CS (2011) Hormone signalling crosstalk in plant growth regulation. Curr Biol 21(9):R365–R373. https://doi.org/10.1016/j.cub.2011.03.013

Edwards JA, Saran UB, Bonnette J, MacQueen A, Yin J, uyen Nguyen T, Schmutz J, Grimwood J, Pennacchio LA, Daum C, Del Rio TG (2023) Genetic determinants of switchgrass-root-associated microbiota in field sites spanning its natural range. Curr Biol 33(10):1926–1938. https://doi.org/10.1016/j.rser.2024.114369

Egorshina AA, Khairullin RM, Sakhabutdinova AR, Luk'Yantsev MA (2012) Involvement of phytohormones in the development of interaction between wheat seedlings and endophytic Bacillus subtilis strain 11BM. Russ J Plant Physiol 59:134–140. https://doi.org/10.1134/S1021443711050062

Elbeltagy A, Nishioka K, Sato T, Suzuki H, Ye B, Hamada T, Isawa T, Mitsui H, Minamisawa K (2001) Endophytic colonization and in planta nitrogen fixation by a Herbaspirillum sp. isolated from wild rice species. Appl Environ Microbiol 67(11):5285–5293. https://doi.org/10.1128/AEM.67.11.5285-5293.2001

Eldakak M, Milad SIM, Nawar AI, Rohila JS (n.d.) Proteomics: a biotechnology tool for crop improvement. Department of Biology and Microbiology, South Dakota State University

Fahad S, Nie L, Chen Y, Wu C, Xiong D, Saud S et al (2014) Crop plant hormones and environmental stress. Sustain Agric Rev:371–400. https://doi.org/10.1007/978-3-319-0

Harfouche A, Meilan R, Altman A (2010) Tree genetic engineering and applications to sustainable forestry and biomass production. Trends Biotechnol. https://doi.org/10.1016/j.tibtech.2010.09.003

Heijo G, Taulé C, Mareque C, Stefanello A, Souza EM, Battistoni F (2021) Interaction among endophytic bacteria, sweet sorghum (Sorghum bicolor) cultivars and chemical nitrogen fertilization. FEMS Microbiol Ecol 97(2):fiaa245. https://doi.org/10.1093/femsec/fiaa245

Hönig M, Roeber VM, Schmülling T, Cortleven A (2023) Chemical priming of plant defense responses to pathogen attacks. Front Plant Sci 14:1146577. https://doi.org/10.3389/fpls.2023.1146577

Iqbal A, Hasnain S (2013) Auxin producing Pseudomonas strains: biological candidates to modulate the growth of Triticum aestivum beneficially. Am J Plant Sci 4(09):1693. https://doi.org/10.4236/ajps.2013.49206

Kim S, Lowman S, Hou G, Nowak J, Flinn B, Mei C (2012) Growth promotion and colonization of switchgrass (Panicum virgatum) cv. Alamo by bacterial endophyte Burkholderiaphytofirmans strain PsJN. Biotechnol Biofuels 5:1–10. https://doi.org/10.1186/1754-6834-5-37

Li Q, Song J, Peng S, Wang JP, Qu GZ, Sederoff RR, Chiang VL (2014) Plant biotechnology for lignocellulosic biofuel production. Plant Biotechnol J 12(9):1174–1192

Liu Y, Zhang M, Meng Z, Wang B, Chen M (2020) Research progress on the roles of cytokinin in plant response to stress. Int J Mol Sci 21(18):6574. https://doi.org/10.3390/ijms21186574

Luo S, Xu T, Chen L, Chen J, Rao C, Xiao X, Wan Y, Zeng G, Long F, Liu C, Liu Y (2012) Endophyte-assisted promotion of biomass production and metal-uptake of energy crop sweet sorghum by plant-growth-promoting endophyte Bacillus sp. SLS18. Appl Microbiol Biotechnol 93:1745–1753. https://doi.org/10.1007/s00253-011-3483-0

Luo J, Xia W, Cao P, Xiao ZA, Zhang Y, Liu M, Zhan C, Wang N (2019) Integrated transcriptome analysis reveals plant hormones jasmonic acid and salicylic acid coordinate growth and defense responses upon fungal infection in poplar. Biomolecules 9(1):12. https://doi.org/10.3390/biom9010012

Mehmood MA, Ibrahim M, Rashid U, Nawaz M, Ali S, Hussain A, Gull M (2017) Biomass production for bioenergy using marginal lands. Sustain Prod Consump 9:3–21. https://doi.org/10.1016/j.spc.2016.08.003

Nair SS, Kang S, Zhang X, Miguez FE, Izaurralde RC, Post WM, Dietze MC, Lynd LR, Wullschleger SD (2012) Bioenergy crop models: descriptions, data requirements, and future challenges. Glob Change Biol Bioenergy 4(6):620–633. https://doi.org/10.1111/j.1757-1707.2012.01166.x

Nature (2017, Dec 13) The transcriptome of Arabidopsis thaliana during systemic acquired resistance

Nature (2023, Oct 27) N-hydroxypipecolic acid triggers systemic acquired resistance through extracellular NAD(P)

Nature (2024a, Feb 12) Marker and readout genes for defense priming in Pseudomonas cannabina pv. alisalensis interaction aid

Nature (2024b, Feb 14) Choreographing root architecture and rhizosphere interactions through synthetic biology

Naveed M, Mitter B, Reichenauer TG, Wieczorek K, Sessitsch A (2014) Increased drought stress resilience of maize through endophytic colonization by BurkholderiaphytofirmansPsJN and Enterobacter sp. FD17. Environ Exp Bot 97:30–39. https://doi.org/10.1016/j.envexpbot.2013.09.014

Ndimba BK, Ndimba RJ, Johnson TS, Waditee-Sirisattha R, Baba M, Sirisattha S, Shiraiwa Y, Agrawal GK, Rakwal R (2013) Biofuels as a sustainable energy source: an update of the applications of proteomics in bioenergy crops and algae. J Proteome 93:234–244. https://doi.org/10.1016/j.jprot.2013.05.041

Olatoye MO, Clark LV, Wang J, Yang X, Yamada T et al (2019) Evaluation of genomic selection and marker-assisted selection in Miscanthus and energycane. Mol Breeding, Springer. https://doi.org/10.1007/s11032-019-1081-5

Ouyang F, Su W, Zhang Y, Liu X (2014) Ecological control service of the predatory natural enemy and its maintaining mechanism in rotation-intercropping ecosystem via wheat-maize-cotton. Ecosyst Reg Serv J. https://doi.org/10.1016/j.ecoserv.2014.05.003

Özyazıcı G, Açıkbaş S, Özyazıcı MA (2023) Effects of salicylic acid priming application in some switchgrass (Panicum virgatum L.) cultivars. Int J Nat Life Sci 7(2):137–146. https://doi.org/10.47947/ijnls.1400366

Parmar N, Singh KH, Sharma D, Singh L, Kumar P, Nanjundan J, Khan YJ, Chauhan DK, Thakur AK (2017) Genetic engineering strategies for biotic and abiotic stress tolerance and quality enhancement in horticultural crops: A comprehensive review. 3 Biotech 7:239. https://doi.org/10.1007/s13205-017-0943-x

Pathak RK, Baunthiyal M, Pandey D, Kumar A (2018) Augmentation of crop productivity through interventions of omics technologies in India: challenges and opportunities. 3 Biotech 8:454. https://doi.org/10.1007/s13205-018-1473-y

Paungfoo-Lonhienne C, Lonhienne TG, Yeoh YK, Webb RI, Lakshmanan P, Chan CX, Lim PE, Ragan MA, Schmidt S, Hugenholtz P (2014) A new species of B urkholderia isolated from sugarcane roots promotes plant growth. Microb Biotechnol 7(2):142–154. https://doi.org/10.1111/1751-7915.12105

Pieterse CM, Zamioudis C, Berendsen RL, Weller DM, Van Wees SC, Bakker PA (2014) Induced systemic resistance by beneficial microbes. Annu Rev Phytopathol 52:347–375. https://doi.org/10.1146/annurev-phyto-082712-102340

Puliga GA, Thiele J, Ahnemann H, Dauber J (2019) Effects of temporal crop diversification of a cereal-based cropping system on generalist predators and their biocontrol potential. Thünen Institute of Biodiversity. https://doi.org/10.1016/j.biocontrol.2019.03.007

Rial RC (2024) Biofuels versus climate change: Exploring potentials and challenges in the energy transition. Renew Sust Energ Rev 196:114369. https://doi.org/10.1016/j.rser.2024.114369

Rizvi A, Ahmed B, Khan MS, Umar S, Lee J (2021) Sorghum-phosphate solubilizers interactions: crop nutrition, biotic stress alleviation, and yield optimization. Front Plant Sci 12:746780. https://doi.org/10.3389/fpls.2021.746780

Rogers A, McDonald K, Muehlbauer MF, Hoffman A, Koenig K, Newman L, Taghavi S, Van Der Lelie D (2012) Inoculation of hybrid poplar with the endophytic bacterium E nterobacter sp. 638 increases biomass but does not impact leaf level physiology. GCB Bioenergy 4(3):364–370. https://doi.org/10.1111/j.1757-1707.2011.01119.x

ScienceDirect.com (2014, Apr 24) Report free radicals mediate systemic acquired resistance

ScienceDirect.com (2018, Jan 3) Signaling mechanisms underlying systemic acquired resistance to microbial pathogens

ScienceDirect.com (2019, Mar 19) Chemical elicitors of systemic acquired resistance—Salicylic acid and its functional analogs

ScienceDirect.com (2021, Apr 21) Computational modeling of metabolism in microbial communities on a genome-scale

ScienceDirect.com (2022a, May 25) Protein glycosylation changes during systemic acquired resistance in Arabidopsis thaliana

ScienceDirect.com (2022b, Mar 4) Division of labor for substrate utilization in natural and synthetic microbial communities

ScienceDirect.com (2023, Jun 15) Design and optimization of artificial light-driven microbial consortia for the sustainable growth and biosynthesis of 2 …

Sharma E, Sharma R, Borah P, Jain M, Khurana JP (2015) Emerging roles of auxin in abiotic stress responses. In: Pandey G (ed) Elucidation of abiotic stress signaling in plants. Springer, New York, NY, pp 299–328. https://doi.org/10.1007/978-1-4939-2211-6_11

Shyu C, Brutnell TP (2015) Growth–defence balance in grass biomass production: the role of jasmonates. J Exp Bot 66(14):4165–4176. https://doi.org/10.1093/jxb/erv011

Smith SE, Read DJ (2010) Mycorrhizal Symbiosis. Academic Press

Songsomboon K, Crawford R, Crawford J, Hansen J, Cummings J, Mattson N, Bergstrom GC, Viands DR (2022) Genome-wide associations with resistance to bipolaris leaf spot (Bipolaris oryzae (Breda de Haan) Shoemaker) in a northern Switchgrass population (Panicum virgatum L.). Plants 11(10):1362. https://doi.org/10.3390/plants11101362

Spaepen S (2014) Plant hormones produced by microbes. In: Lugtenberg B (ed) Principles of Plant-Microbe Interactions. Springer, pp 247–256. https://doi.org/10.1007/978-3-319-08575-3_26

Stoof CR, Richards BK, Woodbury PB, Fabio ES, Brumbach AR, Cherney J, Das S, Geohring L, Hansen J, Hornesky J, Mayton H (2015) Untapped potential: opportunities and challenges for sustainable bioenergy production from marginal lands in the Northeast USA. BioEnergy Res 8:482–501. https://doi.org/10.1007/s12155-014-9515-8

Straub D, Yang H, Liu Y, Tsap T, Ludewig U (2013) Root ethylene signalling is involved in Miscanthus sinensis growth promotion by the bacterial endophyte Herbaspirillumfrisingense GSF30T. J Exp Bot 64(14):4603–4615. https://doi.org/10.1093/jxb/ert276

Taktek S, Trabelsi D, Ben Ammar H, Ben Salah I (2021) Combined inoculation with arbuscular mycorrhizal fungi and phosphate solubilizing bacteria for improving phosphorus availability and plant growth promotion. J Agric Sci Technol 23(5):1251–1266

van Butselaar T, Van den Ackerveken G (2020) Salicylic acid steers the growth–immunity tradeoff. Trends Plant Sci 25(6):566–576. https://doi.org/10.1016/j.tplants.2020.02.002

Vanneste S, Friml J (2009) Auxin: a trigger for change in plant development. Cell 136:1005–1016. https://doi.org/10.1016/J.CELL.2009.03.001

Wang Y, Mostafa S, Zeng W, Jin B (2021a) Function and mechanism of jasmonic acid in plant responses to abiotic and biotic stresses. Int J Mol Sci 22(16):8568. https://doi.org/10.3390/ijms22168568

Wang Z, Peng X, Xia A, Shah AA, Huang Y, Zhu X, Zhu X, Liao Q (2021b) The role of machine learning to boost the bioenergy and biofuels conversion. Bioresour Technol 319:126099. https://doi.org/10.1016/j.biortech.2021.126099

Welfle A, Thornley P, Röder M (2020) A review of the role of bioenergy modelling in renewable energy research & policy development. Biomass Bioenergy 136:105542. https://doi.org/10.1016/j.biombioe.2020.105542

Wen Z, Yao L, Singer SD, Muhammad H, Li Z, Wang X (2017) Constitutive heterologous overexpression of a TIR-NB-ARC-LRR gene encoding a putative disease resistance protein from wild Chinese Vitis pseudoreticulata in Arabidopsis and tobacco enhances resistance to phytopathogenic fungi and bacteria. Plant Physiol Biochem 112:346–361. https://doi.org/10.1016/j.plaphy.2017.01.017

Yadav P, Priyanka P, Kumar D, Yadav A, Yadav K (2019) Bioenergy crops: recent advances and future outlook. In: Rastegari AA, Yadav AN, Gupta A (eds) Prospects of renewable bioprocessing in future energy systems. Springer, pp 315–335. https://doi.org/10.1007/978-3-030-14463-0_12

Ye B, Saito A, Minamisawa K (2005) Effect of inoculation with anaerobic nitrogen-fixing consortium on salt tolerance of Miscanthus sinensis. Soil Sci Plant Nutr 51(2):243–249. https://doi.org/10.1111/j.1747-0765.2005.tb00028.x

Zaynab M, Sharif Y, Fatima M, Afzal MZ, Aslam MM, Raza MF, Anwar M, Raza MA, Sajjad N, Yang X, Li S (2020) CRISPR/Cas9 to generate plant immunity against pathogen. Microb Pathog 141:103996. https://doi.org/10.1016/j.micpath.2020.103996

Zhang X, Ouyang F, Su J, Li Z, Yuan Y, Sarkar SC, Xiao Y, Ge F (2021) Intercropping flowering plants facilitate conservation, movement and biocontrol performance of predators in insecticide-free apple orchard. Ecol Appl. https://doi.org/10.1002/eap.2241

Zhou M, Luo H (2013) MicroRNA-mediated gene regulation: potential applications for plant genetic engineering. Plant Mol Biol 83:59–75. https://doi.org/10.1007/s11103-013-0049-7

11 Exogenous Elicitors as Inducers of Environmental Stress Tolerance in Wheat (*Triticum aestivum* L.) Crop

Anjali Yadav and Shachi Singh

Abstract

The most significant factors that limit crop productivity and yield are abiotic and biotic stresses. Abiotic stress, such as high temperature, drought, soil salinity, and biotic stress in the form of bacterial, fungal, and viral pathogen impose challenges in crop production. To combat these challenges, crops exhibit a range of morphological, physiological, biochemical, and molecular responses. Numerous types of external stressors or stimuli cause plant cells to go through a complicated web of reactions, which eventually result in the synthesis and buildup of defensive metabolites. Exogenous application of these chemical stimuli, called as elicitors have shown promising results in crop protection. These molecules help the plants to develop resistance against environmental stress. Wheat, an important staple crop cultivated worldwide, is also affected by various stresses. Numerous methods have been applied to reduce the harmful effect of environmental stress on wheat productivity. Among which, elicitors have shown to play a significant role in inducing tolerance in wheat crop against biotic and abiotic stress. This chapter summarizes the research work carried out in the past few years, demonstrating the role of elicitors in wheat production.

Keywords

Abiotic stress · Biotic stress · Crop protection · Elicitors · Wheat

A. Yadav · S. Singh (✉)
Department of Botany, MMV, Banaras Hindu University, Varanasi, UP, India

S. Singh, R. Mehrotra (eds.), *Plant Elicitor Peptides*,
https://doi.org/10.1007/978-981-97-6374-0_11

11.1 Introduction

A healthy plant may suffer number of physical (flooding, drought, salinity, heavy metal toxicity, chilling, and heat), chemical (fertilizer, pesticides), and biological (pathogen and parasites) stressful events in its environments, which affect their metabolism and homeostasis, ultimately having negative impact on the crop growth, development, and yield (Nadeem et al. 2023) (Fig. 11.1). Wheat is an important component of the human diet especially in developing country like India. Wheat crop is strongly affected by several biotic and abiotic stresses, which decrease its production in terms of quality and quantity (Yadav and Singh 2023). In the coming year demand for the stress-tolerant crop varieties is going to increases due to the pressure on global food productivity and changing climatic conditions (Yadav et al. 2024). It has been reported that wheat genome can be modified to improve their tolerance against stress. For example, transcription factor (HaHB4) of *Helianthus annuus* (sunflower) was inoculated in wheat to increase water use efficiency, yield and drought, and salt tolerance (González et al. 2019). GmDREB1 gene was isolated from soybean and transferred in wheat to secure salt tolerance (Shiqing et al. 2005). Although, transgenic plants were developed to solve specific problems of agriculture and food security, but they have also raised many concerns in recent years. Issues for the transgenic crops are uncertain and controversy remains constant (Rani and Usha 2013). Secondary metabolites, synthesized by plants in response to various chemical stimuli, called elicitors, have emerged as promising method for enhancing crop resilience against biotic and abiotic stresses (Singh 2014). These compounds exhibit diverse chemical structures and functions, playing pivotal roles in scavenging reactive oxygen species (ROS), regulating osmotic balance, modulating antioxidant systems, and orchestrating stress-responsive signaling pathways within plants (Upadhyay et al. 2012) (Fig. 11.2). Therefore, this study is

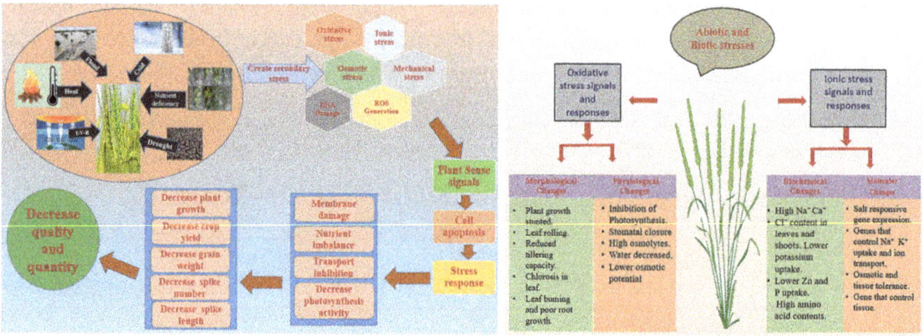

Fig. 11.1 Schematic representation of the impact of environmental stress (*Left*) and their effect on oxidative and ionic stress signals (*Right*)

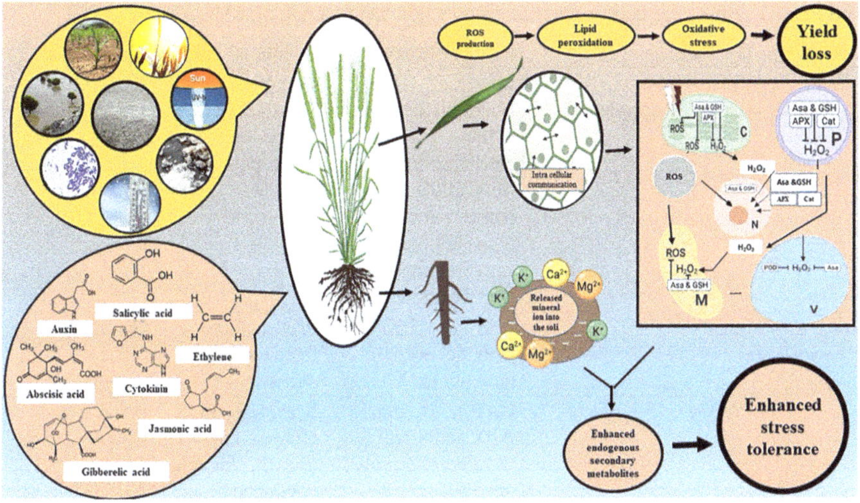

Fig. 11.2 Defense strategies adopted by the wheat plants through various phytohormones

aimed to elucidate the potential of exogenous elicitors as agents against biotic and abiotic stresses in wheat crops.

11.2 Salt Stress

Although it has been established that soil salinity existed before humans and agriculture, the issue has only recently been brought about by agricultural activities like irrigation through saline water (Zhu 2001). Salt stress is linked to decrease germination percentage, reduced growth, altered reproductive behavior, degrade enzymatic activity, disrupted photosynthesis, damage to the ultrastructure of cellular components, hormonal imbalance, and oxidative stress, which frequently has a negative impact on wheat productivity. Under salt stress, soil becomes hyperosmotic and hyperionic (ion toxicity), causing lower water potential and excessive sodium ions formation (Hajihashemi et al. 2009). Convenient methods for enhancing plant performance is to use exogenous phytoprotectants, seed priming, nutrient management, and application of plant hormones (Hasanuzzaman et al. 2017). The chemical compounds proline, glycine betadine (GB), amino acids, and sugars protect plant organelles by scavenging ROS from the osmotic equilibrium of cells (ur Rehman et al. 2021) (Table 11.1). It was found that exogenous spraying of GB benefits wheat crops under salt stress. When GB (100 mM) was sprayed, wheat plants' relative water content under salt stress was improved (Khedr et al. 2022). The inhibition of wheat crop development caused by salt was partially alleviated by exogenous melatonin pretreatment, as determined by biomass, IAA content, leaf photosynthesis rate, maximal photochemistry efficiency, and reduced H_2O_2 buildup (Ke et al. 2018). Salicylic acid (SA) also has been studied for its potential to mitigate salt stress in

Table 11.1 Effect of elicitors, their mode of action and function on wheat crop

Stress	Elicitors	Mode of action	Function	References
Salt	Chitooligosaccharides (COSs) + acetylation (DAs)	Lowered Malondialdehyde (MDA) concentrations, Intracellular ion concentration and boost antioxidant enzymes activities were regulated	Improved photosynthetic efficiency and chlorophyll content	Zhu (2001)
	Sugars	CAT and APX activities were upregulated	Protected crop from oxidative stress	Guo et al. (2015)
	Proline	CAT, SOD, and APX activities were considcrably upregulated	Scavenged ROS, increased plant growth, stomatal conductance and increased CO_2 assimilation rate	Bhaskar and Bingru (2014)
	Glycine betain	Activities of POD, SOD, and CAT significantly increased	Water use efficiency and membrane stability was Increased	
	Trehalose	POD, CAT, APX, and GR activities were considerably increased	Maintained osmotic balance and, K^+ accumulation and K^+/Na^+ ratio	Turan et al. (2012)
	Auxin	Improved ion homeostasis, Increased $CO_{2,}$ and net assimilation rate	Productive tiller, grain yield, and grain weight were Increased	Shaddad et al. (2013)
	Abscisic acid (ABA)	Decreased Na uptake	Percentage germination Increased tiller number, biological yield	Gurmani et al. (2013)
	Potassium sulphate (K_2SO_4),	Increased K^+ uptake and decreased Na^+ uptake	Increased plant height and biomass	Kausar and Gull (2014)
	K_2O	Increased SOD and POD activity	Increased photosynthetic pigments	El-Lethy et al. (2013)
	$CaSO_4$	Increase K^+ and Ca^{++} uptake	Increased root and leaf area	Zaman et al. (2005)

(continued)

Table 11.1 (continued)

Stress	Elicitors	Mode of action	Function	References
	Ca $(NO_3)_2$	Decreased lipid peroxidation, electrolyte leakage, O_2^- and H_2O_2	Increased spike number and length	Tian et al. (2005)
	Ascorbic acid (AsA)	Decreased Na^+ content, and increased K^+ and Ca_2^+ content	Improved photosynthesis machineries	Athar et al. (2007)
	Ascorbin (ascorbic and citric acid 2:1)	Decreased MDA content and Increased carotenoid contents	Enhanced tolerance	Elhamid et al. (2014)
Heat	Salicylic acid (SA)	Heat shock proteins (HSPs), Adenosine triphosphate (ATP) synthase, Rubisco activase (RCA), and Calcium dependent protein kinase (CDPK), were increased, superoxide dismutase (SOD) and osmolyte were accumulated	Protected the nascent proteins from denaturation, Enhanced carbon assimilatory process, regulated signaling process and increased yield	Murakami et al. (2002), Clarke et al. (2004)
	Ascorbic acid	Accumulated osmolyte (proline), decreased H_2O_2 content and increased endogenous ascorbic acid	Protected pollen development, embryogenesis Inhibit lipid peroxidation	Leja et al. (2007)
	Arginine	Decreased peroxidase (POD) and polyphenol-oxidase (PPO) enzyme activities reduced MDA level	Increased water use potential	Khalil et al. (2009)
	Putrescine	Reduced thiobarbituric acid reactive substances (TBARS), increased ascorbate and tocopherol in grains	Protected membrane integrity in root and shoot	Asthir et al. (2012)

(continued)

Table 11.1 (continued)

Stress	Elicitors	Mode of action	Function	References
	α-tocopherol	Increased superoxide dismutase	Prevented dehydration, protected cellular membranes	Kumar et al. (2013)
	PGPR (Bacillus amyloliquefaciens UCMB5113 and Azospirillum brasilense NO40 strains)	Reduced ROS generation	Increased heat tolerance in wheat seedlings, enhanced yield	Abd El-Daim et al. (2014)
Drought	Chitosan nanoparticles (NPs) (especially 90 ppm)	Increased catalase (CAT), and superoxide dismutase (SOD) activities	Increased leaf area, relative water content, chlorophyll content, yield, and biomass	Behboudi et al. (2019)
	Silicon (seed priming)	Increased antioxidant activities and non-enzymatic antioxidants. Facilitated erectness of leaves, thus increased the light interception	Increased water uptake by roots, decreased water loss from leaves plant growth and ultimately yield	Ahmed et al. (2016)
	PEG-6000 (Polyethylene glycols)	Decreased available water required for seed germination and plant growth causes osmotic stress and drought stimulator	Increased biomass and shoot length	Datta et al. (2011)
	Polyamines (Putrescin, spermine, and spermidine)	Promoted ethylene, nucleic acids and protein production Produced conjugated and bound polyamines and stimulated polyamine oxidation	Provided resistance to drought stress	Bouchereau et al. (1999)
	Hydropriming and Osmopriming, in aerated $CaCl_2$ (1.5% solution)	Reduced MDA accumulation lipid peroxidation and osmolytes, scavenged ROS, accumulated stress responsive proteins	Provided adverse effects of drought stresses and increased yield and production	Tabassum et al. 2018

(continued)

Table 11.1 (continued)

Stress	Elicitors	Mode of action	Function	References
	Arbuscular mycorrhizal fungi (*Glomus etunicatum, Glomus mosseae, Glomus etunicatum,* and *Glomus mosseae*)	Maintained soil moisture conditions. Increased phosphorus and iron content	Improved growth, yield and nutrient uptake by plant	Al-Karaki et al. (2004)
	PGPR (*Bacillus amyloliquefaciens* 5113 and *Azospirillum brasilense* NO40)	Regulated ascorbate cycle	Improved growth, yield and production	Kasim et al. (2013)
	Pseudomonas azotoformans FAP5	Improved photosynthetic pigment efficiency and other physiological attributes	Improved growth attributes provide benefit to plant in drought stress condition	Ansari et al. (2021)
	Foliar application of Salicylic acid (SA) and zinc (Zn)	Increased the abscisic acid content, proline accumulation	Improved drought tolerance, regulated stomata and Reduced the detrimental effects of drought	Yavas and Unay (2016)
Cold	PGPR (Psychrophilic Bacillus spp. CJCL2, RJGP41) and temperate Bacillus. velezensis FZB42	Regulated abscisic acid, lipid peroxidation and proline accumulation pathways	Provided resistance to cold stress	Zubair et al. (2019)
	Selenium	Increased antioxidant capacity and biomass accumulation Reduced production of free radicals	Reduced chilling effect and enhanced crop yield	Chu et al. (2010)

(continued)

Table 11.1 (continued)

Stress	Elicitors	Mode of action	Function	References
	Betaine	Improved tolerance to photo-inhibition of Photosystem II (PSII) and the steady-state yield of electron transport over PSII in a manner that mimicked cold-acclimated plants	Improved freezing tolerance by more than 5 °C	Allard et al. (1998)
	Melatonin	Elevated SOD activity, enhanced total glutathione content, increased the ratio of *Glutathione/ Glutathione disulphide*	Increase Cold tolerance scavenged ROS by modulating redox balance and other defense mechanisms	Marta et al. (2016)
	Abscisic acid (ABA)	Decreased the amount of H_2O_2 and relative conductivity Decreased ROS production	Regulated cold tolerance Increased yield	Yu et al. (2020)
	Salicylic acid (SA)	Increased the abscisic acid content, proline accumulation	Alleviated the damaging effects Improved seedlings growth and increased masses and lengths	Rakhmankulova et al. (2010)
	Methyl jasmonate (MeJA)	Enhanced antioxidant and soluble protein content	Improved cold tolerance Reduced moisture	Qi et al. (2005).
	Jasmonic acid (JA)	Motivated the germination of dormant seeds and stimulate a transient rise	Decreased ABA content, improve tolerance	Xu et al. (2016)
Water logging	Ethylene	Increased photosynthesis	Improved stomatal conductance	Iqbal et al. (2011)

(continued)

Table 11.1 (continued)

Stress	Elicitors	Mode of action	Function	References
	Salicylic acid (SA)	Developed shallow root system able to survive in water logging condition	Lateral roots development was enhanced, emergence of surface adventitious roots	
	Foliar Nitrogen application	Increased enzymes activities such as nitrate reductase (NR), glutamine synthetase, and glutamic-pyruvic transaminase	Improved respiratory activity, photosynthetic rate, stomatal conductance, and transpiration rate	Wu et al. (2014)
Heavy metal	PGPR (Pseudomonas moraviensis + Bacillus cereus)	Produced indole-acetic acid (IAA)	Decreased cadmium (Cd) content of the rhizosphere soil by 21%. Decreased Ni content by 38%, Cu by 35%, and Cr by 51% in wheat root Decreased Cd by 24%, Co by 44%, Cr by 45% and Cu by 25% in wheat leaves	Hassan et al. (2017)
	PGPR (Bacillus subtilis BM2)	Significant drop in proline and MDA content, and antioxidant enzymes activity, like CAT, SOD, and GR	Enhanced the length by 14% dry biomass of shoots by 23% and grain yield by 49% (at 195 mg Pb kg^{-1}, 870 mg Ni kg^{-1} and 585 mg Pb kg^{-1})	Rizvi et al. (2019)

(continued)

Table 11.1 (continued)

Stress	Elicitors	Mode of action	Function	References
	PGPR (Pseudomonas aeruginosa CPSB1)	Declined the levels of CAT, GR, and SOD, proline and MDA, and reduced metal uptake by wheat	Root dry biomass of inoculated plants was enhanced by 44% at 2007 mg Cu kg^{-1}, by 28% and at 36 mg Cd kg^{-1} and by 48% at 204 mg Cr kg^{-1}, respectively. This bioinoculant enhanced number of spikes by 25%, grain by 17% and straw yields by 12%	Rizvi and Khan (2017)
	Polyamines (Especially Putrescine)	Improved activities of POD, GR, AAO, PPO, SOD, and CAT, DPPH, and reducing power capacity	Relieved the adverse effects of heavy metals in wheat crop	Taie et al. (2019)
Biotic stress	Chitosan	Elevated expression of chitinase gene	Provided protection against *Fusarium graminearum* and *Microdochium nivale*	Xing et al. 2015), Hofgaard et al. (2005)
	Ulvans (sulfated heteropolysaccharides)	Generation of hydrogen peroxide (H2O2)	Heightened resistance against powdery mildew	Paulert et al. (2010)
	Trichoderma harzianum (Fungus)	Generation of hydrogen peroxide (H2O2)	Stimulate resistance against *Fusarium graminearum*	Saravanakumar et al. (2017)

(continued)

Table 11.1 (continued)

Stress	Elicitors	Mode of action	Function	References
	Iodus40 (Laminarins, derived from brown marine algae)	Accumulation of hydrogen peroxide (H2O2) at the fungal penetration site, leading to a reduction in haustorium formation	Shielding wheat plants against powdery mildew	Renard-Merlier et al. (2007)
	Ozone	Activating the SA/ jasmonic acid (JA) signaling pathway	Exposure has been found to induce resistance to and *Blumeria graminis f.* sp. *tritici*	Pazarlar et al. (2017)
	Aluminum (Al)	Promoting better balance of reactive oxygen species (ROS) and salicylic acid (SA) signaling	Enhances tolerance against Fusarium oxysporum, respectively	Banerjee and Roychoudhury (2017)

wheat crops. The synergistic effects of salinity and seed soaking in varying salicylic acid concentrations on the growth and physiological characteristics of wheat was investigated by Kaydan et al. (2007). Soaking the seeds in SA enhanced the percentage of emergence, dry weight of the shoots and roots, osmotic potential, photosynthetic pigments (Chlorophyll a, b, and carotenoids), and K+/Na+ ratio. Another study found that applying selenium and humic acid mixtures to saline soil increased plant resistance to salinity conditions and increased the percentage of germination as well as the length of the shoot and root and their dry matter contents. Pretreatment of H_2O_2 directly to roots or seeds prior to the germination had shown to induce resistance against salinity in wheat crop (Li et al. 2011). Sodium nitroprusside (SNP), acting as a nitric oxide (NO) donor, facilitated improved salt tolerance in wheat by boosting levels of antioxidant enzymes, proline, ascorbic acid, phenolic compounds and stimulating the activity of beta-amylase (Duan et al. 2007). Priming seeds with sodium hydrosulfide (NaHS) notably alleviated the inhibition of seed germination and seedling growth in a salt-sensitive wheat cultivar, as evidenced by improvements in germination rate, germination index, vigor index, and seedling growth (Bao et al. 2011). Tian et al. (2022) used the TaCEPID peptide elicitor to boost the tolerance of wheat plants to salt, revealing potential functions for small secreted peptides (SSPs) in regulating wheat stress responses (Table 11.1).

11.3 Heat Stress

One of the most notable environmental issues in agriculture is the rise in global temperatures (Deryng et al. 2014). Drought and heat stresses pose significant limitations to plant growth and productivity, surpassing the impact of other environmental factors. For example, a mere 1 °C increase in temperature can result in a reduction of global wheat production by approximately 6% (Asseng et al. 2015). Researchers conducted a climate impact assessment to direct their thinking about various adaptation strategies. In cases of extreme heat stress, they investigated cultivar selection and adjusting sowing dates as adaptation strategies. Other adaption strategies include osmo protectants, antioxidant defense (Deryng et al. 2014), and surface cooling via irrigation (Lobell et al. 2008). In accordance with Singh et al. (2011b), the enhancement of wheat productivity under elevated temperature conditions was achieved through various interventions. These interventions included foliar application of potassium nitrate (KNO3) at concentrations of 0.5% during the 50% flowering stage and 1.0% during the anthesis stage. Additionally, the application of 2.5 mM arginine, zinc foliar spray, supplementation of additional irrigation water during the grain filling stage, and the utilization of potassium fertilizers in conjunction with municipal wastewater were employed. Foliar spray of potassium orthophosphate (KH_2PO_4) was found to be an alternative method to delay the senescence of leaves and increases grain production under heat stress (Dias and Lidon 2010). The benefits of NO_3^- in delaying the production of abscisic acid and stimulating cytokinin activity and potassium ion, increasing photosynthetic activity and the storage of assimilates are well known for increasing grain yield in heat-stressed environments (Singh et al. 2011a). Afzal et al. (2020) demonstrated that the exogenous application of natural herbal extract of moringa leaf extract along with sorghum water extract ameliorated heat stress effects in wheat plants. Such applications improved growth and development, metabolite accumulation, relative water content, and minimize oxidative damage. Conversely, Khan et al. (2013) demonstrated that SA treatment under heat stress conditions enhances proline metabolism, nitrogen assimilation, and photosynthesis. Notably, SA's interaction with proline metabolism and ethylene formation emerged as crucial mechanisms in alleviating heat stress-induced inhibition of photosynthesis in wheat. In another study conducted by Clarke et al. (2004), it was observed that the application of SA resulted in the upregulation of various molecular components crucial for heat stress tolerance, specifically, heat shock proteins (HSPs), oxygen evolving enhancer protein, and calcium dependent protein kinase (CDPK). Furthermore, the study revealed an accumulation of superoxide dismutase (SOD) and osmolytes, which collectively contributed to the preservation of nascent proteins by preventing denaturation. Moreover, SA treatment was found to modulate the photosynthetic activity, enhance carbon assimilatory processes, regulate signaling pathways, and ultimately augment yield. Kumar et al. (2011) investigated the effect of pre-anthesis treatment with 400 mM ascorbic acid on wheat pollens subjected to heat stress (42 °C, 2 h). They observed that ascorbic acid treatment significantly enhanced thermotolerance capacity, as evidenced by biochemical markers. Additionally, Almeselmani et al. (2009)

demonstrated that heat stress-induced growth inhibition and chlorosis in wheat were correlated with decreased leaf water status and heightened oxidative stress. Remarkably, exogenous application of ascorbic acid partly alleviated these effects, suggesting its role in protecting against heat stress.

11.4 Drought Stress

Drought stress has a significant impact on the vegetative and reproductive phases of plants (Zhu 2001). There are many factors that can affect crop performance when it comes to tolerance to water stress such as plant genotype, growth stage (Ingram and Bartels 1996), stress intensity and duration, physiological growth processes, various gene expression patterns respirational activity patterns of the photosynthesis machinery (Flexas et al. 2004), and environmental factors (McDonald and Davies 1996). Additionally, drought can result in pollen sterility, grain loss, and abscisic acid accumulation in spikes and synthesis in the anthers (Ji et al. 2010). In the pursuit of enhancing drought stress tolerance in wheat (*Triticum aestivum* L.), researchers have employed various methods to simulate drought conditions for experimental studies. An earlier study found that silica gel with a 1.5% silicon concentration would be a viable priming choice to grow plants under drought stress (Ahmed et al. 2016). Several significant studies have highlighted the resilience to drought stress achieved through pre-exposure to H_2O_2 (Choudhary et al. 2024). Seed priming of wheat with salicylic acid (SA) resulted in significant enhancements in moisture content, dry matter yield, antioxidant activity, and total chlorophyll content under conditions of drought stress (Singh and Usha 2003). Conversely, selenium priming was observed to augment drought tolerance in rapeseed through the regulation of both enzymatic and non-enzymatic antioxidant activities. Hydrogen sulfide (H_2S) enhances the drought tolerance of wheat seedlings by regulating various biochemical pathways, including those related to energy and carbon metabolism, signal transduction, antioxidant capacity, and protein synthesis (Ding et al. 2018). Datta et al. (2011) investigated the efficacy of polyethylene glycol (PEG-6000) as a means to induce drought stress tolerance. Bacterial priming has demonstrated a beneficial effect on enhancing drought stress tolerance in wheat, resulting in boosted plant biomass and improved survival rates of plants cultivated under severe drought conditions (Timmusk et al. 2014). Haggag Wafaa et al. (2017) used bio-elicitors; *Acremonium coenophiulum, Streptomyces griseus, Trichoderma harzanium, Trichoderma viride, Rhodotorula glutinis, Paenibacillus polymyxa, Bacillus subtilis, B. megaterium, Pseudomonas putida, and P. fluorescens* and proved that wheat plants can display improved stress tolerance through bio-elicitors.

11.5 Cold Stress

Low temperatures also affect the crop, causing a number of changes in its morphological physiological, biochemical, and molecular makeup. Crop plants use their cold-tolerant systems to adapt these changes, which include storing soluble carbohydrates, signaling molecules, and cold tolerance gene expressions (Hassan et al. 2021). An essential plant hormone, abscisic acid (ABA), plays a pivotal role in modulating cold tolerance. Treatment with exogenous ABA significantly enhanced the rate of regreening, suggesting that ABA treatment may ameliorate cold tolerance by augmenting the activity of antioxidant enzymes (Yu et al. 2020). Previous investigations have illustrated that dormant embryo, upon transfer to 20 °C, exhibit upregulation of jasmonic acid (JA) biosynthesis genes and demonstrate cold-stimulated germination of dormant wheat grains. Acetylsalicylic acid (ASA), an inhibitor of JA production, impedes cold-stimulated germination in a dose-dependent manner. Moreover, exploration into the interplay between JA and the well-established dormancy promoter, ABA, in cold-induced dormancy regulation reveals a negative correlation between JA and ABA levels in dormant wheat embryos post-stratification. Upon stratification, ABA content precipitously declines, with this decline being arrested upon supplementation with ascorbic acid (AsA) (Xu et al. 2016). The application of exogenous melatonin mitigates oxidative damage caused by cold stress by increasing the activity of antioxidant enzymes in wheat crop (Li et al. 2018). Furthermore, plant growth-promoting rhizobacteria (PGPR) have been utilized to mitigate cold stress and confer tolerance to wheat crops (Mishra et al. 2011).

11.6 Water Logging Stress

Water logging is an occurrence that negatively impacts upland crops like wheat in downstream area regions, significantly reducing their production. Waterlogging can slow down photosynthesis, cause oxidative stress, speed up the aging process in leaves, impede plant growth and ultimately lower agricultural yields (Yamauchi et al. 2014). Earlier research demonstrated that pre-anthesis waterlogging priming might significantly reduce wheat yield loss caused by post-anthesis waterlogging. According to Mirshel et al. (2005), a shoot's limited access to nitrogen caused waterlogging-related reduced vegetative growth. Waterlogging during the early stages of the crop delayed the tillering period in wheat, favoring the emergence of higher-order tillers, as claimed by Robertson et al. (2009). SA has been recognized as a potential agent for enhancing waterlogging stress tolerance in wheat crops. Application of SA (1 mM) improved wheat tolerance to waterlogging and its promotion of axile root formation (Koramutla et al. 2022). Wu et al. (2014) tested foliar application of nitrogen and their results demonstrate a significant increase in the activities of nitrate reductase (NR), glutamine synthetase, and glutamic-pyruvic transaminase. Moreover, enhanced wheat root respiratory activity, photosynthetic rate, stomatal conductance, and transpiration rate were observed following foliar N treatment.

11.7 Heavy Metal Stress

Industrial wastes, which come from industries like mining, chemistry, metal processing, etc. are a major cause of soil pollution. These wastes contain a wide range of compounds, including organics, heavy metals, etc. (Van Assche and Clijsters 1990). In India, it is a normal practice to use sewage sludge and industrial effluent on agricultural land, which allows these harmful metals to move from the soil and concentrate in plant tissues. As a result of the long-term application of huge amounts of bovine dung, mineral fertilizers, composts, contaminated marine sludge, or sewage sludge, numerous tiny regions around the nation exhibit increasing levels of various heavy metals (Boekhold 1992). Wheat, the major cereal crop in many countries, is sensitive to heavy metal toxicity. In wheat, exudation of acids (malate) in roots is associated with the tolerance against heavy metals (Taie et al. 2019). It was reported that the growth and yield characteristics as well as the activity of the enzymes leaf peroxidase, glutathione reductase, ascorbic acid oxidase, and polyphenol oxidase in wheat plants were significantly reduced when Cd^{2+} or Pb^{2+} were present in the growth medium. As opposed to controls, significant increases were seen for the Cd^{2+} level in roots, leaves, and grains, superoxide dismutase, ascorbate peroxidase and catalase activities, radical scavenging activity α, α-diphenyl-β-picrylhydrazyl (DPPH), reducing power capacity, and DNA fragmentation. Foliar spraying or seed soaking with polyamines Putrescine, spermine and spermidine, considerably improved the growth and yield characteristics as well as the activities of enzymes in stressed wheat plants. Two bacterial strains, namely, *Enterobacter ludwigii* and *Klebsiella pneumoniae*, showed better growth promotion of wheat seedlings under metal stress conditions (Gontia-Mishra et al. 2016). Another study investigated the efficacy of live and dead cell biomasses of *Variovorax paradoxus* and *Arthrobacter viscosus* in removing Zinc (II) from aqueous solutions (Rizvi et al. 2020).

11.8 Biotic Stress

Various living organisms, including fungi, viruses, insects, nematodes, arachnids, and weeds, contribute to biotic stress in plants. Among these stressors, pathogenic fungi pose a notable threat to global wheat cultivation. Common diseases affecting wheat include stripe rust, stem rust, leaf rust, powdery mildew, and head blight. Additionally, wheat faces challenges from insect pests such as aphids, hessian flies, green bugs, and borers (Bakala et al. 2021). Research indicates that chitin, with a special emphasis on chitosan, has the potential to trigger disease resistance in plants against *Fusarium graminearum* in wheat (Xing et al. 2015). Chitosan application also provided protection against *Microdochium nivale*, (snow mold) by stimulation of host-defense responses, as evidenced by elevated expression of chitinase gene, along with a direct inhibitory impact on the pathogen growth (Hofgaard et al. 2005). Ulvans, which are water-soluble sulfated heteropolysaccharides derived from the cell walls of green marine macroalgae belonging to the genus Ulva, have been observed to trigger defense reactions in wheat against powdery mildew (Paulert et al. 2010). In greenhouse experiments, *Trichoderma harzianum* has been employed to stimulate resistance against *Fusarium graminearum* in wheat crops (Saravanakumar et al. 2017). The exopolysaccharides (EPSs) purified from different PGPR (plant growth-promoting rhizobacteria) have demonstrated to enhance resistance against fungal or bacterial pathogens (Ortmann et al. 2006). Laminarins, derived from brown marine algae such as *Laminaria digitata*, are utilized in various formulations like Iodus 40 and Vacciplant to safeguard plants from diverse pathogens (Renard-Merlier et al. 2007). This protection was attributed to the induction of host-natural defense mechanisms, including the accumulation of hydrogen peroxide at the fungal penetration site, leading to a reduction in haustorium formation. Exposing plants to ozone or UV radiation can prompt responses resembling those to biotic stresses, consequently enhancing plants' ability to resist pathogens. Ozone exposure has been found to induce resistance against Tobacco mosaic virus (TMV) in tobacco and *Blumeria graminis f.* sp. *tritici* (Bgt) in wheat by activating the SA/JA signaling pathway (Pazarlar et al. 2017). The accumulation of metals to high levels provides resistance against pathogens. For instance, in *Thlaspi caerulescens*, a hyperaccumulator of heavy metals, the accumulation of zinc (Zn), nickel (Ni), and cadmium (Cd) inhibits the growth of the bacterial pathogen *Pseudomonas syringae pv. maculicola* through a direct ionic effect. In an experiment, pre-treating plants with aluminum (Al) enhances their tolerance against *Phytophthora infestans* and *Fusarium oxysporum*, by promoting better balance of reactive oxygen species and SA signaling (Banerjee and Roychoudhury 2017).

11.9 Conclusion

In summary, the utilization of various elicitors represents a promising strategy for enhancing stress tolerance in wheat crops. These elicitors trigger the production of secondary metabolites that play pivotal roles in combating both abiotic and biotic stresses. This approach offers a sustainable and eco-friendly alternative to traditional chemical-based methods, contributing to enhanced crop productivity and food security. As we confront the challenges posed by climate change and escalating pest pressures, the exploration and optimization of elicitor-induced secondary metabolites hold significant promise for ensuring the long-term viability of wheat production systems. Continued research efforts aimed at elucidating the underlying mechanisms and refining elicitor applications will be crucial for realizing the full potential of this approach in wheat crop management.

References

Abd El-Daim IA, Bejai S, Meijer J (2014) Improved heat stress tolerance of wheat seedlings by bacterial seed treatment. Plant Soil 379:337–350. https://doi.org/10.1007/s11104-014-2063-3

Afzal I, Akram MW, Rehman HU, Rashid S, Basra SMA (2020) Moringa leaf and sorghum water extracts and salicylic acid to alleviate impacts of heat stress in wheat. S Afr J Bot 129:169–174. https://doi.org/10.1016/j.sajb.2019.04.009

Ahmed M, Qadeer U, Ahmed ZI, Hassan FU (2016) Improvement of wheat (*Triticum aestivum*) drought tolerance by seed priming with silicon. Arch Agron Soil Sci 62(3):299–315. https://doi.org/10.1080/03650340.2015.1048235

Al-Karaki G, McMichael BZAKJ, Zak J (2004) Field response of wheat to arbuscular mycorrhizal fungi and drought stress. Mycorrhiza 14:263–269. https://doi.org/10.1007/s00572-003-0265-2

Allard F, Houde M, Kröl M, Ivanov A, Huner NP, Sarhan F (1998) Betaine improves freezing tolerance in wheat. Plant Cell Physiol 39(11):1194–1202. https://doi.org/10.1093/oxfordjournals.pcp.a029320

Almeselmani M, Deshmukh P, Sairam R (2009) High temperature stress tolerance in wheat genotypes: role of antioxidant defence enzymes. Acta Agronomica Hungarica 57(1):1–14. https://doi.org/10.1556/AAgr.57.2009.1.1

Ansari FA, Jabeen M, Ahmad I (2021) Pseudomonas azotoformans FAP5, a novel biofilm-forming PGPR strain, alleviates drought stress in wheat plant. Int J Environ Sci Technol 18:3855–3870. https://doi.org/10.1007/s13762-020-03045-9

Asseng S, Ewert F, Martre P, R€ Otter RP, Lobell D, Cammarano D et al (2015) Rising temperatures reduce global wheat production. Nat Clim Chang 5(2):143. https://doi.org/10.1038/nclimate2470

Asthir B, Koundal A, Bains NS (2012) Putrescine modulates antioxidant defense response in wheat under high temperature stress. Biol Plant 56:757–761. https://doi.org/10.1007/s10535-012-0209-1

Athar HUR, Khan A, Ashraf M (2007) Exogenously applied ascorbic acid alleviates salt-induced oxidative stress in wheat. Environ Exp Bot 63:224–231. https://doi.org/10.1016/j.envexpbot.2007.10.018

Bakala HS, Mandahal KS, Sarao LK, Srivastava P (2021) Breeding wheat for biotic stress resistance: achievements, challenges and prospects. Curr Trends Wheat Res 12:11–34. https://doi.org/10.5772/intechopen.97359

Banerjee A, Roychoudhury A (2017) Abiotic stress, generation of reactive oxygen species, and their consequences: an overview. In: Reactive oxygen species in plants: boon or bane-revisiting the role of ROS. Wiley, pp 23–50. https://doi.org/10.1002/9781119324928.ch2

Bao J, Ding TL, Jia WJ et al (2011) Effect of exogenous hydrogen sulfiocde on wheat seed germination under salt stress. Modern Agric Sci Technol 20:40–42

Behboudi F, Tahmasebi-Sarvestani Z, Kassaee MZ, Modarres-Sanavy SAM, Sorooshzadeh A, Mokhtassi-Bidgoli A (2019) Evaluation of chitosan nanoparticles effects with two application methods on wheat under drought stress. J Plant Nutr 42(13):1439–1451. https://doi.org/10.1080/01904167.2019.1617308

Bhaskar G, Bingru H (2014) Mechanism of salinity tolerance in plants: physiological, biochemical, and molecular characterization. Int J Genomic 18:701596. https://doi.org/10.1155/2014/701596

Boekhold AE (1992) Field scale behaviour of cadmium in soil. Wageningen University and Research, Wageningen, pp 1–24

Bouchereau A, Aziz A, Larher F, Martin-Tanguy J (1999) Polyamines and environmental challenges: recent development. Plant Sci 140(2):103–125. https://doi.org/10.1016/S0168-9452(98)00218-0

Choudhary R, Rajput VD, Ghodake G, Ahmad F, Meena M, Rehman RU, Prasad R, Sharma RK, Singh R, Seth CS (2024) Comprehensive journey from past to present to future about seed priming with hydrogen peroxide and hydrogen sulfide concerning drought, tempera-

ture, UV and ozone stress-a review. Plant and soil 500:351-373. https://doi.org/10.1007/s11104-024-06499-9
Chu J, Yao X, Zhang Z (2010) Responses of wheat seedlings to exogenous selenium supply under cold stress. Biol Trace Elem Res 136:355–363. https://doi.org/10.1007/s12011-009-8542-3
Clarke SM, Mur LA, Wood JE, Scott IM (2004) Salicylic acid dependent signaling promotes basal thermotolerance but is not essential for acquired thermotolerance in Arabidopsis thaliana. Plant J 38(3):432–447. https://doi.org/10.1111/j.1365-313X.2004.02054.x
Datta JK, Mondal T, Banerjee A, Mondal NK (2011) Assessment of drought tolerance of selected wheat cultivars under laboratory condition. J Agric Technol 7(2):383–393
Deryng D, Conway D, Ramankutty N, Price J, Warren R (2014) Global crop yield response to extreme heat stress under multiple climate change futures. Environ Res Lett 9:1–13. https://doi.org/10.1088/1748-9326/9/3/034011
Dias AS, Lidon FC (2010) Bread and durum wheat tolerance under heat stress: a synoptical overview. Emir J Food Agric 22:412–436
Ding H, Han Q, Ma D (2018) Characterizing physiological and proteomic analysis of the action of H2Sto mitigate drought stress in young seedling of wheat. Plant Mol Biol Report 36:45–57. https://doi.org/10.1007/s11105-017-1055-x
Duan P, Ding F, Wang F, Wang BS (2007) Priming of seeds with nitric oxide donor sodium nitroprusside (SNP) alleviates the inhibition on wheat seed germination by salt stress. J Plant Physiol Mol Biol 33:244–250
Elhamid EMA, Sadak MS, Tawfik MM (2014) Alleviation of adverse effects of salt stress in wheat cultivars by foliar treatment with antioxidant to changes in some biochemical aspects, lipid peroxidation, antioxidant enzymes and amino acid contents. Agric Sci 5:1269–1280
El-Lethy SR, Abdelhamid MT, Reda F (2013) Effect of potassium application on wheat (*Triticum aestivum* L.) cultivars grown under salinity stress. World Appl Sci J 26:840–850. https://doi.org/10.5829/idosi.wasj.2013.26.07.13527
Flexas J, Bota F, Loreto G, Cornic, Sharkey TD (2004) Diffusive and metabolic limitations to photosynthesis under drought and salinity in C3 plants. Plant Biol 6(03):269–279. https://doi.org/10.1055/s-2004-820867
Gontia-Mishra I, Sapre S, Sharma A, Tiwari S (2016) Alleviation of mercury toxicity in wheat by the interaction of mercury-tolerant plant growth-promoting rhizobacteria. J Plant Growth Regul 35:1000–1012. https://doi.org/10.1007/s00344-016-9598-x
González FG et al (2019) Field-grown transgenic wheat expressing the sunflower gene HaHB4 significantly outyields the wild type. J Exp Bot 70(5):1669–1681. https://doi.org/10.1093/jxb/erz037
Guo R et al (2015) Comparative metabolic responses and adaptive strategies of wheat (*Triticum aestivum* L.) to salt and alkali stress. BMC Plant Biol 15:1–13. https://doi.org/10.1186/s12870-015-0546-x
Gurmani AR, Bano A, Najeeb U, Zhang J, Khan SU, Flowers TJ (2013) Exogenously applied silicate and abscisic acid ameliorates the growth of salinity stressed wheat (*Triticum aestivum* L.) seedlings through Na+ exclusion. Aust J Crop Sci 7:1123–1130
Haggag Wafaa M, Tawfik M, Abouziena H, Abd El Wahed M, Ali R (2017) Enhancing wheat production under arid climate stresses using bio-elicitors. Gesunde Pflanz 69(3):149. https://doi.org/10.1007/s10343-017-0399-3
Hajihashemi S et al (2009) Effect of paclobutrazol on wheat salt tolerance at pollination stage. Russ J Plant Physiol 56:251–257. https://doi.org/10.1134/S1021443709020149
Hasanuzzaman M et al (2017) Approaches to enhance salt stress tolerance in wheat. Wheat Improv Manag Utiliz:151–187. https://doi.org/10.5772/67247
Hassan TU, Bano A, Naz I (2017) Alleviation of heavy metals toxicity by the application of plant growth promoting rhizobacteria and effects on wheat grown in saline sodic field. Int J Phytoremediation 19(6):522–529. https://doi.org/10.1080/15226514.2016.1267696
Hassan MA et al (2021) Cold stress in wheat: plant acclimation responses and management strategies. Front Plant Sci 12:676884. https://doi.org/10.3389/fpls.2021.676884

Hofgaard IS, Ergon A, Wanner LA et al (2005) The effect of chitosan and bion on resistance to pink snow mould in perennial ryegrass and winter wheat. J Phytopathol 153:108–119. https://doi.org/10.1111/j.1439-0434.2005.00937.x

Ingram, Bartels D (1996) The molecular basis of dehydration tolerance in plants. Annu Rev Plant Physiol Plant Mol Biol 47:377–403. https://doi.org/10.1146/annurev.arplant.47.1.377

Iqbal N, Nazar R, Syeed S, Masood A, Khan NA (2011) Exogenously-sourced ethylene increases stomatal conductance, photosynthesis, and growth under optimal and deficient nitrogen fertilization in mustard. J Exp Bot 62:4955–4963. https://doi.org/10.1093/jxb/err204

Ji X, Shiran B, Wan J et al (2010) Importance of pre-anthesis anther sink strength for maintenance of grain number during reproductive stage water stress in wheat. Plant Cell Environ 33(6):926–942. https://doi.org/10.1111/j.1365-3040.2010.02130.x

Kasim WA, Osman ME, Omar MN, Abd El-Daim IA, Bejai S, Meijer J (2013) Control of drought stress in wheat using plant-growth-promoting bacteria. J Plant Growth Regul 32:122–130. https://doi.org/10.1007/s00344-012-9283-7

Kausar A, Gull M (2014) Effect of potassium sulphate on the growth and uptake of nutrients in wheat (*Triticum aestivum* L.) under salt stressed conditions. J Agric Sci 6:1–12. https://doi.org/10.5539/jas.v6n8p101

Kaydan D, Yagmur M, Okut N (2007) Effects of salicylic acid on the growth and some physiological characters in salt stressed wheat (*Triticum aestivum L.*). Tarim Bilimleri Dergisi 13(2):114–119

Ke Q et al (2018) Melatonin mitigates salt stress in wheat seedlings by modulating polyamine metabolism. Front Plant Sci 9:914. https://doi.org/10.3389/fpls.2018.00914

Khalil SI, El-Bassiouny HMS, Hassanein RA, Mostafa HA, El-Khawas SA, Abd El-Monem AA (2009) Antioxidant defense system in heat shocked wheat plants previously treated with arginine or putrescine. Aust J Basic Appl Sci 3:1517–1526

Khan MIR, Iqbal N, Masood A, Per TS, Khan NA (2013) Salicylic acid alleviates adverse effects of heat stress on photosynthesis through changes in proline production and ethylene formation. Plant Signal Behav 8(11):e26374. https://doi.org/10.4161/psb.26374

Khedr RA et al (2022) Alleviation of salinity stress effects on agro-physiological traits of wheat by auxin, glycine betaine, and soil additives. Soudi J Biol Sci 29(1):534–540. https://doi.org/10.1016/j.sjbs.2021.09.027

Koramutla MK, Tuan PA, Ayele BT (2022) Salicylic acid enhances adventitious root and aerenchyma formation in wheat under waterlogged conditions. Int J Mol Sci 23(3):1243. https://doi.org/10.3390/ijms23031243

Kumar RR, Goswami S, Kumar N, Pandey SK, Pandey VC, Sharma SK, Pathak H, Rai RD (2011) Expression of novel ascorbate peroxidase isoenzymes of wheat (*Triticum aestivum* L) in response to heat stress. Int J Plant Physiol Biochem 3(11):188–194

Kumar RR, Sharma SK, Goswami S, Singh GP, Singh R, Singh K, Pathak H, Rai RD (2013) Characterization of differentially expressed stress associated proteins in starch granule development under heat stress in wheat (*Triticum aestivum* L.). Indian J Biochem Biophys 50:126–138

Leja M, Wyzgolik G, Kaminska I (2007) Some parameters of antioxidant capacity of red cabbage as related to different forms of nutritive nitrogen. Folia Hort 19(1):15–23

Li JT, Qiu ZB, Zhang XW, Wang LS (2011) Exogenous hydrogen peroxide can enhance tolerance of wheat seedlings to salt stress. Acta Physiol 33:835–842. https://doi.org/10.1007/s11738-010-0608-5

Li X et al (2018) Melatonin alleviates low PS I-limited carbon assimilation under elevated CO 2 and enhances the cold tolerance of offspring in chlorophyll b-deficient mutant wheat. J Pineal Res 64(1):e12453. https://doi.org/10.1111/jpi.12453

Lobell DB, Bonfils CJ, Kueppers LM, Snyder MA (2008) Irrigation cooling effect on temperature and heat index extremes. Geophys Res Lett 35:L09705. https://doi.org/10.1029/2008GL034145

Marta B, Szafrańska K, Posmyk MM (2016) Exogenous melatonin improves antioxidant defense in cucumber seeds (*Cucumis sativus* L.) germinated under chilling stress. Front Plant Sci 7:575. https://doi.org/10.3389/fpls.2016.00575

McDonald AJS, Davies WJ (1996) Keeping in touch: responses of the whole plant to deficits in water and nitrogen supply. Adv Bot Res 22:229–300

Mirshel W, Wenkel K-O, Schultz A, Pommerening J, Verch G (2005) Dynamic phenological model for winter rye and winter barley. Eur J Agron 23(2):123–135. https://doi.org/10.1016/j.eja.2004.10.002

Mishra PK et al (2011) Alleviation of cold stress in inoculated wheat (*Triticum aestivum* L.) seedlings with psychrotolerant pseudomonads from NW Himalayas. Arch Microbiol 193:497–513. https://doi.org/10.1007/s00203-011-0693-x

Mona IN, Gawish SM, Taha T, Mubarak M (2017) Response of wheat plants to application of selenium and humic acid under salt stress conditions. Egypt. J Soil Sci 57(2):175–187. https://doi.org/10.21608/EJSS.2017.3715

Murakami R, Ifuku K, Takabayashi A, Shikanai T, Endo T, Sato F (2002) Characterization of an Arabidopsis thaliana mutant with impaired psbO, one of two genes encoding extrinsic 33-kDa proteins in photosystem II. FEBS Lett 523:138–142. https://doi.org/10.1016/S0014-5793(02)02963-0

Nadeem H, Amir KHAN et al (2023) Stress combination: when two negatives may become antagonistic, synergistic or additive for plants? Pedosphere 33(2):287–300. https://doi.org/10.1016/j.pedsph.2022.06.031

Ortmann I, Conrath U, Moerschbacher BM (2006) Exopolysaccharides of *Pantoea agglomerans* have different priming and eliciting activities in suspension-cultured cells of monocots and dicots. FEBS Lett 580:4491–4494. https://doi.org/10.1016/j.febslet.2006.07.025

Paulert R, Ebbinghaus D, Urlass C et al (2010) Priming of the oxidative burst in rice and wheat cell cultures by ulvan, a polysaccharide from green macroalgae, and enhanced resistance against powdery mildew in wheat and barley plants. Plant Pathol 59:634–642. https://doi.org/10.1111/j.1365-3059.2010.02300.x

Pazarlar S, Cetinkaya N, Bor M, Ozdemir F (2017) Ozone triggers different defence mechanisms against powdery mildew (*Blumeria graminis DC. Speer f. sp. tritici*) in susceptible and resistant wheat genotypes. Funct Plant Biol 44:1016–1028. https://doi.org/10.1071/FP17038

Qi F, Li J, Duan L, Li Z (2005) Inductions of coronatine and MeJA to low-temperature resistance of wheat seedlings. Acta Botan Boreali-Occiden Sin 26:1776–1780

Rakhmankulova ZF, Fedyaev VV, Rakhmatulina SR, Ivanov CP, Gilvanova IR, Usmanov IY (2010) The effect of wheat seed presowing treatment with salicylic acid caon its endogenous content, activities of respiratory pathways, and plant antioxidant status. Russ J Plant Physiol 57:778–783. https://doi.org/10.1134/S1021443710060051

Rani SJ, Usha R (2013) Transgenic plants: types, benefits, public concerns, and future. J Pharm Res 6(8):879–883. https://doi.org/10.1016/j.jopr.2013.08.008

Renard-Merlier D, Randoux B, Nowak E et al (2007) Iodus 40, salicylic acid, heptanoyl salicylic acid and trehalose exhibit different efficacies and defence targets during a wheat/powdery mildew interaction. Phytochemistry 68:1156–1164. https://doi.org/10.1016/j.phytochem.2007.02.011

Rizvi A, Khan MS (2017) Biotoxic impact of heavy metals on growth, oxidative stress and morphological changes in root structure of wheat (*Triticum aestivum* L.) and stress alleviation by Pseudomonas aeruginosa strain CPSB1. Chemosphere 185:942–952. https://doi.org/10.1016/j.chemosphere.2017.07.088

Rizvi A, Ahmed B, Zaidi A, Khan MS (2019) Heavy metal mediated phytotoxic impact on winter wheat: oxidative stress and microbial management of toxicity by *Bacillus subtilis BM2*. RSC Adv 9(11):6125–6142. https://doi.org/10.1039/C9RA00333A

Rizvi A, Zaidi A, Ameen F, Ahmed B, AlKahtani MD, Khan MS (2020) Heavy metal induced stress on wheat: phytotoxicity and microbiological management. RSC Adv 10(63):38379–38403. https://doi.org/10.1039/D0RA05610C

Robertson D, Zhang H, Palta JA, Colmer T, Turner NC (2009) Waterlogging affects the growth, development of tillers, and yield of wheat through a severe, but transient, N deficiency. Crop Pasture Sci 60(6):578–586. https://doi.org/10.1071/CP08440

Saravanakumar K, Li Y, Yu C et al (2017) Effect of *Trichoderma harzianum* on maize rhizosphere microbiome and biocontrol of *fusarium* stalk rot. Sci Rep 7:1771. https://doi.org/10.1038/s41598-017-01680-w

Shaddad MAK, Abd El-Samad HM, Mostafa D (2013) Role of gibberellic acid (GA3) in improving salt stress tolerance of two wheat cultivars. Int J Plant Physiol Biochem 5:50–57. https://doi.org/10.5897/IJPPB11.055

Shiqing GAO et al (2005) Improvement of wheat drought and salt tolerance by expression of a stress-inducible transcription factor GmDREB of soybean (Glycine max). Chin Sci Bull 50(23):2714–2723. https://doi.org/10.1360/982005-1234

Singh S (2014) A review on possible elicitor molecules of cyanobacteria: their role in improving plant growth and providing tolerance against biotic or abiotic stress. J Appl Microbiol 117(5):1221–1244. https://doi.org/10.1111/jam.12612

Singh B, Usha K (2003) Salicylic acid induced physiological and biochemical changes in wheat seedlings under water stress. Plant Growth Regul 39:137–141

Singh A, Singh D, Gill BS et al (2011a) Planting time, methods, and practices to reduce the deleterious effects of high temperature on wheat. In: The proceedings of international conference on preparing agriculture for climate change. Punjab Agricultural University, Ludhiana, pp 338–339

Singh A, Singh D, Kang JS, Aggarwal N (2011b) Management practices to mitigate the impact of high temperature on wheat: a review. IIOABJ 2:11–22

Tabassum T, Farooq M, Ahmad R, Zohaib A, Wahid A, Shahid M (2018) Terminal drought and seed priming improves drought tolerance in wheat. Physiol Mol Biol Plants 24:845–856. https://doi.org/10.1007/s12298-018-0547-y

Taie HA, Seif El-Yazal MA, Ahmed SM, Rady MM (2019) Polyamines modulate growth, antioxidant activity, and genomic DNA in heavy metal-stressed wheat plant. Environ Sci Pollut Res 26:22338–22350. https://doi.org/10.1007/s11356-019-05555-7

Tian X, He M, Wang Z, Zhang J, Song Y, He Z, Dong Y (2005) Application of nitric oxide and calcium nitrate enhances tolerance of wheat seedlings to salt stress. Plant Growth Regul 77:343–356. https://doi.org/10.1007/s10725-015-0069-

Tian D et al (2022) Small secreted peptides encoded on the wheat (*Triticum aestivum* L.) genome and their potential roles in stress responses. Front Plant Sci 13:1000297. https://doi.org/10.3389/fpls.2022.1000297

Timmusk S, Abd El-Daim IA, Copolovici L, Tanilas T, Kannaste A et al (2014) Drought-tolerance of wheat improved by rhizosphere bacteria from harsh environments: enhanced biomass production and reduced emissions of stress volatiles. PLoS One 9:e96086. https://doi.org/10.1371/journal.pone.0096086

Turan S, Cornish K, Kumar S (2012) Salinity tolerance in plants: breeding and genetic engineering. Aust J Crop Sci 6:1337

Upadhyay SK, Singh JS, Saxena AK, Singh DP (2012) Impact of PGPR inoculation on growth and antioxidant status of wheat under saline conditions. Plant Biol 14(4):605–611. https://doi.org/10.1111/j.1438-8677.2011.00533.x

ur Rehman H et al (2021) Sequenced application of glutathione as an antioxidant with an organic biostimulant improves physiological and metabolic adaptation to salinity in wheat. Plant Physiol Biochem 158:43–52. https://doi.org/10.1016/j.plaphy.2020.11.041

Van Assche F, Clijsters H (1990) Effects of metals on enzyme activity in plants. Plant Cell Environ 13:195–206. https://doi.org/10.1111/j.1365-3040.1990.tb01304.x

Wu JD, Li JC, Wei FZ, Wang CY, Zhang Y, Sun G (2014) Effects of nitrogen spraying on the post-anthesis stage of winter wheat under waterlogging stress. Acta Physiol Plant 36:207–216. https://doi.org/10.1007/s11738-013-1401-z

Xing K, Zhu X, Peng X et al (2015) Chitosan antimicrobial and eliciting properties for pest control in Agri culture: a review. Agron Sustain Dev 35:569–588. https://doi.org/10.1007/s13593-014-0252-3

Xu Q, Truong TT, Barrero JM, Jacobsen JV, Hocart CH, Gubler F (2016) A role for jasmonates in the release of dormancy by cold stratification in wheat. J Exp Bot 67:3497–3508. https://doi.org/10.1093/jxb/erw172

Yadav A, Singh S (2023) Effect of exogenous phytohormone treatment on antioxidant activity, enzyme activity and phenolic content in wheat sprouts and identification of metabolites of control and treated samples by UHPLC-MS analysis. Food Res Int 169:1–14. https://doi.org/10.1016/j.foodres.2023.112811

Yadav A, Singh S, Yadav V (2024) Screening herbal extracts as biostimulant to increase germination, plant growth and secondary metabolite production in wheatgrass. Sci Rep 14:607. https://doi.org/10.1038/s41598-023-50513-6

Yamauchi T, Watanabe K, Fukazawa A, Mori H, Abe F, Kawaguchi K, Oyanagi A, Nakazono M (2014) Ethylene and reactive oxygen species are involved in root aerenchyma formation and adaptation of wheat seedlings to oxygen deficient conditions. J Exp Bot 65:261–273. https://doi.org/10.1093/jxb/ert371

Yavas I, Unay A (2016) Effects of zinc and salicylic acid on wheat under drought stress. JAPS: J Anim Plant Sci 26(4):1012–1018

Yu J, Cang J, Lu Q, Fan B, Xu Q, Li W, Wang X (2020) ABA enhanced cold tolerance of wheat 'dn1'via increasing ROS scavenging system. Plant Signal Behav 15(8):1780403. https://doi.org/10.1080/15592324.2020.1780403

Zaman B, Niazi BH, Athar M, Ahmad M (2005) Response of wheat plants to sodium and calcium ion interaction under saline environment. Int J Environ Sci Technol 2:7–12

Zhu JK (2001) Plant salt tolerance. Trends Plant Sci 6(2):66–71

Zubair M, Hanif A, Farzand A, Sheikh TMM, Khan AR, Suleman M, Gao X (2019) Genetic screening and expression analysis of *psychrophilic bacillus spp.* reveal their potential to alleviate cold stress and modulate phytohormones in wheat. Microorganisms 7(9):337. https://doi.org/10.3390/microorganisms7090337

Recent Advancement on Peptide Research and their Application in Eco-agriculture

12

Jyotsna Setty, Pavan Singh, Girish Tantuway, and P. Vijai

Abstract

Plant interactions with pathogens or microbes generate various signals within plants that trigger defense mechanisms. Some of these signals can also be prompted by elicitors, which are protective molecules. The signaling initiated by elicitors guides a series of internal processes in plants, culminating in hormonal signal transduction cascade activation. This activation leads to induced resistance (IR), ultimately enhancing plant immunity against environmental stresses. Therefore, it is crucial to comprehend the role and mechanisms of elicitors in the cellular defense of crops to enhance sustainable crop protection and management. Peptides, recognized as significant physiological regulators, have found applications in varied branches such as medicine, cosmetics, healthcare products, animal nutrition, and plant protection from biotic stresses. Peptides have gained significant attention in recent years as a focal point of research in plant protection due to their roles as antimicrobial and immune inducers, as well as plant growth regulators, insecticides, and herbicides. This chapter provides a concise overview of the advancements in peptide research, outlines their applica-

J. Setty
Department of Plant Physiology, Rajiv Gandhi South Campus- Banaras Hindu University, Mirzapur, India

P. Singh
Department of Soil Science and Agricultural Chemistry, JNKVV, College of Agriculture, Tikamgarh, India

G. Tantuway
Department of Genetics and Plant Breeding, Rajiv Gandhi South Campus- Banaras Hindu University, Mirzapur, India

P. Vijai (✉)
Department of Plant Physiology, Institute of Agricultural Sciences, Banaras Hindu University, Varanasi, India

S. Singh, R. Mehrotra (eds.), *Plant Elicitor Peptides*,
https://doi.org/10.1007/978-981-97-6374-0_12

tions in plant protection, and discusses their potential as environmentally friendly agrochemicals.

Keywords

Elicitors · Peptides · Eco-agriculture · Immunity · Pathogen

12.1 Introduction

Broadly distributed signaling molecules in plants, plant elicitor peptides (Peps) help provide broad-spectrum defenses against insects and diseases. These bioactive Peps are produced from the carboxyl termini of larger propeptide precursors (PROPEPs), and their usual length is between 23 and 36 amino acids. A glycine-enriched motif, (S/G)(S)Gxx(G/P)xx(N), is frequently present in them (Tavormina et al. 2015). Based on amino acid sequence similarity, orthologues of the eight Peps that were first found in Arabidopsis thaliana have been found in many angiosperm genomes, including those of important crops including maize, wheat, rice, potatoes, and soybeans (Lori et al. 2015). In response to oral secretions from insects and pathogen infections, certain PROPEP genes show elevated expression levels (Huffaker et al. 2006, 2013) (Fig. 12.1). Subsequently, mature Peps interact with plant elicitor peptide receptors (PEPRs), which starts calcium signaling, releases nitric oxide and reactive oxygen species (ROS), produces phytohormones, reprograms transcription, and produces defensive proteins and metabolites (Klauser et al. 2013; Ma et al. 2012).

Pesticides are crucial for safeguarding plants and ensuring food security in agriculture. Research shows that without them, there would be significant losses in fruit,

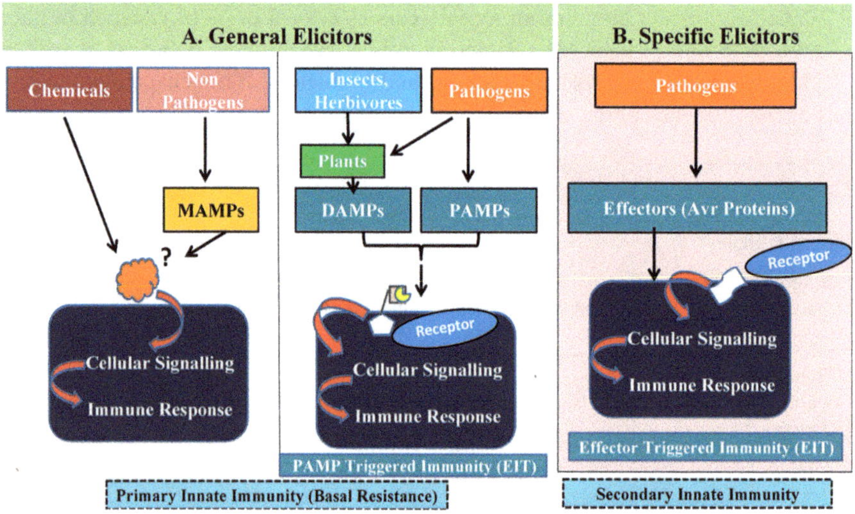

Fig. 12.1 An overview of the origins, categorization, optimization, manufacturing, and applications of peptide research. (Source: Modified from Zhang et al. (2023))

vegetable, and cereal production (fruit output has decreased by 78%, vegetable production has decreased by 54%, and cereal production has decreased by 32%) (Tudi et al. 2021). In today's era of ecological agriculture, which prioritizes sustainability, there is a pressing demand for pesticides that are both effective against pests and environmentally friendly, posing minimal risk to non-target organisms.

In recent years, plant elicitor peptides have gained significant attention in the realm of plant protection. This surge in interest is attributed to their abundant raw materials, demonstrated by studies like those by Mita and Sato (2019), Hedges and Ryan (2019), and Aguilar Troyano et al. (2021). Moreover, their efficacy, and their environmentally friendly nature evidenced by research such as that by Bomgardner (2017) and Zhou et al. (2022), further contribute to their appeal. These peptides serve various roles in plant defense, acting as antimicrobials, immune boosters, growth regulators, insecticides, and herbicides. They offer protection against a range of threats including bacteria, viruses, insects, pests, and obnoxious weeds. Presently, 18 peptides have been introduced into the market as eco-friendly solutions for safeguarding plants. In the USA, Spear®, a bioinsecticide made from a neuropeptide discovered in the venom of the Blue Mountains funnel-web spider, won the Best New Biological Agent Award in 2021 and the Presidential Green Chemistry Challenge Award (Small Business Award) in 2020. Peptides of high caliber and achievement, such as Spear®, are recognized as pivotal advancements in plant protection. Consequently, they have garnered substantial interest in the exploration and creation of environmentally friendly agrochemicals, as noted by Zhang et al. (2023).

Despite extensive research endeavors aimed at discovering and formulating innovative plant protection solutions, pesticide companies have shown limited enthusiasm in introducing novel pesticides to growers. This reluctance stems from the relatively low return on investment, attributed to the market value, and the challenges associated with obtaining regulatory approval, given the stringent requirements imposed by regulatory frameworks. Additionally, numerous novel strategies for disease management, such as RNAi and defense elicitors, are still under development. However, they face safety concerns, such as novel formulations involving nanoparticles, or lack comprehensive evaluation and validation at the field level, as highlighted by Montesinos (2023). Functional peptides have attracted considerable research attention within the realm of crop protection, mirroring similar interest seen in the veterinary, medical, and food industries (Montesinos 2007; Marcos et al. 2008; Cary et al. 2012). Peptides, typically defined as polypeptides consisting of up to 50–60 amino acids, encompass a range of structures, bringing together modified or non-natural amino acids and pseudopeptides with peptide linkages. These peptides are mostly found in living things, and they work by way of processes like microorganisms' antibiosis or antagonism. They play crucial roles as the initial immune defense barrier and are significant contributors to stress mitigation in both animals and plants (Huan et al. 2020).

To uncover novel peptide pesticides, it is essential to comprehend the research and utilization of current peptide compounds. This chapter provides a concise overview of the progress in peptide research, outlines their role in plant protection, elucidates the origins and modes of action of functional peptides, and encapsulates the utilization of peptides in safeguarding plants. Furthermore, it discusses the future prospects of environmentally friendly agrochemicals and the platforms for their production, along with the challenges and opportunities in developing innovative biopesticides.

12.2 Progress in Peptide Research

Peptides are biomolecules consisting of short chains of amino acids, typically ranging from 2 to 50, bound together via bonds of peptides. They can additionally originate from the byproducts that are in between when proteins break down. Peptides are categorized based on their composition as either homomeric, composed solely of amino acids, or heteromeric, which contain a mixture of amino acids and non-amino acids like glycopeptides.

With the exception of cyclic peptides, an N-terminal (amine group) and a C-terminal (carboxyl group) residue are present in every peptide. Peptides are further divided into naturally occurring and synthetically produced categories according to where they come from. Animals, plants, and microbes are the main sources of natural peptides. Peptides can be produced either synthetically or naturally by a variety of methods, such as chemical synthesis, biological fermentation, gene recombination, among others (Zhang et al. 2023). Given their prevalence in living organisms and their role in modulating numerous physiological processes, peptides are extensively studied in fields such as medicine, cosmetics, agriculture, and beyond (Zhang et al. 2023).

Secretin was the first peptide to be discovered; it was found in the gastrointestinal system of animals by Bayliss and Starling in 1902 (Tam et al. 2014). Subsequent breakthroughs included the unveiling of bioactive peptides including oxytocin, which causes uterine contractions, and insulin, which is well known for its ability to lower blood glucose levels. Peptide synthesis was transformed in 1963 with the introduction of solid phase peptide synthesis (SPPS) offering a more efficient and streamlined process compared to traditional liquid phase methods. Merrifield, credited with inventing SPPS, was honored with the Nobel Prize in Chemistry in 1984 for this pioneering advancement. Since then, peptide research has undergone rapid expansion, with ten Nobel Prizes recognizing contributions to peptide-related studies, underscoring their profound significance in the realms of science and technology.

Research into peptides encompasses various aspects including their sources, structures, optimization, production methods, functions, and applications (Fig. 12.2). These molecules, serving as crucial physiological regulators with diverse functions, discover extensive applications in the fields of medicine, skin care products, animal health and sustenance, and crop development and preservation. As of right now, more than 80 peptide medications are available on the market to treat ailments including cancer, diabetes, osteoarthritis, multiple sclerosis (MS), infections with HIV, and persistent pain (Muttenthaler et al. 2021). Additionally, 400–600 peptide therapies are being investigated in preclinical settings, while around 50 peptide medications are presently undergoing clinical development (Muttenthaler et al.

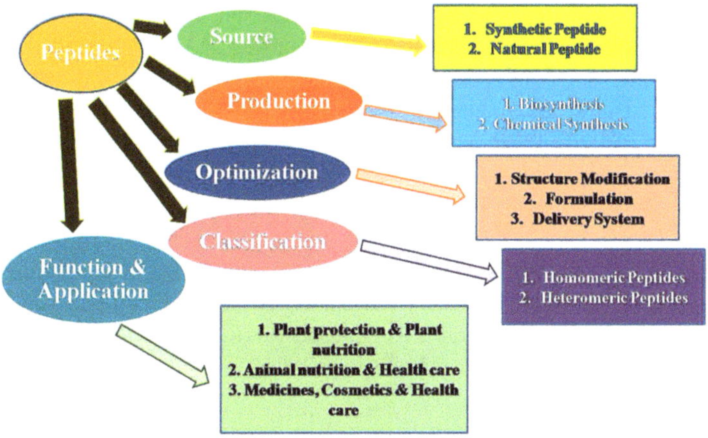

Fig. 12.2 Classification of elicitors. *MAMPs* microbes-associated molecular patterns, *DAMPs* damage-associated molecular patterns, *PAMPs* pathogen-associated molecular patterns. (Source: Modified from Henry et al. (2012)

2021). Given their extensive applications in medicine, researchers are increasingly exploring their potential utilization in enhancing plant protection within modern agriculture.

12.3 Mechanisms of Action

Functional peptides exhibit various mechanisms that impact plant pathogens. For instance, within antimicrobial peptides, lytic peptides induce damage to the cell membrane, leading to its lysis or deactivation. Another subset of antimicrobial peptides influences internal cell processes (Le et al. 2017) by penetrating the target cell. Sometimes they breach the membrane, sometimes they don't, but they obstruct things like cell division, proteinase inhibition, and the production of proteins or nucleic acids. This pattern is shown in magainins, which influence the way that fungi and bacteria produce DNA, as well as in their metabolic pathways (McMillan and Coombs 2020); in cathelicidins, which block ion channels, translation, and replication; and in the antifungal PAF26 (Muñoz et al. 2013). Certain cell-penetrating peptides (CPPs) do not, however, interfere with intracellular functions or damage cell membranes. Rather, they have been employed to move cargo molecules into different eukaryotic cells' cytoplasms, including plants, fungi, and humans (Numata et al. 2014). An alternative mechanism involves priming the plant to induce defense responses against pathogens. Plant cells respond to a variety of peptides through a process known as effector-triggered immunity (ETI). Certain peptides connect with pattern recognition receptors (PRRs), whereas inherent elicitor peptides collaborate with receptors such as endogenous elicitor peptide receptors (PEPR). This mechanism of interaction mimics the recognition observed through as-yet-unidentified interaction mechanisms (peptides) or with MAMPs, microbe-associated molecular patterns, DAMPs, damage-associated molecular patterns, PAMPs, pathogen-associated molecular patterns, and HAMPs, herbivore-associated molecular patterns. The PEP/PROPEP cycle produces effectors and receptors, which starts the process of self-induction. The upregulation of genes linked to the ethylene, jasmonic acid, and salicylic acid pathways (systemic signals), the phenylalanine ammonium lyase pathway (phenylpropanoids), or the synthesis of pathogenesis-related (PR) proteins (antimicrobials) is the consequence of these signaling processes, which also initiate kinase cascades and transcription factors. Reactions also follow activation involving reactive oxygen species (ROS), reinforcement of the cell wall, and alterations in ion flux (Montesinos 2023).

12.4 Peptide Application for Plant Protection

Peptides are advantageous for using effective and environment friendly crop pest management techniques in sustainable agriculture. It has been seen that varied functions have been exhibited, including controlling plant growth, acting as pesticides and herbicides, and stimulating plant immunity and antimicrobial activity.

12.4.1 Peptides with Antimicrobial and Immuno-Stimulant Properties

Pathogens that affect plants pose a significant threat to crop health, resulting in detrimental effects on their growth. While traditional chemical fungicides effectively combat infections brought on by these causal organisms, their persistent use has resulted in resistance, and their leftovers are dangerous for the environment and human health. Consequently, there is an urgent need for management of several plant diseases by more sustainable methods (Donley 2019). Naturally occurring antimicrobial peptides (AMPs) play a crucial role in innate host defense mechanisms and can serve as immune boosters. AMPs are expected to be a potential first line of defense against bacteria, viruses, and fungi because of their quick breakdown, efficacy, and specificity (Yadav et al. 2019).

12.4.1.1 AMPs

The antimicrobial peptide database (APD3) documents a total of 3425 AMPs. Artificially synthesized peptides constitute 49.33% of the total peptides listed, with AMPs sourced from plants, animals, and microorganisms contributing 29.33, 16.00, and 5.33%, respectively (Table 12.1). Natural AMPs are derived from microorganisms, plants, and animals (Li et al. 2021). AMPs including defensins, cecropins, drosocins, attacins, diptericins, ponericins, metchnikowins, and melittin are also produced by insects (Wu et al. 2018). Li et al. (2021) have classified plant antimicrobial peptides (AMPs) based on their sequences and structures. These include

Table 12.1 The origin of peptides that have antimicrobial, herbicidal, insecticidal, and plant growth regulator properties for the protection of plants

	Source of peptides (%)			
Uses of peptides	Animals	Plants	Microorganisms	Synthesis
Antimicrobial activity	16	29.33	5.33	49.33
Plant growth regulator activity	7.14	58.93	14.29	19.64
Insecticidal activity	63.1	19.64	5.95	11.31
Herbicidal activity	–	37.5	43.75	18.75

thionins, defensins, hevein-like peptides, knottins, stable-like peptides, lipid transfer proteins, snakins, and cyclotides. Nevertheless, bacterial AMPs are not generated to protect against infections; rather, they function as a means of competition to drive out other microbes contending for resources in the same biological forte, guaranteeing the subsistence of a solitary bacterial cell (Jenssen et al. 2006).

12.4.1.2 Immune-Inducing Peptides

AMPs play essential roles in innate host defense, functioning as immune stimulators, initiating defense signals, or boosting native immunity in plants (Bende et al. 2015). Among these peptides, three types have been investigated for their potential as commercial immune stimulators. For instance, the peptide maSAMP, developed by Invaio Sciences, is employed in combating citrus Huanglongbing, a highly damaging disease. This peptide not only eradicates *Liberobacter asianticum* Jagoueix, the bacterium responsible for Huanglongbing, but also stimulates the plant's immune system, thereby averting subsequent infections. Plant Health Care (PHC), an American firm, invented the PREtec technology, which was granted a patent in the USA in 2019. This innovative technology, featuring immune-inducing peptides and their combinations with other substances, has been acknowledged for its ability to enhance plant resilience against diseases and stress factors, while also stimulating plant growth.

Numerous immune-inducing peptides are currently undergoing development. For example, in Arabidopsis, PIP1 and PIP2 have been found to augment pathogen resistance and immune responses (Hou et al. 2014). Another peptide, NbPPI1 from *Nicotiana tabacum*, has been shown to stimulate the defense response and enhance plant resistance against *Pytophthora* (Wen et al. 2021). The maize Zip1 immune signal peptide has demonstrated ability to decrease the virulence of maize smut fungus (Segonzac and Monaghan 2019). Furthermore, via inducing an increase in plant hormones related to defense activities including salicylic acid and jasmonic acid, inceptin has been found to contribute to the defense against herbivores in cowpea and kidney bean (Schmelz et al. 2012).

12.4.2 Peptides Regulating Plant Growth

Cellular communication throughout plant embryonic phases, control of plant growth and development are under the control of phytohormones auxins, cytokinins, and gibberellins. However, recent investigations suggest that even peptide signaling molecules tend to play a significant role in various processes of plant developmental and further even responses of plants environmental conditions (Ryan et al. 2002; Boller 2005). These include fruit ripening, abscission, tissue and organ creation, meristematic stem cell differentiation (Vanyushin et al. 2017) under both biotic and abiotic stressors (Chen et al. 2020). PGRPs (plant growth-regulating peptides) represent a novel category of plant hormones (Matsubayashi and Sakagami 2006) possessing signaling properties and hormone-like characteristics (Vanyushin et al. 2017). Notably, their activities are exhibited at minute concentrations ranging from 10^{-7} to 10^{-9} M. These results highlight the role that peptides play in controlling plant development. PGRPs are involved in many aspects of plant development and growth. The PY91, a functional peptide, affects crop growth, while meristem size is regulated by CLAVATA3 peptide (Lay and Anderson 2005). In the case of cruciferous pollen SCR peptide serves as a recognition element for the self-incompatibility mechanism (Fletcher et al. 1999). RALFs represent a peptide family involved in regulating plant cell growth (Okuda et al. 2009). Another peptide found in roots called CLE25 helps plants deal with drought stress by regulating NCED3 expression, which raises abscisic acid (ABA) levels (Endo et al. 2008). This encourages stomatal closure and hence tends to maintains water balance of the plants. The effects of conventional plant hormones either strengthened or weakened thus regulated by the peptide hormones. For example, control of root formation and root growth has been found to be brought about the action of auxin, BR peptides, and CLE41/TDIF.

12.4.3 Peptides with Insecticidal Properties

Management of crop pests is a significant issue in agriculture, as they can result in crop losses ranging from 13 to 16% (Rivera-de-Torre et al. 2022). Chemical insecticides have been the primary method of controlling insect pests. However, their extensive application has resulted in pest resistance and posed risks to both human health and the environment (Windley et al. 2012). Therefore, there is a pressing need to explore bioinsecticides as environmentally safe and sustainable method to pest control (Park et al. 2011).

The majority of insecticidal peptides derived from plant species, such as cyclic peptide, pea albumin, defense, and recombinant peptide (Grover et al. 2021), originate from botanical families such as Rubiaceae, Leguminosae, Violaceae, Solanaceae, and Cucurbitaceae (Craik 2010). Cyclic peptides, over 47 of them sourced from *Clitoria ternatea* (Gilding et al. 2016), have demonstrated insecticidal properties. Additionally, several mimic peptides with promising insecticidal efficacy (Hofmann et al. 2021) have been developed by altering native peptides to enhance bio-stability. By incorporating unnatural amino acids at enzymatic sites, numerous insect kinin mimics have been produced, resulting in products significantly more resistant to enzymatic degradation. The identification of these peptide-mimicking properties of compounds presents an entirely modern method for developing new environmentally friendly agents that can be used in agriculture to control crop pests and reduce losses in crop yields.

12.4.4 Peptides with Herbicidal Properties

Typically, weeds pose one of the greatest impending threats, accounting for an yield loss of 34%, in contrast to other threats including pathogens and insect pests which cause 18 and 16% of losses, respectively. Weed management can also be achieved through mechanical or chemical means, often achieving greater efficacy compared to controlling animal pests or diseases (Oerke et al. 2012). While traditional herbicides play a crucial role in maintaining crop yields, their excessive use has resulted in adverse effects, including residues in crops and the environment. Consequently, there is an increasing demand for new environmentally friendly herbicides (Shi LiQiao et al. 2020). On a commercial scale, two peptides have been successfully tried as herbicides. The important active ingredient of Opportune™ is Thaxomin A (manufactured by Marrone Bio Innovation) has been shown to function as an inhibitor of cellulose synthesis in cells distinctively. It received approval from the USPA (US Environmental Protection Agency) in 2013 as an environmentally friendly bio-herbicide for managing weeds in cereal crops such as paddy (Zhang et al. 2023).

12.5 Opportunities and Obstacles in the Development of Peptide-Based Agrochemicals

12.5.1 Obstacles

Although peptides have demonstrated effectiveness in plant protection, they encounter various challenges due to limitations like systemic instability, little oral action, and expensive manufacture. Typically, naturally occurring peptides are prone to enzymatic degradation within the organism and are susceptible to exterior environmental factors including pH and light, leading to instability and limited bioavailability. In contrast with drugs based on peptides, if peptide-based pesticides are excessively expensive, their acceptance in the commercial market will be restricted. Hence, for peptide-based agrochemicals to gain wider acceptance, they must exhibit improved steadiness, biological activity, and reduced price.

12.5.2 Opportunities in Agrochemicals that are Peptide Based

Enhancing the chemical stability and greater availability of more efficient peptides from natural sources stands as a crucial objective in the quest for new peptide-based medications and agrochemicals. Genetic engineering offers a means to modify natural peptides, yielding bioengineered peptides with preferred characteristics. For example, the bioinsecticide named Spear® is genetically engineered by incorporating a dipeptide, namely glycine-serine into ω/κ-HXTX-Hv1a, the peptide naturally found in spider venom. This resulting product exhibits heightened activity, reduced risk, and prolonged persistence compared to the original compound, rendering it a sustainable and efficient solution for pest management in both agricultural and public health domains (TONG 2022).

12.5.3 Delivery System

Drug delivery systems (DDS) facilitate the targeted delivery of drugs by employing controlled release technologies like hydrogels, cubosomes, and nanocarriers, thereby enhancing the efficiency of drug utilization (Martin-Serrano et al. 2019). By combining easily degradable peptides with innovative DDS strategies, precision agriculture applications can be achieved.

12.5.4 The Process of Peptide Biosynthesis

Various methods including chemicals are implied in synthesizing peptides such as "liquid phase synthesis" and "solid phase synthesis." Nevertheless, chemical implied method of synthesis proves costly also challenging for mass production. As a result, numerous investigations aim to achieve more cost-effective peptide synthesis methods. Peptides synthesis via various approaches including genetic engineering, fermentation, and enzymes are preferred due to their benefits including, abundant availability of raw material and lower expenses (Akbarian et al. 2022). Thus, exploring the preparation of peptide agrochemicals through biosynthesis merits further investigation.

12.6 Conclusion

Over the past century, peptides have emerged as significant entities, with rapid advancements observed in the development of peptide-based drugs and agrochemicals worldwide. This pattern indicates a growing interest in goods known for their effectiveness, low toxicity, and eco-friendliness. In the field of peptide-based agricultural chemicals, significant advancements are noticeable in research endeavors spanning the utilization, production, and investigation of diverse agricultural peptides, positioning them as promising candidates for agricultural applications. Peptides that full fill criteria for both high efficacy and safety play a crucial role in plant protection within eco-agricultural practices. The advancements in the related fields of molecular biology, biochemistry, synthetic biology, and genetic engineering will accelerate the industrialization of peptides. Moreover, alongside traditional methods like X-ray crystallography, the emergence of innovative techniques such as AlphaFold2 and RoseTTAFold enables the acquisition of 3D protein structures. Additionally, innovative methods such as phage peptide library, mRNA display, and virtual screening provide avenues for anticipating useful peptide ligands. Exploring the interaction between proteins and ligands holds significant importance in drug design. Engineered peptides exhibit a strong affinity for specific targets, necessitating only a small quantity to effectively manage weeds, pathogens, and insects. Despite the rapid evolution of peptides, this chapter could not encompass all facets of their role as novel instruments in eco-agricultural plant protection. Nonetheless, our aim is to encourage biochemists, molecular biologists, and agronomists to delve deeper into this emerging field, fostering comprehensive and thorough investigations into agrochemicals based on peptides aim to foster the sustainable progression of eco-friendly agriculture.

References

Aguilar Troyano FJ, Merkens K, Anwar K, Gómez-Suárez A (2021) Radical-based synthesis and modification of amino acids. Angew Chem Int Ed 60(3):1098–1115

Akbarian M, Khani A, Eghbalpour S, Uversky VN (2022) Bioactive peptides: synthesis, sources, applications, and proposed mechanisms of action. Int J Mol Sci 23(3):1445

Bende NS, Dziemborowicz S, Herzig V, Ramanujam V, Brown GW, Bosmans F, Nicholson GM, King GF, Mobli M (2015) The insecticidal spider toxin SFI 1 is a knottin peptide that blocks the pore of insect voltage-gated sodium channels via a large β-hairpin loop. FEBS J 282(5):904–920

Boller T (2005) Peptide signalling in plant development and self/non-self perception. Curr Opin Cell Biol 17(2):116–122

Bomgardner MM (2017) Spider venom: an insecticide whose time has come. Chem Eng News 95(11):30–31

Cary J, Jaynes J, Montesinos E (2012) Small wonders: peptides for disease control, vol 1095. In: Rajasekaran K (ed). American Chemical Society

Chen YL, Fan KT, Hung SC, Chen YR (2020) The role of peptides cleaved from protein precursors in eliciting plant stress reactions. New Phytol 225(6):2267–2282

Craik DJ (2010) Discovery and applications of the plant cyclotides. Toxicon 56(7):1092–1102

Donley N (2019) The USA lags behind other agricultural nations in banning harmful pesticides. Environ Health 18:1–12

Endo A, Sawada Y, Takahashi H, Okamoto M, Ikegami K, Koiwai H, Seo M, Toyomasu T, Mitsuhashi W, Shinozaki K, Nakazono M, Nambara E (2008) Drought induction of Arabidopsis 9-cis-epoxycarotenoid dioxygenase occurs in vascular parenchyma cells. Plant Physiol 147(4):1984–1993

Fletcher JC, Brand U, Running MP, Simon R, Meyerowitz EM (1999) Signaling of cell fate decisions by CLAVATA3 in Arabidopsis shoot meristems. Science 283(5409):1911–1914

Gilding EK, Jackson MA, Poth AG, Henriques ST, Prentis PJ, Mahatmanto T, Craik DJ (2016) Gene coevolution and regulation lock cyclic plant defence peptides to their targets. New Phytol 210(2):717–730

Grover T, Mishra R, Gulati P, Mohanty A (2021) An insight into biological activities of native cyclotides for potential applications in agriculture and pharmaceutics. Peptides 135:170430

Hedges JB, Ryan KS (2019) Biosynthetic pathways to nonproteinogenic α-amino acids. Chem Rev 120(6):3161–3209

Henry G, Thonart P, Ongena M (2012) PAMPs, MAMPs, DAMPs and others: an update on the diversity of plant immunity elicitors. BASE

Hofmann A, Minges A, Groth G (2021) Interfering peptides targeting protein–protein interactions in the ethylene plant hormone signaling pathway as tools to delay plant senescence. Plant Chem Genom Methods Protocols:71–85

Hou S, Wang X, Chen D, Yang X, Wang M, Turrà D, Di Pietro A, Zhang W (2014) The secreted peptide PIP1 amplifies immunity through receptor-like kinase 7. PLoS Pathog 10(9):e1004331
Huan Y, Kong Q, Mou H, Yi H (2020) Antimicrobial peptides: classification, design, application and research progress in multiple fields. Front Microbiol 11:582779
Huffaker A, Pearce G, Ryan CA (2006) An endogenous peptide signal in Arabidopsis activates components of the innate immune response. Proc Natl Acad Sci 103(26):10098–10103
Huffaker A, Pearce G, Veyrat N, Erb M, Turlings TC, Sartor R, Shen Z, Briggs SP, Vaughan MM, Alborn HT, Teal PE, Schmelz EA (2013) Plant elicitor peptides are conserved signals regulating direct and indirect antiherbivore defense. Proc Natl Acad Sci 110(14):5707–5712
Jenssen H, Hamill P, Hancock RE (2006) Peptide antimicrobial agents. Clin Microbiol Rev 19(3):491–511
Klauser D, Flury P, Boller T, Bartels S (2013) Several MAMPs, including chitin fragments, enhance at pep-triggered oxidative burst independently of wounding. Plant Signal Behav 8(9):e25346
Lay FT, Anderson MA (2005) Defensins-components of the innate immune system in plants. Curr Protein Peptide Sci 6(1):85–101
Le CF, Fang CM, Sekaran SD (2017) Intracellular targeting mechanisms by antimicrobial peptides. Antimicrob Agents Chemother 61(4):10–1128
Li J, Hu S, Jian W, Xie C, Yang X (2021) Plant antimicrobial peptides: structures, functions, and applications. Bot Stud 62(1):5
Lori M, Van Verk MC, Hander T, Schatowitz H, Klauser D, Flury P, Gehring CA, Boller T, Bartels S (2015) Evolutionary divergence of the plant elicitor peptides (peps) and their receptors: interfamily incompatibility of perception but compatibility of downstream signalling. J Exp Bot 66(17):5315–5325
Ma Y, Walker RK, Zhao Y, Berkowitz GA (2012) Linking ligand perception by PEPR pattern recognition receptors to cytosolic Ca2+ elevation and downstream immune signaling in plants. Proc Natl Acad Sci 109(48):19852–19857
Marcos JF, Munoz A, Pérez-Payá E, Misra S, López-García B (2008) Identification and rational design of novel antimicrobial peptides for plant protection. Annu Rev Phytopathol 46:273–301
Martin-Serrano Á, Gómez R, Ortega P, de la Mata FJ (2019) Nanosystems as vehicles for the delivery of antimicrobial peptides (AMPs). Pharmaceutics 11(9):448
Matsubayashi Y, Sakagami Y (2006) Peptide hormones in plants. Annu Rev Plant Biol 57:649–674
McMillan KA, Coombs MRP (2020) Examining the natural role of amphibian antimicrobial peptide magainin. Molecules 25(22):5436
Mita T, Sato Y (2019) Syntheses of α-amino acids by using CO2 as a C1 source. Chem An Asian J 14(12):2038–2047
Montesinos E (2007) Antimicrobial peptides and plant disease control. FEMS Microbiol Lett 270(1):1–11
Montesinos E (2023) Functional peptides for plant disease control. Annu Rev Phytopathol 61:301–324
Muñoz A, Gandía M, Harries E, Carmona L, Read ND, Marcos JF (2013) Understanding the mechanism of action of cell-penetrating antifungal peptides using the rationally designed hexapeptide PAF26 as a model. Fungal Biol Rev 26(4):146–155
Muttenthaler M, King GF, Adams DJ, Alewood PF (2021) Trends in peptide drug discovery. Nat Rev Drug Discov 20(4):309–325
Numata K, Ohtani M, Yoshizumi T, Demura T, Kodama Y (2014) Local gene silencing in plants via synthetic ds RNA and carrier peptide. Plant Biotechnol J 12(8):1027–1034
Oerke EC, Dehne HW, Schönbeck F, Weber A (2012) Crop production and crop protection: estimated losses in major food and cash crops. Elsevier
Okuda S, Tsutsui H, Shiina K, Sprunck S, Takeuchi H, Yui R, Kasahara RD, Hamamura Y, Mizukami A, Susaki D, Kawano N, Higashiyama T (2009) Defensin-like polypeptide LUREs are pollen tube attractants secreted from synergid cells. Nature 458(7236):357–361
Park SC, Park Y, Hahm KS (2011) The role of antimicrobial peptides in preventing multidrug-resistant bacterial infections and biofilm formation. Int J Mol Sci 12(9):5971–5992

Rivera-de-Torre E, Rimbault C, Jenkins TP, Sørensen CV, Damsbo A, Saez NJ, Duhoo Y, Hackney CM, Ellgaard L, Laustsen AH (2022) Strategies for heterologous expression, synthesis, and purification of animal venom toxins. Front Bioeng Biotechnol 9:811905

Ryan CA, Pearce G, Scheer J, Moura DS (2002) Polypeptide hormones. Plant Cell 14(suppl_1):S251–S264

Schmelz EA, Huffaker A, Carroll MJ, Alborn HT, Ali JG, Teal PE (2012) An amino acid substitution inhibits specialist herbivore production of an antagonist effector and recovers insect-induced plant defenses. Plant Physiol 160(3):1468–1478

Segonzac C, Monaghan J (2019) Modulation of plant innate immune signaling by small peptides. Curr Opin Plant Biol 51:22–28

Shi LiQiao SL, Wu ZhaoYuan WZ, Zhang YaNi ZY, Zhang ZhiGang ZZ, Fang Wei FW, Wang YueYing WY, Wan ZhongYi WZ, Wang KaiMei WK, Ke ShaoYong KS (2020) Herbicidal secondary metabolites from actinomycetes: structure diversity, modes of action, and their roles in the development of herbicides

Tam JK, Lee LT, Jin J, Chow BK (2014) Secretin/secretin receptors. J Mol Endocrinol 52:T1–T14

Tavormina P, De Coninck B, Nikonorova N, De Smet I, Cammue BP (2015) The plant peptidome: an expanding repertoire of structural features and biological functions. Plant Cell 27(8):2095–2118

TONG Y (2022) Progress of research, development and application on GS-omega/kappa-HXTX-Hv1a, a new polypeptide biological insecticide. World Pestic 44(7):29–40

Tudi M, Daniel Ruan H, Wang L, Lyu J, Sadler R, Connell D, Chu C, Phung DT (2021) Agriculture development, pesticide application and its impact on the environment. Int J Environ Res Public Health 18(3):1112

Vanyushin BF, Ashapkin VV, Aleksandrushkina NI (2017) Regulatory peptides in plants. Biochem Mosc 82:89–94

Wen Q, Sun M, Kong X, Yang Y, Zhang Q, Huang G, Lu W, Li W, Meng Y, Shan W (2021) The novel peptide NbPPI1 identified from Nicotiana benthamiana triggers immune responses and enhances resistance against Phytophthora pathogens. J Integr Plant Biol 63(5):961–976

Windley MJ, Herzig V, Dziemborowicz SA, Hardy MC, King GF, Nicholson GM (2012) Spider-venom peptides as bioinsecticides. Toxins 4(3):191–227

Wu Q, Patočka J, Kuča K (2018) Insect antimicrobial peptides, a mini review. Toxins 10:461

Yadav PK, Kumar S, Yadav S, Kumar S (2019) Role of aptamers in plant defense mechanism against viral diseases. Aptamers Biotechnolo Appl Next Gener Tool:169–174

Zhang YM, Ye DX, Liu Y, Zhang XY, Zhou YL, Zhang L, Yang XL (2023) Peptides, new tools for plant protection in eco-agriculture. Adv Agrochem 2(1):58–78

Zhou YL, Li XL, Zhang YM, Shi Y, Li HH, Zhang Z, Iqbal C, Ye DX, Li XS, Zhao YR, Xu WL, Yang XL (2022) A novel bee-friendly peptidomimetic insecticide: synthesis, aphicidal activity and 3D-QSAR study of insect kinin analogs at Phe2 modification. Pest Manag Sci 78(7):2952–2963

Plant Immunity Inducers: Strategies to Identify and Isolate Them to Boost Defense Responses in Plants

13

Ragiba Makandar

Abstract

Plants are constantly challenged by pathogens and pests during their growth period. Application of chemical pesticides to control pests and pathogens has been the most common method of controlling them worldwide. Long-term use of the chemical pesticides not only affects human and cattle health but also produces potential hazardous effects on the environment. Furthermore, due to improper usage of pesticides, the pathogens are evolving resistance to these pesticides. Taking into consideration the health concerns of the stakeholders, food safety, and environmental protection from chemical pesticides, it necessitates devising environmentally effective and cost-efficient strategies to control pathogens by means of boosting plant's own immunity. Understanding the principles of plant immunity, plant–microbe interactions, and the molecular mechanism/s underlying constitutive and systemic resistance induced by microbial or synthetic elicitors in plants would enable successful detection and isolation of plant immunity inducers. Priming plants with defense inducers will not only provide suitable, cost-effective control measures to control pathogens but also aid in facilitating an integrated pest management system in crop plants. Moreover, combination of chemical and biological agents could provide a wide spectrum of systemic resistance and defense responses against various pests and diseases for sustainable agriculture and crop protection. This chapter highlights the various aspects of plant innate immunity, constitutive and inducible defenses, compatible and incompatible interactions, host responses, and defense signaling leading to plant immunity inducers detection, their mechanism of action and strategies to

R. Makandar (✉)
Department of Plant Sciences, School of Life Sciences, University of Hyderabad, Hyderabad, India
e-mail: mragiba@uohyd.ac.in

S. Singh, R. Mehrotra (eds.), *Plant Elicitor Peptides*,
https://doi.org/10.1007/978-981-97-6374-0_13

be devised for their effective utilization to enhance protection against various pests and pathogens in crop plants.

Keywords

Effectors · Elicitors · Induced systemic resistance · Plant immunity inducers · Plant–microbe interaction · Systemic acquired resistance

13.1 Introduction

Plants undergo both abiotic and biotic stresses during their growth and development. The biotic component which includes pathogens and insect pests pose a serious concern to agricultural crops and economically important plants. Fungi, bacteria, virus, and nematodes are the chief causal agents of several systemic diseases, leading to poor quality and yield losses in crop plants (Baker et al. 1997).

Pesticide usage has been a norm to control pest and pathogens in agricultural crops (Thakur and Sharma 2024). However, the adverse effects caused by the usage of the chemical pesticides for prolonged periods on health and environment due to their hazardous nature, is a major concern thereby necessitating alternate approaches to mitigate crop losses by biotic agents. To minimize the harmful effects of these chemicals, integrated pest management (IPM) is being practiced globally by combining the available pest control methods with the cultural practices and a minimal usage of highly targeted pesticides. However, the emergence of pest resistance and unavailability of active substances are the limiting factors for implementing IPM in crop plants.

In order to overcome these limitations, the most feasible and effective strategy would be to explore and strengthen the host immune responses to confer broad spectrum resistance to pathogens. Furthermore, an insight into the interacting molecular components of the host and the pathogen would provide a basis to devise strategies to determine the plant immunity inducers (PII) that could be utilized to boost immunity in crop plants.

Identification, development, and application of plant immunity inducers serve the modern-day requirement to not only boost crop heath but also ensure food security and environmental safety. The development of PIIs is an emerging field of plant immunity to mitigate losses caused by pests and pathogens in crop plants with the prospects of commercialization of the PIIs as biopesticides (Qiu et al. 2017). Furthermore, enhancing resistance by application of PIIs to control pests and diseases systemically with minimum intervention of chemical pesticides lies within the framework of integrated pest management. This chapter highlights the significance of host–pathogen interactions in identifying the plant immunity inducers (PIIs) and the strategies to be employed to detect, isolate, and utilize PIIs to bolster defenses against pathogens in crop plants.

13.2 Significance of Host–Pathogen Interactions in Identifying Plant Immunity Inducers

To unravel the strategies evolved by pathogens, both adapted and non-adapted to a host, it is important to distinguish between constitutive and inducible defenses, genetic and molecular basis for resistance in the host and virulence in the pathogen, plant immunity mechanism/s, compatible and incompatible interactions, host defense responses, and defense signaling by analyzing host–pathogen interactions. It is also essential to compare the signaling pathways activated by an invading pathogen to those of a beneficial/ plant growth promoting microbes and utilize the information in devising effective strategies to develop PIIs for plant protection and growth promotion.

13.2.1 Constitutive Defenses in Plants

Plants, though sessile, have evolved a strong surveillance system to detect and ward off the invading pathogens. Plants perceive and respond by establishing preformed physical barriers and multi-layered inducible defense responses (Narváez-Barragán et al. 2022). Preformed physical or structural barriers such as multi-layered waxy layers and cuticle and, thickened epidermal, endodermal, and cortical cell walls and chemical barriers which include synthesis of anticipins, phenolic compounds, etc. are considered as "constitutive defenses" and a "first layer of defense" to block pathogen penetration in host cells. The constitutive defense mechanism contributes to the partial resistance against a broad spectrum of pathogens. An example of partial resistance observed due to elevated constitutive expression of defense-related genes against the rice blast fungal pathogen, *Magnaporthe oryzae* has been reported previously (Vergne et al. 2010). Another example of physiological resistance to fungal penetration was demonstrated by silicon (Si) accumulation in rice leaves on leaf surface, in epidermal cell walls, middle lamellae, and intercellular spaces (Kim et al. 2002). Glucosinolates, a class of secondary metabolites produced in Brassica members, are reported to possess antimicrobial and antioxidant properties to boost defense responses not only against a broad spectrum of pathogens (Singh 2017) but also against insect pests (Ajaharuddin et al. 2024).

13.2.2 Inducible Defenses in Plants

The inducible defenses get activated in a host in response to pathogen attack as an immediate defense response against the invading pathogen. It is also referred as

"basal resistance" or "general resistance" or non-specific resistance or "residual resistance' owing to its wide applicability against pathogens. The "innate immunity," which is an inbuilt defense mechanism in a plant, gets activated upon pathogen attack as an immediate defense response against the invading pathogens. The basal resistance is induced in response to the pathogen in the host, regardless of its genotype—resistance or susceptibility, however, its efficiency is determined by the level of pathogen virulence.

The non-aggressive or non-adapted pathogen could trigger strong basal resistance which is synonymous to non-host resistance. Niks and Marcel (2009) opined that non-host resistance and basal resistance may rest on the same or similar principles while analyzing the specificity of the pathogen to the host. Moreover, the non-host resistance is controlled by polygenes and show strong similarity and association with the basal resistance (Niks and Marcel 2009). Furthermore, the constitutive or preformed defenses when augmented by the inducible defenses would provide a broad-spectrum resistance against a variety of pathogens.

13.2.3 Monogenic Gene-for-Gene Concept

According to the classical monogenic gene-for-gene concept of Flor (1971), for every gene that conditions "resistance" in the host, there is a matching or corresponding gene that conditions "virulence" in the pathogen. The basis for such an interaction emphasizes interaction of the protein products of the two complementary genes, a resistance (R) from the host and the avirulence gene (Avr or AVR) of the pathogen resulting in resistance of the host. The matching Avr and R proteins confer a gene-for-gene resistance in plants (Vergne et al. 2010). It is also referred as race-specific resistance since single or individual R gene controls and provides complete resistance to some of the races of a pathogen species. In contrast, the non-race-specific resistance provides mostly partial, quantitative resistance (qR) to all races of a pathogen species and is independent of specific Avr genes (Sánchez-Martín and Keller 2021).

An example of direct interaction between *Pi-ta* (R) and *AVR-Pita* (Avr) gene products demonstrated the gene-for-gene interaction in rice conferring resistance against rice blast disease caused by *M. grisea* (Jia et al. 2000). The study also proposed a receptor–ligand model wherein the R gene product acts as a receptor to bind to a ligand or an elicitor which is either directly or indirectly produced by the Avr gene of the pathogen. The *Pi-Ta* gene in rice encodes a cytoplasmic receptor while the *AVR-Pita* gene in *M. grisea* encoded a metalloprotease protein with an N-terminal signal peptide. The transient expression of *AVR-Pita* gene in the cells of the *Pi-Ta* resistant genotype produced hypersensitive (HR) responses indicating the functional R-Avr mediated resistance in rice against *M. grisea* (Jia et al. 2000).

13.2.4 Receptor and Ligand Recognition

The molecular components of the host commonly termed as "pattern recognition receptors (PRR)" are found on the surface of the cell that are involved in the recognition process. However, the readiness with which plants respond to a variety of intruders largely attributes to the fact that the cell surface is instated with diverse and innumerable number of PRRs. The PRRs comprise both receptor-like kinases

(RLKs) and receptor-like proteins (RLPs) (Shiu and Bleecker 2003). While the RLKs with an extracellular ligand binding domain, a transmembrane domain, and intracellular protein kinase domain are embedded in the plasma membrane, RLPs possess extracellular and transmembrane domains but lack intracellular kinase domains. As a result, RLPs bind to other intracellular co-receptor kinases to transmit the signals downstream via defense signaling pathways (Boutrot and Zipfel 2017).

Earlier studies have demonstrated that plants could perceive diverse molecules via high-affinity cell surface PRRs and transduce the signals to activate induced defenses against the pathogens (Gohre and Robatzek 2008). The PRRs recognize the conserved molecular determinants, such as ligands or elicitors and subsequently, and transduce the signals to activate immune responses in the host. An elicitor is any molecule such as a DNA or a RNA, a peptide, an oligosaccharide, a chemical compound, a phytohormone, etc. capable of eliciting an immune response in the host. The recognition of such non-self-molecules by the host occurs based on the conserved molecular patterns known as PAMPs (pathogen associated molecular patterns) and MAMPs (microbe associated molecular patterns) (Jones and Dangl 2006). Whereas the damage associated molecular patterns (DAMPs) are related to the damage inflicted to the host plants as wounds or cuts by herbivores (Bent and Mackey 2007). The immune responses activated based on the recognition of PAMPs, MAMPs, and DAMPs in plants are collectively known as PAMP-triggered immunity or pattern triggered immunity (PTI).

The diversification of immunogenic patterns and the pattern recognition receptors results from the selective evolutionary pressure. Identification of an increasing number of immunogenic patterns indicated that PTI appears to be a universal defense mechanism and the evolutionary selection pressure drives the diversification of molecular patterns and PRRs. The experiments on Arabidopsis mutants lacking the flagellin receptor FLS2, a transmembrane protein with extracellular leucine-rich repeats (LRR) and an intracellular protein kinase provided the first evidence of the recognition of MAMPs in plants. Absence of the functional FLS2 receptor showed an increased susceptibility to *Pseudomonas syringae* pv. tomato strain DC3000 (Zipfel et al. 2004). In rice, transmembrane glycoprotein CEBiP induced basal resistance against the fungal pathogen, *M. oryzae* (Kaku et al. 2006).

13.2.5 Plant Immunity: A Two Pronged-Defense Strategy

Unlike animals, plants do not possess a systematized immune system and memory cells to recognize and respond to the pathogen. Lack of mobile defender cells and somatic adaptive immune system renders them to solely rely on the immune responses of host cells and the systemic signals generated from the sites of infection (Jones and Dangl 2006). In the race to "acquire arms," the pathogen and the host coevolve to outdo each other and thereby establish a dynamic state in which the pathogen evolves into an aggressive form with more virulence factors to overcome the counter defenses of the host (Tan et al. 1999). The co-evolutionary model proposed by Jones and Dangl (2006) is a two-pronged inducible defense strategy against both adapted and non-adapted pathogens of the host plant. The branch of immunity that functions via recognition of an elicitor such as MAMPs, PAMPs or DAMPs by the PRRs present on the surface of plant cells is known as PTI. Perception of PAMPs constitutes the first layer of innate immunity which is also referred as the basal immune resistance in plants. Plants compromised in PAMP perception showed enhanced disease susceptibility, demonstrating the significance of PAMP perception in activating PTI against pathogens (Zipfel 2008).

The branch of immunity that functions via recognition of microbial effectors or virulence factors that suppress PTI either directly through cognate R proteins or indirectly via interacting intermediate products in plants is known as effector-triggered immunity (ETI) (Jones and Dangl 2006; Bent and Mackey 2007). ETI is an immune reaction induced in response to a virulence-associated activity of a pathogen. Any pathogen-mediated activity that promotes pathogen replication or transmission is considered as virulence-associated activity. The pathogen-encoded virulence factors are also known as effectors. The effectors are proteinaceous molecules secreted by the pathogen to suppress immune responses and to promote pathogen growth in the host tissues. An effector molecule could also elicit immune responses upon recognition by a cognate PRR. The effector PvRxLR16 could elicit basal defense responses including reactive oxygen species (ROS) accumulation, cell death, and defense associated gene expression to confer resistance against *Phytophthora capsici* in *Nicotiana benthamiana* (Xiang et al. 2017). In response to effectors, the host also modifies effector-targeted proteins to resist pathogen invasion in the host cells (Jwa and Hwang, 2017). Both PTI and ETI activate immune responses resulting in a cascade of complex signaling events to combat pathogen attack (Saijo et al. 2018).

13.2.6 Incompatible and Compatible Host Interactions

Host–pathogen interaction is a hostile relationship and the outcome of the interaction is determined by the ability of one organism to outdo the other. It is a dynamic state in which both the host and the pathogen compete for survival leading to molecular reprograming of the cellular and biological processes (Manoharachary and Kunwar 2014). The outcome is either susceptibility or resistance depending on the level of virulence of the pathogen and resistance of the host.

A virulent pathogen is one which causes disease in susceptible host while the avirulent pathogen would elicit a strong defense response to restrict pathogenesis (Nimchuk et al. 2003). The microbial effectors when recognized by the matching R genes are referred as "Avirulence" genes and the outcome such an interaction would result in resistance reaction. The R and Avr interaction conferred resistance against adapted pathogens (Baker et al. 1997) and the interaction is considered as "incompatible" owing to the incompatibility of the pathogen to produce susceptible reaction in the host carrying the matching R gene. Furthermore, the R–Avr interaction activated HR to restrict pathogen spread in plants (Nimchuk et al., 2001).

On the contrary, absence of the matching R-Avr genes produces susceptible reaction in the host also known as effector-triggered susceptibility (ETS). The interaction is considered as "compatible" owing to the vulnerability of the host to the virulent pathogen. In the absence of the cognate R genes, effectors promote ETS pathogen growth or spread in host tissues. ETS could be attributed to the promotion of pathogen growth and spread in host tissues, in the absence of cognate R genes. The ETS could also be attributed to the suppression of PTI by the effectors or virulence proteins of the pathogen (Bhosle and Makandar 2021).

13.2.7 Resistance to Adapted and Non-adapted Pathogens

An adapted pathogen develops high specificity to the host due to its arsenal of pathogenicity and virulence genes and recurring ability to incite and cause infection on the host in an evolutionary time scale. On the contrary, a non-adapted pathogen could be any random pathogen other than a regular pathogen of the host. The blast disease causing pathogen, *M. grisea* is an adapted pathogen of rice (*Oryza sativa* L.). Whereas the same pathogen when inoculated on a different host such as the garden pea (*Pisum sativum* L.), it is considered as a non-adapted pathogen to the host garden pea and the resistance induced by a non-adapted pathogen in the host is referred as non-host resistance (NHR). The terminology NHR and race-specific resistance have been frequently used in plant breeding to distinguish the nature of resistance conferred by the host against non-adapted and adapted pathogens, respectively.

The NHR induced against non-adapted pathogens is manifested by a variety of layered defenses—both constitutive and induced defense responses. Whereas the incapability of the non-adapted pathogen to break these barriers results from molecular incompatibility between virulence factors of the host and their targets in the host (Ayliffe and Sørensen 2019). Though PTI could be elicited by both adapted and non-adapted pathogens, the ETI is contributed by the Avr gene product upon interaction with R gene products either directly or indirectly against adapted pathogens.

Conversely, a virulence factor, phevamine A is demonstrated to promote bacterial growth and virulence by suppressing plant immune responses using the model pathosystem, *P. syringae* pv. tomato DC3000 (Pto) and *Arabidopsis thaliana* (O'Neill et al. 2018). Another study has shown that an endogenous suppressor might suppress the elicitor-induced defense responses and promote susceptibility in plants even against non-adapted fungal and bacterial pathogens (Toyoda et al. 2023).

13.2.8 Regulation of Host Responses against Pathogens

Previous studies suggest that some of the underlying mechanisms of defense against pathogens by plant, invertebrate, and mammalian hosts may be similar suggesting a conserved signal transduction pathway to activate defense response in the host. The conserved or universal host response suggests the deployment of a common strategy for microbial pathogenesis. Several common features about the animal and plant bacterial pathogens have been noticed such as the secretory apparatuses used by both plant and animal bacterial pathogens to export virulence factors into the host cell, gene networks in signal transduction, and virulence mechanism along with production of antimicrobial peptides to mitigate the host defense responses (Tan et al. 1999).

So far the receptor proteins with a role in the recognition of PAMPs, MAMPs, and DAMPs in plants have been identified as cell surface-localized receptors. The surface localized receptors could be PRRs and nucleotide-binding domain and leucine-rich repeat (NLR)-containing proteins. The receptors, FLAGELLIN SENSING2 (*FLS2*) and EF-Tu RECEPTOR (*EFR*), recognize the PAMP molecules comprising a 22 amino acid length bacterial flagellin protein peptide (flg22) and a 18 amino acid length bacterial Elongation Factor Tu protein peptide (elf18), respectively, and these receptors belong to the leucine-rich repeat-receptor kinase (LRR-RK) family XII in *A. thaliana* (Zipfel et al. 2006). The NLR proteins are the plant immune receptors that are generally involved in ETI (Cui et al. 2015). The PPRs and NLRs are regulated at the transcriptional, post-transcriptional, and post-translational levels. Upon recognition of the elicitor molecules, an active receptor-kinase complex is formed at the plasma membrane which triggers phosphorylation events leading to downstream signaling processes. These processes involve formation of ion fluxes across the plasma membrane, the generation of ROS, activation of mitogen-activated protein kinase (MAPK) cascades, hormone signaling and subsequently defense gene expression accompanied by the production of secondary metabolites and proteins with a role in defense response (Boller and Felix 2009).

Studies in Arabidopsis have shown that both cell surface receptors as well as intracellular receptors are essential in activating immunity in plants. While the cell surface receptors recognize and activate immunity upon detection of the microbial elicitor or pathogen-derived molecules, the intracellular receptors activate immunity upon detection of pathogen-secreted effector proteins that function inside the plant cell. Pathogen recognition by surface receptors activates multiple protein kinases and NADPH oxidases, which are potentiated primarily by the intracellular receptors by increasing their abundance levels. Likewise, the HR, which depends on intracellular receptors, is strongly enhanced by the activation of surface receptors. While activation of either PTI or ETI alone was found to be insufficient to provide effective resistance against the bacterial pathogen *P. syringae*, PTI and ETI together could mutually potentiate and produce strong defense response in the host (Ngou et al. 2021). Moreover, earlier studies demonstrated that PTI and ETI share largely overlapping signaling networks and downstream responses (Qi et al. 2011).

Two transcription factors (TFs), WRKY53 and HSPRO2, were shown to function as positive regulators of basal resistance in Arabidopsis through gene knockout studies against *P. syringae* pv. tomato (Murray et al. 2007). Both WRKY53 and HSPRO2 appear to function downstream of salicylic acid (SA) and to be negatively regulated by jasmonic acid (JA) and ethylene (ET) indicating SA's vital role in mediating basal resistance in plants. Earlier studies also demonstrated that Arabidopsis NPR1 (At*NPR1*), a positive regulator of both basal resistance and systemic acquired resistance (SAR) in plants that also functions downstream of SA (Ryals et al. 1996; Cao et al. 1997). These studies emphasize the involvement of SA, an essential hormone in plant immunity. Further recent studies demonstrated the binding of SA to the transcriptional co-regulator, NPR1 is essential for SA-dependent defense responses. The findings also reveal AtNPR1 protein to function as a receptor to SA (Wu et al. 2012).

The SA-bound NPR1 dimer regulates transcription by associating with two TGA transcription factor dimers, forming an 'enhanceosome', a protein complex that congregates at the enhancer region of a DNA to regulate a gene target. With the recent discovery of the NPR1 as a SA-binding receptor, the missing link between NPR1 as a regulator of basal resistance and SAR and, SA's role in activation of defense in plants has been established (Wu et al. 2012). NPR1 also plays a key role in the regulation of the crosstalk between SA and other defense and growth hormones, such as JA by transcriptional reprogramming, stress protein homeostasis, and cell survival (Zavaliev and Dong 2024). Moreover, NPR1 coordinates its manifold functions through the formation of the nuclear and cytoplasmic biomolecular condensates (Zavaliev and Dong 2024). These cellular bodies are non-membrane bound compartments in the cell that serve as active sites for biological processing. Therefore, the findings suggest that manipulating NPR1 levels through genetic engineering will serve as a potential strategy to confer resistance to both biotic and abiotic stresses.

Research studies carried out to investigate the impact on primary metabolism by plant–pathogen interactions revealed upregulation of several genes associated with primary metabolic pathways to modulate signal transduction cascades that resulted in plant defense responses similar to exposure to pathogens or elicitors. These responses involved synthesis or degradation of carbohydrates, amino acids, and lipids (Rojas et al. 2014). Previous studies also observed that the major conversion of carbon from primary metabolism into secondary metabolism was directed in the biosynthesis of phenylpropanoid pathway products, also referred as phytoalexins, due to their antimicrobial properties and induction in response to pathogen attack (Bolton et al. 2008) Likewise, the shikimic acid pathway was found to be upregulated in response to pathogen infection which is mostly driven by phosphoenolpyruvate (PEP) from glycolysis and erthrose-4-phosphate from the stromal oxidative pentose phosphate (OPP) pathway and producing chorismate as an end product along with other intermediates possessing a role in plant defense (Bolton 2009). Moreover, the NADPH (nicotinamide adenine dinucleotide phosphate hydrogen) consumption is chiefly utilized by the ROS-producing NADPH oxidase, a major source of ROS in plant cells (Lamb and Dixon 1997).

The enzymes that regulate defense responses in host include Phenylalanine Ammonia-Lyase (PAL), Polyphenol oxidases (PPO), class III peroxidases (Poxs), chitinases, and β-1, 3-glucanase. PAL is involved in the biosynthesis of phenolic compounds with antimicrobial properties is essential disease resistance responses in plant (Waewthongrak et al. 2015). PPO mainly found in cytosol is required for oxidation of phenols and synthesis defense-related phenolic compounds as well as lignin biosynthesis in response to pathogen attack. Another enzyme, chitinase is one of the defense-related enzymes that hydrolyze the cell wall in most phytopathogenic fungi which possess chitin in their cell walls. Gene expression of defense-related enzymes such as PAL, PPO, Poxs, chitinases, and β-1, 3-glucanases correlated with decrease in the disease development in plants (Kumari and Vengadaramana 2017).

13.3 Plant Immunity Inducers (PIIs): Recognition, Application, and Prospects

The importance of PIIS is felt globally not only for the minimal adverse effects they produce in comparison to the chemical pesticides, but also with immense prospects for their industrial production for commercialization purpose. Besides contributing to increased crop production, PIIs should serve the purpose of effective, efficient, and eco-friendly biopesticides with high cost-benefit effects. Therefore, to identify and isolate effective PIIs, it is vital to understand their nature, mechanism of action, signaling elicited by their application in the host to devise strategies for developing and utilizing them to boost protection against various pests and pathogens in crop plants.

13.3.1 Nature and Types of PIIs

Plant immunity inducers (PIIs) are mostly derived from biological entities such as microbial organisms, plants, and animals and their active molecules, biochemical compounds, metabolites, and other products formed during their interaction with the host. The PIIs also include both natural and synthetic compounds that are applied exogenously as immunity boosters in plants. Based on their chemical properties, the PIIs are grouped into nucleic acid molecules, oligosaccharides, proteins/peptides, microbial effector proteins, glycopeptides, lipids, lipopeptides, small metabolite molecules, phytohormones, chemical analogs of phytohormones, and other chemical substances. These PIIs could serve as valuable resources to develop potential biopesticides or boosters of plant immunity against broad range of pathogens by further experimentation to validate their efficacy in both controlled and field-based plant growth conditions.

The harpin protein isolated from a fire blight causing phytopathogenic bacterium *Erwinia amylovora* was demonstrated as an elicitor of HR (Wei et al. 1992). Harpins are small glycine-rich heat stable secretory proteins targeted to extracellular region by phytopathogenic gram-negative bacteria to modulate host cellular functions. The plant immunity-inducing protein, PeaT1 which was obtained from the fungus, *Alternaria alternata* enhanced resistance to viral pathogens in plants. The PeaT1-treated plants exhibited enhanced systemic resistance against tobacco mosaic virus (TMV) thereby activating SAR pathway mediated by SA and the *NPR1* gene indicating that SA accumulation was essential (Zhang et al. 2011b).

Chitosan, a linear polysaccharide derived from chitin causes hydrolysis of peptidoglycans of harmful bacteria which in turn results in leakage of intracellular electrolytes thereby causing death of those bacterial cells. In addition, chitosan application in plants activated several defense signaling genes such pathogenesis related (PR) proteins and phytoalexins to defend the intruding phytopathogens via nitric oxide (NO) pathway (Zhang et al. 2011a). While analyzing the defense responses in chitosan-elicited tomato cells, it was found out that NO is essential in the generation of the lipid secondary messenger, phosphatidic acid (PA) which could be produced via both phospholipase-D (PLD) and phospholipase-C (PLC)/diglycerol kinase (DGK) pathways (Raho et al. 2011).

Yang and associates (2022) described PIIs as the biological factors which are present in plants, animals, and microorganisms that can activate immune responses in host against invading pests and pathogens. However, a few other studies also indicate that not only the biological factors, any molecule or a compound that is capable of eliciting an immune response or defense response can also act as a PII (Bektas and Eulgem 2015). Burketova et al. (2015) while reviewing the bio-based resistance inducers for sustainable crop protection highlighted the role of several bacterial, fungal, synthetic elicitors as well as algal and plant extracts in activating immune responses in plants. A list of elicitors of bacteria, fungi, and insects that could activate plant defense responses were described by Malik et al. (2020). The microbial or plant-based elicitors such as proteins, peptides, carbohydrates, lipids, lipoproteins, nucleotides, and other chemicals identified that could serve as potential PIIs have been provided (Yang et al. 2022). The potential bacterial elicitors (Orozco-Mosqueda et al. 2023) and oligosaccharides (Guarnizo et al. 2020) identified with a role in plant immunity were presented.

Burketova et al. (2015) reviewed several bacterial and fungal derived molecules with a role in eliciting immune responses in plants. Of the several mobile signals that were found to be crucial in activating SAR in plants post-pathogen infection, methylated SA (MeSA), a volatile organic compound (VOC) is an important mobile signal (Park et al. 2007). MeSA along with a non-proteinaceous amino acid, pipecolic acid (Pip) are involved in long-distance signaling and promoting SA accumulation and priming, and SA signaling (Navarova et al. 2012). Kachroo and Robin (2013) while reviewing the defense signaling involved in SAR in plants pointed that acid diazelaic acid (AzA), glycerol-3-phoshate (G3P), abietane diterpenoid or dehydroabietinal (DA) also serve as mobile SAR inducers along with SA and MeJA. The plant hormone, SA has been demonstrated as the endogenous signaling molecule that gets accumulated in response to the infection by a necrotizing pathogen causing redox changes in infected plant cells triggering SAR in plants. Among the synthetic compounds that trigger SAR, SA and its analog, BTH (benzo[1,2,3]thiadiazole), and MeSA have been reported as potential chemical activators of immunity in plants.

A variety of plant growth promoting rhizobacteria (PGPR)-based elicitors with a putative role in systemic resistance have been previously reported (Pršić and Ongena 2020). One of those elicitors, rhizobacterial CLPs (cyclic lipopeptides) are secondary metabolites that contribute to the biocontrol property and antimicrobial ability of the biocontrol bacteria of the genus, *Bacillus*. Another PGPR metabolite with a role in plant defense is secreted as high-affinity iron-chelating molecules termed as siderophores by the biocontrol agents (BCA). Their primary role is to function as $Fe3^{+}$-acquisition systems which not only enables the BCA to propagate in high Fe (iron)-limited conditions but also to antagonize other pathogenic microbial organisms by depleting the Fe-content and creating a competition for the nutrient. The siderophores produced by various fluorescent pseudomonads known as pyoverdines or pseudobactines were considered as the first PGPR metabolites with a role in eliciting systemic resistance by *Pseudomonas* sp. in different plant species against a wide range of microbial pathogens (Meziane et al. 2005).

A few other secondary metabolites best identified for their antimicrobial function which could serve elicitors were the non-ribosomal peptides, polyketides, bacteriocins, terpenes, phenazines, quinolones, and rhamnolipids that can stimulate induced systemic resistance (ISR) as promising tools for biocontrol against pathogen populations (Pršić and Ongena 2020). Some of the chemicals or synthetic elicitors which when applied as exogenous treatments to plants have resulted in priming the systemic resistance in the way similar to that activated by micronbial pathogens. The natural plant hormones such as SA, JA, and systemin, the chemical elicitors such as DL-β-aminobutyric acid (BABA), oxalic acid, 2,6-dichloro isonicotinic acid (INA) and its derivatives, BTH and its derivatives have been effective in inducing basal and systemic resistance in plants (Henry et al. 2012). Therefore, based on the principles of plant immunity and the information and technology available, it is possible to identify the elicitors which could enhance immunity against pests and pathogens. The techniques used in crop breeding to deploy resistance against pests and pathogens could be complemented by the application of PIIs to confer enhanced resistance against a broad spectrum of pathogens.

13.3.2 Mechanism of PIIs

The recognition of the role of extracellular DNA (exDNA), a key component of root exudates from the border cells of roots in plant defense was revealed to function as "extracellular trapping" of pathogens by Hawes et al. (2012). The exDNA produced and transported into the surrounding mucilage was demonstrated to attract, trap, and immobilize the pathogens in a host–microbe specific manner. Furthermore, the resistance of the root cap to infection was abolished when this plant exDNA was degraded. Recent studies have established plant immune response to a DNA molecule as highly self- or non-self-specific. A plant's self-DNA triggered stronger responses by eliciting early immune signals such as H_2O_2 formation when compared to non-self-DNA from closely related plant species (Vega-Muñoz et al. 2023). Studies have shown that extracellular fragmented self-DNA (esDNA) acts as a signaling molecule (or second messenger) to trigger inhibitory effects on conspecific plants. Application of common bean esDNA to common bean leaves induced ROS and defense responses, which were not elicited by extracellular heterologous DNA (etDNA) confirming the specificity of esDNA to induce plant cell responses and to trigger early signaling events (Duran-Flores and Heil 2014, 2018). These defense responses to the esDNA were similar to plant responses to biotic attack, where ROS production, along with calcium signaling and plasma membrane potential (Vm) depolarization were observed as early signaling events preceding the buildup of chemical defense (Barbero et al. 2016). Furthermore, recent findings also indicated that esDNA could serve as a DAMP molecule to trigger immunity in plants against pathogen attack. The esDNA-induced defense responses were activated in the host species, Arabidopsis (*A. thaliana*) and tomato (*Solanum lycopersicum* L.) by JA production and the expression of JA-responsive genes in a concentration- and species-specific manner against the pathogens, *P. syringae* pv. tomato DC3000 and *Botrytis cinerea,* respectively. The esDNA-mediated growth inhibition and ROS production were achieved through the JA signaling pathway which was impaired in the JA-related mutants (Zhou et al. 2023).

The recognition of the highly conserved microbe-specific molecules has been previously demonstrated to activate PTI in plants. The most common bacterial and fungal elicitors, namely flg22, elf18, peptidoglycans (PGNs), lipopolysaccharides (LPS), Ax21 (Activator of XA21-mediated immunity in rice), fungal chitin, and *β*-glucans from oomycetes have been well studied and applied for activating innate immunity in plants (Newman et al. 2013). With the discovery that PTI is active against bacterial and fungal pathogens, antiviral immunity was also tested through viral suppressors in plants. Cross protection is phenomenon to control diseases caused by viral pathogens by priming with a mild or attenuated viral strain. However, due to non-availability of mild strains or attenuated mutants of plant viruses, the application of cross protection to control viral pathogens is limited (Huang et al. 2019). The possible role of double-stranded RNAs (dsRNA) as well as other viral nucleic acids was identified as candidates for viral PAMPs in plants (Amari and Niehl 2020). The viral PAMPs might be recognized by the plasma-membrane-associated receptor and co-receptor proteins namely BAK1, BKK1 (BAK1-Like 1), Serk1, and NIK1 with a potential role in antiviral PTI (Guzma´ n-Benito et al. 2019). The viral PAMPs might be perceived by the receptors that are recruited to viral replication compartments (VRC) or after relocalization of receptors to the endosomes since viruses could rewire host membrane and transport pathways (Beck et al. 2012; Frescatada-Rosa et al. 2015).

Most of the studies to understand plant immunity have been focussed at investigating the transcriptional reprogramming induced by pathogen infection. As a consequence, little is known about post-transcriptional processes controlling disease resistance. The regulatory role of the non-coding RNAs (ncRNAs) in plant immune responses has been emphasized (Wang et al. 2017; Zaynab et al. 2018; Song et al. 2021). A number of plant ncRNAs directly target the expression of signaling components downstream of the nucleotide-binding domain and NLR receptors that include kinases, transcription factors, and other defense-related proteins (Song et al. 2021). On the contrary, a long-noncoding RNA, designated as ELF18-INDUCED LONG-NONCODING RNA1 (ELENA1) from *A. thaliana*, showed a positive role in regulating plant innate immunity against *P. syringae* pv tomato DC3000. The ELENA1 knockout mutants showed decreased levels of pathogenesis related gene1 (*PR1*) gene expression and were susceptible to the pathogen, while overexpressing ELENA1 plants showed enhanced resistance accompanied with elevated *PR1* gene expression upon treatment with elf18. Furthermore, the studies demonstrated that ELENA1 is required by another mediator gene, MED19a to regulate *PR1* expression implicating the role of long-noncoding RNAs in the transcriptional regulation of immune responses.

Another class of ncRNAs designated as microRNAs (miRNAs) comprises small ncRNAs that direct gene silencing at the post-transcriptional level. The miRNAs, which are 22 nt in length, are shown to target multiple NLR motifs, such as TIR1, kinase-2, and P-loop motifs. These miRNAs trigger the production of phased-small interference RNAs (pha-siRNAs) of the NLRs to target those NLRs (Fei et al. 2013). Furthermore, miRNAs guide cleavage of R genes to repress ETI-based resistance. Previously, their role as negative regulators of plant defense has been demonstrated in case of TIR-NB-LRR immune receptor gene, N that confers resistance against TMV pathogen in tobacco (Li et al. 2012). On the contrary, a member of the miR812 family of rice miRNAs was identified as a positive regulator of defense responses against rice blast fungus *M. oryzae*. The miR812w, a DCL3- dependent 24 nt long miRNA, derived from the Stowaway type of rice MITEs (miniature inverted-repeat transposable elements) directs cytosine DNA methylation in target genes as well as its own locus to regulate disease resistance (Campo et al. 2021). The association between miR812w and Stowaway MITEs suggested the role of miRNAs in not only post- transcriptional regulation but also a regulatory function at the transcriptional level. The CRISPR/Cas9-induced mutations in MIR812w lead to enhanced susceptibility, while its overexpression conferred resistance to the rice blast fungus *M. oryzae*.

13.3.3 PII Induced Defense Signaling

Earlier studies have demonstrated that prior inoculation with pathogens can induce resistance in plants to subsequent infection. The microbial pathogen triggered immunity in plant has been characterized as a complex sequential process that induces defense responses not only at the site of infection but also systemically in all parts of the plants. A prior exposure to a necrotising pathogen would lead to a long-lasting resistance even in the distantly located part of the plant and is known as SAR. SA and MeSA have been shown to act as the long-distance signals, which accumulate in infected tissues (Park et al. 2007). Likewise, SA methyltransferase (SAMT) transcript accumulation in the local infected tissues is central for TMV-induced SAR in tobacco. In addition to SAR in tobacco, SAR in potato and Arabidopsis were associated with SAMT and MES, with MeSA as a potential long-distance signal (Park et al. 2009; Chen et al. 2019). SAR is SA-dependent defense response and NPR1 gene has been demonstrated to regulate SAR in plants via its nuclear role activating the expression of *PR1* that serves as a SAR marker (Conrath et al. 2015). Likewise, NPR1 is also involved in the regulation of ISR in plants but via its cytosolic role in plants thereby activating JA signaling. Both SA and JA signaling pathways have been shown to antagonize each other but with certain key players like NPR1 which regulate the crosstalk to fine tune the signaling events in plants.

The SAR-based immune responses were found to be effective against biotrophic pathogens and also against hemibiotrophs but during early phases of infection as shown against Fusarium head blight (FHB) caused by the hemibiotrophic fungal pathogen, *Fusarium graminearum* in wheat and Arabidopsis (Makandar et al. 2006). While SA and MeSA activated immune responses during the early phases of infection by *F. graminearum* in wheat and Arabidopsis plants, the signaling molecules JA and methyl jasmonate (MeJA) were involved in defense responses at the later stages of infection (Makandar et al. 2010, 2012; Nalam et al. 2015).

While SAR is conferred by a pathogen infection in plants, the ISR is induced by non-pathogenic or beneficial microbes such as PGPRs. ISR resembles SAR considering its mechanism of operation that involves stimulation of early cellular immune-related events upon recognition at the plant cell surface by PRRs followed by activating systemic signaling via a fine-tuned hormonal crosstalk and expression of PR proteins. The activation of the octadecanoid pathway induces the expression of several genes with the end product as JA (Shah et al. 2014). Together SAR and ISR signaling events are modulated by transcriptional reprogramming of genes central to regulation of defense such as expression of PR proteins with antimicrobial activity such as chitinases and β-1,3-glucanases (Xie et al. 2015). Changes in antioxidant systems along with phytoalexins, induction of phenolic compounds, deposition of callose and lignin for cell wall strengthening were observed in manifestation of systemic resistance against pathogen attack (Mandal and Mitra 2007; Davar et al. 2013).

Furthermore, SAR confers resistance to biotrophic and hemibiotrophic pathogens, while ISR is effective against necrotrophic pathogens by activating cell signaling pathways (Pieterse et al. 1996, 2012, 2014). This in turn regulates the synthesis and accumulation of several cellular products, such as phytohormones, enzymes, proteins, and secondary metabolites to restrict pathogen invasion and spread in plant cells. The crosstalk between different phytohormones allows fine-tuning of immune responses toward the specific invading pathogen. Using the model plant, Arabidopsis and its interaction with different pathogens, a mutual antagonism between SA, and JA-ET-dependent defenses and its effect on immune responses against biotrophic and necrotrophic pathogens have been previously reported (Spoel et al. 2007). It suggests that plant deploys a common defense machinery but regulated by a complex network of pathways which exhibit trade-offs between plant defenses against pathogens with different lifestyles for efficient biocontrol potential of plants.

13.3.4 Recognition of Plant Immunity Inducers

Plant protection through plant immunity inducers is gaining more attention with focus on HR proteins, secondary metabolites, oligosaccharides, sea-weed/ plant, and endophytic extracts or molecules (Dewen et al. 2017). Understanding how specific transcriptional, proteome, and metabolome changes are associated with biotic stress adaptations and identifying the central hubs controlling those interactions will be central to detect the PIIs. Also, utilizing plant–microbe interactions, it is possible to detect PIIs by analyzing the biological processes, cellular responses, and molecular mechanism underlying the specific interaction. The plant–microbial interactions may be classified as positive interactions comprising symbiotic or non-symbiotic associations and negative interactions comprising competition or parasitism associations (Bais et al. 2006).

The microbial elicitors upon recognition by the plant cell receptors result in the activation of signaling pathways resulting in the production of secondary messengers and activation of TFs to regulate PR gene expression as well as enzymes involved in secondary metabolites synthesis. The study on microbial elicitors inducing the accumulation of secondary metabolites in host plants has immense potential to induce or increase the synthesis of these metabolites along with novel bioactive molecules with a role in plant immunity. To date, only a few of the secondary messengers such as NO, JA, and hydrogen peroxide (H_2O_2) with a role in secondary metabolite synthesis have been reported to be elicited by fungal elicitors (Chamkhi et al. 2021). Furthermore, in response to the microbial elicitors, the phytochemical classes such as terpenoids, phenolic acids, and flavonoids were either synthesized and/ or increased in medicinal plants indicating their role in drug discovery (Chamkhi et al. 2021). Glyceollins, a class of antimicrobial prenylated pterocarpans were produced in response to fungus elicitation in soybean seedlings and also when primed with ROS prior to elicitation with *Rhizopus oligosporus or R. oryzae* (Kalli et al. 2020).

Plants produce an array of secondary metabolites and some of them possess antimicrobial activities. As a pre-existing or constitutive defense mechanism, plants produce preformed antimicrobial compounds also known as "phytoanticipins" such as phenols and phenolic glycosides which could be explored as potential PIIs. These constitutive plant compounds are also known to exhibit antifungal properties namely sulfur compounds, saponins, cyanogenic glycosides, and glucosinolates (Osbourn 1996). The toxic action of saponins to fungi is associated with the ability of these compounds to complex with membrane sterols and cause pore formation while plants may protect themselves from their own saponins by compartmentalizing them in the vacuole or in other organelles. To counter the effect of saponins, the fungi either modify their membrane composition as in case of certain phytopathogenic fungi to tolerate saponins or perform enzymatic detoxification through specific glycosyl hydrolase enzymes. These enzymes remove sugar moiety from saponin backbone rendering it ineffective to form a complex with membrane sterols and thereby prevent its antifungal activity. These fungal proteins could also serve as effector proteins or molecular determinants of fungal pathogenicity in crop plants. Screening plant metabolites to detect antimicrobial properties could be performed to utilize them as potential elicitors to activate defense responses in plants.

Transcription factors (TFs) function as either activators or repressors in regulating gene expression by binding to cis-acting elements in the promoter regions of the target genes. They modify the expression of the target promoters based on their interaction with other cofactor and corepressor molecules. Out of the 67 TF families identified in potato, 9 TF families namely ARF, NAC, WRYK, AP2/ERF-ERF, AP2/ERF-DREB, ZFP, TCP, bZIP, and BELL are reported to regulate biotic stress responses (Chacón-Cerdas et al. 2020). The defense responses activated by these TFs result in the regulation of PR gene expression, cell wall reinforcement, regulation of SA pathway genes, and activation of antioxidant enzyme activity against the invading pathogens. Furthermore, the oxidative stress responses enable plants more resistant to biotrophic or hemibiotrophic pathogens. Whereas, oxidative burst renders plant susceptible to necrotrophic pathogens since necrotrophs derive their nourishment from dead and decaying host tissues. Considering the diversity among the pathogens in their mode of nourishment/ infection, it necessitates to detect the regulatory components and the gene network pathways conferring resistance in crop plants.

Silicon (Si) is the second most abundant mineral element in soil after oxygen and shown to reduce both biotic and abiotic stresses in plants. Several lines of evidence has been reported indicating that Si application enhanced resistance to fungal diseases in plants. Si application resulted in silicified cells in rice leaves and stems (Kim et al. 2002), other plant structures such as trichomes thereby forming a physical barrier to restrict the growth of invading pathogens (Pozza et al. 2015). Besides providing structural defense, Si also enhances biochemical responses such as antioxidant properties and accumulation of antimicrobial secondary metabolites indicating its potential as plant immunity inducer. A recent study revealed that Si conferred resistance to brown stripe disease caused by *Bipolaris setariae* in sugarcane by inducing ROS scavengers, hormone signaling, antifungal metabolites, and silicon deposition (Chen et al. 2024).

The algal polysaccharides are reported to function as physiological stimulants and resistance elicitors in crop plants. The polysaccharides derived from the seaweeds (macroalgae) such as carrageenans, fucans, laminarans, and ulvans possess enormous potential as elicitors of defense responses in plants (Stadnik and de Freitas 2014). The carrageenans obtained from red algae and fucans found in the cell walls of brown algae and vulvans (both poly- and oligosaccharides) from the cell walls of green seaweeds of *Ulva* spp. are reported to induce resistance in plants. Furthermore, carrageenans and fucans are being utilized as constituents of several commercial fertilizers and biostimulants. The laminarans from *Laminaria digitata* are the most commercially explored to protect plants against a broad spectrum of pathogens. These sea-weed based polysaccharides could be exploited for wider applicability in crop plants as potential new resistance elicitors in agriculture.

The red maple (*Acer rubrum*) leaves are rich sources of polyphenols (PPs) and numerous other phenolic compounds which include gallate derivatives and gallotannins (Zhang et al. 2015). The red maple leaf extracts (RME) when tested for their plant defense induction (PDI) properties, induced HR reaction-like lesions with several upregulated gene expression of antimicrobial markers such as *PR1*, $\beta-$ 1,3-glucanase *PR2*, chitinase *PR3*, and osmotin *PR5* in tobacco leaves following RME treatment. Furthermore, galloyl glucose was detected as the bioactive molecule that might be contributing to the PDI activity of RME (Peghaire et al. 2020).

The communication between plants and the associated beneficial microbes such as PGPRs has been reported to alleviate plant diseases, both directly and indirectly through microbial metabolites and signaling components. PGPR-mediated induced resistance against biotic or abiotic stresses show minimum fitness cost caused by inducible defense mechanism on the wellness of the host. The anti-pathogenic metabolites such as antibiotics, lytic enzymes, hydrogen cyanide, and several other microbial synthesized molecules directly impact pathogen growth in plant tissues and indirectly through activation of ISR (Khoshru et al. 2023). The PGPR association with roots promotes not only plant growth, systemic resistance to pathogen and pests, but also establishes a plant-specific "core root microbiome" (Bukhat et al. 2020). The molecular determinants of the core root microbiome could be detected and utilized to develop synthetic bacterial or fungal consortia that are eco-friendly, non-hazardous, and cost-effective. These microbial consortia may serve as PGPR-based crop biofertilizers for integrated pest management and sustainable agriculture.

The members of the genus, *Trichoderma* are non-virulent symbiotic fungi known to promote plant growth besides functioning as promising biological control agents (Hermosa et al. 2012). The key strategies employed by *Trichoderma* in direct conflict with pathogens include competition, mycoparasitism, and antibiosis (Karuppiah et al. 2019). The *Trichoderma* fungal strains could colonize the rhizosphere and establish symbiotic association with plant roots while exhibiting antibiosis to other fungal pathogens infecting castor by inducing systemic resistance (Pradhan et al. 2023a,b). Previous studies revealed that a hydrophobin-like elicitor, Sm1 of the beneficial soil-borne fungus, *T. virens* induced systemic resistance in maize (Djonović et al. 2007). Two elicitors, Sm1 from *T. virens* and Epl1 from *T. atroviride*, have different specific targets in the same pathway to counteract pathogens with different life styles (Salas-Marina et al. 2015). Likewise, peptaibols, the linear peptide antibiotics produced by *Trichoderma* spp. showed systemic resistance response in cucumber plants against the pathogen, *P. syringae* pv. *lachrymans* (Viterbo et al. 2007). Two synthetic 18-amino-acid peptaibol isoforms when applied to cucumber seedlings also induced systemic protection suggesting their role as potential chemical elicitors. However, *A. thaliana* cell cultures and leaves when treated with peptaibol, alamethicin induced cell death/ lesions indicating phytotoxicity in other plant species (Rippa et al. 2010). The secondary metabolites, harzianolide and pentyl-pyranone from *Trichoderma* species activated defense responses suggesting their putative role as plant immunity inducers (Vinale et al. 2008).

Majority of differentially expressed genes (DEGs) that were downregulated in *Trichoderma* treated castor were related to stress, HR response, ROS, phenylpropanoid pathway, secondary metabolite biosynthesis, etc., indicating that host perceives *Trichoderma* as no obvious threat in order to allow the symbiotic relationship to be established between them (Pradhan and Makandar, 2023). Upregulation of TFs such as *WRKY71*, *RPV*, *ZAT5,* and *RPS2* with a putative role in defense suggests that these genes might be crucial for *Trichoderma*-mediated responses in castor.

The biocontrol ability of the endophytic fungal (Nisa et al. 2015) and bacterial (Li et al. 2018) strains have been emphasized as promising control measures of plant diseases. The endophytic fungal extract, ZhiNengCong (ZNC) is widely used in China as an efficient, environment-friendly plant immunity inducer not only protecting crops against *P. syringae* pv. tomato (Pst) DC3000 but also against viruses at very low concentration (Peng et al. 2020). The secondary metabolites, N-acyl-homoserine lactones (AHLs) produced by certain gram-negative bacteria are involved in cell-to-cell communication to monitor their behavior and serve as quorum-sensing molecules. Some AHLs besides their role in communication between bacterial cells are also involved in signaling inter-kingdom interactions thereby promoting growth and immunity in plants (Schikora et al. 2016). Similarly, some volatile organic compounds (VOCs) are reported to induce systemic resistance in the host plants by functioning as infochemicals involved in inter-kingdom communication (Kai et al. 2016).

13.3.5 Strategies for Isolating and Analyzing the Role of PIIs

Many plant immune elicitors have been successfully identified such as flagellin from *P. syringae* (Felix et al. 1999) and glycosyl hydrolase PsXEG1 from *P. sojae* (Ma et al. 2015). Isolation and purification of these plant immunity inducers from the culture extracts of microorganisms could be achieved through biochemical approaches. Using ion exchange chromatography chromatography (IEC) and high-performance liquid chromatography (HPLC), high-throughput detection of differentially expressing metabolites could be performed. Furthermore, the compositions of those single components could be identified by mass spectrometry (MS). Studies also showed that using nuclear magnetic resonance spectroscopy and fast atom bombardment analysis, cerebroside, a glycosphingolipid from *M. oryzae* was detected (Koga et al. 1998).

The elicitors could serve as potential biostimulants that are environmentally sustainable with no impact on human health. The PAMPs, MAMPs, and DAMPs detected through plant–pathogen interaction or plant–microbe interaction or plant–insect/herbivore interaction studies respectively contribute to the repertoire of conserved structural patterns or molecules for boosting plant immunity. Since these conserved structural patterns specifically bind to the receptor proteins, it is a promising approach to use them as baits for the detection of candidate PRRs or receptors. The most active of the elicitor compounds can be isolated and purified by biotechnological intervention and mass production for commercialization purpose. Ectopic expression of the genes that encode elicitors in plants is also an effective strategy. For instance, the overexpression of harpin-encoding genes confers enhanced resistance to pathogens in crop plants (Du et al. 2018). Alternatively, new structural derivatives of these compounds with higher activity and lower susceptibility to degradation and toxicity could be obtained.

Using computational and bioinformatic studies, new PIIs have been detected based on sequence similarity of existing PIIs as query sequences in BLAST search. A necrosis and ethylene-inducing peptide (Nep) was initially identified from a necrotrophic fungal pathogen, *Fusarium oxysporum* (Bailey 1995). The orthologs of Nep1-like proteins (NLP) were identified based on nucleotide similarity in bacteria, fungi, and oomycetes (Oome et al. 2014). A highly conserved sequence of 24 amino acids (nlp24) of NLP protein was found to be sufficient to induce immune responses in plants and subsequently, the NLP receptor, RLP23 has also been detected in *A. thaliana* (Albert et al. 2015). Likewise, other microbial elicitors, oomycete elicitin and bacterial flagellin were also detected based on sequence homology studies (Derevnina et al. 2016).

The chemical inducers identified through large-scale screening could be utilized in the development of novel PIIs including natural elicitor based chemical derivation, bifunctional combination, and computer-aided design. Chemosynthesis of PIIs is gaining importance with the advent of high-throughput screening systems and comprehensive in silico and in vivo analysis of synthetic plant immunity inducers. With the availability of a diverse collection of known synthetic and nature plant immune inducers, comparison between known elicitors may help in identifying specific moiety critical to the immune inducing ability. The specificity of functional moiety and the conserved pattern of known immune elicitors could be used as leads to design new elicitors. These simple chemical derivatives of known PIIs could be developed as potent immune elicitors (Zhou and Wang 2018).

The combination of a known synthetic plant immunity inducer with another functional compound could result in a new synthetic compound with potential elicitor activity. The combination of 3,4-dichloroisothiazoles with fungicidal strobilurins produced a new synthetic compound with high fungicide activity (Chen et al. 2017). The natural conjugation of JA and isoleucine (IL)—JA-Ile was previously identified as an endogenous bioactive JA molecule that enhances resistance to pathogens. The bifunctional combination approach which combines a known synthetic PII with another compound resulting in a highly effective final product resulted in several endogenous bioactive JA molecules such as JA-Leu, JA-Val, JA-Met, and JA-Ala. By integrating the structural information of the bioactive molecules such as covalent combination, ionic pairing, etc., it may be possible to develop effective elicitors. Using this strategy, several immunity inducers including SA, BTH, INA, BABA were paired with the cholinium cation to form ionic liquids (Kukawka et al. 2018).

Computer-aided design (CAD) of elicitors is considered a new strategy to generate new PIIs. Advances in high-performance computing have made it possible to screen innumerable lead-like molecules computationally for pesticide discovery and property analysis (Burden et al. 2016). Furthermore, based on structural recognition, a large-scale virtual screening of new leading compounds could be carried out to detect the best ligand and their target receptor molecules. The CAD-based screening would not only enable screening of infinite number of elicitors in less time but also minimize the actual experimentation costs to test for their antimicrobial properties in the host plants to a select few putative elicitors.

Identifying and analyzing host genes with a role in defense enabled detection of gene targets in the host as well as defense signaling pathways in crop plants. The host components comprise the putative receptors, TFs, resistance factors, genes encoding defense-related enzymes, secondary metabolites, ROS, HR, and PR proteins. Gene knockouts and targeted gene silencing were used as a potential approach to identify the gene variants for resistance or susceptibility to pathogens. With the discovery of CRISPR-Cas gene editing tools, it is now possible to alter expression of the specific gene/s with precision for developing resistance in crop plants. While majority of the studies carried out to enhance resistance in crop plants were focussed on positive regulators, it will be worth explore negative regulators—that have the potential to provide more stable and durable resistance.

Utilizing the sophisticated next-generation sequencing platforms, the host molecular components modulated by the pathogen could be revealed. These platforms will also reveal the genes coding for defense-related enzymes and PR proteins induced in the host to confer resistance to the pathogen. Priming by various abiotic compounds such as silicon, salicylic acid, acetyl salicylic acid (aspirin), β-amino Butyric Acid (BABA), polyacrylic acid, oxalic acid, benzothidiazole, and biotic inducers from fungal and bacterial endophytes also increased the level of defense-related enzymes in the primed plants against a wide variety of pathogens.

Identifying and analyzing novel effector proteins would also provide a wealth of knowledge to use them as baits to detect cognate receptors or resistance genes in plants. Analyzing compatible and incompatible interaction of adapted and non-adapted would enhance plant immunity against pests and diseases. Further studies using model pathosytems with pathogens of diverse lifestyles would provide insights into the repertoire of effectors, elicitors, receptors, and other defense-related genes involved in those host–pathogen interactions.

13.3.6 Promising New Strategies for the Utilization of Plant Immunity Inducers

A new strategy has been proposed to control plant diseases by combining two effective methods: use of the plant Pep elicitor to enhance resistance to plant parasitic nematodes and use of *B. subtilis* for efficient delivery of these elicitors (Hiltl and Siddique 2020). This strategy will provide new avenues for the combined application of plant immunity inducers and beneficial microorganisms to protect plant health (Zhang and Gleason 2020). Disease resistance could also be improved by overexpressing antibodies fused with antimicrobial peptides and that recognize specific pathogen surface components (Li et al. 2008).

Molecular interference between PAMPs and their cognate plasma membrane sensing systems may also provide the basis for novel strategies to engineer durable plant disease resistance. The host's potential to recognize a broad range of PAMPs against a variety of pathogens could be enhanced successfully via heterologuous expression or overexpression of putative PRRs or LRR-RKs in plants (Gust et al. 2007). Manipulating the expression of key positive regulators of systemic resistance such as NPR1 that confers broad-spectrum resistance against pathogens, may enhance immunity in plants (Makandar et al. 2006).

Defense regulation mediated by endogenous small RNAs such as siRNAs and miRNAs have been reported to confer resistance to viral diseases. In plants, a direct connection between endogenous siRNAs and defense responses has been demonstrated by the miRNA (miR393)-based regulation of plant basal defense by targeting auxin signaling (Jiang et al. 2022). Therefore, the endogenous siRNA-mediated gene silencing may serve as an important strategy to enhance defense responses by gene expression reprogramming against pathogens in plants.

Oligosaccharides as a disease management strategy could be adopted by the construction of new molecules or assays with β-glucans or other oligosaccharides from different sources to activate immunity in plants in response to pathogens (Guarnizo et al. 2020).

Based on the promising results obtained with beneficial ISR-inducing microorganisms, the development of microbial formulations was promoted for application in conventional agriculture. However, rhizobacterial- or fungal-mediated ISR though long-lasting may not confer resistance to a wide variety of pathogens (Van Loon et al. 1998). Even though, neither SAR nor ISR will serve as a stand-alone method for pest control, activating these systemic resistances via elicitors of SAR and ISR could potentially revolutionize pest management in conventional agriculture.

A lot of new molecules acting as PAMPs and MAMPs will most probably be discovered future along with a detailed understanding the mechanisms by which non-pathogenic microorganisms induce resistance in plants.

The lipid component of eukaryotic plasma membranes may act as efficient sensor system for the perception of various abiotic and biotic external signals. In coordination with specialized protein receptors to recognize non-self-molecules may serve as an alternative mode of microbe sensing in plants.

13.3.7 Prospects for the Development of Plant Immunity Inducers

PIIs offer several advantages in comparison to the commercially available chemical pesticides, namely absence of undesirable effects on the environment, non-hazardous nature on human beings and animals, potential biostimulants, effective even at minimum or low dosage levels, ability to induce durable and broad-spectrum resistance, low risk of microbial resistance, and reduced pesticide usage in accordance with integrated pest management—that ensures eco-friendly farming for sustainable agriculture and crop protection.

In recent years, efforts for environmental protection are on the raise to reduce environmental pollution and protect ecological civilization. The reduced usage of chemical pesticides and fertilizers is advocated along with increased efficiency and urgent need to explore more efficient, safe, and eco-friendly modern farming systems.

High-throughput screening for both microbial and synthetic elicitors as well as plant-based and endophytic extracts offer large avenues to discover the potentiality of those biomolecules as green solutions for disease management. With the advent of technology for computer-aided designing of elicitors and drug targets in combination and interdisciplinary approaches combining chemistry, systemic biology and biological host defense, it is possible to identify and isolate via high-throughput screening systems to detect synthetic PIIs. Using this approach, a repertoire of second-generation synthetic elicitors and derivatives of known elicitor compounds with priming ability have been obtained (Bektas and Eulgem 2015). The proposed approach would also reveal the hidden drug-able targets in plant immune system and enable further discovery of new synthetic immune inducers to enhance our ability to dissect plant immune system in a prospective manner. With the availability of the chemical structures of these leads, it is possible to detect more new and potent chemical inducers that could transform agri-farming into more efficient, eco-friendly and chemical-pollutant free enterprise for the benefit of mankind.

13.3.8 Applications of Plant Immunity Inducers

Research findings from most of the studies have demonstrated that application of PIIs has enhanced immunity in plants. The generation of potential plant immunity inducers have been conducted based on known plant immunity elicitors and the PIIs developed and utilized commercially are predominantly from China, Korea, and Japan. A notable example is the harpin protein isolated from the plant pathogen *E. amylovora* (Wei et al. 1992). The harpin protein was industrially produced in a genetically engineered strain of *Escherichia coli* (Bauer et al. 1997). It is developed into a biological pesticide, designated as "Messenger" which is a well-known protective activity against plant diseases by eliciting immune responses in crop plants. A gram-positive bacterium, *Bacillus subtilis*-mediated expression of a component of harpin protein, HpaGXooc that was isolated from *Xanthomonas oryzae* pv. *oryzicola* showed hypersensitive response as well as enhanced growth in tobacco (Wu et al. 2009). Other PIIs, namely, PeaT1 and Hrip1 isolated from *Alternaria tenuissima* and other pathogenic fungi are also utilized as PIIs (Kulye et al. 2012). Field-based trials with application of the PII, ATaiLing, consisting of plant immunity-inducing proteins and oligosaccharides as a biopesticide showed enhanced crop protection against viral disease (Qiu 2016). Likewise, an immune-activating protein VdAL identified from *Verticillium dahlia* showed significant resistance and storage quality when applied on cucumber seedlings (Sun et al. 2016). ZhiNengCong, an extract of the fungal endophyte *Paecilomyces variotii* serves as an elicitor to induce plant resistance and promote crop growth (Lu et al. 2019). Oligosaccharide-based PIIs have been developed, especially chitosan as pesticide application that provided considerable improvement in crop yield and quality (Qiu 2016).

A chitin-based PII with the active ingredient of chitosan oligosaccharide, derived from the shells of shrimp, crab, and other organisms is recognized as environmentally friendly biopesticide. The research findings have shown that oligochitosan can stimulate defense responses in rice leaves resulting in elevated accumulation of phenolic secondary metabolites thereby imparting resistance against rice blast disease (Agrawal et al. 2002). Furthermore, its application in wheat also promoted resistance to the leaf spot causing fungal pathogen, *B. sorokiniana* (Burkhanova et al. 2007). Application of oligochitosan in tobacco and potato crops induced resistance to *P. parasitica* (Falcón et al. 2008) and *P. infestans* (Ozeretskovskaya et al. 2006), respectively. Other PII-chemical compounds, such as dufulin and Isotianil, an isothiazole-based synthetic plant immunity inducer are widely used in Japan to control rice diseases.

13.3.9 Challenges in the Utilization of Plant Immunity Inducers

The research and application of the PIIs pose several important challenges which need to be addressed with appropriate solutions for their effective usage. The major challenge with the application of PIIs, especially PGPRs in boosting immunity in plants results from the lack of optimization of the production of elicitors for commercial purpose. Another challenge is the limited understanding of the mechanism of action of the elicitors in plants, considering the diverse nature of the hosts. Though ISR triggered by PGPR could provide a long-lasting protection without inflicting any collateral damage to the tissues, it may not be conducive for development of resistance against all the pathogens (Köhl et al. 2019; Van Loon et al. 1998). Furthermore, the success of PGPR as plant defense activators is limited owing to a variety of factors including the absence of knowledge regarding the spectrum of PGPRs as well as field-based evaluation of the known elicitors of ISR. Since the ISR determinants identified so far are from limited number of species such as *Pseudomonas* spp. and *Bacillus* spp., there is a dearth in the diversity among them thereby necessitating the discovery of large repertoire of diverse elicitors from as many PGPRs as possible.

Furthermore, the mechanism/s involving PGPR elicitor recognition at the very first challenge stage, i.e., at the plant plasma membrane level and the accompanying downstream molecular events induced by PGPR-mediated priming remains largely unknown. Since the lipid phase is anticipated to act as docking platform for some of these elicitors, it necessitates an interdisciplinary approach to analyze the physicochemical basis of ligand–receptor interactions (Nishimura and Matsumori 2020). These studies may also enable to understand as to why certain elicitors are effective in certain hosts, since the proposed approach would provide an insight into the composition of the lipids in their domain-structured plasma membranes (Gronnier et al. 2018).

Furthermore, several biotic factors may also influence the efficiency of PGPR-induced ISR under natural conditions which include interactions with other microbial members of the soil microbiome or the molecular crosstalk between those members with the host (Venturi and Keel 2016; Andric et al. 2020). In addition, environmental factors may also affect the PGPR's ability to activate ISR responses (Williams et al. 2018). Notable among them are the rhizosphere specific abiotic parameters such as low temperature, acidic pH, and poor oxygen availability that affect the bacterial physiology, which in turn may modulate the production of secondary metabolites as well as the elicitors of the PGPR. Sustaining the PGR concentration in the rhizosphere at optimum levels accompanied by quantification of the metabolites using most-advanced mass spectrometry-based metabolomics with high sensitivity for their detection, might aid in overcoming any inconsistencies in the detection of potential elicitors. A better evaluation of the impact of all these factors for the application of new elicitors deserve further investigation to avoid any discrepancies in PGPR efficacy of those elicitors when applied under field conditions.

To discover new PIIs from different sources, a systematic approach involving screening, functional analysis, and evaluation of the activities of plant immunity inducers need to be adopted. Though simple chemical derivation of known immune inducers has been a norm, the findings confine only to a few known/ reported PIIs. However, the bifunctional combination adopted to develop more potent immune elicitors with either increased efficiency or reduced phytotoxicity, also exhibited complications due to introduction of the second chemical moiety, as has been the case with Pip, a SAR mobile signal candidate which showed significant reduction in SAR-inducing activity when paired with cholinium (Kukawka et al. 2018). On the contrary, pairing cholinium with isonicotinate compared to isonicotinate alone induced SAR activity (Kukawka et al. 2018). As the bifunctional combination is not mere addition of the biological activities of the two chemical moieties but involves complicated interactions induced by two different moieties, careful characterization is therefore, essential to understand the complexity of the biological activities of the new synthetic immune inducers developed by this approach.

The application of CAD to discover new immunity inducers is an emerging technique yet to be explored to its maximum potential. Presently with the availability of only a few number of known immune inducers, recognition of their potentially critical bioactive substructures and patterns is very limited (He et al. 2017; Luzuriaga-Loaiza et al. 2018). Though lead-like compound databases could provide a basic platform for virtual chemical screening, yet they are lacking in the enormity of chemical diversity which may impede the discovery of potent and novel chemical structures/ scaffolds as future PIIs.

Identifying the interacting plant recognition receptors, key recognition sites, and signaling components altered by the new PIIs would reveal the mechanistic action and signaling pathways altered by these elicitors. It is now possible to discern the signaling networks of plant immune system that were not accessible to genetic screens due to the lethality and gene redundancy with the advent of synthetic immune inducers. However, the synthetic immune activators identified by CAD technique based on the structural information of plant defense signaling components may be effective in that specific plant species studied, owing to the sequence variation among different plant species. Whereas integration of evolutionary conserved information of those genetic components from a diverse set of host species may mitigate this problem.

Furthermore, identification of potential effectors that elicit resistance responses via incompatible interaction with the host could be detected by integrating the evolutionarily conserved sequence information. Despite the potential application of the effector proteins in inducing disease resistance in field trials, their potency can be restrained by the susceptible host genotypes as well as variable environmental conditions indicating that the effector proteins alone are ineffectual compared to the chemical pesticides. It necessitates a comprehensive investigation of effector proteins to include them as plant immunity inducers in crop protection practices on crop plants. A comprehensive investigation through field-based trials is required before the effector proteins can be fully included as regular crop protection practices.

Following the identification of plant immunity inducers from different sources, a large-scale production for utilization as PIIs is a major challenge. For an instance, the target proteins such the genetically engineered harpin produced from *E. coli* must be released from the bacterial cells either by sonication or high pressure accompanied by the protein purification step which is expensive and adds to the production cost of the PII. The other challenges such as inefficient gene expression, protein inclusion body formation, and bacterial toxin production encounters during the production and purification process of the harpin proteins. Therefore, it would be ideal to use the appropriate microbial source such as *Bacillus* species that secrete proteins directly into the fermentation broth which is advantageous for the smooth downstream production processing of those proteins.

The present-day disease management measures primarily involve the application of chemical-based pesticides such as synthetic fungicides and bactericides compared to biological inducers owing to their effectiveness, accessibility, and affordability. As a result, the biological inducers appear to be less attractive and unconvincing to the farmers for their application as biopesticides for crop protection thereby posing another important challenge to convince the adoption of biological inducers in farm practices. The rational combination of plant immunity inducers and chemical pesticides or beneficial microorganisms along with fertilizer application is needed for synergistic enhancement of plant resistance and as well as crop production.

13.4 Conclusion

Plant immunity inducers represent a promising approach to boost crop health and protection from a variety of pests and pathogens by eco-friendly alternatives to limit the usage of chemical pesticides and ensure pest management. There has been an increasing interest in the development and application of plant immunity inducers as bio-solution for crop protection. These products when commercially available could contribute to sustainable agriculture through integrated pest management ensuring crop health, food safety, ecological, and environmental protection. So far only a few of the plant immunity inducers or stimulants are being commercially made available in countries like China to restrict the use of chemical-based pesticides considering food safety and environmental security. Several other synthetic and microbial elicitors, especially from fungi and bacteria are yet to be identified and evaluated to exploit their potential as PIIs for commercial application. However, efficacy of PIIs can only be relied if tested on larger scale especially in field-based conditions, since studies carried out so far with these elicitors have been mostly confined to controlled laboratory conditions.

Acknowledgment This work was supported by University of Hyderabad-Institute of Eminence (UoH-IOE-RC3-21-020), Department of Science and Technology (DST)-SERB POWER (SPG/2021/001819) DST; SR/SO/BB02/2010); Department of Biotechnology (DBT)- BT/PR1264/PBD/16/848/2009) Govt. of India, India; Indian Council of Agricultural Research (ICAR)-NFBSFARA (F. No. NFBSFARA/BS-3007/2012-13); Universities with Potential for Excellence (UPE Phase II; UH/ UGC/UPE Phase-2/Interface Studies/research projects/R-29) and DBT-BUILDER funding received by RM. Facilities at UoH; DBT-CREBB, DST-FIST, UGC-SAP, CIL, DBT-BUILDER, UoH-IOE and Plant culture facility at UoH.

Conflict of Interest There is no conflict of interest.

References

Agrawal GK, Rakwal R, Tamogami S, Yonekura M, Kubo A, Saji H (2002) Chitosan activates defense/stress response (s) in the leaves of Oryza sativa seedlings. Plant Physiol Biochem 40(12):1061–1069. https://doi.org/10.1016/S0981-9428(02)01471-7

Ajaharuddin SKMD, Das KK, Kar P, Bandyopadhyay P, Shah MH, Goswami S (2024) Insect-plant-pathogens: toxicity, dependence, and defense dynamics. In: Kumar A, Santoyo G, Singh J, Biocontrol Agents for Improved Agriculture (eds) Plant and soil microbiome. Academic Press, pp 385–411. https://doi.org/10.1016/B978-0-443-15199-6.00019-1

Albert I, Bohm H, Albert M, Feiler CE, Imkampe J, Wallmeroth N, Brancato C, Raaymakers TM, Oome S, Zhang HQ, Krol E, Grefen C, Gust AA, Chai JJ, Hedrich R, Van den Ackerveken G, Nurnberger T (2015) An RLP23-SOBIR1- BAK1 complex mediates NLP-triggered immunity. Nat Plants 1:15140. https://doi.org/10.1038/Nplants.2015.140

Amari K, Niehl A (2020) Nucleic acid-mediated PAMP-triggered immunity in plants. Curr Opin Virol 42:32–39. https://doi.org/10.1016/j.coviro.2020.04.003

Andric S, Meyer T, Ongena M (2020) Bacillus responses to plant-associated fungal and bacterial communities. Front Microbiol 11:1350. https://doi.org/10.3389/fmicb.2020.01350

Ayliffe M, Sørensen CK (2019) Plant nonhost resistance: paradigms and new environments. Curr Opin Plant Biol 50:104–113

Bailey BA (1995) Purification of a protein from culture filtrates of *fusarium oxysporum* that induces ethylene and necrosis in leaves of *Erythroxylum coca*. Phytopathology 85:1250–1255. https://doi.org/10.1094/Phyto-85-1250

Bais HP, Weir TL, Perry LG, Gilroy S, Vivanco JM (2006) The role of root exudates in rhizosphere interactions with plants and other organisms. Annu Rev Plant Biol 57:233–266. https://doi.org/10.1146/annurev

Baker B, Zambryski P, Staskawicz B, Dinesh-Kumar SP (1997) Signaling in plant-microbe interactions. Science 276(5313):26–733. https://doi.org/10.1126/science.276.5313.726

Barbero F, Guglielmotto M, Capuzzo A, Mafei M (2016) Extracellular self-DNA (esDNA), but not heterologous plant or insect DNA (etDNA), induces plasma membrane depolarization and calcium signaling in lima bean (*Phaseolus lunatus*) and maize (*Zea mays*). Int J Mol Sci 17:1659. https://doi.org/10.3390/ijms17101659

Bauer DW, Zumoff CH, Theisen TM, Bogdanove AJ, Beer SV (1997) Optimized production of *Erwinia amylovora* harpin and its use to control plant disease and enhance plant growth. Phytopathology 87:S7–S7

Beck M, Zhou J, Faulkner C, Mac DL, Robatzek S (2012) Spatiotemporal cellular dynamics of the Arabidopsis flagellin receptor reveal activation status-dependent endosomal sorting. Plant Cell 24:4205–4219. https://doi.org/10.1105/tpc.112.100263

Bektas Y, Eulgem T (2015) Synthetic plant defense elicitors. Front Plant Sci 26(5):122149

Bent AF, Mackey D (2007) Elicitors, effectors and R genes: the new paradigm and a lifetime supply of questions. Annu Rev Phytopathol 45:399–436. https://doi.org/10.1146/annurev.phyto.45.062806.094427

Bhosle SM, Makandar R (2021) Comparative transcriptome of compatible and incompatible interaction of *Erysiphe pisi* and garden pea reveals putative defense and pathogenicity factors. FEMS Microbiol Ecol 97:fiab006. https://doi.org/10.1093/femsec/fiab006

Boller T, Felix G (2009) A renaissance of elicitors: perception of microbe-associated molecular patterns and danger signals by pattern-recognition receptors. Annu Rev Plant Biol 60:379–406. https://doi.org/10.1146/annurev.arplant.57.032905.105346

Bolton MD (2009) Primary metabolism and plant defense—fuel for the fire. Mol Plant-Microbe Interact 22(5):487–497. https://doi.org/10.1094/MPMI-22-5-0487

Bolton MD, Kolmer JA, Xu WW, Garvin DF (2008) Lr34- mediated leaf rust resistance in wheat: transcript profiling reveals a high energetic demand supported by transient recruitment of multiple metabolic pathways. Mol Plant-Microbe Interact 21:1515–1527

Boutrot F, Zipfel C (2017) Function, discovery, and exploitation of plant pattern recognition receptors for broad-spectrum disease resistance. Annu Rev Phytopathol 55:257–286. https://doi.org/10.1146/annurev-phyto-080614-120106

Bukhat S, Imran A, Javaid S, Shahid M, Majeed A, Naqqash T (2020) Communication of plants with microbial world: exploring the regulatory networks for PGPR mediated defense signaling. Microbiol Res 238:126486. https://doi.org/10.1016/j.micres.2020.126486

Burden N, Maynard SK, Weltje L, Wheeler JR (2016) The utility of QSARs in predicting acute fish toxicity of pesticide metabolites: a retrospective validation approach. Regul Toxicol Pharmacol 80:241–246. https://doi.org/10.1016/j.yrtph.2016.05.032

Burketova L, Trda L, Ott PG, Valentova O (2015) Bio-based resistance inducers for sustainable plant protection against pathogens. Biotechnol Adv 33(6):994–1004. https://doi.org/10.1016/j.biotechadv.2015.01.004

Burkhanova G, Yarullina L, Maksimov I (2007) The control of wheat defense responses during infection with *Bipolaris sorokiniana* by chitooligosaccharides. Russ J Plant Physiol 54(1):104–110. https://doi.org/10.1134/S1021443707010153

Campo S, Sanchez-Sanuy F, Camargo-Ramırez R, Gomez-Ariza J, Baldrich P, Campos-Soriano L, Soto-Suarez M, Segundo BS (2021) A novel transposable element-derived microRNA participates in plant immunity to rice blast disease. Plant Biotechnol J 19:1798–1811. https://doi.org/10.1111/pbi.13592

Cao H, Glazebrook J, Clarke JD, Volko S, Dong X (1997) The Arabidopsis NPR1 gene that controls systemic acquired resistance encodes a novel protein containing ankyrin repeats. Cell 88(1):57–63. https://doi.org/10.1016/s0092-8674(00)81858-9

Chacón-Cerdas R, Barboza-Barquero L, Albertazzi FJ, Rivera-Méndez W (2020) Transcription factors controlling biotic stress response in potato plants. Physiol Mol Plant Pathol 112:101527. https://doi.org/10.1016/j.pmpp.2020.101527

Chamkhi I, Benali T, Aanniz T, Menyiy NE, Guaouguaou FE, Omari NE, El-Shazly M, Zengin G, Bouyahya A (2021) Plant microbial interaction: the mechanism and the application of microbial elicitor induced secondary metabolites biosynthesis in medicinal plants. Plant Physiol Biochem 167:269–295. https://doi.org/10.1016/j.plaphy.2021.08.001

Chen L, Guo XF, Fan ZJ, Zhang NL, Zhu YJ, Zhang ZM, Khazhieva I, Yurievich MY, Belskaya NP, Bakulev VA (2017) Synthesis and fungicidal activity of 3, 4- dichloroisothiazole based strobilurins as potent fungicide candidates. RSC Adv 7:3145–3315

Chen L, Wang WS, Wang T, Meng XF, Chen TT, Huang XX, Li YJ, Hou BK (2019) Methyl salicylate glucosylation regulates plant defense signaling and systemic acquired resistance. Plant Physiol 180:2167–2181

Chen J, Li Y, Zeng Z, Zhao X, Zhang Y, Li X, Chen J, Shen W (2024) Silicon induces ROS scavengers, hormone signalling, antifungal metabolites, and silicon deposition against brown stripe disease in sugarcane. Physiol Plant 176(e14313):1–22. https://doi.org/10.1111/ppl.14313

Conrath U, Beckers GJM, Langenbach CJG, Jaskiewicz MR (2015) Priming for enhanced defense. Annu Rev Phytopathol 53:97–119

Cui H, Tsuda K, Parker JE (2015) Effector-triggered immunity: from pathogen perception to robust defense. Annu Rev Plant Biol 66:487–511

Davar R, Darvishzadeh R, Majd A (2013) Changes in antioxidant systems in sunflower partial resistant and susceptible lines as affected by *Sclerotinia sclerotiorum*. Biologia 68:821–829

Derevnina L, Dagdas YF, De la Concepcion JC, Bialas A, Kellner R, Petre B, Domazakis E, Du J, Wu CH, Lin X (2016) Nine things to know about elicitins. New Phytol 212:888–895

Dewen Q, Yijie D, Yi Z, Shupeng L, Fachao S (2017) Plant immunity inducer development and application. Mol Plant-Microbe Interact 30(5):355–360. https://doi.org/10.1094/MPMI-11-16-0231-CR

Djonović S, Pozo MJ, Dangott LJ, Howell CR, Kenerley CM (2007) Sm1, a proteinaceous elicitor secreted by the biocontrol fungus *Trichoderma virens* induces plant defense responses and systemic resistance. Mol Plant-Microbe Interact 19(8):838–853

Du Q, Yang XD, Zhang JH, Zhong XF, Kim KS, Yang J, Xing GJ, Li XY, Jiang ZY, Li QY, Dong YS, Pan HY (2018) Over-expression of the *pseudomonas syringae* harpin-encoding gene hrpZm confers enhanced tolerance to Phytophthora root and stem rot in transgenic soybean. Transgenic Res 27:277–288. https://doi.org/10.1007/s11248-018-0071-4

Duran-Flores D, Heil M (2014) Damaged-self recognition in common bean (*Phaseolus vulgaris*) shows taxonomic specificity and triggers signaling via reactive oxygen species (ROS). Front Plant Sci 5:585

Duran-Flores D, Heil M (2018) Extracellular self-DNA as a damage associated molecular pattern (DAMP) that triggers self-specific immunity induction in plants. Brain Behav Immun 72:78–88

Falcón AB, Cabrera JC, Costales D, Ramírez MA, Cabrera G, Toledo V, MartínezTéllez MA (2008) The effect of size and acetylation degree of chitosan derivatives on tobacco plant protection against *Phytophthora parasitica nicotianae*. World J Microbiol Biotechnol 24:103

Fei Q, Xia R, Meyers BC (2013) Phased, secondary, small interfering RNAs in posttranscriptional regulatory networks. Plant Cell 25:2400–2415

Felix G, Duran JD, Volko S, Boller T (1999) Plants have a sensitive perception system for the most conserved domain of bacterial flagellin. Plant J 18:265–276

Flor HH (1971) Current status of the gene-for-gene concept. Annu Rev Phytopathol 9:275–296

Frescatada-Rosa M, Robatzek S, Kuhn H (2015) Should I stay or should I go? Traffic control for plant pattern recognition receptors. Curr Opin Plant Biol 28:23–29. https://doi.org/10.1016/j.pbi.2015.08.007

Gohre V, Robatzek S (2008) Breaking the barriers: microbial effector molecules subvert plant immunity. Annu Rev Phytopathol 46:189–215. https://doi.org/10.1146/annurev.phyto.46.120407.110050

Gronnier J, Gerbeau-Pissot P, Germain V, Mongrand S, Simon-Plas F (2018) Divide and rule: plant plasma membrane organization. Trends Plant Sci 23:899–917. https://doi.org/10.1016/j.tplants.2018.07.007

Guarnizo N, Oliveros D, Murillo-Arango W, Bermúdez-Cardona MB (2020) Oligosaccharides: defense inducers, their recognition in plants, commercial uses and perspectives. Molecules 25:5972. https://doi.org/10.3390/molecules2524597

Gust AA, Biswas R, Lenz HD, Rauhut T, Ranf S, Kemmerling B, Gotz F, Glawischnig E, Lee J, Felix G, Nurnberger T (2007) Bacteria-derived peptidoglycans constitute pathogen-associated molecular patterns triggering innate immunity in Arabidopsis. J Biol Chem 282:32338–32348. https://doi.org/10.1074/jbc.M704886200

Guzma´ n-Benito I, Donaire L, Amorim-Silva V, Vallarino JG, Esteban A, Wierzbicki AT, Ruiz-Ferrer V, Llave C (2019) The immune repressor BIR1 contributes to antiviral defense and undergoes transcriptional and post-transcriptional regulation during viral infections. New Phytol 224:421–438. https://doi.org/10.1111/nph.15931

Hawes MC, Curlango-Rivera G, Xiong Z, Kessler JO (2012) Roles of root border cells in plant defense and regulation of rhizosphere microbial populations by extracellular DNA 'trapping'. Plant Soil 355:1–6

He XR, Chen X, Lin SB, Mo XC, Zhou PY, Zhang ZH et al (2017) Diversity-oriented synthesis of natural-product-like libraries containing a 3-methylbenzofuran moiety for the discovery of new chemical elicitors. ChemistryOpen 6:102–111. https://doi.org/10.1002/open.201600118

Hedrich R, Van den Ackerveken G, Nurnberger T (2015) An RLP23-SOBIR1- BAK1 complex mediates NLP-triggered immunity. Nat Plants 1:15140. https://doi.org/10.1038/Nplants.2015.140

Henry G, Thonart P, Ongena M (2012) PAMPs, MAMPs, DAMPs and others: an update on the diversity of plant immunity elicitors. Biotechnol Agron Soc Environ 16(2):257–268

Hermosa R, Viterbo A, Chet I, Monte E (2012) Plant-beneficial effects of *Trichoderma* and of its genes. Microbiol 158:17–25

Hiltl C, Siddique S (2020) New allies to fight worms. Nat Plants 6(6):598–599. https://doi.org/10.1038/s41477-020-0699-y

Huang XD, Fang L, Gu QS, Tian YP, Geng C, Li XD (2019) Cross protection against the watermelon strain of papaya ringspot virus through modification of viral RNA silencing suppressor. Virus Res 265:166–171. https://doi.org/10.1016/j.virusres.2019.03.016

Jia Y, McAdams SA, Bryan GT, Hershey HP, Valent B (2000) Direct interaction of resistance gene and avirulence gene products confers rice blast resistance. EMBO J 19(15):4004–4014. https://doi.org/10.1093/emboj/19.15.4004

Jiang J, Zhu H, Li N, Batley J, Wang Y (2022) The miR393-target module regulates plant development and responses to biotic and abiotic stresses. Int J Mol Sci 23(16):9477. https://doi.org/10.3390/ijms23169477

Jones J, Dangl J (2006) The plant immune system. Nature 444:323–329. https://doi.org/10.1038/nature05286

Jwa NS, Hwang BK (2017) Convergent evolution of pathogen effectors toward reactive oxygen species signaling networks in plants. Front Plant Sci 8:1687. https://doi.org/10.3389/fpls.2017.01687

Kachroo A, Robin GP (2013) Systemic signaling during plant defense. Curr Opin Plant Biol 16:527–533. https://doi.org/10.1016/j.pbi.2013.06.019

Kai M, Effmert U, Piechulla B (2016) Bacterial-plant-interactions: approaches to unravel the biological function of bacterial volatiles in the rhizosphere. Front Microbiol 7:108. https://doi.org/10.3389/fmicb.2016.00108

Kaku H, Nishizawa Y, Ishii-Minami N, Akimoto-Tomiyama C, Dohmae N, Takio K, Minami E, Shibuya N (2006) Plant cells recognize chitin fragments for defense signaling through a plasmamembrane receptor. Proc Natl Acad Sci USA 103(29):11086–11091. https://doi.org/10.1073/pnas.0508882103

Kalli S, Araya-Cloutier C, Lin Y, de Bruijn WJ, Chapman J, Vincken JP (2020) Enhanced biosynthesis of the natural antimicrobial glyceollins in soybean seedlings by priming and elicitation. Food Chem 317:126389

Karuppiah V, Li T, Vallikkannu M, Chen J (2019) Co-cultivation of *Trichoderma asperellum* GDFS1009 and *bacillus amyloliquefaciens* 1841 causes differential gene expression and improvement in the wheat growth and biocontrol activity. Front Microbiol 1068:1–16

Khoshru B, Mitra D, Joshi K, Adhikari P, Rion MS, Fadiji AE, Alizadeh M, Priyadarshini A, Senapati A, Sarikhani MR, Panneerselvam P (2023) Decrypting the multi-functional biological activators and inducers of defense responses against biotic stresses in plants. Heliyon 9(3):e13825. https://doi.org/10.1016/j.heliyon.2023.e13825

Kim SG, Kim KW, Park EW, Choi D (2002) Silicon–induced cell wall fortification of rice leaves: a possible cellular mechanism of enhanced host resistance to blast. Phytopathology 92:1095–1103

Koga J, Yamauchi T, Shimura M, Ogawa N, Oshima K, Umemura K, Kikuchi M, Ogasawara N (1998) Cerebrosides a and C, sphingolipid elicitors of hypersensitive cell death and phytoalexin accumulation in rice plants. J Biol Chem 273:31985–31991. https://doi.org/10.1074/jbc.273.48.31985

Köhl J, Kolnaar R, Ravensberg WJ (2019) Mode of action of microbial biological control agents against plant diseases: relevance beyond efficacy. Front Plant Sci 10:845. https://doi.org/10.3389/fpls.2019.00845

Kukawka R, Czerwoniec P, Lewandowski P, Pospieszny H, Smiglak M (2018) New ionic liquids based on systemic acquired resistance inducers combined with the phytotoxicity reducing cholinium cation. New J Chem 42:11984–11990. https://doi.org/10.1039/C8NJ00778K

Kulye M, Liu H, Zhang YL, Zeng HM, Yang XF, Qiu DW (2012) Hrip1, a novel protein elicitor from necrotrophic fungus, *Alternaria tenuissima*, elicits cell death, expression of defence-related genes and systemic acquired resistance in tobacco. Plant Cell Environ 35:2104–2120. https://doi.org/10.1111/j.1365-3040.2012.02539.x

Kumari YSMAI, Vengadaramana A (2017) Stimulation of defense enzymes in tomato (*Solanum lycopersicum* L.) and Chilli (*Capsicum annuum* L.) in response to exogenous application of different chemical elicitors. Universal J Plant Sci 5(1):10–15

Lamb C, Dixon RA (1997) The oxidative burst in plant disease resistance. Annu Rev Plant Physiol Plant Mol Biol 48:251–275

Li HP, Zhang JB, Shi RP, Huang T, Fischer R, Liao YC (2008) Engineering fusarium head blight resistance in wheat by expression of a fusion protein containing a fusarium-specific antibody and an antifungal peptide. Mol Plant-Microbe Interact 9:1242–1248. https://doi.org/10.1094/MPMI-21-9-1242

Li F, Pignatta D, Bendix C, Brunkard JO, Cohn MM, Tung J, Sun H, Kumar P, Baker B (2012) microRNA regulation of plant innate immune receptors. Proc Natl Acad Sci USA 109:1790–1795

Li H, Guan Y, Dong Y, Zhao L, Rong S, Chen W, Lv M, Xu H, Gao X, Chen R, Li L, Xu Z (2018) Isolation and evaluation of endophytic *bacillus tequilensis* GYLH001 with potential application for biological control of *Magnaporthe oryzae*. PLoS One 13(10):e203505

Lu C, Liu H, Jiang D, Wang L, Jiang Y, Tang S, Hou X, Han X, Liu Z, Zhang M (2019) *Paecilomyces variotii* extracts (ZNC) enhance plant immunity and promote plant growth. Plant Soil 441:383–397

Luzuriaga-Loaiza WP, Schellenberger R, De Gaetano Y, Akong FO, Villaume S, Crouzet J et al (2018) Synthetic Rhamnolipid Bolaforms trigger an innate immune response in *Arabidopsis thaliana*. Sci Rep 8:8534. https://doi.org/10.1038/s41598-018-26838-y

Ma ZC, Song TQ, Zhu L, Ye WW, Wang Y, Shao YY, Dong SM, Zhang ZG, Dou DL, Zheng XB, Tyler BM, Wang YC (2015) A *Phytophthora sojae* glycoside hydrolase 12 protein is a major virulence factor during soybean infection and is recognized as a PAMP. Plant Cell 27:2057–2072. https://doi.org/10.1105/tpc.15.00390

Makandar R, Essig JS, Schapaugh MA, Trick HN, Shah J (2006) Genetically engineered resistance to fusarium head blight in wheat by expression of Arabidopsis NPR1. Mol Plant-Microbe Interact 19(2):123–129. https://doi.org/10.1094/MPMI-19-0123

Makandar R, Nalam V, Chaturvedi R, Jeannotte R, Sparks AA, Shah J (2010) Involvement of salicylate and jasmonate signaling pathways in Arabidopsis interaction with *fusarium graminearum*. Mol Plant-Microbe Interact 23(7):861–870. https://doi.org/10.1094/MPMI-23-7-0861

Makandar R, Nalam VJ, Lee H, Trick HN, Dong Y, Shah J (2012) Salicylic acid regulates basal resistance to fusarium head blight in wheat. Mol Plant-Microbe Interact 25(3):431–439. https://doi.org/10.1094/MPMI-09-11-0232

Malik NAA, Kumar IS, Nadarajah K (2020) Elicitor and receptor molecules: orchestrators of plant defense and immunity. Int J Mol Sci 21:963. https://doi.org/10.3390/ijms21030963

Mandal S, Mitra A (2007) Reinforcement of cell wall in roots of *Lycopersicon esculentum* through induction of phenolic compounds and lignin by elicitors. Physiol Mol Plant Pathol 71:201–209

Manoharachary C, Kunwar IK (2014) Host–pathogen interaction, plant diseases, disease management strategies, and future challenges. In: Future challenges in crop protection against fungal pathogens. Springer, New York, NY, pp 185–229

Meziane H, Van Der Sluis I, Va Loon LC, Höfte M, Bakker PA (2005) Determinants of *pseudomonas putida* WCS358 involved in inducing systemic resistance in plants. Mol Plant Pathol 6:177–185. https://doi.org/10.1111/j.1364-3703.2005.00276

Murray SL, Ingle RA, Petersen LN, Denby KJ (2007) Basal resistance against *pseudomonas syringae* in Arabidopsis involves WRKY53 and a protein with homology to a nematode resistance protein. Mol Plant-Microbe Interact 20(11):1431–1438. https://doi.org/10.1094/MPMI-20-11-1431

Nalam VJ, Alam S, Keereetaweep J, Venables B, Burdan D, Lee H, Trick HN, Sarowar S, Makandar R, Shah J (2015) Facilitation of *fusarium graminearum* infection by 9-lipoxygenases in Arabidopsis and wheat. Mol Plant-Microbe Interact 10:1142–1152. https://doi.org/10.1094/MPMI-04-15-0096

Narváez-Barragán DA, Tovar-Herrera OE, Guevara-García A, Serrano M, Martinez-Anaya C (2022) Mechanisms of plant cell wall surveillance in response to pathogens, cell wall-derived ligands and the effect of expansins to infection resistance or susceptibility. Front Plant Sci 13:969343. https://doi.org/10.3389/fpls.2022.969343

Navarova H, Bernsdorff F, Doring AC, Zeier J (2012) Pipecolic acid, an endogenous mediator of defense amplification and priming, is a critical regulator of inducible plant immunity. Plant Cell 24:5123–5141

Newman MA, Sundelin T, Nielsen JT, Erbs G (2013) MAMP (microbe-associated molecular pattern) triggered immunity in plants. Front Plant Sci 4:50369. https://doi.org/10.3389/fpls.2013.00139

Ngou BPM, Ahn HK, Ding P, Jones JDG (2021) Mutual potentiation of plant immunity by cell-surface and intracellular receptors. Nature 592:110–115. https://doi.org/10.1038/s41586-021-03315-7

Niks RE, Marcel TC (2009) Nonhost and basal resistance: how to explain specificity? New Phytol 182:817–828. https://doi.org/10.1111/j.1469-8137.2009.02849.x

Nimchuk Z, Rohmer L, Chang JH et al (2001) Knowing the dancer from the dance: R gene products and their interactions with other proteins from host and pathogen. Curr Opin Plant Biol 4:288–294

Nimchuk Z, Eulgem T, Holt BF III, Dangl JL (2003) Recognition and response in the plant immune system. Annu Rev Genet 37:579–609. https://doi.org/10.1146/annurev.genet.37.110801.142628

Nisa H, Kamili AN, Nawchoo IA, Shafi S, Shameem N, Bandh SA (2015) Fungal endophytes as prolific source of phytochemicals and other bioactive natural products: a review. Microb Pathog 82:50–59

Nishimura S, Matsumori N (2020) Chemical diversity and mode of action of natural products targeting lipids in the eukaryotic cell membrane. Nat Prod Rep 37:677–702. https://doi.org/10.1039/c9np00059c

O'Neill EM, Mucyn TS, Patteson IB, Finkel OM, Chung EH, Baccile JA, Massolo E, Schroeder FC, Dangl JL, Li B (2018) Phevamine a, a small molecule that suppresses plant immune responses. PNAS 115(41):E9514–E9522. www.pnas.org/cgi/doi/10.1073/pnas.1803779115

Oome S, Raaymakers TM, Cabral A, Samwel S, Bohm H, Albert I, Nurnberger T, Van den Ackerveken G (2014) Nep1-like proteins from three kingdoms of life act as a microbe-associated molecular pattern in Arabidopsis. Proc Natl Acad Sci USA 111(47):16955–16960. https://doi.org/10.1073/pnas.1410031111

Orozco-Mosqueda MDC, Fadiji AE, Babalola OO, Santoyo G (2023) Review: bacterial elicitors of the plant immune system: an overview and the way forward. Plant Stress 7:100138

Osbourn AE (1996) Preformed antimicrobial compounds and plant defense against fungal attack. Plant Cell 8(10):1821–1831. https://doi.org/10.1105/tpc.8.10.1821

Ozeretskovskaya O, Vasyukova N, Panina YS, Chalenko G (2006) Effect of immunomodulators on potato resistance and susceptibility to *Phytophthora infestans*. Russ J Plant Physiol 53(4):488–494. https://doi.org/10.1134/S1021443706040091

Park SW, Kaimoyo E, Kumar D, Mosher S, Klessig DF (2007) Methyl salicylate is a critical mobile signal for plant systemic acquired resistance. Science 318:113–116

Park SW, Liu PP, Forouhar F, Vlot AC, Tong L, Tietjen K, Klessig DF (2009) Use of a synthetic salicylic acid analog to investigate the roles of methyl salicylate and its esterases in plant disease resistance. J Biol Chem 284:7307–7317

Peghaire E, Hamdache S, Galien A, Sleiman M, ter Halle A, El Alaoui H, Kocer A, Richard C, Goupil P (2020) Inducing plant defense reactions in tobacco plants with phenolic-rich extracts

from red maple leaves: a characterization of Main active ingredients. Forests 11(6):705. https://doi.org/10.3390/f11060705
Peng C, Zhang A, Wang Q, Song Y, Zhang M, Ding X, Li Y, Geng Q, Zhu C (2020) Ultrahigh-activity immune inducer from endophytic fungi induces tobacco resistance to virus by SA pathway and RNA silencing. BMC Plant Biol 20(1):169. https://doi.org/10.1186/s12870-020-02386-4
Pieterse CM, van Wees SC, Hoffland E, van Pelt JA, van Loon LC (1996) Systemic resistance in Arabidopsis induced by biocontrol bacteria is independent of salicylic acid accumulation and pathogenesis-related gene expression. Plant Cell 8:1225–1237
Pieterse CM, Van der Does D, Zamioudis C, Leon-Reyes A, Van Wees SC (2012) Hormonal modulation of plant immunity. Annu Revof Cell Develop Biol 28:489–521
Pieterse CM, Zamioudis C, Berendsen RL, Weller DM, VanWees SC, Bakker PA (2014) Induced systemic resistance by beneficial microbes. Annu Rev Phytopathol 52:347–375
Pozza EA, Pozza AAA, Botelho DMDS (2015) Silicon in plant disease control. Revista Ceres 62(3):323–331
Pradhan DA, Makandar R (2023) Delineating host responses induced by *Trichoderma* in castor through comparative transcriptome analysis. Rhizosphere 27:100745. https://doi.org/10.1016/j.rhisph.2023.100745
Pradhan DA, Bagagoni P, Makandar R (2023a) Assessing rhizosphere *Trichoderma asperellum* strains for root colonizing and antagonistic competencies against fusarium wilt through molecular and biochemical responses in castor. Biol Control 184:105280. https://doi.org/10.1016/j.biocontrol.2023.105280
Pradhan DA, Bagagoni P, Slathia S, Prasad RD, Makandar R (2023b) Characterization of *Trichoderma* strains for novel species-specific markers by multiplex PCR and antagonistic property against *Alternaria ricini* in castor (*Ricinus communis* L.). Biocatal Agric Biotechnol 54:102945. ISSN 1878–8181. https://doi.org/10.1016/j.bcab.2023.102945
Pršić J, Ongena M (2020) Elicitors of plant immunity triggered by beneficial bacteria. Front Plant Sci 9(11):594530
Qi Y, Tsuda K, Glazebrook J, Katagiri F (2011) Physical association of pattern-triggered immunity (PTI) and effector-triggered immunity (ETI) immune receptors in Arabidopsis. Mol Plant Pathol 12:702–708. https://doi.org/10.1111/j.1364-3703.2010.00704.x
Qiu DW (2016) Research status and trend analysis of plant immune induction technology in China. Plant Prot 42:10–14
Qiu DW, Dong YJ, Zhang Y, Li SP, Shi FC (2017) Plant immunity inducer development and application. Mol Plant-Microbe Interact 30(5):355–360. https://doi.org/10.1094/Mpmi-11-16-0231-Cr
Raho N, Ramirez L, Lanteri ML, Gonorazky G, Lamattina L, ten Have A, Laxalt AM (2011) Phosphatidic acid production in chitosan-elicited tomato cells, via both phospholipase D and phospholipase C/diacylglycerol kinase, requires nitric oxide. J Plant Physiol 168(6):534–539. https://doi.org/10.1016/j.jplph.2010.09.004
Rippa S, Eid M, Formaggio F, Toniolo C, Béven L (2010) Hypersensitive-like response to the pore-former peptaibol alamethicin in Arabidopsis thaliana. Chembiochem 11(14):2042–2049. https://doi.org/10.1002/cbic.201000262. PMID: 20818637
Rojas CM, Senthil-Kumar M, Tzin V, Mysore KS (2014) Regulation of primary plant metabolism during plant-pathogen interactions and its contribution to plant defense. Front Plant Sci 5(17):1–12. https://doi.org/10.3389/fpls.2014.00017
Ryals JA, Neuenschwander UH, Willits MG, Molina A, Steiner HY, Hunt MD (1996) Systemic acquired resistance. Plant Cell 8:1809–1819
Saijo Y, Loo EP, Yasuda S (2018) Pattern recognition receptors and signaling in plant–microbe interactions. Plant J 93:92–613. https://doi.org/10.1111/tpj.13808
Salas-Marina MA, Isordia-Jasso MI, Islas-Osuna MA, Delgado-Sánchez P, Jiménez-Bremont JF, Rodríguez-Kessler M, Rosales-Saavedra MT, Herrera-Estrella A, Casas-Flores S (2015) The Epl1 and Sm1 proteins from *Trichoderma atroviride* and *Trichoderma virens* differentially modulate systemic disease resistance against different life style pathogens in *Solanum lycopersicum*. Front Plant Sci 6:77. https://doi.org/10.3389/fpls.2015.00077

Sánchez-Martín J, Keller B (2021) NLR immune receptors and diverse types of non-NLR proteins control race-specific resistance in Triticeae. Current Opin Plant Biol 62:102053. https://doi.org/10.1016/j.pbi.2021.102053

Schikora A, Schenk ST, Hartmann A (2016) Beneficial effects of bacteriaplant communication based on quorum sensing molecules of the N-acyl homoserine lactone group. Plant Mol Biol 90:605–612. https://doi.org/10.1007/s11103-016-0457-8

Shah J, Chaturvedi R, Chowdhury Z, Venables B, Petros RA (2014) Signaling by small metabolites in systemic acquired resistance. Plant J 79(4):645–658. https://doi.org/10.1111/tpj.12464

Shiu SH, Bleecker AB (2003) Expansion of the receptor-like kinase/Pelle gene family and receptor-like proteins in Arabidopsis. Plant Physiol 132:530–543. https://doi.org/10.1104/pp.103.021964

Singh A (2017) Glucosinolates and plant defense. Glucosinolates. Springer International Publishing, pp 237–246

Song L, Fang Y, Chen L, Wang J, Chen X (2021) Role of non-coding RNAs in plant immunity. Plant Commun 2(3):100180

Spoel SH, Johnson JS, Dong X (2007) Regulation of tradeoffs between plant defenses against pathogens with different lifestyles. Proc Natl Acad Sci USA 104:18842–18847

Stadnik MJ, de Freitas MB (2014) Algal polysaccharides as source of plant resistance inducers. Trop Plant Pathol 39(2):111–118. https://doi.org/10.1590/S1982-56762014000200001

Sun FQ, Li JQ, Qi JS, Gao LH (2016) Effect of foliar spraying *Verticillium dahiliae* allergen asp f2-like (VdAL) during seedling stage on storage quality of commercial cucumber seedlings. Chin Veg 3:48–52

Tan MW, Mahajan-Miklos S, Ausubel FM (1999) Killing of *Caenorhabditis elegans* by *Pseudomonas aeruginosa* used to model mammalian bacterial pathogenesis. Proc Natl Acad Sci USA 96:715–720

Thakur R, Sharma S (2024) Impact of pesticides used in agriculture: their benefits and hazards. AIP Conf Proc 2986:030163. https://doi.org/10.1063/5.0193979

Toyoda K, Fitrianti AN, Itoh C, Hasegawa H, Matsui H, Noutoshi Y, Yamamoto M, Ichinose Y, Shiraishi T (2023) CEP peptide, a family of conserved, secreted small peptides acts as an endogenous suppressor in Arabidopsis. Physiol Mol Plant Pathol 125:102019. https://doi.org/10.1016/j.pmpp.2023.102019

Van Loon LC, Bakker PAHM, Pieterse CMJ (1998) Systemic resistance induced by rhizosphere bacteria. Annu Rev Phytopathol 36:453–483

Vega-Muñoz I, Herrera-Estrella A, Martínez-de la Vega O, Heil M (2023) ATM and ATR, two central players of the DNA damage response, are involved in the induction of systemic acquired resistance by extracellular DNA, but not the plant wound response. Front Immunol 14:1175786

Venturi V, Keel C (2016) Signaling in the rhizosphere. Trends Plant Sci 21:187–198. https://doi.org/10.1016/j.tplants.2016.01.005

Vergne E, Grand X, Ballini E, Chalvon V, Saindrenan P, Tharreau D, Nottéghem J-L, Morel JB (2010) Preformed expression of defense is a hallmark of partial resistance to rice blast fungal pathogen *Magnaporthe oryzae*. BMC Plant Biol 10:206. https://doi.org/10.1186/1471-2229-10-206

Vinale F et al (2008) A novel role for *Trichoderma* secondary metabolites in the interactions with plants. Physiol Mol Plant Pathol 72:80–86

Viterbo A et al (2007) The 18mer peptaibols from *Trichoderma virens* elicit plant defence responses. Mol Plant Pathol 8:737–746

Waewthongrak W, Pisuchpen S, Leelasuphakul W (2015) Effect of *Bacillus subtilis* and chitosan applications on green mold *(Penicillium digitatum* Sacc.) decay in citrus fruit. Postharvest Biol Technol 99:44–49

Wang M, Weiberg A, Dellota E, Yamane D, Jin H (2017) Botrytis small RNA Bc-siR37 suppresses plant defense genes by cross-kingdom RNAi. RNA Biol 14:421–428. https://doi.org/10.1080/15476286.2017.1291112

Wei ZM, Laby RJ, Zumoff CH, Bauer DW, He SY, Collmer A, Beer SV (1992) Harpin, elicitor of the hypersensitive response produced by the plant pathogen *Erwinia amylovora*. Science 257:85–88. https://doi.org/10.1126/science.1621099
Williams A, Pétriacq P, Beerling DJ, Cotton TEA, Ton J (2018) Impacts of atmospheric CO_2 and soil nutritional value on plant responses to rhizosphere colonization by soil bacteria. Front Plant Sci 871:1493. https://doi.org/10.3389/fpls.2018.01493
Wu HJ, Wang SA, Qiao JQ, Liu J, Zhan J, Gao XW (2009) Expression of HpaG (Xooc) protein in *Bacillus subtilis* and its biological functions. J Microbiol Biotechnol 19:194–203. https://doi.org/10.4014/jmb.0802.154
Wu Y, Zhang D, Chu JY, Boyle P, Wang Y, Brindle ID, De Luca V, Després C (2012) The Arabidopsis NPR1 protein is a receptor for the plant defense hormone salicylic acid. Cell Rep 1(6):639–647. https://doi.org/10.1016/j.celrep.2012.05.008
Xiang J, Li X, Yin L, Liu Y, Zhang Y, Qu J, Lu J (2017) A candidate RxLR effector from *Plasmopara viticola* can elicit immune responses in *Nicotiana benthamiana*. BMC Plant Biol 17:75. https://doi.org/10.1186/s12870-017-1016-4
Xie YR, Raruang Y, Chen ZY, Brown RL, Cleveland TE (2015) ZmGns, a maize class I β-1,3-glucanase, is induced by biotic stresses and possesses strong antimicrobial activity. J Integr Plant Biol 57:271–283
Yang B, Yang S, Zheng W, Wang Y (2022) Plant immunity inducers: from discovery to agricultural application. Stress Biol 2:5. https://doi.org/10.1007/s44154-021-00028-9
Zavaliev R, Dong X (2024) NPR1, a key immune regulator for plant survival under biotic and abiotic stresses. Mol Cell 84(1):131–141. https://doi.org/10.1016/j.molcel.2023.11.018
Zaynab M, Fatima M, Abbas S, Umair M, Sharif Y, Raza MA (2018) Long non-coding RNAs as molecular players in plant defense against pathogens. Microb Pathog 121:277–282
Zhang L, Gleason C (2020) Enhancing potato resistance against root-knot nematodes using a plant-defence elicitor delivered by bacteria. Nat Plants 6:625–629. https://doi.org/10.1038/s41477-020-0689-0
Zhang H, Zhao X, Yang J, Yin H, Wang W, Lu H, Du Y (2011a) Nitric oxide production and its functional link with OIPK in tobacco defense response elicited by chitooligosaccharide. Plant Cell Rep 30:1153–1162
Zhang W, Yang X, Qiu D, Guo L, Zeng H, Mao J, Gao Q (2011b) PeaT1-induced systemic acquired resistance in tobacco follows salicylic acid-dependent pathway. Mol Biol Rep 38:2549–2556
Zhang Y, Ma H, Yuan T, Seeram NP (2015) Red maple (*Acer rubrum*) aerial parts as a source of bioactive phenolics. Nat Prod Commun 10:1409–1412
Zhou M, Wang W (2018) Recent advances in synthetic chemical inducers of plant immunity. Front Plant Sci 9:1613
Zhou X, Gao H, Zhang X, Khashi u Rahman M, Mazzoleni S, Du M, Wu F (2023) Plant extracellular self-DNA inhibits growth and induces immunity via the jasmonate signaling pathway. Plant Physiol 192(3):2475–2491. https://doi.org/10.1093/plphys/kiad195
Zipfel C (2008) Pattern-recognition receptors in plant innate immunity. Current Opin Immunol 20(1):10–16. https://doi.org/10.1016/j.coi.2007.11.003
Zipfel C, Robatzek S, Navarro L, Oakeley EJ, Jones JD et al (2004) Bacterial disease resistance in Arabidopsis through flagellin perception. Nature 428:764–767
Zipfel C, Kunze G, Chinchilla D, Caniard A, Jones JDG et al (2006) Perception of the bacterial PAMP EF-Tu by the receptor EFR restricts agrobacterium-mediated transformation. Cell 125:749–760

Deep Learning Approaches for Off-targets Prediction in CRISPR-Cas9 Genome Editing to Improve Resistant in Plants

14

Awadhesh Kumar

Abstract

Genome editing enables the modification, insertion of mutations, and alteration of living organisms' genomes. This breakthrough technology expands the possibilities for genetics, molecular biology, and biomedical research. CRISPR-Cas9 has revolutionized genome editing across various organisms, including plants. Deep learning techniques have been increasingly utilized for off-target prediction in CRISPR/Cas9 gene editing, aiming to enhance the accuracy and efficiency of identifying potential off-target sites. To analyze biological data and predict off-target sites with superior performance, I compared differed variants of deep learning algorithm. In this chapter, I used different variants of three deep learning models, namely feedforward neural network (FNN), convolutional neural network (CNN), and recurrent neural network (RNN) which significantly improved the prediction of off-target cleavage sites and genome vulnerability, achieving high accuracies of up to 99.5%. In three variants of FNN models, FNN5 outperforms from FNN3 and FNN7 with highest accuracy, low loss, good off-target and on-target prediction, and better F1 score. In two variants of CNN, CNN3 performs better than CNN5 in terms of overall evaluating parameters, and in two variants of RNN, RNN-GRU performs better than RNN-LSTM with high accuracy of 0.995, low loss of 0.0195, best off-target, and on-target prediction of CRISPR/Cas9. When compared the performance of all the discussed models, found that RNN-GRU outperforms all other models. The performance of the said models are evaluated based on several evaluation metrics such as confusion matrix, precision, recall, support, F1-Score, microaverage, macroaverage, and accuracy. Overall, the application of deep learning in off-target prediction for CRISPR-Cas9 gene editing showcases promising advancements in enhancing

A. Kumar (✉)
Department of Computer Science, MMV, Banaras Hindu University, Varanasi, India
e-mail: akmcsmmv@bhu.ac.in

S. Singh, R. Mehrotra (eds.), *Plant Elicitor Peptides*,
https://doi.org/10.1007/978-981-97-6374-0_14

the precision and safety of genome editing techniques to develop resistance against various diseases in plants.

Keywords

Deep learning · Neural networks · FNN · CNN · RNN · CRISPR/Cas9 · Gene editing

14.1 Introduction

Genome editing allows us to insert mutations, indels, and manipulates genomes of living organisms thus increasing the potential for biomedicine, molecular biology, and genetics research. Researchers in labs use genome editing in plant cells and animal models to identify various diseases (Charlier et al. 2021). It shows promising results in major domains like plant–pathogen interactions, personalized medicine, biofuel, gene therapy, HIV, cancer, and creating bio-models by mimicking complex genetic makeup. Some of the effective genome editing techniques, namely clustered regularly interspaced palindromic repeats (CRISPR/Cas), zinc finger nucleases, pentatricopeptide repeat proteins (PPRs), homologous recombination (HR), and transcription activator-like effector nucleases (TALENs), out of these, CRISPR being the latest and most widely used (Manghwar et al. 2020).

The CRISPR/Cas9 gene editing technique has two key components: a guide RNA that matches a particular target gene and Cas9 (CRISPR-associated protein 9), an endonuclease that breaks double strands of DNA to allow alterations to the genome (Redman et al. 2016). Targeting specific DNA sequences, CRISPR-Cas9 is an accurate and effective genome editing technique that makes use of a bacterial defense mechanism. With the help of RNA molecules, the Cas9 enzyme cuts DNA exactly at the targeted places, functioning as molecular scissors. This makes it possible to precisely alter the plant genome (Mengstie and Wondimu 2021). During evolution, it occurs naturally in bacteria as an adaptive defense mechanism against plasmids and viruses, majorly in archaea. This system can efficiently distinguish between foreign genomes and genome of its own. CRISPR are regions in the bacterial genome composed of groups of short repeat sequences that are highly conserved, implying great importance for bacterial survival. There can be more than one CRISPR site in the bacterial genome. These repeat sequences are separated by unique sequences called spacer sequencing, which can vary greatly in length, originating from the invader genome during the course of infection (Redman et al. 2016). These spacer sequences even vary greatly among closely related species used as molecular markers by the bacterial system to recognize the organism. An AT-rich leader sequence is present at one end of CRISPR loci which leads to the initiation of the transcription. Cas are found near the CRISPR loci which results in cas proteins having nuclease, helicase, and primase activity. Cas protein plays a crucial role in non-self-genome fragment recognition and inhibition of pathogens. The general mechanism can be broken into three major steps namely initiation, maturation, and interference. In the first step, cas protein collects genome fragments of invader pathogens after its first infection, these fragments are known as protospacer. Cas proteins integrate this protospacer as new spacers in the CRISPR loci. In the second step, pre-CRISPR RNA (pre-cr-RNA) is transcribed from the CRISPR region and sliced in a way that every fragment contains a spacer and some part of the repeat sequence, now called cr-RNA. This cr-RNA creates complexes with cas proteins and guides the complex to the target. In the last step, upon second exposure between host and pathogen, the cr-CAS complex recognizes complementary sequence in the pathogen genome and cleaves at the target site.

There are basically three types of CRISPR/Cas systems on the basis of genetic material recognition and cleavage, also the type of Cas protein involved varies in all three types of CRISPR/Cas system (Redman et al. 2016). Out of the three systems, type II has been effectively used by researchers for genome editing in eukaryotic systems. The Type II system contains Cas 9 protein which is derived from Streptococcus pyogenes and is a PAM-dependent system. In order to enable genome editing in mammalian cells and induce desired genes through type II CRISPR/Cas system three components are essential that are Cas9, a CRISPR-associated nuclease; cr-RNA for target recognition and tracr-RNA which serves as the binding scaffold for cas9 protein. Cr-RNA and tracr-RNA together compose the single guide RNA (sgRNA) of about 20 nucleotides in length, complementary to the target DNA segment of interest, which is responsible for specific binding to the target. The target segment is followed by a protospacer adjacent motif (PAM) which is a short sequence of three nucleotides in length. PAM helps sgRNA recognize the binding sequence and the absence of PAM prevents targeting of the self-genome.

Albeit the presence of PAM sequence for recognition and specificity provided by 20 nucleotides sgRNA, sgRNA sometimes binds and cleaves other DNA sites giving rise to the off-targets (Manghwar et al. 2020). This could be occurring because of two reasons, high probability of finding thousands of three base pair PAM in large DNA composed of only four nucleotides (ATGC) repeating and binding between sgRNA and target can tolerate three to five mismatches at the distal end of the sgRNA-DNA complex. Off-targets are a major limitation in applying genome editing techniques as it can mutate necessary genes, hindering proper gene function and expression, and it can be fatal for the biological system. To apply genome editing tools for clinical analysis and to enhance the safety of biological systems, knowledge of potential off-targets is essential during experimental designing and choosing the sgRNA in such a way that minimizes the chances of targets. For that, we need a robust method to predict off-targets caused by the sgRNAs in a genome of study. Machine learning has shown good results on sequential data be it genome or polypeptides in biological domains such as protein structure prediction, transcription binding site prediction, evolutionary relationship, etc. not long-ago machine learning and deep learning algorithms are applied in an attempt to accurately predict off-targets in CRISPR-cas9 genome editing tool (Sharkawy 2020). Deep learning neural network models thrive on large amounts of data to extract features from them and train, the large amount of genomic data of CRISPR-cas off-targets came from experiments involving expression profiling followed by sequencing off-targets by researchers and published for open access.

The objective of this article is also one such attempt by building different CNN, RNN, and FNN models with different architectures and training them to accurately predict off-targets in CRISPR-case9, then comparing and evaluating these models to improve the resistant in plants.

14.2 Literature Review

Support vector machine (SVM) was the first ever machine learning (ML) model used for the prediction of sgRNA efficacy. At that time, biological data was very limited and the algorithm used targeting ribosomal genes as the indicators (Sherkatghanad et al. 2023). Some researchers also applied linear regression but Elastic-net outperformed all the other models at that time, Elastic-net authors proposed ways to achieve better performance in the CRISPR-cas technique and gives insight into CRISPR-cas9 gene knockout (Ogutu et al. 2012). CRISPR targeting assessment (CRISTA) algorithm was proposed, which depends on random forest machine learning framework to determine of targets, he analyzed five traditional ML models on human and mouse genomes (Dimauro et al. 2022). A ML model named inDelphi predicts insertion–deletion mutations and phenotypes using again mouse and the human genome (Li et al. 2021). With the advancement in data availability of CRISPR-cas due to genome sequencing method discoveries, researchers started publishing experimental data using techniques like guide-seq, and CIRCLE-seq, and this data was widely used to construct large repositories of CRISPR data like GenomeCRISPOR datasae and CRISPR (Bin Moon et al. 2019). The major libraries contributing to CRISPOR are CHANGE-seq, SITE-seq, guide-seq, CIRCLE-seq, etc. Recently, resemble learning (Trivedi et al. 2020) is also applied in CRISPR for potential off- and on-target classification, these models contained five scoring methods, namely CCTop, MITwebsite, CFD, MIT, AND Cropit. One-dimensional CNN design showed significant improvement in sequence analysis and achieve accuracy of 99% in predicting genome vulnerabilities (Khan et al. 2023). The development of disease- and pest-resistant crop lines is one of the many agricultural concerns that can be effectively addressed by recent advances in genome editing technology (Jhu et al. 2023). Research in agriculture is excited by the revolutionary promise of CRISPR/Cas tools in creating disease resistance in plants.

14.3 Methodology

14.3.1 Data Retrieval and Preprocessing

Data preprocessing is the most crucial as well as critical part of developing a deep learning (DL) model construction. Data need to be cleaned and processed in a form that can be fed into the DL model without increasing the complexity of the model too much. The first step is data retrieval followed by preprocessing, model building, and ultimately evaluation.

14.3.1.1 Retrieval of Large Amounts of Biological Data

DL models need a huge amount of data to train and for accurate prediction all for real-life applications where vagueness or not precise predictions can be proved fatal. The availability of valid biological data is the major limitation in the application of DL in a biological system, but due to boom in computer science and molecular biology made it possible. One such sequencing process used by researchers is the GUIDE-seq technique, creating an important off-target database providing validated data, but it was sourced from very few sgRNA sites providing not sufficient data to train the model with good accuracy (Rodríguez et al. 2020). Ultimately CRISPR database (Hodgkins et al. 2015) created by combining validated experimental work is created which provides approx 26,000 validated off-target sequences, enough for DL. The CRISPOR data was retrieved from the GitHub repository of the author for this study (Usluer et al. 2023). The data contains some validated on-targets sequences, used to perform binary classification and train models to distinguish between off-targets and on-targets. The data contains some duplicates which are needed to be removed before feeding the models.

14.3.1.2 Data Preprocessing and One-hot Encoding

CRISPR targeting specificity, which is the site of binding sgRNA ensured by two components, first 20 nucleotide sgRNA sequence itself and 3 nucleotide PAM sequences present near the target site, a total of 23 nt length of sgRNA-DNA sequence is used as input for DL models. The sequence ATGC cannot be fed; it should be vectorized before training. The one-hot encoding (Lv et al. 2021) is used after removing duplicate data to represent categorical values of nucleotide sequence ATGC in 0 and 1; [1,0,0,0], [0,1,0,0], [0,0,1,0] and [0,0,0,1] for A, T, G, and C respectively. They are resulting in the matrix of 4 × 23. The out of the model is also categorical, binary that is input sequence is either off-target or on-target. Off-targets represented by [1,0] and on-target by [0,1]. Pyplot: Pyplot module of matplotlip python library is used for visualizing 4 × 23 encoding as images shown in Fig. 14.1. The figure shows the summation of both sgRNA and DNA metrics which were used by us as input. Pickle: We have used the pickling method to save the 4 × 23 one-hot encoding for organized storage and easy access. Pickle Python library is used for pickling.

14.3.1.3 Test: Train Split

The encoded data is divided into three groups, namely training data, validation data, and test data. 80% of the data i.e. 20841 off-targets were used for training the ML model, of which 10% of the data was kept aside for validation of models. Validation is done after every epoch, by comparing the accuracy of prediction on data never seen by the model. The rest 20% of the data which is 5211 off-targets was used for testing the model and calculating various evaluation scores.

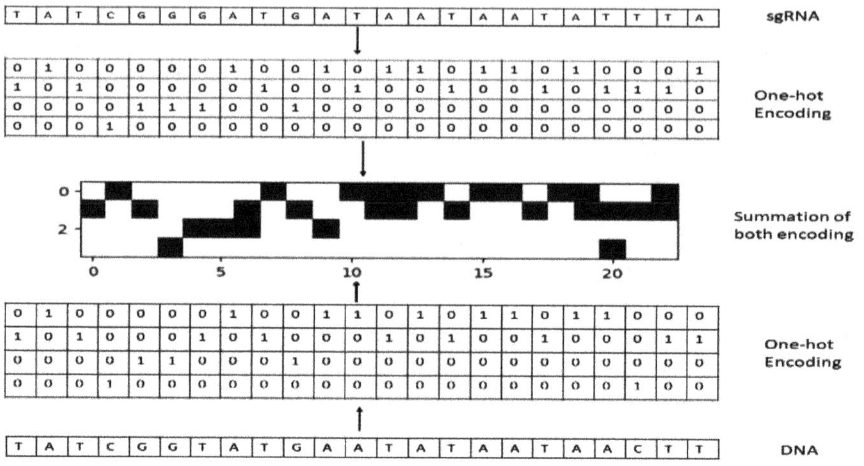

Fig. 14.1 One-hot encoding for sgRNA and DNA of size 4 × 23

14.3.2 Neural Network Model Building

To produce accurate results, human input is required, and a data scientist manually selects the relevant features that the software needs to assess in machine learning techniques. Therefore, developing and maintaining unstructured and large amounts of data can be a tedious and laborious task. On the other hand, in deep learning techniques, the data scientist only provides raw data, and the deep learning network learns and generates features on its own. Deep learning techniques have become an effective tool for off-target prediction in CRISPR-Cas9. A neural network, which is a type of deep learning process, teaches computers to process data in a way inspired by the human brain. It uses interconnected nodes or neurons arranged in layers that resemble the human brain. A simple neural network consists of interconnected neurons in three layers: input, hidden, and output layers. A neuron is the basic information processing unit of a neural network, and its simple architecture is shown in Fig. 14.2. Neural network consists of inputs x_1, x_2, x_3 .....x_n, weights associated with each inputs w_1, w_2, w_3 .... w_n, bias value (b), and then computed the weighted sum and added bias to it. An activation function determines the range of activation values for an artificial neuron. Apply activation function for nonlinear transformation of inputs for making complex task. There are different types of activation function such as linear, tanh, sigmoid, RELU, SoftMax, etc. which adds nonlinearity features in neural network (Feng and Lu 2019). Depending upon the data flow from input node to output node, neural networks are classified into different categories as follows:

This paper uses three types of neural networks such as feedforward neural network (FNN), convolutional neural network (CNN), and recurrent neural network (RNN) with slightly different architectures. Two RNN models were also build, one with GRU and the other with LSTM, both with the same activation functions and

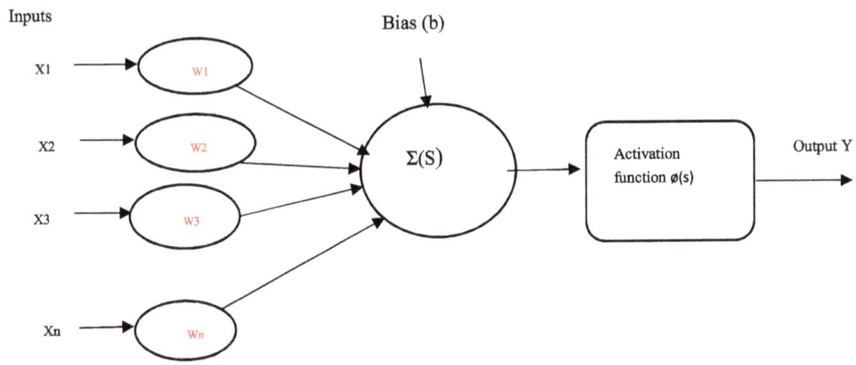

Fig. 14.2 General architecture of neural network

regularization layers (dropout and batch normalizations). We created DL models using Python programming with the help of TensorFlow and Keras. Short descriptions of different types of neural networks are presented here:

14.3.2.1 Feed Forward Neural Networks (FFN)

In this network, data moves from every node in the previous layer to every node of the next layers and this transformation process continues in only forward direction until the data outputs are not generated (Choi et al. 2020). Perceptrons are a type of ML algorithm that takes a set of features and their corresponding targets as input and searches for a line, plane, or hyperplane that separates classes in two-, three-, or hyperdimensional space. It has input, hidden, and output layers while feedback loops are absent. When several perceptrons are connected, then model is called multilayer perceptron which consists of many number of hidden layers between input and output layers (Plastiras et al. 2018). The output layer of an ANN typically has a single node for binary classification and a number of nodes for multiclass classification, depending on the number of classes to be predicted. Depending on the number of connected layers, there are several variants of FNN such as FNN3, FNN5, and FNN7 which contains three connected dense layers, five connected dense layers and seven connected dense layers, respectively, with occasional batch optimization and dropout layers (dropout set to 0.5). The activation function uses in dense layers are rectified linear unit (ReLU) and softmax for the output layer. ReLU is a nonlinearity function with only positive values, negatives are truncated to 0, while softmax squeezes everything between 0 and 1, classifying off-targets from on-targets (Feng and Lu 2019). Pattern recognition, classification tasks, regression analysis, Image recognition, time series prediction, etc. are several applications of FNN.

14.3.2.2 Convolutional Neural Network (CNN)

There is a limitation of feedforward ANN (FANN) in image recognition and computer vision task because each input into a FANN corresponds to a pixel in the image, hence spatial features in the image are lost. CNN is a special type of FNN that solve the problem of this issue by maintaining the spatial link between pixels in an image. A CNN preserves the spatial context from which a feature was derived by feeding patches of an image—rather than individual pixels—to particular nodes in the subsequent layer of nodes. These node patches extract specific information and are referred to as convolutional filters (Plastiras et al. 2018; Zhang et al. 2021). Convolutional layers, which form the foundation of the structure of CNNs, are created by convolving images with multiple small kernels. These many kernels function as feature identifiers to classify different features of the input data, typically images. Activation functions and pooling are necessary for the complex utilization of these functions. Once the features are extracted, fully connected neural networks will be used to connect them to the output layer. In this paper, build two CNN models, one with 2 convolutions and 1 dense layer while the other with 3 convolutions and 2 dense layers, both architectures have batch normalization, dropout, and 2 × 2 max-pooling layers to pass local maxima towards the next level.

14.3.2.3 Recurrent Neural Network (RNN)

It is a specialized type of ANN made to handle sequential data. Instead of FNN networks that process information in a single pass, RNNs can leverage their internal memory to process sequences and take into account dependencies between elements. The neural network consists of input, hidden, and output units. The hidden units are connected through feedback (FB) loops (IntroductiontoSequenceLearningModels.pdf.crdownload n.d.). RNNs possess an internal state, known as the hidden state, which captures information from preceding elements in the sequence. This hidden state is updated as the network processes each element, enabling it to learn from the contextual information (Schmidt 2019). The decision made at time $t-1$ in an RNN has an impact on the decision made at time t. Therefore, the network's response to new data depends on two factors: (1) the input at that specific moment and (2) the output from a recent past event (Zhang et al. 2021; Shewalkar et al. 2019). Output of RNN is computed in two steps by the following Eqs. (14.1) and (14.2), where x, y represent the inputs and outputs sequence, h represents the hidden layer sequence at time $t = 1$ to $t = n$, w and b represent the weight and bias metrices, and f represents the activation functions.

$$h_t = f\left(w_{xh} x_t + w_{hh} x_{t-1} + b_h\right) \tag{14.1}$$

$$y_t = w_{hy} h_t + b_y \tag{14.2}$$

This paper uses RNN with long short term memory unit (LSTM) and RNN with gated recurrent unit (GRU). In LSTM, special memory cell unit make it simple to retain data and information for long duration of time. GRU has a fewer feature then LSTM but it has an additional gated unit, which regulate information flow inside the unit without the need for independent memory cells. Unlike an LSTM, all the content of GRU is visible because of lack of output gate in GRU (Shewalkar et al. 2019). In this paper, I discuss two RNN models, one with GRU and the other with

LSTM, both with the same activation functions and regularization layers (dropout and batch normalizations).

I created these deep learning models using Python programming with the help of TensorFlow and Keras. TensorFlow (Weber et al. 2021) is the open-source platform provides tools for building ML models with mathematical aspects of models while Keras (Chicho and Sallow 2021) is an open-source library, provided by TensorFlow. It provided high-level programming for ML and fast performance. We build all the DL models using Keras' sequential, all the layers with their attributes can be easily lined up, compiled, and trained. Keras also has features like callbacks for feature refinement.

14.3.3 Training and Regularization

The encoded matrix is exported and unpickled to train the model in Jupyter notebook. We used adam optimizer to control the learning rate and weights, for all of the models and binary cross entropy as a loss function for DL training. While training history callback from Keras is used to store loss and accuracy measures per epoche and later used for model evaluation. The training was set up for 300 epotched in batches with batch size 50. Overfitting is the major problem when training the DL models. We tried to control overfitting by using dropout layers, batch optimization, and using Keras callback, named early stopping, which halts the training process when the model starts to overfit, decided by observing validation loss.

14.4 Model Evaluation Metrics

Model evaluation is necessary to compare the performance of the models I build on the training data. Following are the scoring I have chosen for model evaluation (Feng and Lu 2019; Shewalkar et al. 2019).

14.4.1 Confusion Matrix

Confusion matrix (Shewalkar et al. 2019; Konstantakos et al. 2022) helps to visualize the performance of the classifier, in my case, I am performing binary classification, thus confusion matrix is 2 × 2. It shows the prediction made by the classifier using for measures: true positive (TP), false positive (FP), true negative (TN), and false negative (FN).

14.4.2 Accuracy Score

It tells us by what percentage of data our classifier can classify correctly.

$$Accuracy\ Score = (TP + TN) / (TP + FN + TN + FP)$$

14.4.3 Loss

Loss gives cumulative errors done by the classifier while training, the higher the value of loss more errors are done by the model.

14.4.4 Precision Score

It is the measure to know how much proportion of positively predicted labels are truly positive in all the positive labels predicted by the classifier.

$$Precision\ Score = TP / \left(FP + TP\right)$$

14.4.5 Recall Score

It is also known as the true positive rate which measures the proportion of predicted positive labels out of true positives.

$$Recall\ Score = TP / \left(FN + TP\right)$$

14.4.6 F1 Score

Consider both precision and recall to compare a classifier and measures the accuracy of the ML model. Sklearn provides a classification report function that evaluates the classifier and produces scores for Precision, Recall, F1 score, and Support for each class and then averages them. Support is the ground truth, showing the actual classification of labels. It also provides average, macroaverage, and weighted average. The macroaverage is the classical mathematical mean of the F1 scores of two classes. Weighted avg normalizes the F1 score with the support, it is very helpful in case of imbalanced data like ours where on-targets are present in less quality.

$$F1\,Score = 2 * Precision\ Score * Recall\ Score / \left(Precision\ Score + Recall\ Score\right)$$

14.5 Results and Discussion

The CRISPR-cas9 data collected from CRISPR used to train our neural network contains 26,000 off-targets, out of that only 177 are validated on-targets, making our data asymmetrical. 80% of the data is used to train our seven neural network models and rest 20% is used to test them. The 23 nucleotides are stored as images after one-hot encoding, thus every sequence is a sparse matrix off 4 × 23. The input data structure, architecture of the model, its training, and prediction on the test data are discussed in this section. It has been found that the CRISPR-cas9 data obtained from CRISPR was imbalanced and consisted mostly of off-targets, data portion used for training contains 5181 off-targets and only 30 on-targets. The results are as follows:

14.5.1 Analysis of FNN

For an input in FNN, the 4 × 23 one-hot encoded sequence is flattened and linearized before feeding to the DL model. It is the simplest model with only three fully connected dense layers, without any backward pass. The layers contain the rectified linear unit (ReLU) activation function and the output layer contains the sigmoid function. FNN3 has 12,507 total parameters and 12,307 trainable parameters. The results shown in Table 14.1 showed that the model could classify with an accuracy of 0.9948 and a loss of 0.0194. The model also has good score for off-targets and very good macro precisions and weighted F1 is 0.99. Confusion matrix shown in Fig. 14.3a found that TP = 5178, TN = 7, FP = 3, and FN = 23.

FNN5 contains five fully connected dense layers with 22,582 trainable parameters out of 22,932 parameters. From Table 14.1, I found that the model could classify with 0.9944 accuracy and 0.024 loss. The model has 0.7 precision against on-targets. The model also has good score for off-targets and very good macro precisions and weighted F1 is 0.99. Confusion matrix shown in Fig. 14.3b found that TP = 5178, TN = 4, FP = 3, and FN = 26.

FNN7 contains seven fully connected dense layers with few intermediate batch optimization layers. It is also a bad classifier compared to others. This model showed very poor prediction for on-targets, and there are 0 true positives for class 1 (on-targets) prediction in the confusion matrix. From Table 14.1, an accuracy of 0.9942 and a loss score of 0.0251 are found. The model also has good score for off-targets but F1 and precision of label 1 is zero making it the worst model in comparisons of FNN3 and FNN4. Confusion matrix shown in Fig. 14.3c found that TP = 5176, TN = 0, FP = 5, and FN = 30 showed predicting all the on-target wrongly.

Table 14.1 Performance evaluation scores of all the DL models tested on CRISPR data

							RNN	
Model	Label	CNN3	CNN5	FNN3	FNN5	FNN7	GRU	LSTM
Loss		0.0159	0.0295	0.0194	0.024	0.0251	0.0159	0.0724
Accuracy		0.995	0.9942	0.9948	0.9944	0.9942	0.995	0.9946
Recall	Label 0	0.9998	1	0.9994	0.9996	0.999	0.9986	0.9982
	Label 1	0.1366	0	0.2333	0.2	0	0.3666	0.3666
Precision	Label 0	0.9951	0.9942	0.9955	0.9953	0.9942	0.9963	0.9963
	Label 1	0.8333	0	0.7	0.75	0	0.6111	0.55
F1	Label 0	0.9974	0.9971	0.9974	0.9974	0.9966	0.9974	0.9973
	Label 1	0.2777	0	0.35	0.3157	0	0.4583	0.44

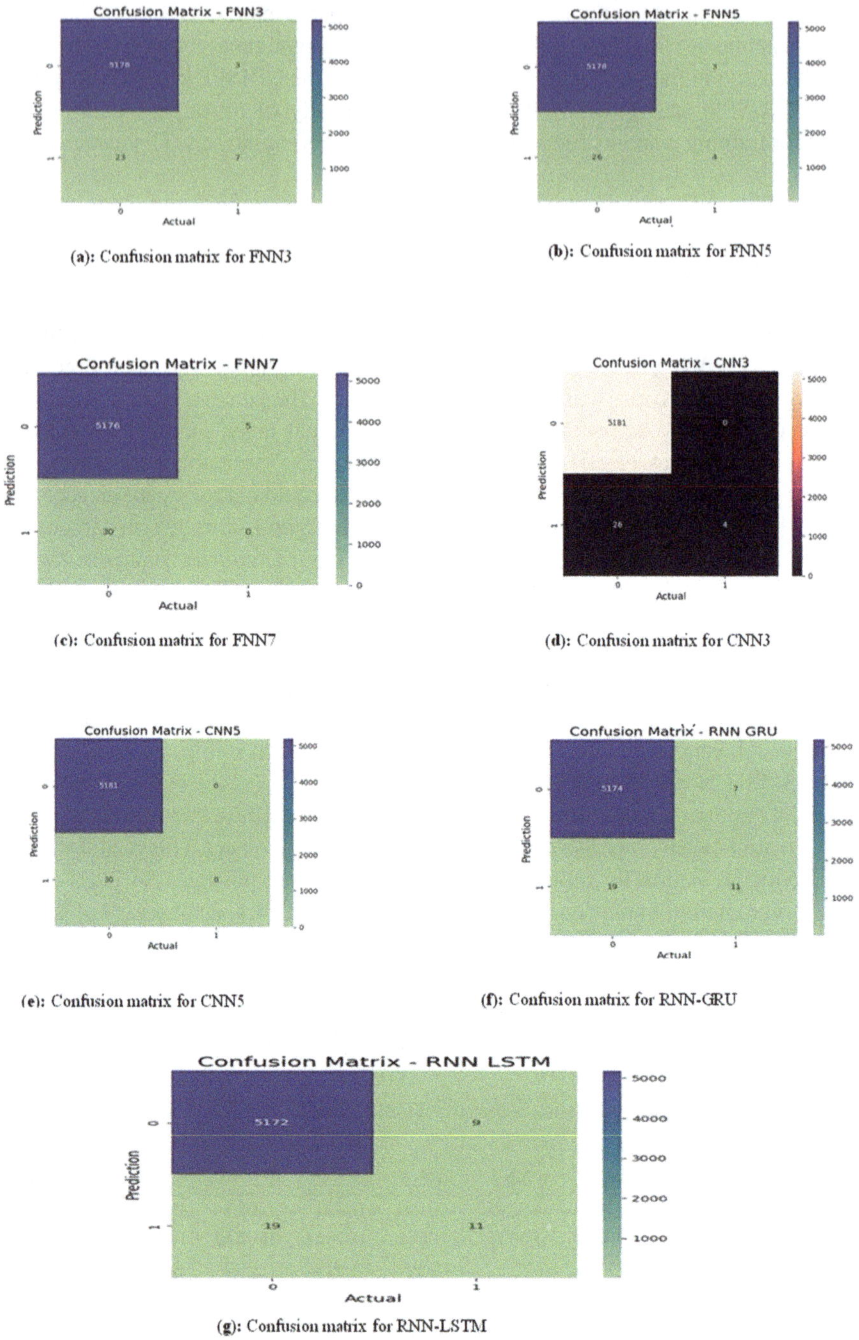

(**a**): Confusion matrix for FNN3

(**b**): Confusion matrix for FNN5

(**c**): Confusion matrix for FNN7

(**d**): Confusion matrix for CNN3

(**e**): Confusion matrix for CNN5

(**f**): Confusion matrix for RNN-GRU

(**g**): Confusion matrix for RNN-LSTM

Fig. 14.3 Confusion metrics for different variants FNN, CNN, and RNN neural network models. (**a**) Confusion matrix for FNN3, (**b**) Confusion matrix for FNN5, (**c**) Confusion matrix for FNN7, (**d**) Confusion matrix for CNN3, (**e**) Confusion matrix for CNN5, (**f**) Confusion matrix for RNN-GRU, (**g**) Confusion matrix for RNN-LSTM

After analyzing the results of FNN3, FNN5, and FNN7, I found that FNN3 performs best in overall parameters for predicting on-targets and off-targets CRISPRR Cas/9 but in some parameters FNN5 performs slightly better than FNN3 while FNN7 performs worst in comparisons to other two variants of FNN.

14.5.2 Analysis of CNN

CNN is usually used for image classification. The encoded sparse matrix of 4 × 23 is inserted into the CNN models as 4 × 23 × 1, which is as images of 4 × 23 with only one grayscale. The CNN3 contains three hidden layers, two convolution layers one with a 3 × 3 kernel and the other with 1 × 1, and one fully connected layer before the output node. There are a total of 84,738 trainable parameters. From Table 14.1, I found that the CNN3 model predicted off-targets with a 0.995 accuracy rate, comparatively, it has the lowest loss score (0.0159) from the other models. It is the best classifier. The F1 score for off-target classification (label 0), 0.9974 is the highest in CNN3. The confusion matrix shown in Fig. 14.3d found that TP = 5181, TN = 4, FP = 0, and FN = 26, determined it perform well on an off-target prediction but not so well for on-targets.

CNN5 contains three convolution layers first one with a 3 × 3 kernel and other two with a 1 × 1 kernel and two fully connected layers, it also contains dropout layers and batch normalization later like CNN3 but in addition, contains a flatten layer, with a total of 97,026 trainable parameters. From Table 14.1, its accuracy is 0.9942 which is less than the CNN3 model, and its loss score is 0.0295, which is highest, this model showed very poor prediction for on-targets, there are 0 true positives for class 1 (on-targets) prediction in the confusion matrix. This model is most precisely classified as off-targets but this model is one of the worst models among others in terms of accuracy and loss. The confusion matrix shown in Fig. 14.3e found that TP = 5181, TN = 0, FP = 0, and FN = 30 showed predicting all the on-target wrongly. Hence CNN3 model outperforms than CNN5 model among overall evaluating parameters.

14.5.3 Analysis of RNN

The 4 × 23 encoded matrix is directly fed to the model. RNN-GRU model contains one GRU layer, two dense layers, one batch normalization layer, and one batch layer. There are a total of 49,774 trainable parameters. From Table 14.1, it is the best model with 0.9950 accuracy and 0.0159 loss score. It does have a good score for class 1 (off-targets) but is also better for on-target prediction. The classifier has a 0.73 mean average F1 score for both classes. The confusion matrix shown in Fig. 14.3f found that TP = 5174, TN = 11, FP = 7, and FN = 19, determined it perform well on an off-target prediction as well as for on-targets prediction. This model is the best in our experimental setup.

The RNN-LSTM model contains one LSTM layer, two dense layers, one batch normalization layer, and one batch layer, with 60,170 trainable parameters. It results are very similar as RNN-GRU as accuracy of 0.9946 with loss of 0.0724. It does have a good score for class 1 (off-targets) but is also better for on-target prediction similar as RNN-GRU. The confusion matrix shown in Figure 14.3g revealed that TP = 5172, TN = 11, FP = 9, and FN = 19, determines that it performed well on off-target prediction as well as on-target prediction as well as for on-targets prediction. The overall performance of RNN-GRU is better than RNN-LSTM.

When I compared all the above-discussed variants of FNN, CNN, and RNN models, found that RNN-GRU outperforms all the discussed variants of the models.

14.6 Conclusion

In this article, three variants of FNN, two variants of CNN, and two variants of RNN Neural network models are used to predict off-target in CRISPR-cas9, a genome editing technique. We cleaned the data and encode it using one-hot encoding retrieved from the CRISPR repository on GitHub. We then trained our model and performed binary classification on the data. Out of the discussed variants of FNN, CNN, and RNN neural network models, FNN3, CNN3, and RNN-GRU are best-performing classifiers based on accuracy, loss score, and F1 score. All of the discussed neural network models have good accuracy with very slight differences, and very close F1, precision, and recall scores for off-target prediction. However, these scores for on-target prediction differ significantly as FNN7 and FNN5 showed zero scores for ontarget predictions as well as 0 true negatives in their confusion matrix, both of which are the worst classifiers. The poor prediction for on-targets is because CRISPOR data is highly imbalanced. There were only 177 on-targets out of total of 26,000 samples, and a 20% split of this data is kept for training and evaluation. In training data, only 30 samples belonged to on-targets out of 5211. This results in training bias toward on-targets. The results analyzed from Table 14.1 and Fig. 14.3 showed that RNN-GRU variant outer perform from rest of the discussed variants. In these results, some of the models perform better but RNN-GRU outperform from rest of the models for CRISPR/Cas9 gene editing to improve the resistant in plants.

References

Bin Moon S, Kim DY, Ko JH, Kim YS (2019) Recent advances in the CRISPR genome editing tool set. Exp Mol Med 51(11). https://doi.org/10.1038/s12276-019-0339-7

Charlier J, Nadon R, Makarenkov V (2021) Accurate deep learning off-target prediction with novel sgRNA-DNA sequence encoding in CRISPR-Cas9 gene editing. Bioinformatics 37(16):2299–2307. https://doi.org/10.1093/bioinformatics/btab112

Chicho BT, Sallow AB (2021) A comprehensive survey of deep learning models based on keras framework. J Soft Comput Data Min 2(2):49–62. https://doi.org/10.30880/jscdm.2021.02.02.005

Choi RY, Coyner AS, Kalpathy-Cramer J, Chiang MF, Peter Campbell J (2020) Introduction to machine learning, neural networks, and deep learning. Transl Vis Sci Technol 9(2):1–12. https://doi.org/10.1167/tvst.9.2.14

Dimauro G, Barletta VS, Catacchio CR, Colizzi L, Maglietta R, Ventura M (2022) A systematic mapping study on machine learning techniques for the prediction of CRISPR/Cas9 sgRNA target cleavage. Comput Struct Biotechnol J 20:5813–5823. https://doi.org/10.1016/j.csbj.2022.10.013

Feng J, Lu S (2019) Performance analysis of various activation functions in artificial neural networks. J Phys Conf Ser 1237(2):111–122. https://doi.org/10.1088/1742-6596/1237/2/022030

Hodgkins A et al (2015) WGE: A CRISPR database for genome engineering. Bioinformatics 31(18):3078–3080. https://doi.org/10.1093/bioinformatics/btv308

IntroductiontoSequenceLearningModels.pdf.crdownload (n.d.)

Jhu MY, Ellison EE, Sinha NR (2023) CRISPR gene editing to improve crop resistance to parasitic plants. Front Genome Ed 5(October):1–11. https://doi.org/10.3389/fgeed.2023.1289416

Khan Z et al (2023) Genome editing in cotton: challenges and opportunities. J Cotton Res 6(1). https://doi.org/10.1186/s42397-023-00140-3

Konstantakos V, Nentidis A, Krithara A, Paliouras G (2022) CRISPR-Cas9 gRNA efficiency prediction: an overview of predictive tools and the role of deep learning. Nucleic Acids Res 50(7):3616–3637. https://doi.org/10.1093/nar/gkac192

Li Y, Brian Golding G, Ilie L (2021) DELPHI: Accurate deep ensemble model for protein interaction sites prediction. Bioinformatics 37(7):896–904. https://doi.org/10.1093/bioinformatics/btaa750

Lv Z, Ding H, Wang L, Zou Q (2021) A convolutional neural network using dinucleotide one-hot encoder for identifying DNA N6-methyladenine sites in the rice genome. Neurocomputing 422:214–221. https://doi.org/10.1016/j.neucom.2020.09.056

Manghwar H et al (2020) CRISPR/Cas systems in genome editing: methodologies and tools for sgRNA design, off-target evaluation, and strategies to mitigate off-target effects. Adv Sci 7(6). https://doi.org/10.1002/advs.201902312

Mengstie MA, Wondimu BZ (2021) Mechanism and applications of crispr/ cas-9-mediated genome editing. Biol Targets Ther 15:353–361. https://doi.org/10.2147/BTT.S326422

Ogutu JO, Schulz-Streeck T, Piepho HP (2012) Genomic selection using regularized linear regression models: ridge regression. BMC Proc 6(Suppl 2):S10

Plastiras G, Kyrkou C, Theocharides T (2018) Efficient convnet-based object detection for unmanned aerial vehicles by selective tile processing. ACM Int Conf Proceeding Ser. https://doi.org/10.1145/3243394.3243692

Redman M, King A, Watson C, King D (2016) What is CRISPR/Cas9? Arch Dis Child Educ Pract Ed 101(4):213–215. https://doi.org/10.1136/archdischild-2016-310459

Rodríguez TC, Pratt HE, Liu PP, Amrani N, Zhu LJ (2020) GS-Preprocess: containerized GUIDE-seq data analysis tools with diverse sequencer compatibility. bioRxiv:26–28

Schmidt RM (2019) Recurrent neural networks (RNNs): a gentle introduction and overview. 1:1–16. [Online] http://arxiv.org/abs/1912.05911

Sharkawy A-N (2020) Principle of neural network and its main types: review. J Adv Appl Comput Math 7:8–19

Sherkatghanad Z, Abdar M, Charlier J, Makarenkov V (2023) Using traditional machine learning and deep learning methods for on- and off-target prediction in CRISPR/Cas9: a review. Brief Bioinform 24(3):1–25. https://doi.org/10.1093/bib/bbad131

Shewalkar A, Nyavanandi D, Ludwig SA (2019) Performance evaluation of deep neural networks applied to speech recognition: RNN, LSTM and GRU. J Artif Intell Soft Comput Res 9(4):235–245. https://doi.org/10.2478/jaiscr-2019-0006

Trivedi TB et al (2020) Crispr2vec: machine learning model predicts off-target cuts of CRISPR systems. bioRxiv:1–19. [Online] https://doi.org/10.1101/2020.10.28.359885

Usluer S et al (2023) Optimized whole-genome CRISPR interference screens identify ARID1A-dependent growth regulators in human induced pluripotent stem cells. Stem Cell Rep 18(5):1061–1074. https://doi.org/10.1016/j.stemcr.2023.03.008

Weber M et al (2021) DeepLab2: A TensorFlow library for deep labeling. 1–7. [Online] http://arxiv.org/abs/2106.09748

Zhang W, Li H, Li Y, Liu H, Chen Y, Ding X (2021) Application of deep learning algorithms in geotechnical engineering: a short critical review. 54(8), Springer Netherlands. https://doi.org/10.1007/s10462-021-09967-1

Peps, Pathogens, and Pests: Challenges and Opportunities for Usage of Pep Signaling in Sustainable Farming

15

Alice Kira Zelman and Gerald Alan Berkowitz

Abstract

Peps (plant elicitor peptides) are important messengers in plant stress responses and activate plant immune as well as in some cases anti-predation defense responses. These signaling molecules have predominantly been studied in their roles in biotic interactions. While early research focused on model species' Peps and Pep receptors (PEPRs), Pep signaling components are now often investigated in crop species. Much of this interest owes to a desire to use Peps in sustainable agriculture as an alternative to applying environmentally harmful pesticides. Peps, as polypeptides native to plant species, are easily degraded in the environment after treatment and are native to the crop plant to which they are applied; factors favoring their benign effects on the environment. However, there are challenges to overcome in utilizing Peps as tools in agriculture. This chapter will discuss what problems must be overcome to effectively make use of Pep signaling in sustainable agriculture, such as incorporating Peps into pest management strategies and disease mitigation. Activating Pep signaling can only be beneficial when it is an effective pathway for dealing with a given biotic threat(s). Future directions for research into Peps will be discussed in the context of overcoming these challenges and the opportunities for using Peps in different applications in sustainable agriculture.

Keywords

Plant elicitor peptides · Damage-associated molecular pattern · Plant biotic stress response · Organic agricultural strategies

A. K. Zelman (✉) · G. A. Berkowitz
Department of Plant Science and Landscape Architecture, University of Connecticut, Storrs, CT, USA
e-mail: alice.zelman@uconn.edu

S. Singh, R. Mehrotra (eds.), *Plant Elicitor Peptides*,
https://doi.org/10.1007/978-981-97-6374-0_15

15.1 Introduction

Agriculture has always been faced with disease burden and herbivore pressure. From the potato famine to plagues of locusts, farmers have contended with dire threats from biotic stresses as well as less severe yield losses from more quotidian pests and pathogens. The advent of modern pesticides, fungicides, and other chemicals for controlling these problems has alleviated the plight of conventional crop producers. However, sustainable farming faces the challenge of controlling losses to biotic stressors without synthetic chemicals or genetically bioengineered plants in a world of increasingly complex threats. The arguments for sustainable agriculture are numerous. Of all human activity, agriculture uses the greatest amount of land (Mader et al. 2011), and therefore changes to agricultural practices can have an outsize impact. Sustainable agriculture can offer improvements for neighboring ecosystem health compared with "conventional" farming by reducing runoff and groundwater contamination, enhancing carbon sequestration, and retaining populations of beneficial insects such as pollinators, among many other benefits. Sustainable practices like diversifying crops with intercropping can aid with the suppression of pathogens and encouragement of beneficial insects in an economically advantageous fashion (Huss et al. 2022). Sustainable farming practices can result in better microbial diversity (Ishaq 2017; Blundell et al. 2020) and higher populations of beneficial insects in the field and neighboring habitat such as hedgerows and roadside banks (Mader et al. 2011). Global consumer demand for sustainable food and acreage devoted to sustainable farming are high and shows no signs of slowing (Bonciu 2022). Popular watchdog *Consumer Reports* suggests consumers choose organic produce when possible, despite its higher cost at the grocery store, especially when buying blueberries, bell peppers, hot peppers, strawberries, potatoes, green beans, kale and mustard greens, and watermelons, because of higher levels of potentially dangerous pesticide residues in these crops (Roberts 2024).

Naturally occurring compounds are an alternative (to synthetic pesticides) that can complement other sustainable agricultural practices. Plant elicitor peptides (Peps) are a type of signaling molecule produced by plants to spread danger signals around the organs of an individual plant and activate cellular level immune and other defensive responses. Peps were discovered two decades ago in the model plant *Arabidopsis thaliana* (Huffaker et al. 2006) and since that time, a body of fascinating research has uncovered their involvement in defense against fungi, bacteria, oomycetes, herbivorous insects, and nematodes in several crop plants. The aim of this chapter is to examine the case for using Pep signaling to protect plants in sustainable farming, the challenges that lie ahead for sustainable farmers hoping to use Peps for control of biotic threats to crops, and suggestions for research that may enable Pep signaling to be effectively used in crop production.

Others have reviewed Peps and Pep signaling (for example, Bartels and Boller 2015); this chapter will review Pep signaling only in brief. Peps are a type of damage-associated molecular pattern (DAMP), which activate components of the plant immune response. DAMPs are an endogenously produced class of molecules, but foreign molecules can evoke a similar response: microbe-associated molecular patterns (MAMPs) are compounds of uniquely microbial origin that also activate immune response components at a cellular level, and nematodes produce unique molecules that activate the innate immune system in plant cells as well (Choi and Klessig 2016). Pep signaling is potentiated by Pep-activated transcription of PROPEP genes, as demonstrated in *Arabidopsis* (Huffaker and Ryan 2007; Safaeizadeh and Boller 2019) and *Prunus* (Ruiz et al. 2018), in a positive feedback loop. PROPEPs detach from the tonoplast and enter the cytosol in response to a calcium signal (which could be a secondary cytosolic response to perception of MAMPs such as flagellin peptides corresponding to portions of the motility organ of bacteria); PROPEPs are cleaved into mature signaling peptide Peps by metacaspases (Hander et al. 2019). Peptides cleaved from progenitor polypeptides are common in stress signaling in plants (Chen et al., 2020). Peps are perceived by plasmalemma-bound Pep receptors (PEPRs) (Yamaguchi et al. 2006). Pep signaling requires rapid reversible influx of Ca^{2+} into the cytosol after perception by PEPRs (Qi et al. 2010; Ma et al., 2012). Thus Ca^{2+} is both upstream and downstream of Pep perception. Reactive oxygen species (ROS) are also both upstream and downstream of Pep signaling: RBOHD/F generate ROS after *Arabidopsis* Pep1 (AtPep1) perception, via phosphorylation and activation of BOTRYTIS-INDUCED KINASE1 by PEPR, and ROS is a positive regulator of Pep signaling (Jing et al. 2020).

PROPEPs are expressed at different levels in disparate organs (Mehdi and Boller, 2019). For example, in tomato unchallenged with any stimulus, SlPROPEP mRNA was detected in roots, stems, leaves, fruits, and flowers, with highest levels in stems and flowers and lowest level in roots (Trivilin et al. 2014). In soybean, only GmPROPEP1 was constitutively expressed in all tested organs: seeds, flowers, pods, leaves, roots, and root nodules, while GmPROPEP4 mRNA was found in seeds (Lee et al. 2018). In Arabidopsis, the AtPROPEPs are induced by the jasmonic acid (JA) derivative methyl jasmonate, a phytohormone associated with defense against herbivores and necrotrophic pathogens as well as AtPep1 itself (Safaeizadeh and Boller 2019).

Pep sequences are not highly conserved between plant families. Crops come from diverse plant families, and there appears to be little immune activation when a Pep from another family is applied to a plant (Lori et al. 2015). Therefore, no single Pep peptide can serve as an anti-herbivore or disease prevention treatment across crop species. Arbitrary peptide sequences can be synthesized easily, so this does not necessarily pose a manufacturing hurdle. However, it does mean that Pep sequences must be identified for at least each specific crop family, and preferably species.

Natural predators of herbivorous insects can be recruited to help control damage. Unfortunately, climate change, habitat loss, diseases, parasites, and competition from invasive species can reduce populations of beneficial insects (Mader et al. 2011). A growing number of non-native pests have no natural enemies in areas of heavy infestation. Organic farmers often have little recourse when the limited number of approved control substances are ineffective. In some cases when approved substances are effective at control, they are not cost-effective. For example, even though spinosad and azadirachtin performed better than control in a trial of spinosad, pyrethrin, and azadirachtin on tomato for tomato hornworm (*Manduca quinquemaculata*) control, none of the microbial biopesticides was economically advantageous (Koirala et al. 2021). There are also cases where approved organic treatments are harmful to the environment. For example, the spinosad Entrust is an Organic Materials Review Institute-approved anti-herbivore organic pesticide that is used on many different crops. Entrust was the only organic treatment that reduced tobacco hornworm (*Manduca sexta*) sufficiently to warrant use during infestation of tobacco (Toennisson and Burrack 2018). Entrust and other spinosad compounds have deleterious effects on other organisms; counterproductively, this includes beneficial arthropods (Biondi et al. 2012). According to the manufacturer label, "This product is toxic to aquatic invertebrates. Do not apply directly to water, to areas where surface water is present or to intertidal areas below the mean high water mark. Do not contaminate water when disposing of equipment washwaters. Do not apply where runoff is likely to occur. Do not apply when weather conditions favor drift from treated areas. Drift and runoff from treated areas may be hazardous to aquatic organisms in neighboring areas. Applying this product when rain is not predicted for the next 24 hours will help reduce potential risk to aquatic invertebrates by reducing pesticide runoff from the treatment area into water bodies. This product is highly toxic to bees and other pollinating insects exposed to direct treatment, or to residues in/on blooming crops or weeds." (Environmental Protection Agency website https://www3.epa.gov/pesticides/chem_search/ppls/062719-00621-20200214.pdf, accessed 24 April 2024). Peps, on the other hand, are unlikely to have any such effects, and could be used for crops in water, crops that are blooming, and before rain, for example.

Natural peptides of a comparable size to Peps have direct effects on deleterious organisms. Insects make multiple 20–40 amino acid antibacterial peptides (Otvos 2000), while bacteria make insecticidal peptides. While surprisingly few studies have evaluated whether there is a direct biocidal effect of Peps on pathogens and insects, because Peps are endogenous plant signaling molecules (i.e., in order to activate downstream defense responses, a cell must have a cognate receptor; thus there would be no reason to expect such receptors to be present in pathogens or insects). Limited evidence suggests that Peps do not have antimicrobial activity against *Pseudomonas syringae* pv. *tomato* (by SlPep, Zelman, unpublished data; by AtPep3, Berkowitz, unpublished data). Deleterious effects on non-target organisms during use on field crops are unlikely compared with other treatments. Peps would also presumably be degraded in the field quickly. For this reason, it is thought that Peps would be harmless to ecosystems. If treatments are effective enough, even conventional farmers might use sustainable methods for pest and disease control, protecting beneficial insect populations that can also benefit their crops and the planet. For example, in the southeastern USA, a single native digger bee can pollinate 6000 blueberry flowers in its lifetime (Mader et al. 2011). While an endogenous plant peptide hormone is unlikely to have off-target effects on invertebrates, sensitivity of benign and beneficial invertebrates could be checked using a typical ecotoxicological approach that characterizes the toxicity of Peps in a few key representative native species and applying the results to a species sensitivity distribution (Spurgeon et al. 2020).

15.2 Applications of Peps in Sustainable Crop Production

One possible application for Pep signaling in sustainable agriculture is post-harvest treatment to reduce losses to pathogenic organisms. Economic consequences of post-harvest disease are severe: 30% of fruits are lost post-harvest globally, and losses can reach up to 50% (Noam and Fortes, 2015; Poveda 2020). Pep application as a spray or drench treatment might be a good choice for post-harvest on fruits, for example, pepper, eggplant, tomato, strawberry, peach, cherry, and others for which Peps have been discovered and validated. Defensive compounds cease to accumulate in harvested fruits once they are separated from the plant that bore them (Alkan and Forte 2015) and plant hormone treatment has been demonstrated to reduce mold growth and post-harvest losses in numerous crops (reviewed in Poveda 2020). Typically, methyl jasmonate (a derivate of JA that activates JA signaling) or salicylic acid (SA) are used depending on the type of pathogen that is most damaging to the crop in question.

A new exciting application of Peps is regeneration in tissue culture. Tomato Pep (termed REGENERATION FACTOR1) was shown to aid in shoot regeneration from callus (Yang et al. 2023). Shoot regeneration is a critical step when using tissue culture in plant breeding and micropropagation. Cannabis is an example of a crop that has a need for improved shoot regeneration protocols, as current regeneration of transgenic cannabis plants is low (Hesami et al. 2021) and there is a great need for an efficient system for micropropagation (Monthony et al., 2021). Cannabis is also a crop for which consumers demand pesticide-free production (Monthony et al. 2021). The MAMP harpin has been shown to increase cannabis seedling growth in the absence of a pathogen and enhance defense against a necrotrophic pathogen (Sands et al. 2022), and Peps merit study for these applications in this increasingly important crop.

In a perfect environment, plants could grow and reproduce without needing to worry about disease or herbivory. However, a plant without defenses quickly succumbs to biotic threats. Plants must therefore make decisions about how to balance defense and growth. The existence of the Pep/PEPR system and the responses evoked by Pep signaling are testament to this need: signaling is activated when a threat is detected so that the balance of resource utilization is shifted toward defense. Plants successfully defend themselves against most such dangers, but this comes at a cost: metabolic resources that could otherwise have resulted in growth, photosynthetic product storage, reproductive development, and seed production are diverted to reinforcing cell walls, defensive compound synthesis, cuticle thickening, etc. (Huot et al. 2014). However, the growth-defense tradeoff may not be a significant detraction from using Pep treatments in at least some applications. Both above-ground and below-ground biomass was unaffected by Pep treatment of seeds, and the Pep concentration used was an effective anti-nematode treatment (Lee et al.

2018). AtPep application to seedlings inhibited root growth (Okada et al. 2021; Krol et al. 2010), but overexpressing AtPROPEPs increased root growth, and Peps are probably involved in root formation (Bartels and Boller 2015). It is worth noting that root inhibition occurred when a very high concentration of Pep was used (Okada et al. 2021)—presumably much higher than physiological concentrations would be. Tomato Pep and its receptor are implicated in shoot regeneration after wounding and callus formation (Yang et al. 2023).

CRISPR/Cas editing could be used to alter the promoters or coding sequences of PROPEPs to enhance expression, and such procedures can bring new crops to readiness quickly (Hundleby and Harwood 2019). While plants resulting from such new genomic techniques (NGTs) are currently regulated under the same framework as transgenic plants in many places, they may soon face more relaxed regulatory strictures, for example in the United Kingdom and European Union (Bohle et al. 2024; Menary and Fuller 2024), which historically has forbidden almost any genetically modified crop production.

Unfortunately, whether or not activation of Pep signaling proves to be an effective method of control for biotic threats, significant regulatory barriers remain for utilizing Pep signaling in organic farming. While consumers purchase a vast amount of genetically modified food, they are thought to generally have a negative view of bioengineered foods (Oselinsky et al., 2021; Sendhil et al., 2022). Organic farming trade organizations generally oppose any kind of genetically modified crop in organic agriculture (Hundleby and Harwood 2019). It is not currently possible to label bioengineered products as United States Department of Agriculture (USDA) Organic: the National Organic Standards Board of the United States prohibits "means that are not possible under natural conditions or processes… including gene deletion, gene doubling, introducing a foreign gene, and changing the positions of genes when achieved by recombinant DNA technology" (Reference §205.2 https://www.ams.usda.gov/sites/default/files/media/MSExcludedMethodsDiscDoc.pdf accessed 2 May 2024). However, as in the case of MAMP use in activating crop plant pathogen defense responses, the exogenous application of Peps would be consistent with organic disease and insect control (Ma et al. 2012).

"Pesticide-free" labels are also viewed favorably by consumers, although the regulatory framework for this label is not as stringent as, for example, USDA Organic certification (Zheng et al. 2022). For farms that cannot afford or do not wish to convert to organic production, synthetic Peps for biotic threat mitigation or transgenic high-PROPEP-expressing plants could be feasible for achieving yields without using pesticides. Peps are known to be involved in plant defense against insects (Huffaker 2015). Nine kinds of produce have been reported to have potentially dangerous pesticide residue levels, and therefore these crops have been recommended to consumers as the most important to buy organic (Roberts 2024). Of these crops, there are Peps already discovered in potatoes, bell peppers, and hot peppers of the species *Capsicum annuum* (Huffaker et al. 2013), hot peppers of the species *C. chinensis* (Zelman et al. 2024), and strawberries (Ruiz et al. 2018). Since PROPEPs are conserved within plant families, it would likely be trivial to identify Peps in kale (*B. oleracea*) because a Pep is already known from

Brassica rapa (Lori et al. 2015) and there are genomes available for *B. oleracea* (Liu et al. 2014). An alternative to USDA certified Organic produce production protocols are 'biorational pesticides' (e.g., https://ag.umass.edu/fruit/ne-small-fruit-management-guide/general-information/biorational-disease-control accessed 31 May 2024). Compounds such as harpin, which act to activate plant immune responses in a similar fashion as Peps, have been characterized as useful biorational pesticides and thus provide some degree of benefit as crop aids that reduce the negative impacts of pesticide applications to the environment and non-target organisms. Harpin is highly effective as a foliar treatment of tomato infected with the necrotroph *Phytophthora infestans* (Bourbos and Barbolopoulou 2006) and numerous harpins have been shown to evoke plant immune responses and increased plant growth (Choi et al. 2013). Harpin is sold as a plant growth enhancer under several brand names, such as Axiom and H2Copla. The global compound annual growth rate of harpin products over the period from 2020 and 2023 was 20% and 2024 is showing increased growth over that rate (Plant Health Care website https://www.planthealthcare.com/products/commerical-business, accessed 30 May 2024). Peps are bioactive at low concentrations—20 nM Pep solutions are used to elicit immune responses and prime plant immunity in laboratory studies. This has led us to conclude that they would be economically competitive with MAMP products such as harpin (Zelman and Berkowitz 2023), which is used commercially at ~1 uM. The low concentration at which Peps are effective suggests a possibility for using Peps to grow crops for certified organic products. As previously noted, exogenous Peps have been shown to induce a feedback loop that results in further endogenous PROPEP production *in planta* (Huffaker and Ryan 2007; Safaeizadeh and Boller 2019). Furthermore, plant extracts have been effectively used to increase yields of organic leafy crops (Ciriello et al. 2024). Extracts of plants are considered natural/nonsynthetic in the current NOP 5034 Materials for Organic Crop Production (Reference: 7 CFR 205.105 https://www.ams.usda.gov/sites/default/files/media/NOP-5034.pdf accessed 26 May 2024). This suggests the possibility that conspecific plants could be treated with synthetic Peps, and these plants could be processed into extracts that could then be applied to organic crops as a preventative measure against biotic threats.

The future use of Peps in agriculture will in any case be dependent on the application of patent law. The initial discoverers of Peps patented Pep and PROPEP sequences for cotton, rice, poplar, grape, sunflower, sugarcane, canola, potato, soybean, and medicago (patents US-8686224-B2 and US-20130061352-A1) (Ryan et al. 2013). Commercial applications of synthetic Peps and recombinant PROPEP expression strategies will need to account for this.

15.3 Peps and Differing Biotic Threats

Pests and pathogens can be categorized based on their infection and feeding strategies. Pathogens are classified as biotrophic or necrotrophic. As Laluk and Mengiste elegantly expressed, "biotrophs have evolved intricate infection strategies aimed at maintaining host viability whereas necrotrophs disrupt cellular integrity." Necrotrophs tend to cause widespread cell death, tissue damage, and visible rot while biotrophs use specialized structures to grow inside living tissues without killing cells (Laluk et al. 2010).

Because plants do not have mobile immune cells akin to vertebrate macrophages, each plant cell is equipped to mount an immune response (Jones and Dangl 2006). The plant immune response to disease has canonically been characterized as having two distinct modes or layers, pathogen-associated molecular pattern (PAMP)-triggered immunity (PTI) and effector-triggered immunity (ETI). (MAMPs present in pathogens are termed PAMPs.) PTI is a response to detection of molecules that are common to multiple pathogenic or herbivorous organisms. For example, flagellin is a component of the flagella of many bacterial pathogens, and flagellin is a MAMP that triggers PTI. ETI, in contrast, is a response to the detection of race-specific molecules. ETI is typically a more extreme defense response to the detection of specific effector proteins from pathogens that includes planned cell death at the site of infection (the hypersensitive response) (Thomma et al. 2011). ETI increases expression of defensive genes and synthesis of antimicrobial compounds and other defense components in the cells surrounding a smaller region of cell death. It has been reported that JA and SA signaling both occur in ETI but localized differently: SA is active at and around the infection site, while JA is present in the surrounding area (Betsuyaku et al. 2018). JA is known to be generally associated with defense against necrotrophs, while SA is typically involved with defense against biotrophs (Pieterse et al. 2012). Plant pathogens take advantage of the plant innate defense system by "hijacking" phytohormones. For example, bacterial and fungal necrotrophs may interfere with JA signaling (Laluk et al. 2010). SA and JA inhibit each other's pathways, but in a seeming contradiction, SA in some cases increases JA in ETI, presumably to ward off secondary attack by a necrotroph after primary infection by a biotroph (Liu et al. 2016).

Similarly, different feeding types of herbivores elicit different defense responses in plants. Chewing herbivores and piercing-sucking phloem-feeders have distinct feeding styles that lead to differential damage and transcriptional responses in hosts (Montesinos et al. 2024). Other invertebrate pests of plants include obligate parasites like cyst-forming and root-knot forming nematodes, which are particularly economically damaging in tropical regions and on tomato, cotton, soybean, and cucurbits, in some cases affecting virtually 100% of a plant's root mass (Perry et al. 2009). Chewing insects can eat massive quantities of aerial plant parts, typically leaves. Some are economically damaging pests for sustainable farmers, such as tobacco hornworm (*Manduca sexta*) on solanaceous crops. Piercing-sucking insects cause less wounding than chewing herbivores but can still be devastating to crops. For example, the worst insect threat to rice globally is brown planthopper (Shen et al. 2022). The recent and ongoing sweep of the spotted lanternfly, a piercing-sucking pest, on grape across eastern North America can cause yield losses of 90% (Harner et al. 2022). In Arabidopsis, a chewing insect induced JA and ABA signaling and growth suppression; a piercing-sucking pest induced SA signaling and a longer-term but lower-amplitude change in gene expression overall (Montesinos et al. 2024). Considering insect pests along with pathogens presents a complex picture. In the current understanding, infection by a necrotroph can cause a plant to mount better defenses against piercing-sucking and chewing herbivores, while infection by a biotroph can make a plant mount a poor defense against chewing insects (Lazebnik et al. 2014). It is thought that insects associate with pathogens in order to interfere with host plant defenses (Kazan and Lyons 2014).

Due to the desirability of publishing positive results, studies of microbial and insect threats for which Pep treatment was not promising may have been abandoned. Whatever the cause, we currently have a lack of knowledge about specific Peps' relative efficacy against biotrophs, necrotrophs, hemibiotrophs, chewing herbivores, and piercing-sucking herbivores in crop plants. Sustainable agriculture functions best when a holistic approach to biotic threat mitigation is used. Integrated pest management is one stalwart approach to dealing with herbivore damage. Growers may be able to take advantage of plant innate immune signaling to facilitate disease severity reduction or herbivore damage mitigation. However, in common with using plant growth regulators and taking advantage of innate immune signaling, knowing which threat(s) to respond to is of paramount importance.

15.4 What Do we Need to Learn about Pep Signaling?

Since Peps native to a plant species are purported to be efficacious when applied only to other plants of a similar family, each type of crop species must have a corresponding Pep sequence discovered. More Peps are being identified all the time. Once a Pep has been discovered in a family, the process can be very easy, because within families, PROPEPs are similar enough to be detected without difficulty, even if the Pep sequences themselves are very dissimilar (Lori et al. 2015). If there are multiple Peps in a given crop, research must determine which is the most effective for specific threats. An example of this issue can be found in the work of Huffaker et al. who identified several Pep isoforms in maize, and one Pep was active against insect feeders while a different Pep-activated pathogen immune defense responses in this crop plant (Huffaker et al. 2011, 2013). Biochemical assays such as those used by Miao et al. (2019) will be helpful in identifying Pep signaling components in plant species that are amenable to transgene insertion.

Of course, pathogen strategies to suppress Pep signaling need not target Pep and PEPR directly but might interfere with upstream or downstream processes. For example, S-acylation of ROTUNDIFOLIA4 (ROT4) inhibits its binding to BRASSINOSTEROID-SIGNALING KINASE 5 (BSK5), which interferes with the physical association between BSK5 and PEPR1, and activates immunity by causing the release of PEPR1 from this complex. (Li et al. 2024). Post-translational modification of Peps has been studied some but there may be more to learn. Currently we know that metacaspases are responsible for cleaving PROPEP into Pep (Hander et al. 2019). Other post-translational processes should be investigated to see if we can improve the efficacy of Pep application. S-acylation (S-palmitoylation) of a non-secreted peptide, for example, has been shown to affect bacterial load in an infected plant (Li et al. 2024).

On a related note, it is unknown how activation of Pep signaling would affect the populations and activity of beneficial microbes. Beneficial microorganisms can include fungi and bacteria in the rhizosphere, endophytes that live within plant tissues, nitrogen-fixing bacteria, and microbes that live on the phylloplane (leaf surfaces). Beneficial microorganisms such as endophytes can act as biocontrols for disease-causing fungi and insect infestation, fix nitrogen, and solubilize micronutrients and macronutrients for better uptake (reviewed in Medison et al. 2022). Organic farms are reported to have higher microbial diversity than conventional farms as well as improved soil organic matter and plant–microbe interactions (reviewed in Ishaq 2017). Thus, the interactions of plants and microorganisms may be more complicated in sustainable agriculture. To our knowledge, no studies have examined the effect of Pep application or increasing *PROPEP* expression on beneficial microbe interactions. This is a vastly unknown topic and would be a worthy avenue of research.

Studies of the effects of exogenous Pep application have imbibed Pep aqueous solutions into cut aerial portions of plantlets, infiltrated via syringe into leaf tissue, or sprayed. Obviously, protocols for exogenous Pep application as a disease preventative or herbivory deterrent must be refined to reduce labor and material costs of deployment. On the other hand, some laboratory techniques may be adapted to crop production regimes readily. Pep solutions could be added to water used in hydroponic growth solutions to expose roots to solutions of synthetic Peps, for example. Pep application via spraying is possible. For example, rice seedlings treated with OsPep3 showed a promising reduction in symptoms of brown leafhopper attack (Shen et al. 2022). Seedling-stage plants are particularly vulnerable to disease, and infection during seedling growth may cause more damaging symptoms during mature phases than infection during later stages does, as seen for example in blackleg disease of canola (Li et al. 2006). Treatment of seedlings may be economically more viable than treatment at other phases of plant growth and development, for these reasons, and because a much smaller amount of Pep solution would be needed per plant due to the vastly smaller surface area of seedlings compared with larger plants.

Seed treatments with aqueous Pep has proved effective in reducing root-knot nematode infestation (Lee et al. 2018) and could be scaled up while allowing for planting with industrial farming equipment. Treating seed prior to planting is commonplace but requires drying before placing into planting equipment; different seeds can be treated for different lengths of time before germination rates are affected. The necessary durations of treatment for biotic stress mitigation will be required for any Pep seed treatments. Commercially planted seeds are often treated with aqueous solutions of growth regulators (hydroprimed) or osmolyte solutions (osmoprimed) to increase germination rates and growth rates (Weissmann et al. 2023a) and adding Peps to these solutions should be straightforward to implement. More niche applications can be envisioned. For example, anti-transpirants used on bedding plants and floral crops are effective at delaying wilting (Park et al. 2016); they could also include Peps to retard pathogen growth. Since Pep signaling appears to be important in salinity stress (Nakaminami et al. 2018; Wang et al. 2022), osmopriming with a Pep may also be useful when crops need to be grown in soil with a high salt content. The approach of using Pep as a seed soak to activate pathogen defenses during subsequent growth of a plant is well supported by the work of Dong et al. 2004; they showed that overnight incubation of Arabidopsis seeds in a solution containing the MAMP harpin (which, again, acts similarly to Peps as far as activation of pathogen defense responses) impacted the plant during its entire growth cycle to flowering (Dong et al. 2004).

Of course, treating crops with Peps will require a sufficiently stable Pep in the treatment medium. In medicine, peptides as therapeutics have encountered problems with aggregation and stability in solution (Zapadka et al. 2017). Studies of Pep stability are needed to determine how these peptides will fare during storage and after application, for example, how long aqueous solutions can be stored without refrigeration before significant degradation. Synthetic Peps could incorporate unnatural amino acids to extend viability (Badosa et al., 2022). Alternatively, crops could be bioengineered to have overexpressed PROPEPs or inducible PROPEP expression that could be exploited prior to an expected threat. Another approach to address the stability of exogenous Peps applied to crops as a defense against biotic stresses could involve nanoencapsulation, which has been shown to be an efficacious approach to improve stability and bioactivity of bioactive peptides (Aguilar-Toala et al. 2022).

It is unlikely that a herbivore or pathogen attacking a crop plant could directly develop resistance against these peptides. Peps act within the plant rather than directly against herbivores and pathogens, so the invading organism would need to interfere with the defense pathways activated by Peps *in planta*. The competitive inhibition of PEPR by a Pep analog secreted by a smut fungal pathogen is brought to mind as an example. An interesting case of interference with Pep signaling by a pathogen has been discovered. One pathogen of sugarcane produces a mimic of sugarcane Pep (SsPep1). This sugarcane smut fungus is able to transport the Pep mimic (ScPele1) into the plant where it acts as a ligand of sugarcane PEPR1, inhibiting Pep-induced defense pathways (Ling et al. 2022). The exact mechanism for this inhibition has not yet been elucidated but would be a worthy subject of further study.

Yield loss for major crops such as wheat, rice, potato, soybean, and maize already ranges from 20%-30% globally (Savary et al., 2019). Climate change will present agriculture with increased challenges and can directly affect pathogenesis and disease severity (Singh et al. 2023). Climate change may cause periods of moisture perturbation—either increased humidity or drought. Higher humidity can increase the virulence of a bacterial pathogen and appears overall to increase fungal pathogen virulence (Singh et al. 2023), while one study found that drought increased chickpea dry root rot. As a generalization it appears that necrotrophs cause more severe infection in plant subjected to drought, while biotrophs are less severe in drought due to the lack of productivity of plant tissues that nourish these pathogens (Singh et al. 2023). Anthropogenic increases in carbon dioxide concentration may generally have the opposite effect, since elevated carbon dioxide reduced plants' susceptibility to necrotrophs such as *Peronospora manshurica* and *Botrytis cinerea*, but increased susceptibility to biotrophs such as powdery mildew on cucurbits and rust of aspen (Singh et al. 2023). Clearly climatic shifts will have complex effects on plant–pathogen interactions and forecasts of specific threats are necessary to understand how to slow disease progression. This highlights the need for new sustainable strategies to protect plants from biotic threats. Using Pep peptides could be one such strategy.

Anthropogenic climatic changes include episodes of increased temperature, which can induce heat stress and yield losses in crops. The highest risk of yield damage due to extreme heat incidents will be the Indian subcontinent, central North America, and central and eastern Asia, areas that include vast acreage of crops such as maize and rice (Teixeira et al. 2013). Plants at elevated temperatures appear to have compromised disease resistance. The mechanisms for such temperature-induced susceptibility are largely unknown but some evidence suggests that SA and JA are involved. SA generation is reduced at higher temperatures. Huffaker et al. observed that ZmPep1 did not increase SA production (2011), so the impact of temperature on Pep-induced hormone generation may not be compromised. However, the impact of temperature on JA is unknown. It is important to remember that our knowledge of hormone production after Pep perception is limited to a few plants, and may differ in other plant species, in different plant organs, or in response to different processes that induce Pep signaling. Additionally, biotic and abiotic stresses will increasingly co-occur due to climate change (Carpentier et al., 2022). Components of Pep signaling, and temperature stress have been linked.

15.5 Conclusion

In an ever shifting, escalating landscape of biotic threats to crops, sustainable agricultural producers need multiple concurrent strategies to control these threats. In this chapter, we have clarified some of the major challenges of using Peps in sustainable agriculture, and some of the important directions that research can take to overcome these challenges. Peps have been found in numerous clades of plants, so stimulating Pep signaling may have a role as a disease and herbivory mitigation strategy in many crops. Scientists and crop producers will need to communicate with each other to assess the best ways to incorporate Peps into pest and pathogen management.

Acknowledgment The author thanks Jeff Chandler of the North Carolina Soybean Producers Association for helpful insight on seed treatments. Joseph Heckmann advised on USDA organic agriculture trends. Lorelei McFadden provided valuable editing of the manuscript.

References

Aguilar-Toala J, Quintanar-Guerrero D, Liceaga A, Zambrano- Zaragoza M (2022) Encapsulation of bioactive peptides: A strategy to improve the stability, protect the nutraceutical bioactivity and support their food applications. RSC Adv 12(11):6449–6458. https://doi.org/10.1039/d1ra08590e

Alkan N, Fortes AM (2015) Insights into molecular and metabolic events associated with fruit response to post-harvest fungal pathogens. Front Plant Sci 6(OCTOBER):1–14. https://doi.org/10.3389/fpls.2015.00889

Badosa E, Planas M, Feliu L, Montesinos L, Bonaterra A, Montesinos E (2022) Synthetic peptides against plant pathogenic bacteria. Microorganisms 10(9). https://doi.org/10.3390/microorganisms10091784

Bartels S, Boller T (2015) Quo vadis, Pep? Plant elicitor peptides at the crossroads of immunity, stress, and development. J Exp Bot 66(17):5183–5193. https://doi.org/10.1093/jxb/erv180

Betsuyaku S, Katou S, Takebayashi Y, Sakakibara H, Nomura N, Fukuda H (2018) Salicylic acid and jasmonic acid pathways are activated in spatially different domains around the infection site during effector-triggered immunity in Arabidopsis thaliana. Plant Cell Physiol 59(1):8–16. https://doi.org/10.1093/pcp/pcx181

Biondi A, Mommaerts V, Smagghe G, Viñuela E, Zappalà L, Desneux N (2012) The non-target impact of Spinosyns on beneficial arthropods. Pest Manag Sci 68(12):1523–1536. https://doi.org/10.1002/ps.3396

Blundell R, Schmidt J, Igwe A, Cheung A, Vannette R, Gaudin A, Casteel C (2020) Organic management promotes natural Pest control through altered plant resistance to insects. Nat Plants 6(5):483–491. https://doi.org/10.1038/s41477-020-0656-9

Bohle F, Schneider R, Mundorf J, Zühl L, Simon S, Engelhard M (2024) Where Does the EU-path on new genomic techniques Lead us? Front Genome Ed 6(July 2018):1–7. https://doi.org/10.3389/fgeed.2024.1377117

Bonciu E (2022) Trends in the evolution of organic agriculture at the global level - a brief review. Scientific Papers Series Management Economic Engineering in Agriculture and Rural Development 22(3):81–86

Bourbos V, Barbolopoulou EA (2006) Effect of Harpin Ea on the fruit production and control of Phytophthora Infestans in greenhouse tomato. In: X international symposium on plant bioregulators in fruit production, Orlando, FL, USA, pp 557–660

Carpentier S, Aldon D, Berthomé R, Galaud J (2022) Is there a specific calcium signal out there to decode combined biotic stress and temperature elevation? Front Plant Sci 13(November):1–7. https://doi.org/10.3389/fpls.2022.1004406

Chen Y, Fan K, Hung S, Chen Y (2020) The role of peptides cleaved from protein precursors in eliciting plant stress reactions. New Phytol 225(6):2267–2282. https://doi.org/10.1111/nph.16241

Choi H, Klessig D (2016) DAMPs, MAMPs, and NAMPs in plant innate immunity. BMC Plant Biol 16(1):1–10. https://doi.org/10.1186/s12870-016-0921-2

Choi M, Kim W, Lee C, Oh C (2013) Harpins, multifunctional proteins secreted by gram-negative plant-pathogenic bacteria. Mol Plant-Microbe Interact 26(10):1115–1122. https://doi.org/10.1094/MPMI-02-13-0050-CR

Ciriello M, Campana E, Colla G, Youssef R (2024) An appraisal of nonmicrobial biostimulants' impact on the productivity and mineral content of wild rocket organic conditions. Plan Theory 13(1326)

Dong H, Peng J, Bao Z, Meng X, Bonasera J, Chen G, Beer S, Dong H (2004) Downstream divergence of the ethylene signaling pathway for Harpin-stimulated Arabidopsis growth and insect defense. Plant Physiol 136(3):3628–3638. https://doi.org/10.1104/pp.104.048900

Hander T, Fernández-Fernández A, Kumpf R, Willems P, Schatowitz H, Rombaut D, Staes A, Nolf J, Pottie R, Yao P, Gonçalves A, Pavie B, Boller T, Gevaert K, Van Breusegem F, Bartels S, Stael S (2019) Damage on plants activates ca 2+ -dependent Metacaspases for release of immunomodulatory peptides. Science 363(6433). https://doi.org/10.1126/science.aar7486

Harner A, Leach H, Briggs L (2022) Prolonged phloem feeding by the spotted lanternfly, an invasive Planthopper, alters resource allocation and inhibits gas exchange in grapevines. Plant Direct 6(March):1–18. https://doi.org/10.1002/pld3.452

Hesami M, Baiton A, Alizadeh M, Pepe M, Torkamaneh D, Jones A (2021) Advances and perspectives in tissue culture and genetic engineering of cannabis. Int J Mol Sci 22(11). https://doi.org/10.3390/ijms22115671

Huffaker A (2015) Plant elicitor peptides in induced defense against insects. CurrOpin Insect Sci 9:44–50. https://doi.org/10.1016/j.cois.2015.06.003
Huffaker A, Dafoe N, Schmelz E (2011) ZmPep1, an Ortholog of Arabidopsis elicitor peptide 1, regulates maize innate immunity and enhances disease resistance. Plant Physiol 155(3):1325–1338. https://doi.org/10.1104/pp.110.166710
Huffaker A, Ryan C (2007) Endogenous peptide defense signals in Arabidopsis differentially amplify signaling for the innate immune response. Proc Natl Acad Sci USA 104(25):10732–10736. https://doi.org/10.1073/pnas.0703343104
Huffaker A, Pearce G, Ryan CA (2006) An endogenous peptide signal in Arabidopsis activates components of the innate immune response. Proc Natl Acad Sci USA 103(26):10098–10103. https://doi.org/10.1073/pnas.0603727103
Huffaker A, Pearce G, Veyrat N, Erb M, Turlings T, Sartor R, Shen Z, Briggs S, Vaughan M, Alborn H, Teal P, Schmelz E (2013) Plant elicitor peptides are conserved signals regulating direct and indirect Antiherbivore defense. Proc Natl Acad Sci USA 110(14):5707–5712. https://doi.org/10.1073/pnas.1214668110
Hundleby P, Harwood A (2019) Impacts of the EU GMO regulatory framework for plant genome editing. Food Energy Secur 8(2):1–8. https://doi.org/10.1002/fes3.161
Huot B, Yao J, Montgomery B, He S (2014) Editor's choice: growth–defense tradeoffs in plants: A balancing act to optimize fitness. Mol Plant 7(8):1267. https://doi.org/10.1093/MP/SSU049
Huss C, Holmes K, Blubaugh C (2022) Benefits and risks of intercropping for crop resilience and pest management. J Econ Entomol 115(5):1350–1362. https://doi.org/10.1093/jee/toac045Ishaq S (2017) Plant-microbial interactions in agriculture and the use of farming systems to improve diversity and productivity. AIMS Microbiol 3(2):335–353. https://doi.org/10.3934/MICROBIOL.2017.2.335
Jing Y, Shen N, Zheng X, Fu A, Zhao F, Lan W, Luan S (2020) Danger-associated peptide regulates root immune responses and root growth by affecting ROS formation in Arabidopsis. Int J Mol Sci 21(13):1–16. https://doi.org/10.3390/ijms21134590
Jones J, Dangl J (2006) The plant immune system. Nature 444(7117):323–329. https://doi.org/10.1038/nature05286
Kazan K, Lyons R (2014) Intervention of Phytohormone pathways by pathogen effectors. Plant Cell 26(6):2285–2309. https://doi.org/10.1105/tpc.114.125419
Koirala B, Quarcoo F, Kpomblekou-A K, Mortley D (2021) Organic tomato production in Alabama: host preference of the tomato hornworm ("Manduca Quinquemaculata") and performance of selected biopesticides. Am J Entomol 5(1):10. https://doi.org/10.11648/j.aje.20210501.12
Krol E, Mentzel T, Chinchilla D, Boller T, Felix G, Kemmerling B, Postel S, Arents M, Jeworutzki E, Al-Rasheid KAS, Becker D, Hedrich R (2010) Perception of the Arabidopsis danger signal peptide 1 involves the pattern recognition receptor AtPEPR1 and its close homologue AtPEPR2. J Biol Chem 285(18):13471–13479. https://doi.org/10.1074/jbc.M109.097394
Laluk K, Tesfaye M, Laluk K, Mengiste T (2010) Necrotroph attacks on plants: wanton destruction or covert extortion? In: The Arabidopsis book, the Arabidopsis book. The American Society of Plant Biologists, pp 1–34
Lazebnik J, Frago E, Dicke M, van Loon J (2014) Phytohormone mediation of interactions between herbivores and plant pathogens. J Chem Ecol 40(7):730–741. https://doi.org/10.1007/s10886-014-0480-7
Lee M, Huffaker A, Crippen D, Robbins R, Goggin F (2018) Plant elicitor peptides promote plant Defences against nematodes in soybean. Mol Plant Pathol 19(4):858–869. https://doi.org/10.1111/mpp.12570
Li H, Smyth F, Barbetti M, Sivasithamparam K (2006) Relationship between brassica Napus seedling and adult plant responses to Leptosphaeria Maculans is determined by plant growth stage at inoculation and temperature regime. Field Crop Res 96(2–3):428–437. https://doi.org/10.1016/J.FCR.2005.08.006
Li W, Ye T, Ye W, Liang J, Wang W, Han D, Liu X, Huang L, Ouyang Y, Liao J, Chen T, Yang C, Lai J (2024) S-acylation of a non-secreted peptide controls plant immunity via secreted-peptide signal activation. EMBO Rep. https://doi.org/10.1038/s44319-023-00029-x

Ling H, Fu X, Huang N, Zhong Z, Su W, Lin W, Cui H, Que Y (2022) A sugarcane smut fungus effector simulates the host endogenous elicitor peptide to suppress plant immunity. New Phytol 233(2):919–933. https://doi.org/10.1111/nph.17835

Liu L, Sonbol FM, Huot B, Gu Y, Withers J, Mwimba M, Yao J, He SY, Dong X (2016) Salicylic acid receptors activate Jasmonic acid Signalling through a non-canonical pathway to promote effector-triggered immunity. Nat Commun 7. https://doi.org/10.1038/NCOMMS13099

Liu S, Liu Y, Yang X, Tong C, Edwards D, Parkin IAP, Zhao M, Ma J, Yu J, Huang S, Wang X, Wang J, Lu K, Fang Z, Bancroft I, Yang TJ, Hu Q, Wang X, Yue Z et al (2014) The brassica oleracea genome reveals the asymmetrical evolution of polyploid genomes. Nat Commun 5(May). https://doi.org/10.1038/ncomms4930

Lori M, Van Verk M, Hander T, Schatowitz H, Klauser D, Flury P, Gehring C, Boller T, Bartels S (2015) Evolutionary divergence of the plant elicitor peptides (peps) and their receptors: interfamily incompatibility of perception but compatibility of downstream Signalling. J Exp Bot 66(17):5315–5325. https://doi.org/10.1093/jxb/erv236

Ma Y, Walker R, Zhao Y, Berkowitz G (2012) Linking ligand perception by PEPR pattern recognition receptors to cytosolic Ca2+ elevation and downstream immune signaling in plants. Proc Natl Acad Sci USA 109:19852–19857. https://doi.org/10.1073/pnas.1205448109/-/DCSupplemental. www.pnas.org/cgi/doi/10.1073/pnas.1205448109

Mader E, Shepherd M, Vaughan M, Black SH, LeBuhn G (2011) Attracting Native Pollinators. In: Burns D (ed) North Adams. Storey Publishing, MA

Medison R, Tan L, Medison M, Kenani Edward Chiwina K (2022) Use of beneficial bacterial endophytes: A practical strategy to achieve sustainable agriculture. AIMS Microbiol 8(4):624–643. https://doi.org/10.3934/microbiol.2022040

Miao S, Boller T (2019) Differential and tissue-specific activation pattern of the AtPROPEP and AtPEPR genes in response to biotic and abiotic stress in Arabidopsis thaliana. Plant Signal Behav 14(5):1–17. https://doi.org/10.1080/15592324.2019.1590094

Menary J, Fuller S (2024) New genomic techniques, old divides: stakeholder attitudes towards new biotechnology regulation in the EU and UK. PLoS One 19(3 March):1–16. https://doi.org/10.1371/journal.pone.0287276

Monthony A, Page S, Hesami M, Jones A (2021) The past, present and future of cannabis sativa tissue culture. Plants 10(1):1–29. https://doi.org/10.3390/plants10010185

Montesinos A, Sacristán S, Prado-Polonio P, Arnaiz A, Díaz-González S, Diaz I, Santamaria M (2024) Contrasting plant transcriptome responses between a Pierce-sucking and a chewing herbivore go beyond the infestation site. BMC Plant Biol 24(1):1–17. https://doi.org/10.1186/s12870-024-04806-1

Nakaminami K, Okamoto M, Higuchi-Takeuchi M, Yoshizumi T, Yamaguchi Y, Fukao Y, Shimizu M, Ohashi C, Tanaka M, Matsui M, Shinozaki K, Seki M, Hanada K (2018) AtPep3 is a hormone-like peptide that plays a role in the salinity stress tolerance of plants. Proc Natl Acad Sci USA 115(22):5810–5815. https://doi.org/10.1073/pnas.1719491115

Noam A, Fortes A (2015) Insights into molecular and metabolic events associated with fruit response to post-harvest fungal pathogens. Front Plant Sci 6(OCTOBER):1–14. https://doi.org/10.3389/fpls.2015.00889

Okada K, Kubota Y, Hirase T, Otani K, Goh T, Hiruma K, Saijo Y (2021) Uncoupling root hair formation and Defence activation from growth inhibition in response to damage-associated pep peptides in Arabidopsis Thaliana. New Phytol 229(5):2844–2858. https://doi.org/10.1111/nph.17064

Oselinsky K, Johnson A, Lundeberg P, Holm A, Mueller M, Graham D (2021) Gmo food labels do not affect college student food selection, despite negative attitudes towards Gmos. Int J Environ Res Public Health 18(4):1–19. https://doi.org/10.3390/ijerph18041761

Otvos L (2000) Antibacterial peptides isolated from insects. J Pept Sci 6(10):497–511. https://doi.org/10.1002/1099-1387(200010)6:10<497::AID-PSC277>3.0.CO;2-W

Park S, Mills S, Moon Y, Waterland N (2016) Evaluation of Antitranspirants for enhancing temporary water stress tolerance in bedding plants. HortTechnology 26(4):444–452. https://doi.org/10.21273/horttech.26.4.444

Perry R, Moens M, Starr J (2009) Root-Knot Nematodes. CAB International, London

Pieterse C, Van der Does D, Zamioudis C, Leon-Reyes A, Van Wees S (2012) Hormonal modulation of plant immunity. Annu Rev Cell Dev Biol 28(1):489–521. https://doi.org/10.1146/annurev-cellbio-092910-154055

Poveda J (2020) Use of plant-defense hormones against pathogen-diseases of postharvest fresh produce. Physiol Mol Plant Pathol 111(101521). https://doi.org/10.1016/j.pmpp.2020.101521

Qi Z, Verma R, Gehring C, Yamaguchi Y, Zhao Y, Ryan CA, Berkowitz GA (2010) Ca2+ signaling by plant Arabidopsis thaliana Pep peptides depends on AtPepR1, a receptor with guanylyl cyclase activity, and cGMPactivated Ca2+ channels. Proceedings of the National Academy of Sciences of the United States of America 107(49):21193–21198. https://doi.org/10.1073/pnas.1000191107

Roberts C (2024) Produce without pesticides. Consum Rep:28–35

Ruiz C, Nadal A, Montesinos E, Pla M (2018) Novel Rosaceae plant elicitor peptides as sustainable tools to control Xanthomonas Arboricola Pv. Pruni in Prunus Spp. Mol Plant Pathol 19(2):418–431. https://doi.org/10.1111/mpp.12534

Ryan C, Ryan P, Pearce G, Yamaguchi A, Yamaguchi Y, Ryan P (2013) Plant Defense Signal. Peptides:1–196

Sands L, Cheek T, Reynolds J, Ma Y, Berkowitz G (2022) Effects of Harpin and Flg22 on growth enhancement and pathogen defense in cannabis sativa seedlings. Plan Theory 11(9). https://doi.org/10.3390/plants11091178

Safaeizadeh M, Boller T (2019) Differential and tissue-specific activation pattern of the AtPROPEP and AtPEPR genes in response to biotic and abiotic stress in Arabidopsis thaliana. Plant Signal Behav 14(5). https://doi.org/10.1080/15592324.2019.1590094

Savary S, Willocquet L, Pethybridge S, Esker P, McRoberts N, Nelson A (2019) The global burden of pathogens and pests on major food crops. Nat Ecol Evol 3(3):430–439. https://doi.org/10.1038/s41559-018-0793-y

Sendhil R, Nyika J, Yadav S, Mackolil J, Prashat G, Workie E, Ragupathy R, Ramasundaram P (2022) Genetically modified foods: bibliometric analysis on consumer perception and preference. GM Crops Food 13(1):65–85. https://doi.org/10.1080/21645698.2022.2038525

Shen W, Zhang X, Liu J, Tao K, Li C, Xiao S, Zhang W, Li J (2022) Plant elicitor peptide Signalling confers Rice resistance to piercing-sucking insect herbivores and pathogens. Plant Biotechnol J 20(5):991–1005. https://doi.org/10.1111/pbi.13781

Singh BK, Delgado-Baquerizo M, Egidi E, Guirado E, Leach JE, Liu H, Trivedi P (2023) Climate change impacts on plant pathogens, food security and paths forward. Nat Rev Microbiol 21(10), 640–656. https://doi.org/10.1038/s41579-023-00900-7

Spurgeon D, Lahive E, Robinson A, Short S, Kille P (2020) Species sensitivity to toxic substances: evolution, ecology and applications. Front Environ Sci 8(December):1–25. https://doi.org/10.3389/fenvs.2020.588380

Betsuyaku BP, Nürnberger T, Joosten MH (2011) Of PAMPs and effectors: the blurred PTI-ETI dichotomy. Plant Cell 23(1):4–15. https://doi.org/10.1105/tpc.110.082602

Toennisson A, Burrack H (2018) Efficacy of organically acceptable materials against tobacco pests, 2016. Arthropod Manag Tests 43(1):1–2. https://doi.org/10.1093/amt/tsy055

Trivilin A, Hartke S, Moraes M (2014) Components of different Signalling pathways regulated by a new orthologue of AtPROPEP1 in tomato following infection by pathogens. Plant Pathol 63(5):1110–1118. https://doi.org/10.1111/ppa.12190

Teixeira E, Fischer G, van Velthuizen H, Walter C, Ewert F, Global hot-spots of heat stress on agricultural crops due to climate change (2013). Agric For Meteorol 170:206-215. https://doi.org/10.1016/j.agrformet.2011.09.002

Wang A, Guo J, Wang S, Zhang Y, Lu F, Duan J, Liu Z, Ji W (2022) BoPEP4, a C-terminally encoded plant elicitor peptide from broccoli, plays a role in salinity stress tolerance. Int J Mol Sci 23(6). https://doi.org/10.3390/ijms23063090

Weissmann E, Raja K, Gupta A, Patel M, Buehler A (2023a) Seed quality enhancement. In: Dadlani M, Yadava DK (eds) Seed science and technology. Springer Nature, Singapore, pp 391–414

Yang R, Liu J, Wang Z, Zhao L, Xue T, Meng J, Luan Y (2023) The SlWRKY6-SlPROPEP-SlPep module confers tomato resistance to Phytophthora Infestans. Sci Hortic 318(112117)

Yamaguchi Y, Pearce G, Ryan C (2006) The cell surface leucine-rich repeat receptor for AtPep1, an endogenous peptide elicitor in Arabidopsis, is functional in transgenic tobacco cells. Proc Natl Acad Sci 103(26):10104–10109. https://doi.org/10.1073/pnas.0603729103

Zapadka K, Becher F, Gomes dos Santos A, Jackson S (2017) Factors affecting the physical stability (aggregation) of peptide therapeutics. Interface. Focus 7(6). https://doi.org/10.1098/rsfs.2017.0030

Zelman AK, Berkowitz GA (2023) Plant elicitor peptide (Pep) signaling and pathogen defense in tomato. Plants 12(15):2856. https://doi.org/10.3390/PLANTS12152856

Zelman AK, Ma Y, Berkowitz GA (2024) Pathogen elicitor peptide (Pep), systemin, and their receptors in tomato: sequence analysis sheds light on standing disagreements about biotic stress signaling components. BMC Plant Biol 24(1):1–13. https://doi.org/10.1186/s12870-024-05403-y

Zheng Q, Wen X, Xiu X, Yang X, Chen Q (2022) Can the part replace the whole? A choice experiment on organic and pesticide-free labels. Food Secur 11(17):1–16. https://doi.org/10.3390/foods11172564

Printed in the USA
CPSIA information can be obtained
at www.ICGtesting.com
CBHW052359241124
17936CB00003B/44

9 789819 763733